# Solutions Manual

## Jan William Simek
### *California Polytechnic State University*

# O R G A N I C
# C H E M I S T R Y

### EIGHTH EDITION

## L.G.  WADE,  JR.

**PEARSON**

Boston  Columbus  Indianapolis  New York  San Francisco  Upper Saddle River
Amsterdam  Cape Town  Dubai  London  Madrid  Milan  Munich  Paris  Montréal  Toronto
Delhi  Mexico City  São Paulo  Sydney  Hong Kong  Seoul  Singapore  Taipei  Tokyo

Editor In Chief: Adam Jaworski
Senior Marketing Manager: Jonathan Cottrell
Senior Project Editor: Jennifer Hart
Assistant Editor: Coleen McDonald
Managing Editor, Chemistry and Geosciences: Gina M. Cheselka
Senior Project Manager, Production: Shari Toron
Operations Specialist: Jeffrey Sargent
Supplement Cover Designer: Seventeenth Street Studios
Cover Image: Don Paulson Photography/PureStock/Alamy

1 2 3 4 5 6 7 8 9 10—BRR— 15 14 13 12

www.pearsonhighered.com          ISBN-10: 0-321-77389-6; ISBN-13: 978-0-321-77389-0

# TABLE OF CONTENTS

Preface ..................................................................................................................v

Symbols and Abbreviations ................................................................................ vii

Chapter 1    Introduction and Review ...............................................................1
Chapter 2    Structure and Properties of Organic Molecules .........................31
Chapter 3    Structure and Stereochemistry of Alkanes ..................................52
Chapter 4    The Study of Chemical Reactions ...............................................68
Chapter 5    Stereochemistry ...........................................................................87
Chapter 6    Alkyl Halides:  Nucleophilic Substitution and Elimination .......106
Chapter 7    Structure and Synthesis of Alkenes ...........................................141
Chapter 8    Reactions of Alkenes ..................................................................167
Chapter 9    Alkynes .......................................................................................198
Chapter 10   Structure and Synthesis of Alcohols ..........................................216
Chapter 11   Reactions of Alcohols ................................................................235
Chapter 12   Infrared Spectroscopy and Mass Spectrometry .........................264
Chapter 13   Nuclear Magnetic Resonance Spectroscopy ...............................277
Chapter 14   Ethers, Epoxides, and Thioethers ..............................................306
Chapter 15   Conjugated Systems, Orbital Symmetry, and Ultraviolet Spectroscopy .............326
Chapter 16   Aromatic Compounds .................................................................346
Chapter 17   Reactions of Aromatic Compounds ...........................................371
Chapter 18   Ketones and Aldehydes ..............................................................405
Chapter 19   Amines ........................................................................................439
Chapter 20   Carboxylic Acids ........................................................................467
Chapter 21   Carboxylic Acid Derivatives ......................................................496
Chapter 22   Condensations and Alpha Substitutions of Carbonyl Compounds .....................531
Chapter 23   Carbohydrates and Nucleic Acids ..............................................581
Chapter 24   Amino Acids, Peptides, and Proteins .........................................613
Chapter 25   Lipids .........................................................................................641
Chapter 26   Synthetic Polymers ....................................................................657

Appendix 1:   Summary of IUPAC Nomenclature of Organic Compounds ...........675
Appendix 2:   Summary of Acidity and Basicity ..............................................689
Appendix 3:   Sample Alkene Reaction Summary ............................................701

# NOTES TO THE STUDENT

Chapter 1     How to answer "Explain" questions ....................................................30
Chapter 2     Wedge and dashed bond symbolism.................................................31
Chapter 2     Functional group concept map ........................................................40
Chapter 3     IUPAC nomenclature ....................................................................52
Chapter 3     Wedge and dashed bond symbolism reminder ..............................55
Chapter 3     Creating practice problems in nomenclature ................................67
Chapter 5     Molecular models ........................................................................87
Chapter 6     Substitution and elimination concept map ....................................140
Chapter 7     IUPAC nomenclature II ................................................................142
Chapter 8     Regiochemistry and major products ..............................................168
Chapter 8     Achiral reagents produce racemic products ..................................169
Chapter 8     Predicting stereochemistry of alkene additions ............................178
Chapter 8     Deduction and inference ..............................................................193
Chapter 11    Conventions in oxidation methods ..............................................237
Chapter 12    Spectroscopy web sites ................................................................276
Chapter 13    Representations of benzene ..........................................................277
Chapter 13    Appearance of NMR spectra ........................................................278
Chapter 16    Representations of benzene II ......................................................346
Chapter 16    Anthropomorphizing ....................................................................370
Chapter 18    Reminders about synthesis problems ............................................406
Chapter 18    Mechanism of imine formation has six steps ................................412
Chapter 18    Mechanism of acetal formation has seven steps ..........................415
Chapter 23    Fischer projections ......................................................................581
Chapter 23    Acetal and ketal concept map ......................................................585
Chapter 26    Polymer chain symbolism ............................................................657
Chapter 26    Bon voyage ..................................................................................673

# PREFACE

**Hints for Passing Organic Chemistry**

Do you want to pass your course in organic chemistry? Here is my best advice, based on over thirty-five years of observing students learning organic chemistry:

Hint #1: *Do the problems*. It seems straightforward, but humans, including students, try to take the easy way out until they discover there is no shortcut. Unless you have a measured IQ above 200 and comfortably cruise in the top 1% of your class, *do the problems*. Usually your teacher (professor or teaching assistant) will recommend certain ones; try to do all those recommended. If you do half of them, you will be half-prepared at test time. (Do you want your surgeon coming to your appendectomy having practiced only *half* the procedure?) And when you do the problems, keep this Solutions Manual CLOSED. Avoid looking at *my* answer before you write *your* answer—your trying and struggling with the problem is the most valuable part of the problem. Discovery is a major part of learning. Remember that the primary goal of doing these problems is *not* just getting the right answer, but understanding the material well enough to get right answers to the questions you haven't seen yet.

Hint #2: *Keep up*. Getting behind in your work in a course that moves as quickly as this one is the Kiss of Death. For most students, organic chemistry is the most rigorous intellectual challenge they have faced so far in their studies. Some are taken by surprise at the diligence it requires. Don't think that you can study all of the material in the couple of days before the exam—well, you can, but you won't get a passing grade. Study organic chemistry like a foreign language: try to do some every day so that the freshly trained neurons stay sharp.

Hint #3: *Get help when you need it*. Use your teacher's office hours when you have difficulty. Many schools have tutoring centers (in which organic chemistry is a popular offering). Here's a secret: absolutely the best way to cement this material in your brain is to get together with a few of your fellow students and make up problems for each other, then correct and discuss them. When *you* write the problems, you will gain great insight into what this is all about.

Hint #3.5: When you write answers to problem, *write* them. Use the old-fashioned method of a writing implement on paper. Keep a notebook with your work. Show your instructor; he/she will be impressed.

**Purpose of This Solutions Manual**

So what is the point of this Solutions Manual? First, I can't do your studying for you. Second, since I am not leaning over your shoulder as you write your answers, I can't give you direct feedback on what you write and think—the print medium is limited in its usefulness. What I *can* do for you is: 1) provide correct answers: the publishers, Professor Wade, Professor Palandoken (my reviewer), and I have gone to great lengths to assure that what I have written is correct, for we all understand how it can shake a student's confidence to discover that the answer book flubbed up; 2) provide a considerable degree of rigor: beyond the fundamental requirement of correctness, I have tried to flesh out these answers, being complete but succinct; 3) provide insight into how to solve a problem and into where the sticky intellectual points are. Insight is the toughest to accomplish, but over the years, I have come to understand where students have trouble, so I have tried to anticipate your questions and to add enough detail so that the concept, as well as the answer, is clear.

It is difficult for students to understand or acknowledge that their teachers are human (some are more human than others). Since I am human (despite what my students might report), I can and do make mistakes. If there are mistakes in this book, they are my sole responsibility, and I am sorry. If you find one, PLEASE let me know so that it can be corrected in future printings. Nip it in the bud.

**What's New in This edition?**

Better answers! Part of my goal in this edition has been to add more explanatory material to clarify how to arrive at the answer. In many problems, the possibility of more than one answer to a problem has been noted. Concept maps have been added at appropriate places to demonstrate the logic of particular concepts.

Better graphics! The print medium is very limited in its ability to convey three-dimensional structural information, a problem that has plagued organic chemists for over a century.

Appendix 2 on Acidity has been revised, and Appendix 3 has been added as a suggestion to students on how to organize reaction summaries to make studying more effective.

Better jokes? Too much to hope for.

**Some Web Stuff**

Here I am:  http://www.calpoly.edu/~chem/faculty/simek.html .
The Publisher (Pearson) maintains a web site related to the Wade text:  try
http://www.masteringchemistry.com
Two essential web sites providing spectra are listed on the bottom of p. 276.

## Acknowledgments

No project of this scope is ever done alone.  These are team efforts, and several people who have assisted and facilitated in one fashion or another deserve my thanks.

Professor L. G. Wade, Jr., your textbook author, is a remarkable person.  He has gone to extraordinary lengths to make the textbook as clear, organized, informative, and insightful as possible.  He has solicited and followed many suggestions on his text, and his comments on my solutions have been perceptive and valuable.  We agreed early on that our primary goal is to help the students learn a fascinating and challenging subject, and all of our efforts have been directed toward that goal.  I have appreciated our collaboration.

My friend and colleague, Dr. Hasan Palandoken, has reviewed the entire manuscript for accuracy and style.  His extraordinary diligence, attention to detail, and chemical wisdom have made this a better manual.  Hasan stands on the shoulders of previous reviewers who scoured earlier editions for errors: Dr. Kristen Meisenheimer, Jessica Gilman Ernakovich, Dr. Eric Kantorowski, and Dr. Dan Mattern. Mr. Richard King of Pasadena, Texas, Editorial Adviser, has offered numerous suggestions on how to clarify murky explanations.  I am grateful to them all.

The people at Pearson have made this project possible.  Good books would not exist without their dedication, professionalism, and experience.  Among the many people who contributed are:  Lee Englander, who originally connected me with this project; Jeanne Zalesky, Executive Editor in Chemistry; Jennifer Hart, Senior Project Editor in Chemistry; and Coleen McDonald, Assistant Editor in Chemistry.

The entire manuscript was produced using *ChemDraw*®, the remarkable software for drawing chemical structures developed by CambridgeSoft Corp., Cambridge, MA.

Finally, I appreciate my friends who supported me throughout this project, most notably my wife and friend of over forty-six years, Judy Lang.  The students are too numerous to list, but it is for them that all this happens.

Jan William Simek, Professor Emeritus
Department of Chemistry and Biochemistry
Cal Poly State University
San Luis Obispo, CA  93407
Email:  jsimek@calpoly.edu

## DEDICATION

To my inspirational chemistry teachers:
*Joe Plaskas*, who made the batter;
*Kurt Kaufman*, who baked the cake;
*Carl Djerassi*, who put on the icing;

and to my parents:
*Ervin J. and Imilda B. Simek*,
who had the original concept.

# SYMBOLS AND ABBREVIATIONS

Below is a list of symbols and abbreviations used in this Solutions Manual, consistent with those used in the textbook by Wade; see the inside front cover of the text. (Do not expect all of these to make sense to you now. You will learn them throughout your study of organic chemistry.)

## BONDS

———————  a single bond

═══════  a double bond

≡≡≡≡≡≡  a triple bond

◄━━━━  a bond in three dimensions, coming out of the paper toward the reader

ıııııııııı  a bond in three dimensions, going behind the paper away from the reader

- - - - - - - -  a stretched bond, in the process of forming or breaking

## ARROWS

⟶  in a reaction, shows direction from reactants to products

⇌  signifies equilibrium (not to be confused with resonance)

⟷  signifies resonance (not to be confused with equilibrium)

shows direction of electron movement:
the arrowhead with one barb shows movement of one electron;
the arrowhead with two barbs shows movement of a pair of electrons

⊢⟶  shows polarity of a bond or molecule, the arrowhead signifying the more negative end of the dipole

## SUBSTITUENT GROUPS

Me      a methyl group, $CH_3$

Et      an ethyl group, $CH_2CH_3$

Pr      a propyl group, a three-carbon group (two possible arrangements)

Bu      a butyl group, a four-carbon group (four possible arrangements)

R      the general abbreviation for an alkyl group (or any substituent group bonded at carbon)

Ph      a phenyl group, the name of a benzene ring as a substituent, represented:

     or    

Ar      the general abbreviation for an aromatic group

continued on next page

## SUBSTITUENT GROUPS, continued

Ac    an acetyl group:    $CH_3 - \overset{\overset{\displaystyle O}{\|}}{C} -$

*c*-Hx    a cyclohexyl group:

Ts    tosyl, or *p*-toluenesulfonyl group:    $CH_3 - \overset{\overset{\displaystyle O}{\|}}{\underset{\underset{\displaystyle O}{\|}}{S}} -$

Boc    a *tert*-butoxycarbonyl group (amino acid and peptide chemistry):    $(CH_3)_3C - O - \overset{\overset{\displaystyle O}{\|}}{C} -$

Z, or    a carbobenzoxy (benzyloxycarbonyl) group (amino acid and peptide chemistry):
Cbz
$- CH_2 - O - \overset{\overset{\displaystyle O}{\|}}{C} -$

## REAGENTS AND SOLVENTS

DCC    **dic**yclohexyl**c**arbodiimide    $- N = C = N -$

DMSO    **dim**ethyl**s**ulf**ox**ide    $H_3C \overset{\overset{\displaystyle O}{\|}}{\underset{}{S}} CH_3$

ether    diethyl ether, $CH_3CH_2OCH_2CH_3$

HA or H—A is a generic acid; the conjugate base may appear as:    $A^-$    $\overset{\ominus}{A}$    $:\overset{\ominus}{A}$

LG    leaving **g**roup

MCPBA    *meta*-**c**hloro**p**eroxy**b**enzoic **a**cid    $\overset{Cl}{}$    $\overset{\overset{\displaystyle O}{\|}}{C} - O - OH$

MVK    **m**ethyl **v**inyl **k**etone    $H_3C \overset{\overset{\displaystyle O}{\|}}{C}$

NBS    *N*-**b**romo**s**uccinimide    $O = \underset{\underset{\displaystyle Br}{|}}{N} = O$

continued on next page

<u>Symbols and Abbreviations</u>, continued

<u>REAGENTS AND SOLVENTS</u>, continued

Nuc or :Nuc or Nuc⁻ is a generic nucleophile, a Lewis base; E or E⁺ is a generic electrophile, a Lewis acid

**PCC**      **p**yridinium **c**hloro**c**hromate,    $CrO_3 \cdot HCl \cdot N$

**Sia₂BH**      **disia**mylborane

$$\begin{array}{ccccccccc} & CH_3 & H & & H & & H & & CH_3 \\ & | & | & & | & & | & & | \\ H- & C- & C- & B- & C- & & C- & H \\ & | & | & & & | & & | \\ & CH_3 & CH_3 & & & CH_3 & CH_3 \end{array}$$

**THF**      **t**etra**h**ydro**f**uran

<u>SPECTROSCOPY</u>

| | |
|---|---|
| IR | **i**nf**r**a**r**ed spectroscopy |
| NMR | **n**uclear **m**agnetic **r**esonance spectroscopy |
| MS | **m**ass **s**pectrometry |
| UV | **u**ltra**v**iolet spectroscopy |
| ppm | **p**arts **p**er **m**illion, a unit used in NMR |
| Hz | hertz, cycles per second, a unit of frequency |
| MHz | megahertz, millions of cycles per second |
| TMS | **t**etra**m**ethyl**s**ilane, $(CH_3)_4Si$, the reference compound in NMR |
| s, d, t, q, m | **s**inglet, **d**oublet, **t**riplet, **q**uartet, **m**ultiplet: the number of peaks an NMR absorption gives |
| nm | nanometers, $10^{-9}$ meters (usually used as a unit of wavelength) |
| m/z | mass-to-charge ratio, in mass spectrometry |
| $\delta$ | in NMR, chemical shift value, measured in ppm (Greek lower case delta) |
| $\lambda$ | wavelength (Greek lambda) |
| $\nu$ | frequency (Greek nu) |
| $J$ | coupling constant in NMR |

<u>OTHER</u>

| | |
|---|---|
| •• or ⦂ | unshared electron pair |
| *a, ax* | axial (in chair forms of cyclohexane) |
| *e, eq* | equatorial (in chair forms of cyclohexane) |
| HOMO | **h**ighest **o**ccupied **m**olecular **o**rbital |
| LUMO | **l**owest **u**noccupied **m**olecular **o**rbital |
| *NR* | no reaction |
| *o, m, p* | *ortho, meta, para* (positions on an aromatic ring) |
| $\Delta$ | when written over an arrow: "heat"; when written before a letter: "change in" |
| $\delta^+, \delta^-$ | partial positive charge, partial negative charge |
| h$\nu$ | energy from electromagnetic radiation (light) |
| $[\alpha]_D$ | specific rotation at the D line of sodium (589 nm) |

Students:  Add your own notes on symbols and abbreviations.

# CHAPTER 1—INTRODUCTION AND REVIEW

**1-1**

(a) Nitrogen has atomic number 7, so all nitrogen atoms have 7 protons. The mass number is the total number of neutrons and protons; therefore, $^{13}N$ has 6 neutrons, $^{14}N$ has 7 neutrons, $^{15}N$ has 8 neutrons, $^{16}N$ has 9 neutrons, and $^{17}N$ has 10 neutrons.

(b)

| | | | | |
|---|---|---|---|---|
| Na | $1s^2 2s^2 2p^6 3s^1$ | | P | $1s^2 2s^2 2p^6 3s^2 3p_x^1 3p_y^1 3p_z^1$ |
| Mg | $1s^2 2s^2 2p^6 3s^2$ | | S | $1s^2 2s^2 2p^6 3s^2 3p_x^2 3p_y^1 3p_z^1$ |
| Al | $1s^2 2s^2 2p^6 3s^2 3p_x^1$ | | Cl | $1s^2 2s^2 2p^6 3s^2 3p_x^2 3p_y^2 3p_z^1$ |
| Si | $1s^2 2s^2 2p^6 3s^2 3p_x^1 3p_y^1$ | | Ar | $1s^2 2s^2 2p^6 3s^2 3p_x^2 3p_y^2 3p_z^2$ |

**1-2** Lines between atom symbols represent covalent bonds between those atoms. Nonbonding electrons are indicated with dots.

(a) H—N̈—H with H below

(b) H—Ö—H

(c) H—Ö⁺—H with H below

(d) H—C—C—C—H (propane, each C with H above and below)

(e) H—C—N̈—C—H (each C with H, N with lone pair)

(f) H—C—C—Ö—C—C—H

(g) H—C—C—C—Cl̈:

(h) H—C—C—C—H with :O: above middle C and H above

(i) H—B—H with H below

(j) :F̈—B—F̈: with :F̈: below

The compounds in (i) and (j) are unusual in that boron does not have an octet of electrons—normal for boron because it has only three valence electrons.

**1-3**

(a) :N≡N:

(b) H—C≡N:

(c) H—Ö—N̈=Ö

(d) Ö=C=Ö

(e) H—C—C=N̈—H (each C with H)

(f) H—C—Ö—H with :O: double bonded above

(g) H—C=C—Cl̈: (each C with H)

(h) H—N̈=N̈—H

(i) H—C=C—C—H (each C with H)

(j) H—C=C=C—H (each C with H)

(k) H—C≡C—C—H with H above and below

**1-4** There are no unshared electron pairs in parts (i), (j), and (k).

(a) :N≡N: (with circled dots)

(b) H—C≡N: (with circled dots)

(c) H—O—N=O (with circled dots)

(d) O=C=O (with circled dots)

(e) H—C—C=N—H (with circled dots, each C with H)

(f) H—C—O—H with O double bonded (circled dots)

(g) H—C=C—Cl (with circled dots, each C with H)

(h) H—N=N—H (with circled dots)

**1**

**1-5** The symbols "δ⁺" and "δ⁻" indicate bond polarity by showing partial charge. (In the arrow symbolism, the arrow should point to the partial negative charge.)

(a) $\overset{\delta^+}{C}-\overset{\delta^-}{Cl}$  (b) $\overset{\delta^+}{C}-\overset{\delta^-}{O}$  (c) $\overset{\delta^+}{C}-\overset{\delta^-}{N}$  (d) $\overset{\delta^+}{C}-\overset{\delta^-}{S}$  (e) $\overset{\delta^-}{C}-\overset{\delta^+}{B}$

(f) $\overset{\delta^+}{N}-\overset{\delta^-}{Cl}$  (g) $\overset{\delta^+}{N}-\overset{\delta^-}{O}$  (h) $\overset{\delta^-}{N}-\overset{\delta^+}{S}$  (i) $\overset{\delta^-}{N}-\overset{\delta^+}{B}$  (j) $\overset{\delta^+}{B}-\overset{\delta^-}{Cl}$

**1-6** Non-zero formal charges are shown beside the atoms, circled for clarity.

(a)

(b), (c)

In (b) and (c), the chlorine is present as chloride ion. There is no covalent bond between chloride and other atoms in the formula.

(d) Na⊕  ⊖:Ö—C—H with H's (structure)   (e) H—C⊕—H structure   (f) H—C⊖—H structure   (g) Na⊕  H—B⊖—H structure

(h) Na⊕  H—B⊖—C≡N: structure   (i) structure with F—B⊖—F   (j) H—Ö—N⊕—H structure

(k) K⊕  ⊖:Ö—C—C—H structure   (l) H—C=Ö⊕—H structure

As shown in (d), (g), (h), and (k), alkali metals like sodium and potassium form only ionic bonds, never covalent bonds.

**1-7** Resonance forms in which all atoms have full octets are the most significant contributors. In resonance forms, ALL ATOMS KEEP THEIR POSITIONS—ONLY ELECTRONS ARE SHOWN IN DIFFERENT POSITIONS. (In this Solutions Manual, braces {} are commonly used to denote resonance forms.)

(a) { ⊖:Ö—C(=O)—Ö:⊖  ↔  ⊖:Ö—C=Ö (with O⊖ above)  ↔  Ö=C—Ö:⊖ (with O⊖ above) }

(b) { ⊖:Ö—N(=O)(⊕)—Ö:⊖  ↔  ⊖:Ö—N(⊕)=Ö (with O⊖ above)  ↔  Ö=N(⊕)—Ö:⊖ (with O⊖ above) }

(c) { ⊖:Ö—N=Ö  ↔  Ö=N—Ö:⊖ }

1-7 continued

(d) { H—C=C—C⁺—H  ⟷  H—C⁺—C=C—H }
with H's below

(e) { H—C=C—C⁻—H  ⟷  H—C⁻—C=C—H }

(f) Sulfur can have up to 12 electrons around it because it has d orbitals accessible.

{ :O:⁻ / :O—S—O:⁻ (with :O: below) ⟷ :O:⁻ / :O—S=O (with :O: below) ⟷ :O:⁻ / O=S—O:⁻ (with :O: below) ⟷ :O: / O=S—O:⁻ (with :O:⁻ below) }

{ :O: / :O—S=O (with :O:⁻ below) ⟷ :O:⁻ / O=S=O (with :O:⁻ below) }

(g) { 

structure 1: H—C—H (top, with H above); H—C—C⁺ with :O: groups; H—C—H (bottom, with H below)

⟷

structure 2: H—C—H (top); H—C—C with ⁺O: double bond and :O: ; H—C—H

⟷

structure 3: H—C—H (top); H—C—C with :O: and ⁺O: double bond ; H—C—H

}

(h) { 

H—O: / H—O—B—O—H with :O: groups  ⟷  H—O⁺ / H—O—B—O—H (with ⊖ on O's)  ⟷  H—O: / H—O=B—O—H (with ⊖ and ⊕)  ⟷  H—O: / H—O—B—O—H (with ⊖ and ⊕)

}

1-8  Major resonance contributors would have the lowest energy. The most important factors are:  maximize full octets; maximize pi bonds; put negative charge on electronegative atoms; minimize charge separation—see the Problem-Solving Hint in text Section 1-9B.  Part (a) has been solved in the text.

(b) { 

H\C=C—N⁺=O: with H and :O:⁻ below
  major

⟷

H\C=C—N⁺—O:⁻ with H and :O: (double bond below)
  major

⟷

⁺H/⊕C—C=N⁺—O:⁻ with H and :O:⁻ below
  minor

}

These first two forms have equivalent energy and are major because they have full octets, more bonds, and less charge separation than the minor contributor.

1-8 continued

(c) 

H—C=Ö—H  ⟷  H—C—Ö—H

major        minor

The first structure has full octets and one more pi bond.

(d)

minor   ⟷   minor   ⟷   major

All atoms have octets; same number of pi bonds; third structure has negative charge on the more electronegative oxygen atoms instead of carbon.

(e)

H—C⁻—C≡N:  ⟷  H—C=C=N:⁻

minor        major

The second structure has negative charge on the more electronegative atom.

(f)

minor   ⟷   major

These two forms are major contributors because all atoms have full octets.

minor   ⟷   major

(g)

minor   ⟷   major   ⟷   major

The latter two structures have equivalent energy and are major because they have full octets and more pi bonds.

(h)

minor   ⟷   major   ⟷   major

The latter two structures have equivalent energy and are major because the negative charge is on the more electronegative oxygen atom.

1-8 continued

(i)
$$\left\{ \begin{array}{ccc} \text{:O:} & & \text{:O:}^{\ominus} \\ \parallel & & \vert \\ \text{H}-\text{C}-\overset{..}{\text{N}}-\text{H} & \longleftrightarrow & \text{H}-\text{C}=\overset{\oplus}{\text{N}}-\text{H} \\ \vert & & \vert \\ \text{major} \;\; \text{H} & & \text{minor} \;\; \text{H} \end{array} \right\}$$
(no charge separation)

(j)
$$\left\{ \begin{array}{ccc} \text{H}-\text{C}=\text{C}-\overset{\ominus}{\underset{..}{\text{N}}}-\text{H} & \longleftrightarrow & \text{H}-\overset{\ominus}{\text{C}}-\text{C}=\overset{..}{\text{N}}-\text{H} \\ \quad\vert \;\;\; \vert & & \quad\vert \;\;\; \vert \\ \quad\text{H} \;\; \text{H} & & \quad\text{H} \;\; \text{H} \end{array} \right\}$$
major—negative charge on the       minor
more electronegative atom

1-9  Your Lewis structures may *appear* different from these.  As long as the atoms are connected in the same order and by the same type of bond, they are equivalent structures.  For now, the exact placement of the atoms on the page is not significant.  A Lewis structure is "complete" with unshared electron pairs shown.

(a) a six-carbon chain: $\text{H}-\text{C}-\text{C}-\text{C}-\text{C}-\text{C}-\text{C}-\text{H}$ with a $CH_3$ branch (all carbons bearing H's)

(b) $\text{H}-\text{C}-\text{C}-\text{C}-\overset{..}{\underset{..}{\text{Cl}}}\text{:}$ with a $CH_3$ branch

(c) $\text{H}-\text{C}-\text{C}-\text{C}-\text{C}\equiv\text{N:}$ with a C=O on the third carbon

Always be alert for the implied double or triple bond.  Remember that the normal valence of C is four bonds, nitrogen has three bonds, oxygen has two bonds, and hydrogen has one bond.  The only exceptions to these valence rules are structures with formal charges.  (We will see other unusual exceptions in later chapters.)

(d) $\text{C}=\text{C}-\text{C}-\text{H}$ (with :O: double-bonded to the carbonyl carbon)

(e) $\text{H}-\text{C}-\text{C}-\text{C}-\text{C}=\text{C}$ (with a C=O and branching $CH_3$ groups)

(f) $\text{H}-\text{C}-\text{C}-\text{C}-\overset{..}{\underset{..}{\text{O}}}-\text{H}$ (with two :O: double bonds)

(g) $\text{H}-\text{C}-\text{C}-\text{C}-\text{C}-\text{C}-\text{H}$ (with :O: double-bonded to the central carbon)

(h) $\text{H}-\text{C}-\text{C}-\text{C}-\text{H}$ (with :O: double-bonded to the central carbon and a $CH_3$ branch)

1-10  Complete Lewis structures show all atoms, bonds, and unshared electron pairs.

(a) a six-membered ring containing N with H substituents  $C_6H_{13}N$

(b) a ring structure containing O with H substituents  $C_8H_{16}O$

(c) a five-membered ring containing N with C=C double bonds  $C_4H_5N$

5
Copyright © 2013 Pearson Education, Inc.

(d)  $C_5H_{10}O$

(e) $C_7H_{10}O$

(f) $C_6H_8O$

(g) $C_8H_8O$

(h) $C_5H_{10}O$

1-11 Line-angle structures, sometimes called "stick" figures, usually omit unshared electron pairs.

(a) $C_7H_{16}$

(b) $C_4H_9Cl$

(c) $C_4H_5NO$

(d) $C_3H_4O$

H on C are usually not shown but this an an exception; it clarifies what ends this chain.

(e) $C_7H_{12}O$

better placement

not as good

(f) $C_3H_4O_3$

(g) $C_5H_{10}O$

(h) OR $C_4H_{10}O$

These two structures are equally acceptable.

1-12 If the percent values do not sum to 100%, the remainder must be oxygen. Assume 100 g of sample; percents then translate directly to grams of each element.

There are usually MANY possible structures for a molecular formula. Yours may be different from the examples shown here and they could still be correct.

some possible structures:

(a) $\dfrac{40.0 \text{ g C}}{12.0 \text{ g/mole}}$ = 3.33 moles C ÷ 3.33 moles = 1 C

$\dfrac{6.67 \text{ g H}}{1.01 \text{ g/mole}}$ = 6.60 moles H ÷ 3.33 moles = 1.98 ≈ 2 H

$\dfrac{53.33 \text{ g O}}{16.0 \text{ g/mole}}$ = 3.33 moles O ÷ 3.33 moles = 1 O

empirical formula = $\boxed{CH_2O}$ ⟹ empirical weight = 30.02

molecular weight = 90, three times the empirical weight ⟹

three times the empirical formula = molecular formula = $\boxed{C_3H_6O_3}$

Other structures are possible.

1-12 continued

(b)

$$\frac{32.0 \text{ g C}}{12.0 \text{ g/mole}} = 2.67 \text{ moles C} \div 1.34 \text{ moles} = 1.99 \approx 2 \text{ C}$$

$$\frac{6.67 \text{ g H}}{1.01 \text{ g/mole}} = 6.60 \text{ moles H} \div 1.34 \text{ moles} = 4.93 \approx 5 \text{ H}$$

$$\frac{18.7 \text{ g N}}{14.0 \text{ g/mole}} = 1.34 \text{ moles N} \div 1.34 \text{ moles} = 1 \text{ N}$$

$$\frac{42.6 \text{ g O}}{16.0 \text{ g/mole}} = 2.66 \text{ moles O} \div 1.34 \text{ moles} = 1.99 \approx 2 \text{ O}$$

empirical formula = $\boxed{C_2H_5NO_2}$ $\Longrightarrow$ empirical weight = 75.05

molecular weight = 75, same as the empirical weight $\Longrightarrow$

empirical formula = molecular formula = $\boxed{C_2H_5NO_2}$

some possible structures:

*MANY other structures are possible.*

- - - - - - - - - - - - - - - - - - - - - - - - - - - - - - - - - - - - - - - - - - - - - - - - - - - -

(c)

$$\frac{25.6 \text{ g C}}{12.0 \text{ g/mole}} = 2.13 \text{ moles C} \div 1.07 \text{ moles} = 1.99 \approx 2 \text{ C}$$

$$\frac{4.32 \text{ g H}}{1.01 \text{ g/mole}} = 4.28 \text{ moles H} \div 1.07 \text{ moles} = 4 \text{ H}$$

$$\frac{37.9 \text{ g Cl}}{35.45 \text{ g/mole}} = 1.07 \text{ moles Cl} \div 1.07 \text{ moles} = 1 \text{ Cl}$$

$$\frac{15.0 \text{ g N}}{14.0 \text{ g/mole}} = 1.07 \text{ moles N} \div 1.07 \text{ moles} = 1 \text{ N}$$

$$\frac{17.2 \text{ g O}}{16.0 \text{ g/mole}} = 1.07 \text{ moles O} \div 1.07 \text{ moles} = 1 \text{ O}$$

empirical formula = $\boxed{C_2H_4ClNO}$ $\Longrightarrow$ empirical weight = 93.49

molecular weight = 93, same as the empirical weight $\Longrightarrow$

empirical formula = molecular formula = $\boxed{C_2H_4ClNO}$

some possible structures:

*MANY other structures are possible.*

- - - - - - - - - - - - - - - - - - - - - - - - - - - - - - - - - - - - - - - - - - - - - - - - - - - -

(d)

$$\frac{38.4 \text{ g C}}{12.0 \text{ g/mole}} = 3.20 \text{ moles C} \div 1.60 \text{ moles} = 2 \text{ C}$$

$$\frac{4.80 \text{ g H}}{1.01 \text{ g/mole}} = 4.75 \text{ moles H} \div 1.60 \text{ moles} = 2.97 \approx 3 \text{ H}$$

$$\frac{56.8 \text{ g Cl}}{35.45 \text{ g/mole}} = 1.60 \text{ moles Cl} \div 1.60 \text{ moles} = 1 \text{ Cl}$$

empirical formula = $\boxed{C_2H_3Cl}$ $\Longrightarrow$ empirical weight = 62.45

molecular weight = 125, twice the empirical weight $\Longrightarrow$

twice the empirical formula = molecular formula = $\boxed{C_4H_6Cl_2}$

some possible structures:

*MANY other structures are possible.*

1-13

(a)   $5.00 \text{ g HBr} \times \dfrac{1 \text{ mole HBr}}{80.9 \text{ g HBr}} = 0.0618 \text{ moles HBr}$

$0.0618 \text{ moles HBr} \implies 0.0618 \text{ moles } H_3O^+ \text{ (100\% dissociated)}$

$\dfrac{0.0618 \text{ moles } H_3O^+}{100 \text{ mL}} \times \dfrac{1000 \text{ mL}}{1 \text{ L}} = \dfrac{0.618 \text{ moles } H_3O^+}{1 \text{ L solution}}$

$pH = -\log_{10}[H_3O^+] = -\log_{10}(0.618) = \boxed{0.209}$

(b)   $1.50 \text{ g NaOH} \times \dfrac{1 \text{ mole NaOH}}{40.0 \text{ g NaOH}} = 0.0375 \text{ moles NaOH}$

$0.0375 \text{ moles NaOH} \implies 0.0375 \text{ moles } {}^-OH \text{ (100\% dissociated)}$

$\dfrac{0.0375 \text{ moles } {}^-OH}{50.\, \text{mL}} \times \dfrac{1000 \text{ mL}}{1 \text{ L}} = \dfrac{0.75 \text{ moles } {}^-OH}{1 \text{ L solution}} = 0.75 \text{ M}$

$[H_3O^+] = \dfrac{1 \times 10^{-14}}{[{}^-OH]} = \dfrac{1 \times 10^{-14}}{0.75} = 1.33 \times 10^{-14}$

$pH = -\log_{10}[H_3O^+] = -\log_{10}(1.33 \times 10^{-14}) = \boxed{13.88}$

(The number of decimal places in a pH value is the number of significant figures.)

1-14

(a) By definition, an acid is any species that can donate a proton. Ammonia has a proton bonded to nitrogen, so ammonia can be an acid (although a very weak one). A base is a proton acceptor, that is, it must have a pair of electrons to share with a proton; in theory, any atom with an unshared electron pair can be a base. The nitrogen in ammonia has an unshared electron pair so ammonia is basic. In water, ammonia is too weak an acid to give up its proton; instead, it acts as a base and pulls a proton from water to a small extent.

(b)   water as an acid:   $H_2O + NH_3 \rightleftharpoons {}^-OH + NH_4^+$

water as a base:   $H_2O + HCl \rightleftharpoons H_3O^+ + Cl^-$

(c) Hydronium acting as an acid in water solution will have this chemical equation:

$\underset{HA}{H_3O^+} + H_2O \rightleftharpoons \underset{A^-}{H_3O^+} + H_2O \implies K_a = \dfrac{[H_3O^+][A^-]}{[HA]} = \dfrac{[H_3O^+][H_2O]}{[H_3O^+]} = [H_2O]$

$[H_2O] = \dfrac{1000 \text{ g } H_2O}{1 \text{ L } H_2O} \times \dfrac{1 \text{ mole } H_2O}{18.0 \text{ g } H_2O} = \underline{55.55 \text{ M}} = K_a \implies pK_a = -\log(55.55) = \underline{-1.74}$

(d)   methanol as an acid:  $CH_3OH + NH_3 \rightleftharpoons CH_3O^- + NH_4^+$

methanol as a base:  $CH_3OH + H_2SO_4 \rightleftharpoons CH_3OH_2^+ + HSO_4^-$

1-15

(a)  HCOOH  +  ⁻CN  ⇌  HCOO⁻  +  HCN       FAVORS **PRODUCTS**
     stronger      stronger        weaker       weaker
     acid          base            base         acid
     $pK_a$ 3.76                                $pK_a$ 9.22


(b)  $CH_3COO^-$  +  $CH_3OH$  ⇌  $CH_3COOH$  +  $CH_3O^-$       FAVORS **REACTANTS**
     weaker           weaker          stronger        stronger
     base             acid            acid            base
                      $pK_a \approx 15.9$    $pK_a$ 4.74
                      (actually 15.5)*


(c)  $CH_3OH$  +  $NaNH_2$  ⇌  $CH_3O^- Na^+$  +  $NH_3$       FAVORS **PRODUCTS**
     stronger      stronger        weaker            weaker
     acid          base            base              acid      *Numerical values of $pK_a$ higher than water,*
     $pK_a \approx 15.9$                             $pK_a$ 33  *15.7, can be variable.  See Appendix 2 in*
     (actually 15.5)*                                          *this manual for an explanation.*


(d)  $Na^+$ ⁻$OCH_3$  +  HCN  ⇌  $HOCH_3$  +  NaCN       FAVORS **PRODUCTS**
     stronger            stronger        weaker         weaker
     base                acid            acid           base
                         $pK_a$ 9.22     $pK_a \approx 15.9$
                                         (actually 15.5)*


(e)  HCl  +  $H_2O$  ⇌  $H_3O^+$  +  $Cl^-$       FAVORS **PRODUCTS**
     stronger   stronger      weaker       weaker
     acid       base          acid         base
     $pK_a$ −7                $pK_a$ −1.7

     The first reaction in text Table 1-5 shows the $K_{eq}$ for this reaction is 1 x 10$^7$, favoring products.


(f)  $H_3O^+$  +  $CH_3O^-$  ⇌  $H_2O$  +  $CH_3OH$       FAVORS **PRODUCTS**
     stronger      stronger        weaker      weaker
     acid          base            base        acid
     $pK_a$ −1.7                               $pK_a \approx 15.9$ (actually 15.5)*

     *The ninth reaction in Table 1-5 shows the $pK_a$ of a structure similar to $CH_3OH$ is 15.9, so it is
     reasonable to infer that the $pK_a$ value of $CH_3OH$ is approximately the same.  (Text Appendix 4
     gives a value of 15.5.)  Using either value indicates that $CH_3OH$ is the weaker acid, so products
     are favored.


1-16

Protonation of the double-bonded oxygen gives three resonance forms (as shown in Solved Problem
1-5(c)); protonation of the single-bonded oxygen does not give any significant resonance forms, just
the structure shown; it is not stabilized by resonance.  In general, the more resonance forms a
species has, the more stable it is, so the proton would bond to the oxygen that gives a more stable
species, that is, the double-bonded oxygen.

1-17 In Solved Problem 1-4, the structure of methylamine is shown to be similar to ammonia. It is reasonable to infer that their acid-base properties are also similar.

(a) This problem can be viewed in two ways. 1) Quantitatively, the $pK_a$ values determine the order of acidity. 2) Qualitatively, the stabilities of the conjugate bases determine the order of acidity (see Solved Problem 1-4 for structures): the conjugate base of acetic acid, acetate ion, is resonance-stabilized, so acetic acid is the most acidic; the conjugate base of ethanol has a negative charge on a very electronegative oxygen atom; the conjugate base of methylamine has a negative charge on a mildly electronegative nitrogen atom and is therefore the least stabilized, so methylamine is the least acidic. (The first two $pK_a$ values are from text Table 1-5.)

$$\text{acetic acid} \quad > \quad \text{ethanol} \quad > \quad \text{methylamine}$$
$$pK_a \ 4.74 \qquad pK_a \ 15.9 \qquad pK_a \approx 40 \ \text{(from text Appendix 4)}$$
$$\text{strongest acid} \qquad\qquad\qquad \text{weakest acid}$$

(b) Ethoxide ion is the conjugate base of ethanol, so it must be a stronger base than ethanol; Solved Problem 1-4 and text Table 1-5 indicate ethoxide is analogous to hydroxide in base strength. Methylamine has $pK_b$ 3.36. The basicity of methylamine is between the basicity of ethoxide ion and ethanol.

$$\text{ethoxide ion} \quad > \quad \text{methylamine} \quad > \quad \text{ethanol}$$
$$\text{strongest base} \qquad\qquad\qquad\qquad \text{weakest base}$$

1-18 Curved arrows show electron movement, as described in text Section 1-14.

(a) equilibrium favors PRODUCTS

(b) equilibrium favors PRODUCTS

(c) equilibrium favors PRODUCTS

conjugate base, weaker base
resonance stabilized

Copyright © 2013 Pearson Education, Inc.

1-18 continued

(d) Na⊕ ⊖:Ö—H  +  H—S̈—H  ⟶⟵  H—Ö—H  +  Na⊕ ⊖:S̈—H
    stronger base        stronger acid         conjugate acid    conjugate base
    equilibrium favors PRODUCTS               weaker acid       weaker base
                                                                larger anion—size matters!

(e) CH₃—N⊕H₃  +  CH₃—Ö:⊖  ⟶⟵  CH₃—N̈—H  +  CH₃—Ö—H
    (H above and below N)    stronger base        (H below N)         conjugate acid
    stronger acid                                  conjugate base      weaker acid
    equilibrium favors PRODUCTS                    weaker base

(f) CH₃—C(=Ö)—Ö—H  +  CH₃—Ö:⊖  ⟶⟵  CH₃—Ö—H  +  { CH₃—C(=Ö)—Ö:⊖  ↔  CH₃—C(—Ö:⊖)=Ö }
    stronger acid       stronger base            conjugate acid       conjugate base
    equilibrium favors PRODUCTS                  weaker acid          weaker base
                                                                       more stable anion—
                                                                       resonance

(g) CH₃—C(=Ö)—Ö—H  +  { ⊖:Ö—S(=Ö)(CH₃)—Ö:  ↔  Ö=S(CH₃)(=Ö)—Ö⊖  ↔  ⊖Ö=S(CH₃)(—Ö:⊖)=Ö }
    weaker acid                              stronger base       weaker base      more stable anion—more
                                                                                   resonance
    equilibrium favors REACTANTS

    ⟶⟵  { CH₃—C(=Ö)—Ö:⊖  ↔  CH₃—C(—Ö:⊖)=Ö }  +  H—Ö—S(=Ö)(=Ö)—CH₃
         conjugate base                            conjugate acid
         stronger base                             stronger acid

(h) CF₃—C(=Ö)—Ö—H  +  { CH₃—C(=Ö)—Ö:⊖  ↔  CH₃—C(—Ö:⊖)=Ö }  ⟶⟵
    stronger acid        stronger base
    pKₐ 0.2
    from text
    Appendix 4

    { CF₃—C(=Ö)—Ö:⊖  ↔  CF₃—C(—Ö:⊖)=Ö }  +  CH₃—C(=Ö)—Ö—H
    conjugate base                            conjugate acid
    weaker base                               weaker acid
                                              pKₐ 4.74
    equilibrium favors PRODUCTS

The presence of electronegative atoms like F will make an acid stronger by the inductive effect.

**11**

## 1-18 continued

(i)

$$CH_3CH-\overset{F}{\underset{}{\mid}}\overset{:O:}{\underset{}{\overset{\parallel}{C}}}-\overset{..}{\underset{..}{O}}-H \;+\; \left\{ FCH_2CH_2-\overset{:O:}{\overset{\parallel}{C}}-\overset{..}{\underset{..}{O}}\overset{\ominus}{:} \;\longleftrightarrow\; FCH_2CH_2-\overset{:\overset{\ominus}{O}:}{\underset{}{C}}=\overset{..}{O}\overset{..}{:} \right\} \;\longrightarrow$$

stronger acid        stronger base

$$\left\{ CH_3CH-\overset{F}{\underset{}{\mid}}\overset{:O:}{\overset{\parallel}{C}}-\overset{..}{O}\overset{\ominus}{:} \;\longleftrightarrow\; CH_3CH-\overset{F}{\underset{}{\mid}}\overset{:\overset{\ominus}{O}:}{\underset{}{C}}=\overset{..}{O}\overset{..}{:} \right\} \;+\; FCH_2CH_2-\overset{:O:}{\overset{\parallel}{C}}-\overset{..}{O}-H$$

*equilibrium favors PRODUCTS*      conjugate base      conjugate acid
                                 weaker base         weaker acid

The presence of electronegative atoms like F will make an acid stronger by the inductive effect. The closer the electronegative atom is to the acidic group, the stronger its effect. The acid with the F on the second carbon is a stronger acid than the one with the F on the third carbon. From the point of view of the anions, the anion with the F closer to it is more stable, that is, a weaker base, leading to the same conclusion about which side is favored.

(j)

*equilibrium favors REACTANTS*

$$CF_3CH_2-\overset{..}{\underset{..}{O}}\overset{\ominus}{:} \;+\; FCH_2CH_2-\overset{..}{\underset{..}{O}}-H \;\longrightarrow\; CF_3CH_2-\overset{..}{\underset{..}{O}}-H \;+\; FCH_2CH_2-\overset{..}{\underset{..}{O}}\overset{\ominus}{:}$$

weaker base        weaker acid            conjugate acid      conjugate base
                                        stronger acid       stronger base

The presence of electronegative atoms like F will make an acid stronger by the inductive effect. Three F atoms will make a stronger acid than just one F atom. From the point of view of the anions, the anion with three F atoms is more stable, that is, a weaker base. None of these structures has other resonance forms.

**1-19** Solutions for (a) and (b) are presented in the Solved Problem in the text. Here, the newly formed bonds are shown in bold.

(c)

$$H-\overset{H}{\underset{\mid}{B}}-H \;+\; CH_3-\overset{..}{\underset{..}{O}}-CH_3 \;\rightleftharpoons\; H-\overset{H}{\underset{}{\overset{\mid}{\underset{}{B}}}}\overset{\ominus}{}-H$$

Lewis acid—electrophile    Lewis base—nucleophile          $CH_3-\underset{\oplus}{O}-CH_3$

(d)

$$CH_3-\overset{:O:}{\overset{\parallel}{C}}-H \;+\; \overset{\ominus}{:}\overset{..}{O}-H \;\rightleftharpoons\; CH_3-\overset{:\overset{\ominus}{O}:}{\underset{}{\overset{\mid}{C}}}-H$$

Lewis acid—electrophile    Lewis base—nucleophile          $\overset{..}{\underset{..}{O}}-H$

The problem asks for the *imaginary* movement of electrons in the resonance forms.

(e) Bronsted-Lowry—proton transfer

$$H-\overset{H}{\underset{\overset{\mid}{H}}{\overset{\mid}{C}}}-\overset{:O:}{\overset{\parallel}{C}}-H \;+\; \overset{\ominus}{:}\overset{..}{O}-H \;\rightleftharpoons\; \left\{ H-\overset{\ominus}{\underset{..}{C}}-\overset{:O:}{\overset{\parallel}{C}}-H \;\longleftrightarrow\; H-\overset{H}{C}=\overset{:\overset{\ominus}{O}:}{\underset{}{C}}-H \right\} \;+\; H-\overset{..}{O}-H$$

acid—electrophile      base—nucleophile

(f)

$$CH_3-\overset{..}{\underset{\underset{H}{\mid}}{N}}-H \;+\; CH_3-\overset{..}{\underset{..}{Cl}}: \;\rightleftharpoons\; CH_3-\overset{CH_3}{\underset{\underset{H}{\mid}}{\overset{\mid}{\overset{\oplus}{N}}}}-H \;+\; :\overset{..}{\underset{..}{Cl}}\overset{\ominus}{:}$$

Lewis base—nucleophile     Lewis acid—electrophile

1-19 continued

(g) $CH_3-\underset{\underset{H}{|}}{C}=CH_2$ + $\underset{\underset{F}{|}}{\overset{\overset{F}{|}}{B}}-F$ ⇌ $CH_3-\underset{\underset{\oplus}{|}}{C}-\underset{\underset{H_2}{|}}{C}\!-\!\underset{\underset{F}{|}}{\overset{\overset{F}{|}}{B}}\!\overset{\ominus}{-}\!F$

Lewis base—
nucleophile        Lewis acid—
electrophile

1-20

(a) { $\ddot{O}=\overset{\oplus}{\underset{}{S}}-\ddot{\underset{..}{O}}\!:^{\ominus}$ ⟷ $^{\ominus}:\ddot{\underset{..}{O}}-\overset{\oplus}{S}=\ddot{O}$ ⟷ $\ddot{O}=\overset{}{S}=\ddot{O}$ }

(b) { $\ddot{O}=\overset{\oplus}{\underset{}{O}}-\ddot{\underset{..}{O}}\!:^{\ominus}$ ⟷ $^{\ominus}:\ddot{\underset{..}{O}}-\overset{\oplus}{O}=\ddot{O}$ }

(c) The last resonance form of $SO_2$ has no equivalent form in $O_3$. Sulfur, a third row element, can have more than eight electrons around it because of d orbitals, whereas oxygen, a second row element, must adhere strictly to the octet rule.

1-21 (a) CARBON! (the best element)    (b) oxygen    (c) phosphorus    (d) chlorine

1-22

| valence e⁻ → | 1 | 2 | 3 | 4 | 5 | 6 | 7 | 8 |
|---|---|---|---|---|---|---|---|---|
| | H | | | | | | | He (2e⁻) |
| | Li | Be | B | C | N | O | F | Ne |
| | | | | | P | S | Cl | |
| | | | | | | | Br | |
| | | | | | | | I | |

1-23

(a) ionic only    (b) covalent (H—O⁻) and ionic (Na⁺ ⁻OH)

(c) covalent (H—C and C—Li), but the C—Li bond is strongly polarized

(d) covalent only    (e) covalent (H—C and C—O⁻) and ionic (Na⁺ ⁻OCH₃)

(f) covalent (H—C and C=O and C—O⁻) and ionic (HCO₂⁻ Na⁺)    (g) covalent only

1-24

(a)

(b)

CANNOT EXIST

$NCl_5$ violates the octet rule; nitrogen can have no more than eight electrons (or four atoms) around it. Phosphorus, a third-row element, can have more than eight electrons because phosphorus can use d orbitals in bonding, so $PCl_5$ is a stable, isolable compound.

1-25 Your Lewis structures may look different from these. As long as the atoms are connected in the same order and by the same type of bond, they are equivalent structures. For now, the exact placement of the atoms on the page is not significant.

(a) $H-\ddot{N}-\ddot{N}-H$ (with H below each N)

(b) $H-\ddot{N}=\ddot{N}-H$

(c) $H-\underset{\underset{H}{|}}{\overset{\overset{H}{|}}{C}}-\overset{\oplus}{\underset{\underset{H}{|}}{\overset{\overset{H}{|}}{N}}}-\underset{\underset{H}{|}}{\overset{\overset{H}{|}}{C}}-H$    $:\ddot{\underset{..}{Cl}}\!:^{\ominus}$

**13**

1-25 continued

(d)
```
      H
      |
 H —  C — C ≡ N:
      |
      H
```

(e)
```
      H  :O:
      |   ||
 H — C — C — H
      |
      H
```

(f)
```
      H  :O:  H
      |   ..  |
 H — C — S — C — H
      |   ..  |
      H       H
```

(g)
```
       :O:
        ||
 H — ·O· — S — ·O· — H
        ||
       :O:
```

(h)
```
      H
      |    ..
 H — C — N = C = ·O·
      |
      H
```

(i)
```
      H   :O:    H
      |    ||    |
 H — C — ·O· — S — ·O· — C — H
      |    ||    |
      H   :O:    H
```

(j)
```
            H
            \
      H  :N  H
      |   ||  |
 H — C — C — C — H
      |      |
      H      H
```

(k)
```
          H
        H | H
         \|/
          C
          |      ..
 H — C — C — N = ·O·
          |
          C
        H/|\H
          H
```
(with H— on left carbon having three H's)

1-26

(a)
```
      H   O   H            :O:
      |   ||  |             ||
 H — C — C — C — C = C — C — ·O· — H
      |       |   |   |
      H       H   H   H
```

(b)
```
            H  :O:  H  :O:
            |   ||  |   ||
 :N ≡ C — C — C — C — C — H
            |       |
            H       H
```

(c)
```
       H — ·O:  H  :O:
            |   |   ||
 H — C = C — C — C — C — ·O· — H
      |   |   |   |
      H   H   H   H
```

(d)
```
                       :O:      H
                        ||      |
 H — C = C — C = C — C — ·O· — C — H
      |   |   |   |             |
      H   H   C   H             H
              |
            H/|\H
              H
```

1-27 In each set below, the second structure is a more correct line formula. Since chemists are human (surprise!), they will take shortcuts where possible; the first structure in each pair uses a common abbreviation, either COOH or CHO. Make sure you understand that COOH does not stand for C—O—O—H. Likewise for CHO.

(a)

OR

(b)

OR

(c)

OR

(d)

OR

1-28

(a)

$$H-\overset{\overset{\displaystyle H}{|}}{\underset{\underset{\displaystyle H}{|}}{C}}-\overset{\overset{\displaystyle H}{|}}{\underset{\underset{\displaystyle H}{|}}{C}}-\overset{\overset{\displaystyle H}{|}}{\underset{\underset{\displaystyle H}{|}}{C}}-\overset{\overset{\displaystyle H}{|}}{\underset{\underset{\displaystyle H}{|}}{C}}-H$$

and

These are the only two possibilities, but your structures may appear different—making models will help you visualize these structures.

(b)

and

These are the only two possibilities.

(c)

and

These are the only two possibilities, but your structures may appear different—making models will help you visualize these structures.

(d)

Think of all the ways an oxygen **could** be added to the structures in (c). There are many more!

(e) There are several other possibilities **as we**ll. Your answer may be correct even if it does not appear here. Check with others in your study **group.**

(f)

These are the only three structures with this molecular formula.

1-29

(a) only three possible structures

HOCH₂CH₂CH₃

CH₃CH₂OCH₃

CH₃CH(OH)CH₃

1-29 continued

(b) This is most (maybe all) of the possible structures.

CH₃CH₂CHO     CH₃COCH₃     H₂C═CHCH₂OH     H₂C═C(OH)CH₃

- - - - - - - - - - - - - - - - - - - - - - - - - - - - - - - - - - - - - - - - - - - - - - - -

H₂C═CHOCH₃     HOCH═CHCH₃

- - - - - - - - - - - - - - - - - - - - - - - - - - - - - - - - - - - - - -

1-30   General rule: *molecular formulas of stable hydrocarbons must have an even number of hydrogens.*
The formula CH₂ does not have enough atoms to bond with the four orbitals of carbon.

one carbon:

CH₄

two carbons:   H─C≡C─H     H─C═C─H     H─C─C─H

C₂H₂     C₂H₄     C₂H₆

three carbons:   H─C≡C─C─H     H─C═C─C─H     H─C─C─C─H

C₃H₄     C₃H₆     C₃H₈

1-31

(a)

(b)

(c)

(d)

(e)

1-31 continued

(f)

(g)

(h)

1-32 (a) $C_5H_5N$   (b) $C_4H_9N$   (c) $C_4H_4O$   (d) $C_4H_9NO_2$   (e) $C_{11}H_{19}NO$

(f) $C_6H_{12}O$   (g) $C_7H_8O_3S$   (h) $C_7H_8O_3$

1-33

(a) $100\% - 62.0\%\ C - 10.4\%\ H = 27.6\%$ oxygen

$$\frac{62.0\ \text{g C}}{12.0\ \text{g/mole}} = 5.17\ \text{moles C} \div 1.73\ \text{moles} = 2.99 \approx 3\ \text{C}$$

$$\frac{10.4\ \text{g H}}{1.01\ \text{g/mole}} = 10.3\ \text{moles H} \div 1.73\ \text{moles} = 5.95 \approx 6\ \text{H}$$

$$\frac{27.6\ \text{g O}}{16.0\ \text{g/mole}} = 1.73\ \text{moles O} \div 1.73\ \text{moles} = 1\ \text{O}$$

(b) empirical formula = $\boxed{C_3H_6O}$ $\Rightarrow$ empirical weight = 58

molecular weight = 117, about double the empirical weight

$\Rightarrow$ double the empirical formula = molecular formula =

$$\boxed{C_6H_{12}O_2}$$

(c) some possible structures—MANY other structures are possible:

1-34 Non-zero formal charges are shown by the atoms.

(a)

(b)

(c)   (d)   (e)

1-35 The symbols "$\delta^+$" and "$\delta^-$" indicate bond polarity by showing partial charge. Electronegativity differences greater than or equal to 0.5 are considered large.

(a) $\overset{\delta^+}{C}-\overset{\delta^-}{Cl}$ large

(b) $\overset{\delta^-}{C}-\overset{\delta^+}{H}$ small

(c) $\overset{\delta^-}{C}-\overset{\delta^+}{Li}$ large

(d) $\overset{\delta^+}{C}-\overset{\delta^-}{N}$ small

(e) $\overset{\delta^+}{C}-\overset{\delta^-}{O}$ large

**17**

1-35 continued

(f) $\overset{\delta^-}{C}-\overset{\delta^+}{B}$    (g) $\overset{\delta^-}{C}-\overset{\delta^+}{Mg}$    (h) $\overset{\delta^-}{N}-\overset{\delta^+}{H}$    (i) $\overset{\delta^-}{O}-\overset{\delta^+}{H}$    (j) $\overset{\delta^+}{C}-\overset{\delta^-}{Br}$

    large         large         large         large         small

1-36 Resonance forms must have atoms in identical positions. If any atom moves position, it is a different structure.

(a) Different compounds—a hydrogen atom has changed position.
(b) Resonance forms—only the position of electrons is different.
(c) Different compounds—a hydrogen atom has changed position.
(d) Resonance forms—only the position of electrons is different.
(e) Different compounds—a hydrogen atom has changed position.
(f) Resonance forms—only the position of electrons is different.
(g) Resonance forms—only the position of electrons is different.
(h) Different compounds—a hydrogen atom has changed position.
(i) Resonance forms—only the position of electrons is different.
(j) Resonance forms—only the position of electrons is different.

1-37

(a)

(b)

(c)

(d)

When drawing resonance forms with charges on ring atoms, it helps keep track by writing the C or N or O with the charge.

18
Copyright © 2013 Pearson Education, Inc.

1-37 continued

(e) [resonance structures]

(f) [resonance structures]

(g) [resonance structures]

(h) [resonance structures]

(i)

$$H-C=C-C=C-\overset{+}{C}-CH_3 \longleftrightarrow H-C=C-\overset{+}{C}-C=C-CH_3$$

with H substituents on carbons

$$H-\overset{+}{C}-C=C-C=C-CH_3$$

with H substituents on carbons

(j) No resonance forms—the charge must be on an atom next to a double or triple bond, or next to a non-bonded pair of electrons, in order for resonance to delocalize the charge.

1-38   One of the fundamental principles of acidity is that the strength of the acid depends on the stability of the conjugate base.  The two primary factors governing the strength of organic acids are resonance and inductive effects; of these two, resonance is usually the stronger and more important effect.  For a more complete discussion, see Appendix 2 in this manual, especially section III.A.

Any organic structure with an $-SO_3H$ in it is a very strong acid because the anion has three significant resonance contributors; see the solution to 1-18(g).  An organic structure with $-COOH$ is moderately strong since the conjugate base has two significant resonance contributors, also shown in the solution to 1-18(g).  A structure with a simple $-OH$ does not have any resonance stabilization of the conjugate base, so it is the weakest acid.  Within each group, inductive effects from an electronegative atom like Cl will have a small effect.

$$CH_3CH_2OH \; > \; CH_3CH_2COOH \; > \; ClCH_2CH_2COOH \; > \; CH_3CHClCOOH \; > \; CH_3CH_2SO_3H$$

|  (b)  |  (c)  |  (e)  |  (d)  |  (a)  |
|-------|-------|-------|-------|-------|
| *weakest acid* |  | Cl further from COOH | Cl closer to COOH | *strongest acid* |

**19**

1-39

(a)

(b) Protonation at nitrogen #3 gives four resonance forms that delocalize the positive charge over all three nitrogens and a carbon—a very stable condition. Nitrogen #3 will be protonated preferentially, which we interpret as being more basic.

1-40

(a) $\left\{ \begin{array}{ccc} CH_3-\overset{..}{\underset{H}{C}}-C\equiv N\colon^{\ominus} & \longleftrightarrow & CH_3-\underset{H}{C}=C=\overset{..}{N}\colon^{\ominus} \end{array} \right\}$

minor          major (negative charge on electronegative atom)

(b) $\left\{ \begin{array}{ccccc} \colon\overset{\colon}{O}\colon^{\ominus} & & & & \\ CH_3-C=C-\overset{\oplus}{C}-CH_3 & \longleftrightarrow & CH_3-\overset{\colon O\colon^{\ominus}}{C}-C=C-CH_3 & \longleftrightarrow & CH_3-\overset{\colon O\colon}{C}-C=C-CH_3 \\ \quad\; H \;\; H & & \oplus \;\; H \;\; H & & \qquad\; H \;\; H \end{array} \right\}$

minor                           minor                           major—full octets, no charge separation

(c) $\left\{ \begin{array}{ccccc} \colon O\colon \;\; \colon O\colon & & \colon\overset{\colon}{O}\colon^{\ominus} \;\;\; \colon O\colon & & \colon O\colon \;\;\; \colon\overset{\colon}{O}\colon^{\ominus} \\ CH_3-C-\overset{\ominus}{\underset{H}{C}}-C-CH_3 & \longleftrightarrow & CH_3-C=C-\overset{}{\underset{H}{C}}-CH_3 & \longleftrightarrow & CH_3-C-C=\overset{}{\underset{H}{C}}-CH_3 \end{array} \right\}$

minor                           major                           major

negative charge on electronegative atoms—equal energy

**1-40 continued**

(d)

$$\left\{ \ce{CH3-\underset{H}{\overset{\ominus}{C}}-\underset{H}{\overset{}{C}}=\underset{H}{\overset{}{C}}-\overset{\oplus}{N}=\overset{..}{\underset{..}{O}} \underset{:\overset{..}{\underset{\ominus}{O}}:}{} \longleftrightarrow CH3-C=C-\overset{\ominus}{C}-\overset{\oplus}{N}=\overset{..}{O} \longleftrightarrow CH3-C=C-C=\overset{\oplus}{N}-\overset{..}{\underset{..}{O}}{}^{\ominus} } \right\}$$

minor          minor          major—negative charge on electronegative atoms

*NOTE: The two structures below are resonance forms, varying from the first two structures in part (d) by the different positions of the double bonds in the $NO_2$. Usually, chemists omit drawing the second form of the $NO_2$ group although we all understand that its presence is implied. It is a good idea to draw all of the resonance forms until they become second nature. The importance of understanding resonance forms cannot be overemphasized.*

$$\left\{ \ce{CH3-\overset{\ominus}{C}-C=C-\overset{\oplus}{N}-\overset{..}{O}{}^{\ominus}} \longleftrightarrow \ce{CH3-C=C-\overset{\ominus}{C}-\overset{\oplus}{N}-\overset{..}{O}{}^{\ominus}} \right\}$$

(e)

$$\left\{ \ce{CH3CH2-\underset{\oplus}{\overset{\overset{..}{NH2}}{C}}-\overset{..}{N}H2} \longleftrightarrow \ce{CH3CH2-\underset{}{\overset{\overset{\oplus}{NH2}}{C}}=\overset{..}{N}H2} \longleftrightarrow \ce{CH3CH2-\overset{\overset{..}{NH2}}{C}=\overset{\oplus}{N}H2} \right\}$$

minor       major—full octets       major—full octets

equal energy

**1-41**

(a)

$$\ce{CH3-\underset{H}{\overset{\oplus}{C}}-CH3}$$

no resonance stabilization

$$\left\{ \ce{CH3-\underset{H}{\overset{\oplus}{C}}-\overset{..}{O}-CH3} \longleftrightarrow \ce{CH3-\underset{H}{\overset{}{C}}=\overset{..}{\overset{\oplus}{O}}-CH3} \right\}$$

more stable—resonance stabilized

(b)

$$\left\{ \ce{CH2=\underset{H}{\overset{\oplus}{C}}-\underset{H}{\overset{}{C}}-CH3} \longleftrightarrow \ce{\overset{\oplus}{C}H2-\underset{H}{\overset{}{C}}=\underset{H}{\overset{}{C}}-CH3} \right\}$$

more stable—resonance stabilized

$$\ce{CH2=\underset{H}{\overset{}{C}}-\underset{H}{\overset{\overset{H}{|}}{C}}-\overset{\oplus}{C}H2}$$

no resonance stabilization

(c)

$$\ce{H-\underset{H}{\overset{\ominus}{\overset{..}{C}}}-CH3}$$

no resonance stabilization

$$\left\{ \ce{H-\underset{H}{\overset{\ominus}{\overset{..}{C}}}-C#N:} \longleftrightarrow \ce{H-\underset{H}{\overset{}{C}}=C=\overset{..}{\underset{..}{N}}{}^{\ominus}} \right\}$$

more stable—resonance stabilized

(d)

more stable—resonance stabilized

no resonance stabilization

1-41 continued

(e)

more stable—resonance stabilized                    no resonance stabilization

1-42 These $pK_a$ values from the text, Table 1-5 and Appendix 4, provide the answers. The lower the $pK_a$, the stronger the acid. Water and $CH_3OH$ are very close.

least acidic                                                                                        most acidic

$NH_3$ < $H_2O$ ~ $CH_3OH$ < $CH_3COOH$ < HF < $H_3O^+$ < $H_2SO_4$

33 (or 36)    15.7    15.5       4.74     3.2      −1.7      −5

1-43 Conjugate bases of the weakest acids will be the strongest bases. The $pK_a$ values of the conjugate acids are listed here. (The relative order of some bases was determined from the $pK_a$ values in Appendix 4 of the textbook.)

least basic                                                                                        most basic

$HSO_4^-$ < $H_2O$ < $CH_3COO^-$ < $NH_3$ < $CH_3O^-$ ~ NaOH < $^-NH_2$

from −5    from −1.7    from 4.74    from 9.4    from 15.5   from 15.7    from 33 (or 36)

1-44

(a)   $pK_a = -\log_{10} K_a = -\log_{10}(5.2 \times 10^{-5}) =$ **4.3** for phenylacetic acid

for propionic acid, $pK_a$ 4.87: $K_a = 10^{-4.87} =$ **1.35 x 10$^{-5}$**

(b)   Phenylacetic acid is 3.9 times stronger than propionic acid.

$$\frac{5.2 \times 10^{-5}}{1.35 \times 10^{-5}} = 3.9$$

(c)

weaker acid                                        stronger acid

Equilibrium favors the weaker acid and base. In this reaction, **reactants** are favored.

1-45 The newly formed bond is shown in bold.

(a)

nucleophile     electrophile
Lewis base       Lewis acid

(b)

nucleophile
Lewis base

electrophile
Lewis acid

1-45   continued

(c)

$$H-\underset{\underset{\displaystyle }{\overset{\displaystyle \ddot{O}}{\|}}}{C}-H \;+\; \overset{\displaystyle H}{\underset{\displaystyle H}{:N-H}} \longrightarrow H-\underset{\underset{\displaystyle H-\overset{\displaystyle \oplus}{\underset{\displaystyle H}{N}}-H}{\overset{\displaystyle :\ddot{O}:^{\ominus}}{\overset{\displaystyle |}{C}}}-H$$

electrophile     nucleophile
Lewis acid     Lewis base

(d)   $CH_3-\ddot{N}H_2$     $CH_3CH_2-\ddot{Cl}:$   $\longrightarrow$   $CH_3-\overset{\displaystyle H}{\underset{\displaystyle H}{\overset{\oplus}{N}}}-CH_2CH_3$  +  $:\ddot{\underset{..}{Cl}}:^{\ominus}$

    nucleophile      electrophile
    Lewis base      Lewis acid

(e)   $CH_3-\underset{\overset{\displaystyle \|}{\displaystyle }}{\overset{\displaystyle :O:}{C}}-CH_3$ + $H-\ddot{O}-\underset{\underset{\displaystyle :O:}{\|}}{\overset{\displaystyle :O:}{\overset{\displaystyle \|}{S}}}-\ddot{O}H$ $\longrightarrow$ $CH_3-\underset{\overset{\displaystyle \|}{\displaystyle }}{\overset{\displaystyle :\overset{\oplus}{O}-H}{C}}-CH_3$ + $^{\ominus}:\ddot{O}-\underset{\underset{\displaystyle :O:}{\|}}{\overset{\displaystyle :O:}{\overset{\displaystyle \|}{S}}}-\ddot{O}H$

nucleophile
Lewis base

electrophile
Lewis acid

plus resonance forms

(f)   $(CH_3)_3C-\ddot{\underset{..}{Cl}}:$  +  $AlCl_3$  $\longrightarrow$  $H_3C-\underset{\underset{\displaystyle CH_3}{|}}{\overset{\overset{\displaystyle CH_3}{|}}{C}}{\oplus}$  +  $:\ddot{\underset{..}{Cl}}-\underset{\underset{\displaystyle Cl}{|}}{\overset{\overset{\displaystyle Cl}{|}}{\overset{\ominus}{Al}}}-Cl$

   nucleophile       electrophile
   Lewis base       Lewis acid

This may also be written in two steps:  association of the Cl with Al, and a second step where the C—Cl bond breaks.

(g)   $CH_3-\underset{\underset{\displaystyle H}{|}}{\overset{\overset{\displaystyle :O:}{\|}}{C}}-CH_2$  +  $^{-}:\ddot{O}-H$  $\longrightarrow$  $CH_3-\overset{\displaystyle :\ddot{O}:^{\ominus}}{C}=CH_2$  +  $H-\ddot{O}-H$

              nucleophile
electrophile      Lewis base
Lewis acid

(h)   $F-\underset{\underset{\displaystyle F}{|}}{B}-F$    $CH_2=CH_2$  $\longrightarrow$  $F-\underset{\underset{\displaystyle F}{|}}{\overset{\overset{\displaystyle F}{|}}{\overset{\ominus}{B}}}-CH_2-\overset{\oplus}{C}H_2$

electrophile    nucleophile
Lewis acid     Lewis base

(i)   $^{\ominus}BF_3-CH_2-\overset{\oplus}{C}H_2$ + $CH_2=CH_2$ $\longrightarrow$ $^{\ominus}BF_3-CH_2-CH_2-CH_2-\overset{\oplus}{C}H_2$

electrophile      nucleophile
Lewis acid      Lewis base

1-46

(a) $H_2SO_4$ + $CH_3COO^-$ ⇌ $HSO_4^-$ + $CH_3COOH$

(b) $CH_3COOH$ + $(CH_3)_3N:$ ⇌ $CH_3COO^-$ + $(CH_3)_3\overset{\oplus}{N}-H$

(c) + $^-OH$ ⇌ + $H_2O$

(d) $HO-\overset{\overset{O}{\|}}{C}-OH$ + 2 $^-OH$ ⇌ $^-O-\overset{\overset{O}{\|}}{C}-O^-$ + **2** $H_2O$

(e) $H_2O$ + $NH_3$ ⇌ $HO^-$ + $^+NH_4$

(f) $(CH_3)_3\overset{\oplus}{N}-H$ + $^-OH$ ⇌ $(CH_3)_3N:$ + $H_2O$

(g) $HCOOH$ + $CH_3O^-$ ⇌ $HCOO^-$ + $CH_3OH$

(h) $\overset{\oplus}{N}H_3CH_2COOH$ + 2 $^-OH$ ⇌ $NH_2CH_2COO^-$ + **2** $H_2O$

1-47 The critical principle: *the strength of an acid is determined by the stability of its conjugate base.*
(a) conjugate bases

(b) **X** is a stronger acid than **W** because the more electronegative N in **X** can support the negative charge better than carbon, so the anion of **X** is more stable than the anion of **W**.

(c) **Y** is a stronger acid than **X** because the negative charge in **Y** is stabilized by the *inductive effect* from the electronegative oxygen substituent, the OH.

(d) **Z** is a stronger acid than **Y** because of two effects: O is more electronegative than N and can support the negative charge of the anion better, plus the anion of **Z** has two EQUIVALENT resonance forms which is particularly stable.

**1-48** Basicity is a measure of the ability of an electron pair to form a new bond with $H^+$ of an acid. Availability of electrons is the key to basicity.

(a)

$$\left\{ \begin{array}{c} \overset{H}{\underset{H}{\diagdown}} \ddot{N}-C \overset{\ddot{O}:}{\diagdown} CH_3 \end{array} \longleftrightarrow \begin{array}{c} \overset{H}{\underset{H}{\diagdown}} \overset{\oplus}{N}=C \overset{:\ddot{O}:^{\ominus}}{\diagdown} CH_3 \end{array} \right\}$$

$$\overset{H}{\underset{H}{\diagdown}} \ddot{N}-CH_2CH_3$$

The electron pair in acetamide is *delocalized* over many atoms, not readily available for bonding with $H^+$, making it a much weaker base than ethylamine.

The electron pair in ethylamine is *localized*, not distributed over many atoms. It is readily available for bonding with $H^+$.

(b) Acetamide has two possible sites of protonation, the N and the O. HA symbolizes a generic acid.

Protonation of the N produces an ion with no particular stabilization.

Protonation of the O produces an ion with *resonance stabilization*, a far more stable product than protonation on N. The O is the more basic atom in this structure.

**1-49**

(a) $CH_3CH_2-O-H \ + \ CH_3-Li \longrightarrow CH_3CH_2-O^- \ Li^+ \ + \ CH_4$

(b) The conjugate acid of $CH_3Li$ is $CH_4$. Table 1-5 gives the $pK_a$ of $CH_4$ as $> 40$, one of the weakest acids known. The conjugate base of one of the weakest acids known must be one of the strongest bases known.

**1-50** (a) conjugate bases

The negative charge cannot be delocalized by resonance in either of these two structures.

$$CH_3CH_2-\ddot{O}:^{\ominus}$$

**1-50 continued**

(b)

*most stable—delocalization of negative charge by both resonance and induction (electronegative F)*

*delocalization of negative charge by resonance and a weaker inductive effect as F atoms are farther away*

*delocalization of negative charge by resonance only*

*weak delocalization of negative charge by induction only*

*least stable—no delocalization of negative charge*

(c) The strongest acid will have the most stable conjugate base. The actual $pK_a$ values (some from text Appendix 4) are listed beneath each acid.

F₃C—C(=O)—OH > F₃C—CH₂—C(=O)—OH > H₃C—C(=O)—OH > H₃C—C(=O)—O—OH > CH₃CH₂OH

*strongest acid*
$pK_a$ 0.2

$pK_a$ 3.07

$pK_a$ 4.7

$pK_a$ 8.2

*weakest acid*
$pK_a$ 15.9

**1-51**
(a) conjugate acids

The oxygen is more basic than the nitrogen in this structure. See the solution to 1-48(b) on the previous page.

CH₃CH₂—N⁺H₃

H—OH    CH₃CH₂—O⁺H₂    H—NH₂

*minor*

(b) order of decreasing stability ($pK_a$ values from Appendix 4 in the text)

H—NH₂ > H—OH > CH₃CH₂—N⁺H₃ > [N—C resonance structure] > CH₃CH₂—O⁺H₂

*most stable—neutral molecule with less polar NH bonds, lower electronegativity*

$pK_a$ 36

*neutral molecule*
$pK_a$ 15.7

*positive charge on less electronegative atom*
$pK_a$ 10.7

*positive charge on more electronegative atom but resonance stabilized*
$pK_a$ 0.0

*least stable—positive charge on more electronegative atom*
$pK_a$ −2.4

1-51 continued
(c) The weakest conjugate acid will form the **strongest** conjugate base.

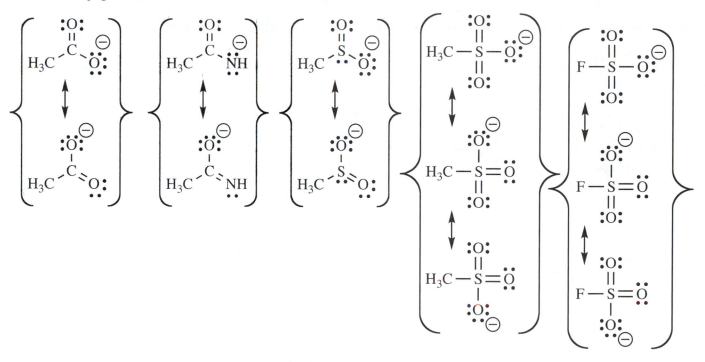

1-52 (a) conjugate bases

(b)

*most stable—*
*delocalization of*
*negative charge by*
*three resonance*
*forms and*
*induction (F and S*
*more electro-*
*negative than C)*

*delocalization of*
*negative charge by*
*three resonance*
*forms and*
*induction (S more*
*electronegative*
*than C)*

*delocalization of*
*negative charge by*
*two resonance forms*
*and induction (S*
*more electronegative*
*than C)*

*delocalization of*
*negative charge by*
*two equivalent*
*resonance forms*

*least stable—*
*delocalization of*
*negative charge by*
*two resonance*
*forms, although*
*non-equivalent*

(c) The strongest acid will have the most **stable** conjugate base. The $pK_a$ values are listed beneath each acid.

*strongest acid*
$pK_a < -5$

$pK_a$ −1.2

$pK_a$ 2.3

$pK_a$ 4.7

*weakest acid*
$pK_a$ 16

27

Copyright © 2013 Pearson Education, Inc.

1-53  In each product, the new bond is shown in bold.

(a)  $CH_3-\overset{\cdot\cdot}{N}-CH_3$  +  $H-\overset{\cdot\cdot}{\underset{\cdot\cdot}{Cl}}:$  $\longrightarrow$  $CH_3-\overset{\oplus}{\underset{H}{N}}-CH_3$  +  $:\overset{\cdot\cdot}{\underset{\cdot\cdot}{Cl}}:^{\ominus}$

Lewis base— nucleophile    Lewis acid— electrophile

(b)  $CH_3-\overset{\cdot\cdot}{N}-CH_3$  +  $H_3C-\overset{\cdot\cdot}{\underset{\cdot\cdot}{Cl}}:$  $\longrightarrow$  $CH_3-\overset{\oplus}{\underset{H}{\overset{CH_3}{N}}}-CH_3$  +  $:\overset{\cdot\cdot}{\underset{\cdot\cdot}{Cl}}:^{\ominus}$

Lewis base— nucleophile    Lewis acid— electrophile

(c)  $CH_3-\overset{\overset{:O:}{||}}{C}-H$  +  $H-\overset{\cdot\cdot}{\underset{\cdot\cdot}{Cl}}:$  $\longrightarrow$  $CH_3-\overset{\overset{\oplus}{\overset{:O}{||}}\,^H}{C}-H$  +  $:\overset{\cdot\cdot}{\underset{\cdot\cdot}{Cl}}:^{\ominus}$

Lewis base— nucleophile    Lewis acid— electrophile

(d)  $CH_3-\overset{\overset{:O:}{||}}{C}-H$  +  $^{\ominus}\overset{\cdot\cdot}{\underset{\cdot\cdot}{O}}-CH_3$  $\longrightarrow$  $CH_3-\overset{\overset{:\overset{\ominus}{O}:}{|}}{\underset{\overset{|}{:O-CH_3}}{C}}-H$

Lewis acid— electrophile    Lewis base— nucleophile

(e)  $H-\overset{\overset{H}{|}}{C}-\overset{\overset{:O:}{||}}{C}-H$  +  $^{\ominus}\overset{\cdot\cdot}{\underset{\cdot\cdot}{O}}-CH_3$  $\longrightarrow$  $\underset{H}{\overset{H}{{>}}}C=\overset{\overset{:\overset{\ominus}{O}:}{|}}{C}-H$  +  $H-\overset{\cdot\cdot}{\underset{\cdot\cdot}{O}}-CH_3$

Lewis acid— electrophile    Lewis base— nucleophile

1-54  From the amounts of $CO_2$ and $H_2O$ generated, the milligrams of C and H in the original sample can be determined, thus giving by difference the amount of oxygen in the 5.00-mg sample.  From these values, the empirical formula and empirical weight can be calculated.

(a)  how much carbon in 14.54 mg $CO_2$

14.54 mg $CO_2$  x  $\dfrac{1 \text{ mmole } CO_2}{44.01 \text{ mg } CO_2}$  x  $\dfrac{1 \text{ mmole C}}{1 \text{ mmole } CO_2}$  x  $\dfrac{12.01 \text{ mg C}}{1 \text{ mmole C}}$  = 3.968 mg C

how much hydrogen in 3.97 mg $H_2O$

3.97 mg $H_2O$  x  $\dfrac{1 \text{ mmole } H_2O}{18.016 \text{ mg } H_2O}$  x  $\dfrac{2 \text{ mmoles H}}{1 \text{ mmole } H_2O}$  x  $\dfrac{1.008 \text{ mg H}}{1 \text{ mmole H}}$  = 0.444 mg H

how much oxygen in 5.00 mg estradiol

5.00 mg estradiol – 3.968 mg C – 0.444 mg H  =  0.59 mg O

*continued on next page*

1-54  continued

calculate empirical formula

$$\frac{3.968 \text{ mg C}}{12.01 \text{ mg/mmole}} = 0.3304 \text{ mmoles C} \div 0.037 \text{ mmoles} = 8.93 \approx 9 \text{ C}$$

$$\frac{0.444 \text{ mg H}}{1.008 \text{ mg/mmole}} = 0.440 \text{ mmoles H} \div 0.037 \text{ mmoles} = 11.9 \approx 12 \text{ H}$$

$$\frac{0.59 \text{ mg O}}{16.00 \text{ mg/mmole}} = 0.037 \text{ mmoles O} \div 0.037 \text{ mmoles} = 1 \text{ O}$$

empirical formula = $\boxed{C_9H_{12}O}$  ⟹  empirical weight = 136

(b)  molecular weight = 272, exactly twice the empirical weight

twice the empirical formula = molecular formula = $\boxed{C_{18}H_{24}O_2}$

1-55  (a)  Ascorbic acid has four OH groups that could act as acids.  The ionization of each shows that one gives a more stable conjugate base.

ascorbic acid

localized negative charge— no resonance forms

localized negative charge— no resonance forms

*continued on next page*

## 1-55 (a) continued

three resonance forms, two
with negative charge on
oxygen

two resonance forms, one with
negative charge on oxygen

(b) The ionization of the OH labeled "3" produces a conjugate base with three resonance forms, two of which have negative charge on oxygen. This OH is the most acidic group in ascorbic acid.

(c) The conjugate base of acetic acid, the acetate ion, $CH_3COO^-$, has two resonance forms (see the solution to problem 1-52(a), page 27 of this manual), each of which has a C=O and a negatively charged oxygen, similar to two of the resonance forms of the ascorbate ion. The acidity of these two very different molecules is similar because the stabilization of the conjugate base is so similar. *The strength of an acid is determined by the stability of its conjugate base.*

Note to the student: Organic chemistry professors will ask you to "explain" questions, that is, to explain a certain trend in organic structures or behavior of an organic reaction. The professor is trying to determine two things: 1) does the student understand the principle underlying the behavior? 2) does the student understand how the principle applies in this particular example?

To answer an "explain" question, somewhere in your answer should be a statement of the principle, like: "*The strength of an acid is determined by the stability of its conjugate base.*" From there, show through a series of logical steps how the principle applies, like drawing resonance forms to show which acid has the most stable conjugate base through resonance or induction. Answering these questions is like crossing a creek on stepping stones. Each phrase or sentence is a step to the next stone. When strung together, the steps bridge the gap between the principle and the observation.

# CHAPTER 2—STRUCTURE AND PROPERTIES OF ORGANIC MOLECULES

2-1 *The fundamental principle of organic chemistry is that a molecule's chemical and physical properties depend on the molecule's structure:* the structure-function or structure-reactivity correlation. It is essential that you understand the three-dimensional nature of organic molecules, and there is no better device to assist you than a molecular model set. You are strongly encouraged to use models regularly when reading the text and working the problems.

(a) Requires use of models.

(b)

The wedge bonds represent bonds coming out of the plane of the paper toward you.
The dashed bonds represent bonds going behind the plane of the paper.

2-2 (a) The hybridization of oxygen is sp³ since it has two sigma bonds and two pairs of nonbonding electrons. The reason that the bond angle of 104.5° is less than the perfect tetrahedral angle of 109.5° is that the lone pairs in the two sp³ orbitals are repelling each other more strongly than the electron pairs in the sigma bonds, thereby compressing the bond angle.

(b) The electrostatic potential map for water shows that the hydrogens have low electron potential (blue), and the area of the unshared electron pairs in sp³ orbitals has high electron potential (red).

high electron potential (red)

low electron potential (blue)

repulsion                    compression

2-3 Each double-bonded atom is sp² hybridized with bond angles about 120°; geometry around sp² atoms is trigonal planar. In (a), all four carbons and the two hydrogens on the sp² carbons are all in one plane. Each carbon on the end is sp³ hybridized with tetrahedral geometry and bond angles about 109°. In (b), the two carbons, the nitrogen, and the two hydrogens on the sp² carbon and nitrogen are all in one plane. The CH₃ carbon is sp³ hybridized with tetrahedral geometry and bond angles about 109°.

(a)

(b)

2-4 The hybridization of the nitrogen and the triple-bonded carbon are sp, giving linear geometry (C—C—N are linear) and a bond angle around the triple-bonded carbon of 180°. The CH₃ carbon is sp³ hybridized, tetrahedral, with bond angles about 109°.

2-5

(a) linear, bond angle 180°

$$sp^2 \quad sp \quad sp^2$$

(b) All atoms are $sp^3$; tetrahedral geometry and bond angles of 109° around each atom.

not a bond—shows lone pair coming out of paper

not a bond—shows lone pair going behind paper

(c) All atoms are $sp^3$; tetrahedral geometry and bond angles of 109° around each atom.

not a bond—shows lone pair going behind paper

(d) trigonal planar around the carbonyl carbon (C=O), bond angles 120°; tetrahedral around the single-bonded oxygen and the $CH_3$, bond angles 109°

all atoms in one plane except for the two H on the $sp^3$ C

(e) tetrahedral around the $sp^3$ carbon; the other two carbons both $sp$, linear, bond angle 180°

all three carbon atoms in a line

(f) trigonal planar around the $sp^2$ carbon and nitrogen, bond angles 120°; around the $sp^3$ carbons, tetrahedral geometry and 109° bond angles

(g) linear, 180° bond angle, around the central carbon; trigonal planar, 120° bond angles, around the $sp^2$ carbon

**2-6** Carbon-2 is sp hybridized. If the p orbitals making the pi bond between C-1 and C-2 are in the plane of the paper (putting the hydrogens in front of and behind the paper), then the other p orbital on C-2 must be perpendicular to the plane of the paper, making the pi bond between C-2 and C-3 perpendicular to the paper. This necessarily places the hydrogens on C-3 in the plane of the paper. (Models will surely help.)

model of *perpendicular* π bonds

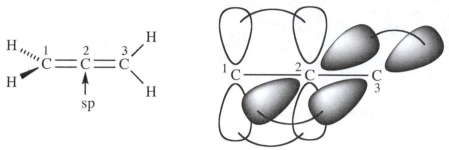

**2-7** For clarity, electrons in sigma bonds **are** not shown. Part (a) has been solved in the text.

(b) Carbon and oxygen are both $sp^2$ hybridized.

One pair of electrons on oxygen is always in an $sp^2$ orbital. The other pair of electrons is shown in a p orbital in the first resonance form, and in a pi bond in the second resonance form.

(c) Oxygen and both carbons are $sp^2$ hybridized.

2-7 continued

(d) All atoms are sp$^2$ hybridized except the C labeled as sp$^3$ (and H which is NEVER hybridized); all bond angles around the sp$^3$ carbon are 109°.

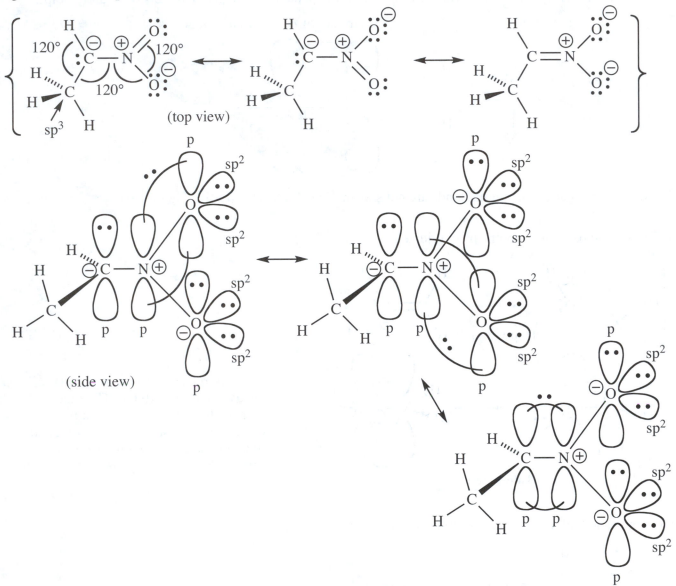

(side view)

(e) The nitrogen and the carbon bonded to it are sp hybridized; the left carbon is sp$^2$.

34

2-7 continued
(f) The boron and the oxygens bonded to it are sp² hybridized.

(g)

All oxygens are sp² with bond angle 120°.

35
Copyright © 2013 Pearson Education, Inc.

**2-8** Very commonly in organic chemistry, we have to determine whether two structures are the same or different, and if they are different, what structural features are different. In order for two structures to be the same, all bonding connections have to be identical, and in the case of double bonds, the groups must be on the same side of the double bond in both structures. (A good exercise to do with your study group is to draw two structures and ask if they are the same; or draw one structure and ask how to draw a different compound.)

(a) Different compounds; H and $CH_3$ on one carbon of the double bond, and $CH_3$ and $CH_2CH_3$ on the other carbon—same in both structures. Drawing a plane through the p orbitals shows the H and $CH_3$ are on the same side of the double bond in the first structure, and the H and the $CH_2CH_3$ are on the same side in the second structure, so they are DIFFERENT compounds.

These are DIFFERENT.

(b) Same compound; in the structure on the right, the right carbon has been rotated, but the bonding is identical between the two structures.

(c) Different compounds; H and Br on one carbon, F and Cl on the other carbon in both structures; H and Cl on the same side of the plane through the C=C in the first structure, and H and F on the same side of the plane through the C=C in the second structure, so they are DIFFERENT compounds.

(d) Same compound: in the structure on the right, the right carbon has been rotated 120°.

**2-9**

(a)

(b)

NOT INTER-CONVERTIBLE

two $CH_3$ on opposite sides of the C=N

two $CH_3$ on the same side of the C=N

(c) The $CH_3$ on the N is on the same side as another $CH_3$ no matter how it is drawn—only one possible structure.

**2-10**

(a)

cis         and         trans

(b) no *cis-trans* isomerism  ⎫
(c) no *cis-trans* isomerism  ⎬ two identical groups on one
(d) no *cis-trans* isomerism  ⎭ carbon of the double bond

(e)

cis         and         trans

(f)

and

These structures show cis-trans isomerism (also known as geometric isomerism), although "*cis*" and "*trans*" are not defined for this particular case.

**2-11** Models will be helpful here.
(a) Constitutional isomers—the carbon skeleton is different.
(b) *Cis-trans* isomers—the first is *trans*, the second is *cis*.
(c) Constitutional isomers—the bromines are on different carbons in the first structure, on the same carbon in the second structure.

2-11 continued

(d) same compound—just flipped over
(e) same compound—just rotated
(f) same compound—just rotated
(g) not isomers—different molecular formulas
(h) Constitutional isomers—the double bond has changed position.
(i) same compound—just reversed
(j) Constitutional isomers—the $CH_3$ groups are in different relative positions.
(k) Constitutional isomers—the double bond is in a different position relative to the $CH_3$ and an H has moved.

2-12

(a)   $2.4\,D = 4.8 \times \delta \times 1.23\,\text{Å}$

   $\delta = 0.41$, or 41% of a positive charge on carbon and 41% of a negative charge on oxygen

(b)

Resonance form **A** must be the major contributor. If **B** were the major contributor, the value of the charge separation would be between 0.5 and 1.0, meaning >50% of a positive charge on C and >50% of a negative charge on O. Even though **B** is "minor", it is quite significant, explaining in part the high polarity of the C=O.

2-13

Both $NH_3$ and $NF_3$ have a pair of nonbonding electrons on the nitrogen. In $NH_3$, the *direction* of polarization of the N—H bonds is *toward* the nitrogen; thus, all three bond polarities and the lone pair polarity reinforce each other. In $NF_3$, on the other hand, the direction of polarization of the N—F bonds is *away* from the nitrogen; the three bond polarities cancel the lone pair polarity, so the net result is a very small *molecular* dipole moment.

polarities reinforce; large dipole moment (1.50)

polarities oppose; small dipole moment (0.20)

2-14 Some magnitudes of dipole moments are difficult to predict; however, the direction of the dipole should be straightforward, in most cases. Actual values of molecular dipole moments are given in parentheses. (Each halogen atom has three nonbonded electron pairs, not shown below.) The C—H is usually considered nonpolar.

(a)   large dipole (1.54)

(b)   large dipole (1.81)

(c)   net dipole = 0

(d)   large dipole (1.70)

2-14 continued

(e) Each end oxygen has one-half negative charge as it is the composite of two resonance forms; see solution to 2-7(g).

net

small dipole (0.52)

(f) large dipole (2.95)

(g) large dipole (2.72)

(h) large dipole

(i) small dipole (0.67)

(j) large dipole (1.45)

(k) net dipole = 0

(l) net dipole = 0

In (k) through (m), the symmetry of the molecule allows the individual bond dipoles to cancel.

(m) net dipole = 0

2-15 With chlorines on the same side of the double bond, the bond dipole moments reinforce each other, resulting in a large net dipole. With chlorines on opposite sides of the double bond, the bond dipole moments exactly cancel each other, resulting in a zero net dipole.

large net dipole (2.40)

net dipole = 0

2-16

(a) (hydrogen bonds shown as wavy bond)

(b)

2-16 continued

(c)

(d)

2-17
(a) $(CH_3)_2CHCH_2CH_2CH(CH_3)_2$ has less branching and boils at a higher temperature than $(CH_3)_3CC(CH_3)_3$ .
(b) $CH_3(CH_2)_5CH_2OH$ can form hydrogen bonds and will boil at a much higher temperature than $CH_3(CH_2)_6CH_3$, which cannot form hydrogen bonds.
(c) $CH_3CH_2CH_2CH_2OH$ can form hydrogen bonds and boils at a higher temperature than $CH_3CH_2OCH_2CH_3$.
(d) $HOCH_2(CH_2)_4CH_2OH$ can form hydrogen bonds at both ends and has no branching; it will boil at a much higher temperature than $(CH_3)_3CCH(OH)CH_3$ .
(e) $(CH_3CH_2CH_2)_2NH$ has an N—H bond and can form hydrogen bonds; it will boil at a higher temperature than $(CH_3CH_2)_3N$, which cannot form hydrogen bonds.
(f) The second compound shown (**B**) has the higher boiling point for two reasons: **B** has a higher molecular weight than **A**; and **B**, a primary amine with two N—H bonds, has more opportunity for forming hydrogen bonds than **A**, a secondary amine with only one N—H bond.

A  ⬡N‑H          B  ⬡—N⟨ H / H

2-18

(a) $CH_3CH_2OCH_2CH_3$ can form hydrogen bonds with water and is more soluble than $CH_3CH_2CH_2CH_2CH_3$ which cannot form hydrogen bonds with water.
(b) $CH_3CH_2CH_2OH$ is more soluble in water because it has one fewer carbon than $CH_3CH_2OCH_2CH_3$ .
(c) $CH_3CH_2NHCH_3$ is more water soluble because it can form hydrogen bonds with water; $CH_3CH_2CH_2CH_3$ cannot form hydrogen bonds.
(d) $CH_3CH_2OH$ is more soluble in water. The polar O—H group forms hydrogen bonds with water, overcoming the resistance of the nonpolar $CH_3CH_2$ group toward entering the water. In $CH_3CH_2CH_2CH_2OH$, however, the hydrogen bonding from only one OH group cannot carry a four-carbon chain into the water; this substance is only slightly soluble in water.
(e) Both compounds form hydrogen bonds with water at the double-bonded oxygen, but only the smaller molecule ($CH_3COCH_3$) dissolves. The cyclic compound has too many nonpolar $CH_2$ groups to dissolve.

2-19

(a) [structure] alkane
(Usually, we use the term "alkane" only when no other groups are present.)

(b) [structure] alkene

(c) [structure] alkyne

continued on page 41

**39**

Note to the student: One of the most fundamental skills in organic chemistry is to identify what functional groups are present in a molecular structure. This page is called a "concept map" that prompts you with a series of questions about a structure, leading to the determination of the functional group. Do not memorize this chart! Use it as an aid as you are becoming familiar with the functional groups.

WHAT IS THE FUNCTIONAL GROUP?

This chart depicts only the most common functional groups. There are many more!

2-19 continued

(d) cycloalkyne and cycloalkene

(e) cycloalkane and alkene
(a "cycloalkene" would
have the C=C in the ring)

(f) aromatic hydrocarbon
and alkyne

(g) cycloalkene and alkene

(h) cycloalkane and alkane

(i) aromatic hydrocarbon
and cycloalkene

2-20

(a) $H-C=C-C-H$   alkene and aldehyde

(b) alcohol

(c) ketone

(d) ether and alkene

(e) carboxylic acid

(f) ether and alkene

(g) alkene and ketone

(h) aldehyde

(i) alcohol

2-21

(a) $H-C-C-C-N-C-H$   amide

(b) amine

(c) ester

(d) $H-C-C=C-C-Cl$   alkene and acid chloride

(e) ether

(f) $H-C-C-C-C\equiv N$   nitrile

**41**

2-21 continued

(g) carboxylic acid

(h) cycloalkene and cyclic ester

(i) cyclic ketone and ether

(j) cyclic amine

(k) cyclic amide

(l) amide

(m) cyclic ester

(n) cyclic amine and aldehyde

(o) cycloalkene and ketone

2-22 When the identity of a functional group depends on several atoms, all of those atoms should be circled. For example, an ether is an oxygen between two carbons, so the oxygen and both carbons should be circled. A ketone is a carbonyl group between two other carbons, so all those atoms should be circled.

(a) CH₂=CHCH₂COOCH₃
alkene, ester, —C-OCH₃

(b) CH₃—O—CH₃
ether

(c) CH₃—C—H aldehyde

(d) CH₃—C—NH₂
amide

(e) CH₃—N—CH₃
amine

(f) R—C—O—H
carboxylic acid

R is the symbol that organic chemists use to represent alkyl groups. Sometimes, when the identity of the group does not matter, aryl groups or others can also be included in the R abbreviation.

(g) aromatic, CH₂OH, alcohol

(h) C≡N nitrile, alkene

(i) ketone

## 2-22 continued

(j)

ketone
(also includes two
adjacent carbons)

alcohol

alcohol

HO

ketone

O

alkene

alcohol

CH₂OH

O

OH

alcohol

(k)

(In later chapters, you will learn that the OH group on a benzene ring is a special functional group called a "phenol". For now, it fits the broad definition of an alcohol.)

aromatic

CH₃

ether

CH₃

CH₃

O

$C_{16}H_{33}$

HO

alcohol

CH₃

*This group is represented by $C_{16}H_{33}$ in the problem.*

## 2-23

(a)
major: all atoms have octets, maximum pi bonds

(b)
major: negative charge on more electronegative atom

(c)
major: all atoms have octets, maximum pi bonds

2-24 The examples here are representative. Your examples may be different and still be correct. What is important in this problem is to have the same functional group.

(a) alkane: hydrocarbon with all single bonds; can be acyclic (no ring) or cyclic

```
    H   H   H
    |   |   |
H — C — C — C — H
    |   |   |
    H   H   H
```

(b) alkene: contains a carbon-carbon double bond

```
        H
        |
H — C — C = C — H
    |   |   |
    H   H   H
```

(c) alkyne: contains a carbon-carbon triple bond

```
        H
        |
H — C — C ≡ C — H
    |
    H
```

2-24 continued

(d) alcohol: contains an OH group on a carbon

H H
| |
H–C–C–O–H
| |
H H

(e) ether: contains an oxygen between two carbons

H H
| |
H–C–O–C–H
| |
H H

(f) ketone: contains a carbonyl group between two carbons

H O H
| || |
H–C–C–C–H
| |
H H

(g) aldehyde: contains a carbonyl group with a hydrogen on one side

H H O
| | ||
H–C–C–C–H
| |
H H

(h) aromatic hydrocarbon: a cyclic hydrocarbon with alternating double and single bonds

(i) carboxylic acid: contains a carbonyl group with an OH group on one side

H O
| ||
H–C–C–O–H
|
H

(j) ester: contains a carbonyl group with an O—C on one side

H O H
| || |
H–C–C–O–C–H
| |
H H

(k) amine: contains a nitrogen bonded to one, two, or three carbons

H   H or R group
|   |
H–C–N–H or R group
|
H

(l) amide: contains a carbonyl group with a nitrogen on one side

H O   H or R group
| ||   |
H–C–C–N–H or R group
|
H

(m) nitrile: contains the carbon-nitrogen triple bond:   $H_3C—C≡N$

2-25 Models show that the tetrahedral geometry of $CH_2Cl_2$ precludes stereoisomers.

2-26

(a)

(b) Cyclopropane must have 60° bond angles compared with the usual $sp^3$ bond angle of 109.5° in an acyclic molecule.

(c) Like a bent spring, bonds that deviate from their normal angles or positions are highly strained. Cyclopropane is reactive because breaking the ring relieves the strain.

2-27

(a)

$sp^3$, ≈ 109°

(b)

$sp^3$, no bond angle because oxygen is bonded to only one atom

(c)

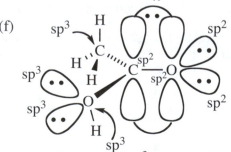

C=C has $sp^2$ carbons, ≈120° angles; nitrile has sp C and N, 180° angle

(d)

all $sp^3$, all ≈ 109°

(e)

both $sp^3$, all ≈ 109°

(f)

angles around $sp^3$ atoms ≈ 109°
angles around $sp^2$ carbon ≈ 120°

44

2-27 continued
(g)

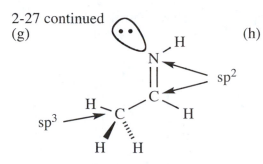

angles around sp³ atom ≈ 109°
angles around sp² atoms ≈ 120°

(h) in front of the plane of the paper    behind the plane of the paper

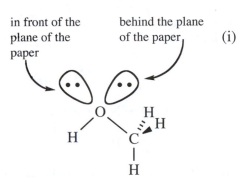

both sp³, all ≈ 109°

(i)

both sp², all ≈ 120°

2-28 For clarity in these pictures, bonds between hydrogen and an sp³ atom are not labeled; these bonds are s-sp³ overlap.

(a)

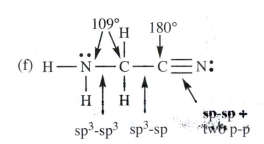

< 109°

(b) sp²-sp²
+ p-p

all angles ≈120°
all bonds to H are sp²-s

(c) 120°   sp²-sp³   109°

sp²-s   sp²-sp²   120°
+ p-p

(d) 109° H sp³-sp²   120°

sp²-s   sp²-sp²
+ p-p

sp²-s

(e) sp-sp + two p-p   120°

s-sp

sp²-sp² + p-p

180°

sp-sp²   sp²-s

(f) 109° H   180°

sp-sp +
two p-p

sp³-sp³   sp³-sp

(g)   120°   sp²-sp²
+ p-p

109° H

sp³-sp²   sp²-sp³   109°

(h) sp³-sp³   sp³-sp²

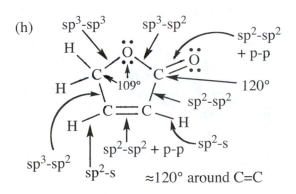

sp²-sp²
+ p-p

120°

sp³-sp²   sp²-s   sp²-sp²+ p-p   sp²-s

≈120° around C=C

(i) sp²-sp³   H   H   sp³-sp²

sp²-s   H

sp²-sp²   sp²-sp²

H   sp²-sp²   H
sp²-s   + p-p   sp²-s

Resonance forms show that the O is sp². All atoms are sp² except the top C which is sp³. Angles ≈120° around sp², ≈109° around the sp³ C.

**45**
Copyright © 2013 Pearson Education, Inc.

**2-29** The second and third resonance forms of urea are minor but still significant. They show that the nitrogen-carbon bonds have some double bond character, requiring that the nitrogens be sp² hybridized with bond angles approaching 120°.

**2-30**

(a) The major resonance contributor shows a carbon-carbon double bond, suggesting that both carbons are sp² hybridized with trigonal planar geometry. The CH₃ carbon is sp³ hybridized with tetrahedral geometry.

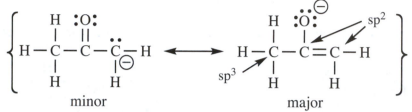

(b) The major resonance contributor shows a carbon-nitrogen double bond, suggesting that all three carbons and the nitrogen are sp² hybridized with trigonal planar geometry.

(c) The nitrogen and the carbon bonded to it are sp hybridized; the other carbon is sp². This ion has linear geometry. See the solution to 2-7(e) in this manual for an orbital picture.

**2-31** In (c), (d), and (e), the unshadowed p orbitals are vertical and parallel. The shadowed p orbitals are perpendicular and horizontal.

2-32
(a)

*cis*

(b) The coplanar atoms in the structures to the left and below are marked with asterisks.

(c)

*trans*

There are still six coplanar atoms.

(d)

H—C

H—C

H

H

C

CH₃

C

C

C

H

H

H

H

2-33 Collinear atoms are marked with asterisks.

H

H

C—C≡C—C

H

*

*

*

*

H

C

H

H

H

2-34
(a)

H₃C          CH₃
   \        /
    C = C          and
   /        \
  H   *cis*   H

H          CH₃
 \        /
  C = C
 /        \
H₃C  *trans*  H

(b) no *cis-trans* isomerism around a triple bond
(c) no *cis-trans* isomerism; two groups on each carbon are the same
(d) Theoretically, cyclopentene could show *cis-trans* isomerism. In reality, the *trans* form is too unstable to exist because of the necessity of stretched bonds and deformed bond angles. *trans*-Cyclopentene has never been detected.

*cis*

"*trans*"—not possible because of ring strain

(e)

CH₃          CH₂CH₃
   \        /
    C = C          and
   /        \
  H   CH₂CH₂CH₃

H          CH₂CH₃
 \        /
  C = C
 /        \
CH₃   CH₂CH₂CH₃

These are *cis-trans* isomers, but the designation of *cis* and *trans* to specific structures is not defined because of four different groups on the double bond.

(f)

H₃C          CH₃
   \        /
    C = N          and
   /        ..
  H   *cis*

H          CH₃
 \        /
  C = N
 /        ..
H₃C   *trans*

2-35

(a) constitutional isomers—The carbon skeletons are different.
(b) constitutional isomers—The position of the chlorine atom has changed.
(c) *cis-trans* isomers—The first is *cis*, the second is *trans*.
(d) constitutional isomers—The carbon skeletons are different.
(e) *cis-trans* isomers—The first is *trans*, the second is *cis*.
(f) same compound—Rotation of the first structure gives the second.
(g) *cis-trans* isomers—The first is *cis*, the second is *trans*.
(h) constitutional isomers—The position of the double bond relative to the ketone has changed (while it is true that the first double bond is *cis* and the second is *trans*, in order to have *cis-trans* isomers, the rest of the structure must be identical).

**2-36**  $CO_2$ is linear; its bond dipoles cancel, so it has no net dipole.  $SO_2$ is bent, so its bond dipoles do not cancel.

net dipole moment = 0          net dipole moment

**2-37**  Some magnitudes of dipole moments are difficult to predict; however, the direction of the dipole should be straightforward in most cases.  Actual values of molecular dipole moments are given in parentheses.  (The C—H bond is usually considered nonpolar.)

(a)

large dipole moment          large dipole moment

(b) $CH_3$—C≡N
net
large dipole moment  (3.96)

(c)
net dipole moment  = 0
Electron pairs on bromines are not shown.

(d)
net
large dipole moment  (2.89)

(e)
net dipole moment  = 0

(f)
net
moderate dipole moment

(g)
net
moderate dipole moment
Electron pairs on chlorine are not shown.

**2-38**  Diethyl ether and butan-1-ol each have one oxygen, so each can form hydrogen bonds with water (water supplies the H for hydrogen bonding with diethyl ether); their water solubilities should be similar. The boiling point of butan-1-ol is much higher because these molecules can hydrogen bond with each other, thus requiring more energy to separate one molecule from another.  Diethyl ether molecules cannot hydrogen bond with each other, so it is relatively easy to separate them.

$CH_3CH_2$—O—$CH_2CH_3$          $CH_3CH_2CH_2CH_2$—OH

diethyl ether          butan-1-ol

Can hydrogen bond with water;          Can hydrogen bond with water;
cannot hydrogen bond with itself.          can hydrogen bond with itself.

N-methylpyrrolidine
b.p. 81 °C

piperidine
b.p. 106 °C

tetrahydropyran
b.p. 88 °C

cyclopentanol
b.p. 141 °C

(a) Piperidine has an N—H bond, so it can hydrogen bond with other molecules of itself. *N*-Methylpyrrolidine has no N—H, so it cannot hydrogen bond and will require less energy (lower boiling point) to separate one molecule from another.

(b) Two effects need to be explained: 1) Why does cyclopentanol have a higher boiling point than tetrahydropyran? and 2) Why do the oxygen compounds have a greater difference in boiling points than the analogous nitrogen compounds?

   The answer to the first question is the same as in (a): cyclopentanol can hydrogen bond with its neighbors while tetrahydropyran cannot.

   The answer to the second question lies in the text, Table 2-1, that shows the bond dipole moments for C—O and H—O are much greater than C—N and H—N; bonds to oxygen are more polarized, with greater charge separation than bonds to nitrogen.

   How is this reflected in the data? The boiling points of tetrahydropyran (88 °C) and *N*-methylpyrrolidine (81 °C) are close; tetrahydropyran molecules would have a slightly stronger dipole-dipole attraction, and tetrahydropyran is a little less "branched" than *N*-methylpyrrolidine, so it is reasonable that tetrahydropyran boils at a slightly higher temperature. The large difference comes when comparing the boiling points of cyclopentanol (141 °C) and piperidine (106 °C). The greater polarity of O—H versus N—H is reflected in a more negative oxygen (more electronegative than nitrogen) and a more positive hydrogen, resulting in a much stronger intermolecular attraction. The conclusion is that hydrogen bonding due to O—H is much stronger than that due to N—H.

(c) Resonance plays a large role in the properties of amides.

*N,N*-dimethylformamide, b.p. 150 °C

*N*-methylacetamide, b.p. 206 °C

The first structure has no hydrogen bonding because it has no O—H or N—H bond, but it is a highly polar structure and has a large dipole moment, so its dipole-dipole intermolecular force is very high. The second structure has hydrogen bonding because of the N—H bond in addition to the strong dipole-dipole interaction, reflected in its higher boiling point. Both amides boil higher than the four at the beginning of the problem because of the dipole-dipole interactions that are accentuated by the resonance forms.

2-40

(a) Can hydrogen bond with itself and with water.　(b) Can hydrogen bond only with water.
(c) Can hydrogen bond with itself and with water.　(d) Can hydrogen bond only with water.
(e) Cannot hydrogen bond: no N or O.　(f) Cannot hydrogen bond: no N or O.
(g) Can hydrogen bond only with water.　(h) Can hydrogen bond with itself and with water.
(i) Can hydrogen bond only with water.　(j) Can hydrogen bond only with water.
(k) Can hydrogen bond only with water.　(l) Can hydrogen bond with itself and with water.

Water solubility depends on the number of carbons and the number of hydrogen-bonding groups in a molecule. A structure with one hydrogen-bonding group can carry in three carbons for sure, called "miscible", meaning completely soluble; four carbons will be partially soluble; five or more carbons will be only slightly soluble, which is sometimes called "insoluble".
• completely soluble, miscible: (c) 3C; (g) 3C; (h) 3C; (i) 3C; (l) 2C;
• moderately soluble, possibly miscible: (a), (j), (k)—each has one H-bonding group and 4C;
• slightly soluble, "insoluble": (b) 6C; (d) 6C; (e) and (f) have no H-bonding group.

**2-41** Higher-boiling compounds are listed.

(a) $CH_3CH(OH)CH_3$ can form hydrogen bonds with other molecules of itself.

(b) $CH_3CH_2CH_2CH_2CH_3$ has a higher molecular weight than $CH_3CH_2CH_2CH_3$ .

(c) $CH_3CH_2CH_2CH_2CH_3$ has less branching than $(CH_3)_2CHCH_2CH_3$ .

(d) $CH_3CH_2CH_2CH_2CH_2Cl$ has a higher molecular weight AND a dipole-dipole interaction compared with $CH_3CH_2CH_2CH_2CH_3$ .

**2-42**

**2-43**

sp$^2$—planar

a misleading picture

sp$^3$—tetrahedral

The key to this problem is understanding that sulfur has a *lone pair of electrons*. The second resonance form shows four pairs of electrons around the sulfur atom, an electronic configuration requiring sp$^3$ hybridization. Sulfur in DMSO cannot be sp$^2$ like carbon in acetone, so we would expect sulfur's geometry to be pyramidal (the four electron pairs around sulfur require tetrahedral geometry, but the three atoms around sulfur define its shape as pyramidal). The first resonance form is a misleading picture because it suggests a p-p pi bond in DMSO that does not exist. Sulfur might use a d orbital for some pi bonding but it is definitely not a p-p pi bond.

**2-44**

(a) penicillin G

2-44 continued

(b) dopamine

(In later chapters, you will learn that the OH group on a benzene ring is a special functional group called a "phenol". For now, it fits the broad definition of an alcohol.)

(c) capsaicin

(In later chapters, you will learn that the OH group on a benzene ring is a special functional group called a "phenol". For now, it fits the broad definition of an alcohol.)

(d) thyroxine

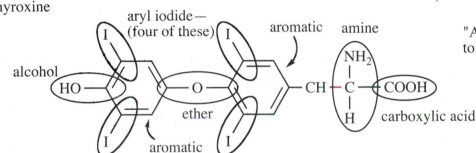

"Aryl" iodides are attached to aromatic rings.

(e) testosterone

ketone

alcohol

alkene

3-1 (a) $C_nH_{2n+2}$ where n = 28 gives $C_{28}H_{58}$    (b) $C_nH_{2n+2}$ where n = 44 gives $C_{44}H_{90}$

> Note to the student: The IUPAC system of nomenclature has a well-defined set of rules determining how structures are named. You will find a summary of these rules as Appendix 1 in this Solutions Manual.

3-2 Use hyphens to separate letters from numbers. Structures might not display their longest chains left to right—you have to search for it. In the name, write substituents in alphabetical order, even though that might be different from the numerical order of position numbers.

(a) 3-methylpentane (Always find the longest chain; it may not be written in a straight line.)
(b) 2-bromo-3-methylpentane (Always find the longest chain.)
(c) 5-ethyl-2-methyl-4-propylheptane ("When there are two longest chains of equal length, use the chain with the greater number of substituents.")
(d) 4-isopropyl-2-methyldecane ("Number from the end closest to the first substituent." —also for (c))

3-3 This Solutions Manual will present line–angle formulas, or simply "line formulas", where a question asks for an answer including a structure. If you use condensed structural formulas instead, be sure that you are able to "translate" one structure type into the other.

3-4 Separate numbers from numbers with commas.

(a) 2-methylbutane          (b) 2,2-dimethylpropane          (c) 3-ethyl-2-methylhexane
(d) 2,4-dimethylhexane    (e) 3-ethyl-2,2,4,5-tetramethylhexane    (f) 4-*tert*-butyl-3-methylheptane

3-5 In some cases of an ambiguous or incorrect name, more than one possible structure might be implied. That is often the problem with a wrong name: it does not describe a unique structure.

(a) incorrect: *2-methylethylpentane*

No position number given for the ethyl, although the only legitimate position would be on carbon-3.

correct: *3-ethyl-2-methylpentane*

(b) incorrect: *2-ethyl-3-methylpentane*

The longest chain was not identified.

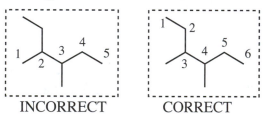

correct: *3,4-dimethylhexane*

(c) incorrect: *3-dimethylhexane*

Two substituents require two position numbers, even if they are both 3; that is the example shown below. It is also possible that a different position number was omitted, like "3,4-".

correct: *3,3-dimethylhexane*

(d) incorrect: *4-isobutylheptane*

This is a more subtle error. If two chains of equal length are possible, select the one that maximizes the number of substituents.

INCORRECT          CORRECT
one substituent     two substituents
correct: *2-methyl-4-propylheptane*

3-5 continued

(e) incorrect: *2-bromo-3-ethylbutane*

The longest chain was not identified.

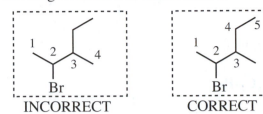

correct: *2-bromo-3-methylpentane*

(f) incorrect: *2-diethyl-3-methylhexane*

This name has two problems: two ethyl groups require two position numbers, presumably both 2 but it is ambiguous; and the longest chain was not identified.

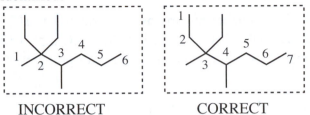

correct: *3-ethyl-3,4-dimethylheptane*

3-6 (Hints: *systematize* your approach to these problems. For the isomers of a six-carbon formula, for example, start with the isomer containing all six carbons in a straight chain, then the isomers containing a five-carbon chain, then a four-carbon chain, *etc*. Carefully check your answers to AVOID DUPLICATE STRUCTURES.)

(a)

hexane
(or *n*-hexane)   2-methylpentane   3-methylpentane   2,2-dimethylbutane   2,3-dimethylbutane

(b)

heptane
(or *n*-heptane)   2-methylhexane   3-methylhexane   2,2-dimethylpentane

3,3-dimethylpentane   2,3-dimethylpentane   2,4-dimethylpentane   3-ethylpentane   2,2,3-trimethylbutane

3-7 For this problem, the carbon numbers in the substituents are indicated in italics. The bold bond shows the point of attachment.

(a)

$CH_3$
—$CHCH_3$
*1*  *2*

1-methylethyl
common name = isopropyl

(b)

$CH_3$
—$CH_2CHCH_3$
*1*  *2*  *3*

2-methylpropyl
common name = isobutyl

(c)

$CH_3$
—$CHCH_2CH_3$
2°    *1*  *2*  *3*

1-methylpropyl
common name = *sec*-butyl, abbreviation for *secondary*-butyl

(d)

$CH_3$
—$C$—$CH_3$
*1*  *2*
3°   $CH_3$

1,1-dimethylethyl
common name = *tert*-butyl or *t*-butyl, abbreviation for *tertiary*-butyl

(e)

$CH_3$
—$CH_2CH_2CHCH_3$
*1*  *2*  *3*  *4*

3-methylbutyl
common name = isopentyl or isoamyl

3-8

(a)

(b)

(c)

3-9 Once the number of carbons is determined, $C_nH_{2n+2}$ gives the formula.

(a) octane = 8C; plus 1,1-dimethylethyl = 4C $\Longrightarrow$ 12 C, $C_{12}H_{26}$

(b) nonane = 9C; plus trimethyl = 3C; plus propyl = 3C $\Longrightarrow$ 15 C, $C_{15}H_{32}$

3-10

(a) (lowest b.p.) hexane < octane < decane (highest b.p.) —molecular weight

(b) $(CH_3)_3C-C(CH_3)_3$ < $CH_3CH_2C(CH_3)_2CH_2CH_2CH_3$ < octane —branching differences
     (lowest b.p.)                                  (highest b.p.)

*some branching*
*actual b.p. 112 °C*

*no branching*
*actual b.p. 126 °C*

*most branching*
*actual b.p. 107 °C*

3-11

1 x 4.2 kJ/mole + 2 x 5.4 kJ/mole = 15.0 kJ/mole (3.6 kcal/mole)
    H–H           H–$CH_3$

3-12

Relative energies on the graph on the next page were calculated using these values from the text: 3.8 kJ/mole (0.9 kcal/mole) for a $CH_3$–$CH_3$ gauche (staggered) interaction; 4.2 kJ/mole (1.0 kcal/mole) for a H–H eclipsed interaction; 5.4 kJ/mole (1.3 kcal/mole) for a H–$CH_3$ eclipsed interaction; 12.5 kJ/mole (3.0 kcal/mole) for a $CH_3$–$CH_3$ eclipsed interaction. These values in kJ/mole are noted on each structure and are summed to give the energy value on the graph. Slight variations between values here and in the text are due to rounding.

3-12 continued

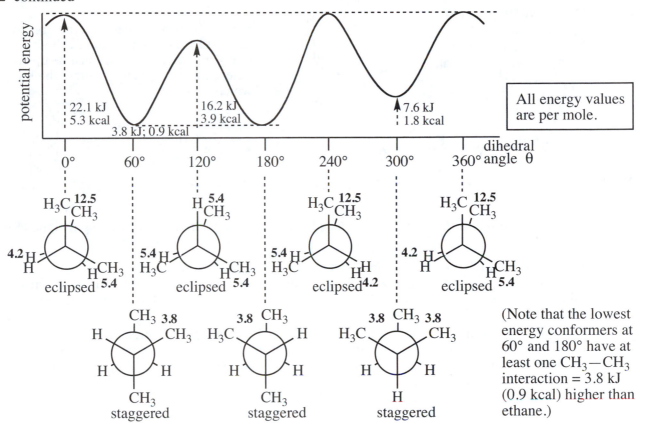

All energy values are per mole.

(Note that the lowest energy conformers at 60° and 180° have at least one CH₃—CH₃ interaction = 3.8 kJ (0.9 kcal) higher than ethane.)

3-13

All bonds are staggered.

Wedge bonds are coming toward the reader above the plane of the paper.

⊪ Dashed bonds are going away from the reader behind the plane of the paper.

3-14 Note that in rings, parts (a) and (b), two substituents on one carbon indicate where to begin numbering.

(a) 3-*sec*-butyl-1,1-dimethylcyclopentane
(b) 3-cyclopropyl-1,1-dimethylcyclohexane
(c) 4-cyclobutylnonane (The chain is longer than the ring.)

3-15 (a) $C_{10}H_{20}$     (b) $C_{10}H_{20}$     (c) $C_8H_{14}$     (d) $C_{10}H_{20}$

(e) $C_{10}H_{22}$     (f) $C_{16}H_{32}$

**3-16** It helps to visualize *cis-trans* isomerism by putting in the H atoms on the carbons with substituents.

(a) no *cis-trans* isomerism possible

(b)

*cis*

and

*trans*

For *cis*, both substituents have to be on the same side; it does not matter if both are up or down. For *trans*, one must be up and the other down; it does not matter which is which.

(c)

*cis*

and

*trans*

(d)

*cis*

and

*trans*

**3-17** In (a) and (b), numbering of the ring is determined by the first group *alphabetically* being assigned to ring carbon 1.

(a) *cis*-1-methyl-3-propylcyclobutane ("M" comes before "p"—practice that alphabet!)
(b) *trans*-1-*tert*-butyl-3-ethylcyclohexane (*Tert*, *sec*, and *n* are ignored in assigning alphabetical priority.)
(c) *trans*-1,2-dimethylcyclopropane (Either carbon with a $CH_3$ could be carbon-1; the same name results.)

**3-18** Combustion of the *cis* isomer gives off more energy, so *cis*-1,2-dimethylcyclopropane must start at a higher energy than the *trans* isomer. The Newman projection of the *cis* isomer shows the two methyls are eclipsed with each other; in the *trans* isomer, the methyls are still eclipsed, but with hydrogens, not each other—a lower energy interaction.

**3-19** *trans*-1,2-Dimethylcyclobutane is more stable than *cis* because the two methyls can be farther apart when *trans*, as shown in the Newman projections.

In the 1,3-dimethylcyclobutanes, however, the *cis* allows the methyls to be farther from other atoms and therefore more stable than the *trans*.

3-20

equatorial only ⇒ showing both axial and equatorial ⇐ axial only

**3-21** The abbreviation for a methyl group, $CH_3$, is "Me". Ethyl is "Et", propyl is "Pr", and butyl is "Bu".

(a)

Me    Me
  Me  H
H                H
H                H
  H   Me
Me    Me

all methyls axial
(all H equatorial)

(b)

    H    H
   H  Me
Me            Me
Me          Me
   H  Me  H
   H    H

all methyls equatorial
(all H axial)

Note that axial groups alternate
up and down around the ring.

**3-22** Carbons 4 and 6 are the back carbons.

anti

**3-23** The isopropyl group can rotate so that its hydrogen is near the axial hydrogens on carbons 3 and 5, similar to a methyl group's hydrogen, and therefore similar to a methyl group in energy. The *tert*-butyl group, however, must point a methyl group toward the hydrogens on carbons 3 and 5, giving severe diaxial interactions, causing the energy of this conformer to jump dramatically.

isopropylcyclohexane

*tert*-butylcyclohexane

**3-24** The most stable conformers have larger substituents equatorial.

(a)

(b)

(c)

The *cis*-1,4 isomer must have one group axial and the other equatorial. The *tert*-butyl group is larger and will take the equatorial position.

**3-25**

**(a)** *cis*

CH₃      CH₃

H      H

H₃C      ⇌      CH₃

H     **EQUAL ENERGY**     H

     equatorial, axial          axial, equatorial

**(b)** *trans*

CH₃      H

H     H₃C

H      ⇌      CH₃

CH₃      H

    axial, axial       equatorial, equatorial

   higher energy       lower energy

**(c)** The *trans* isomer is more stable because BOTH substituents can be in the preferred equatorial positions.

**3-26**

| Positions | cis | trans |
|-----------|-----------|-----------|
| 1,2 | (e,a) or (a,e) | (e,e) or (a,a) |
| 1,3 | (e,e) or (a,a) | (e,a) or (a,e) |
| 1,4 | (e,a) or (a,e) | (e,e) or (a,a) |

**3-27** The more stable conformer places the larger group equatorial.

**(a)**

CH₂CH₃
H
CH₃
H     ⇌     CH₃   CH₂CH₃     more stable—
   H   H     larger group equatorial

**(b)**

CH₂CH₃
H
H
CH₂CH₃     ⇌     H   CH₂CH₃     more stable—
   CH₂     both groups equatorial
   H₃C   H

**(c)** more stable—
larger group equatorial

(CH₃)₂CH
CH₂CH₃
H
H     ⇌     H₃C–CH   H
   H₃C   CH₂CH₃
   H

**(d)**

CH₂CH₃
H
H     ⇌     H   CH₃     more stable—
CH₃     CH₂CH₃     both groups equatorial
   H

**3-28** The key to determining *cis* and *trans* around a cyclohexane ring is to see whether a substituent group is "up" or "down" *relative to the H at the same carbon*. Two "up" groups or two "down" groups will be *cis*; one "up" and one "down" will be *trans*. This works independent of the conformation the molecule is in!

(a) *cis*-1,3-dimethylcyclohexane

(b) *cis*-1,4-dimethylcyclohexane

(c) *trans*-1,2-dimethylcyclohexane

(d) *cis*-1,3-dimethylcyclohexane

(e) *cis*-1,3-dimethylcyclohexane

(f) *trans*-1,4-dimethylcyclohexane

**3-29**

(a)

(b)

(c) Bulky substituents like *tert*-butyl adopt equatorial rather than axial positions, even if that means altering the conformation of the ring. The twist boat conformation allows both bulky substituents to be "equatorial".

**3-30** The nomenclature of bicyclic alkanes is summarized in Appendix 1 in this manual.

(a) bicyclo[3.1.0]hexane    (b) bicyclo[3.3.1]nonane    (c) bicyclo[2.2.2]octane    (d) bicyclo[3.1.1]heptane

**3-31** Using models is essential for this problem.

from text Figure 3-27

**3-32**

(a) Here are eighteen isomers of $C_8H_{18}$. An easy way to compare is to name yours and see if the names match. Note how these isomers were generated systematically: 8C chain (only one), all the possible 7C chains with one methyl, all the 6C chains with two methyls or 1 ethyl, etc.

octane

2-methylheptane

3-methylheptane

4-methylheptane

2,2-dimethylhexane

2,3-dimethylhexane

2,4-dimethylhexane

2,5-dimethylhexane

*continued on the next page*

3-32 (a) continued

3,3-dimethylhexane      3,4-dimethylhexane      3-ethylhexane      2,2,3-trimethylpentane

2,2,4-trimethylpentane   2,3,3-trimethylpentane   2,3,4-trimethylpentane   3-ethyl-2-methylpentane

3-ethyl-3-methylpentane   2,2,3,3-tetramethylbutane

(b)

*constitutional isomers*

ethylcyclopentane      1,1-dimethylcyclopentane
("1" is not necessary)

cis-1,2-dimethylcyclopentane   cis-1,3-dimethylcyclopentane

*geometric isomers*      *geometric isomers*

trans-1,2-dimethylcyclopentane   trans-1,3-dimethylcyclopentane

3-33

(a) The third structure is 2-methylpropane (isobutane). The other four structures are all butane (*n*-butane). Remember that a compound's identity is determined by how the atoms are connected, not by the position of the atoms when a structure is drawn on a page.

(b) The first and fourth structures are both *cis*-but-2-ene. The second and fifth structures are both but-1-ene. The third structure is *trans*-but-2-ene. The last structure is 2-methylpropene.

(c) The first two structures are both *cis*-1,2-dimethylcyclopentane. The next two structures are both *trans*-1,2-dimethylcyclopentane. The last structure, *cis*-1,3-dimethylcyclopentane, is different from all the others.

(d)

Analysis of the structures shows that some double bonds begin at carbon-2 and some at carbon-3 of the longest chain.

The three structures labeled **A** are the same, with the double bond *trans*; **B** is a geometric isomer (*cis*) of **A**. **C** and **D** are constitutional isomers of the others.

**60**

3-33 continued

(e) Naming the structures shows that three of the structures are *trans*-1,4-dimethylcyclohexane, two are the *cis* isomer, and one is *cis*-1,3-dimethylcyclohexane. Although a structure may be shown in two different conformations, it still represents only one compound.

cis-1,4-dimethyl          trans-1,4-dimethyl          cis-1,4-dimethyl

cis-1,3-dimethyl          trans-1,4-dimethyl          trans-1,4-dimethyl

(f) An important and useful skill is the ability to translate Newman projections into other structural formulas and *vice versa*. The way to do this is to start with the two carbons that represent the front and back carbons in the middle of the Newman projections, then attach the three groups emanating from each one.

2,3-dimethylbutane   2,3-dimethylbutane   2,2-dimethylbutane   2,3-dimethylbutane   2,2-dimethylbutane

What may not be clear in the Newman projections is revealed in the structural formulas and especially in the names. There are only two compounds represented here: the first, second, and fourth are the same; the third and fifth are the same.

(g) Names have been incorporated into the other solutions.

3-34 Line–angle formulas are shown.

(a)          (b)          (c)

(d)          (e)          (f) or

(g)          (h)          (i)

(j)          (k)          (l)

3-35 There are often many possible answers to this type of problem, although some can have unique solutions. The ones shown here are examples of correct answers. Your answers may be different AND correct. Check your answers in your study group.

(a) 4-Isopropylheptane is the only possible answer. Why not 1- or 2- or 3- ? Draw these structures in your study group, and name them.

(b)

4,5-diethyldecane             3,5-diethyldecane

(Any combination is correct except using position numbers 1 or 2 or 9 or 10. Why won't these work?)

(c)

CH₂CH₃
CH₂CH₃
cis-1,2-diethylcyclohexane

CH₂CH₃
CH₂CH₃
cis-1,3-diethylcyclohexane

CH₂CH₃
CH₃CH₂
cis-1,4-diethylcyclohexane

NOT 1,5
NOT 1,6

(d) "Halo" is the abbreviation for any of the four halogens (F, Cl, Br, I), and "X" is the symbol used in organic structures. If unspecified, any of the four elements will satisfy the "halo" group. However, on a 5-membered ring, the substitution can be only two possibilities: 1,2- or 1,3- .

"X" stands for any of the four halogen elements.

trans-1,2-dihalocyclopentane  trans-1,3-dihalocyclopentane

(e)

(2,3-dimethylpentyl)cycloheptane        (2,3-dimethylpentyl)cyclooctane

Other ring sizes are possible, although they must have 6 or more carbons to be longer than the 5 carbons of the substituent chain.

(f)

bicyclo[4.3.0]nonane

bicyclo[3.2.2]nonane    bicyclo[3.3.1]nonane

Any combination where the number of carbons in the bridges sums to 7 will work. (Two carbons are the bridgehead carbons.)

3-36    HO−CH₃           HO−CH₂CH₂CH₃           HO−CH₂CH₂CH₂CH₂CH₃

        HO−CH₂CH₃        HO−CH₂CH₂CH₂CH₃        HO−CH₂CH₂CH₂CH₂CH₂CH₃

3-37

(a) 3-ethyl-2,2,6-trimethylheptane
(c) 3,7-diethyl-2,2,8-trimethyldecane
(e) bicyclo[4.1.0]heptane
(g) (1,1-diethylpropyl)cyclohexane

(b) 3-ethyl-2,6,7-trimethyloctane
(d) 1,1-diethyl-2-methylcyclobutane
(f) cis-1-ethyl-3-propylcyclopentane
(h) cis-1-ethyl-4-isopropylcyclodecane

**3-38** The ethyl group, $-CH_2CH_3$, is abbreviated Et. The front carbon is kept constant while the back carbon is rotated. The problem asks for relative energies, not calculations; the ethyl group is slightly larger than a methyl group, causing a higher energy than methyl in eclipsed conformations.

Interestingly, all of the minima and all of the maxima are at different energies; this arises because the three groups on the back carbon (H, $CH_3$, $CH_2CH_3$) are all different.

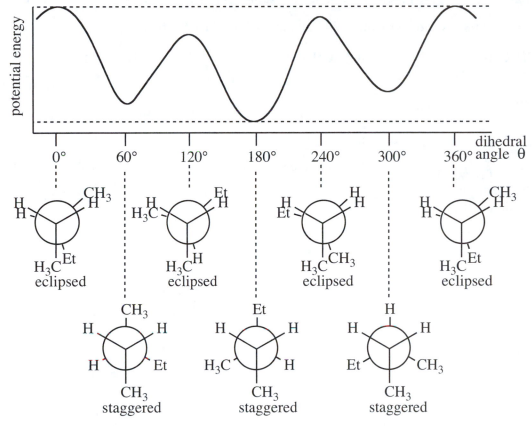

**3-39**

(a)

correct name: 3-methylhexane
(longer chain)

(b)

correct name: 3-ethyl-2-methylhexane
(more branching with this numbering)

(c)

correct name: 2-chloro-3-methylhexane
(Begin numbering at end closest to substituent.)

(d)

correct name: 2,2-dimethylbutane (Include
a position number for each substituent,
regardless of redundancies.)

(e)

correct name: *sec*-butylcyclohexane or
(1-methylpropyl)cyclohexane
(The longer chain or ring is the base name.)

(f)

correct name: 1,2-diethylcyclopentane
(Position numbers are the lowest possible.)

3-40

(a) Octane has a higher boiling point than 2,2,3-trimethylpentane because linear molecules boil higher than branched molecules of the same molecular weight (increased van der Waals interaction).
(b) Nonane has a higher boiling point than 2-methylheptane because nonane has a higher molecular weight and it is a linear structure as opposed to the lighter, branched 2-methylheptane.
(c) Nonane boils higher than 2,2,5-trimethylhexane for the same reason as in (a).

3-41 The point of attachment is shown by the bold bond at the left of each structure.

—$CH_2CH_2CH_2CH_2CH_3$
1°        pentyl
        (old: amyl)

—$CHCH_2CH_2CH_3$
2° $CH_3$    1-methylbutyl

—$CH_2CHCH_2CH_3$
1°    $CH_3$    2-methylbutyl

—$CH_2CH_2CHCH_3$
1°    $CH_3$  3-methylbutyl
          (isopentyl;
          old: isoamyl)

—$CHCH_2CH_3$
2° $CH_2CH_3$  1-ethylpropyl

          $CH_3$
—$C-CH_2CH_3$
3°  $CH_3$  1,1-dimethylpropyl
          (tert-pentyl;
          old: tert-amyl)

     $CH_3$
—$CHCHCH_3$
2° $CH_3$  1,2-dimethylpropyl

          $CH_3$
—$CH_2-C-CH_3$
1°    $CH_3$  2,2-dimethylpropyl
          (neo-pentyl)

3-42 In each case, put the largest groups on adjacent carbons in anti positions to make the most stable conformations.

(a) 3-methylpentane

Newman projections were introduced in text section 3-7. They are used extensively to show the three-dimensional arrangement of atoms in a molecule.

C-2 is the front carbon with H, H, and $CH_3$
C-3 is the back carbon with H, $CH_3$, and $CH_2CH_3$

Carbon-3 is behind carbon-2.

(b) 3,3-dimethylhexane

C-3 is the front carbon with $CH_3$, $CH_3$, and $CH_2CH_3$
C-4 is the back carbon with H, H, and $CH_2CH_3$

Carbon-4 is behind carbon-3.

3-43
(a)

axial H  axial H  $CH_3$ equatorial
$CH_3$ equatorial

equatorial H      H equatorial
$CH_3$ axial  $CH_3$ axial

(b)  more stable
     (lower energy)

less stable
(higher energy)

**3-43 continued**

(c) From Section 3-14 of the text, each gauche interaction raises the energy 3.8 kJ/mole (0.9 kcal/mole), and each axial methyl has two gauche interactions, so the energy is:
2 methyls x 2 interactions per methyl x 3.8 kJ/mole per interaction = 15.2 kJ/mole (3.6 kcal/mole)

(d) The steric strain from the 1,3-diaxial interaction of the methyls must be the difference between the total energy and the energy due to gauche interactions: 23 kJ/mole − 15.2 kJ/mole = 7.8 kJ/mole
(5.4 kcal/mole − 3.6 kcal/mole = 1.8 kcal/mole)

**3-44** The more stable conformer places the larger group equatorial.

(a)

more stable —
larger group equatorial

(b)

Et = ethyl
more stable —
both groups equatorial

(c)

more stable —
both groups equatorial

(d)

more stable —
larger group equatorial

(e)

more stable —
larger group equatorial

(f)

more stable —
both groups equatorial

**3-45** (Using models is essential to this problem.)

In both *cis-* and *trans-*decalin, the cyclohexane rings can be in chair conformations. The relative energies will depend on the number of axial substituents.

*trans*
no axial substituents
MORE STABLE

*cis*
one axial substituent

3-46 Solutions are given in the boxes. The front and back carbons from the Newman projection are denoted by a dark dot.

(a)

H
H   H
1
H   H
$CH_2CH_3$
3   4

butane

(b)
4   5
$CH_2CH_3$
H   $CH_3$
2
H   $CH_3$
1 $CH_3$

2,3-dimethylpentane

(c)
Br
H   $CH_3$
1
H   H
$CH_2CH_3$
3   4

Br 1 2 3 4
1-bromo-2-methylbutane

(d)
5   6
$CH_2CH_3$
H   Cl
3
H   $CH_3$
$CH(CH_3)_2$
2   1

1 2 3 4 5 6
Cl
3-chloro-2,4-dimethylhexane

(e)
2   1
$C(CH_3)_3$
Br   H
3
Cl   $CH_3$
$CH_2CH_3$
5   6

Cl
1 2 3 4 5 6
Br
4-bromo-3-chloro-
2,2,3-trimethylhexane

(f)
5 $CH_3$
H   $CH_3$
3
H   H
$CH(CH_3)_2$
2   1

1 2 3 4 5
2,3-dimethylpentane

This is the same structure as part (b), sighted down a different bond.

(g)
F   1
H   $CBr_3$
3
F   F
H

F F Br Br
F 3 2 1 Br
Br
1,1,1-tribromo-3,3,3-
trifluoropropane

Br comes before F in the alphabet

(h)
H
F   Cl
2
H   Br
$CH_2CH_3$
3   4

Br
4 3 2 1
F   Cl
1-bromo-2-chloro-2-
fluorobutane

(i)
2   1
$CH(CH_3)_2$
Br   H
3
Br   H
$CH_2CH_3$
5   6

Br
1 2 3 4 5 6
Br
3,4-dibromo-2-
methylhexane

(j)
Br
H   $CH_3$
2
Cl   F
$CH_2CH_3$
3   4

Br
Cl
4 3 2 1
F
1-bromo-1-chloro-1-
fluoro-2-methylbutane

**3-47**

3-ethyl-2,4,4-trimethylheptane

C-4 is the back carbon.

most stable conformation —
two largest groups anti in a
staggered conformation

least stable conformation —
two largest groups eclipsed

**3-48** Usually, the anti conformation is more stable than the gauche because the two larger groups are farther away in the anti conformer. There must be something special about the gauche conformer, some attractive force that overcomes the normal repulsion between groups in the gauche conformation. What could it be?

*anti*                *gauche*

Hydrogen bonding! Moving the two OH groups closer together permits the H of one to come close to the O of the other, as can be seen in the picture below. Such intramolecular (within one molecule) hydrogen bonding cannot take place in the anti conformation where the two OH groups are too far apart.

intramolecular hydrogen bonding, a strong attractive force

*gauche*

**3-49** chair form of glucose with all substituents equatorial

(without ring H atoms shown)

Note to the student: Creating practice problems for your study group is very helpful in learning new material, and in some areas, it is not difficult. For example, to create nomenclature problems, draw a cyclopentane, add three substituents, and name it. Then erase one of the ring bonds and name that acyclic structure. Go around the ring, erasing one bond at a time, and you have generated five new nomenclature problems for your group to practice naming!

4-1

(a)

$$H-\underset{\underset{H}{|}}{\overset{\overset{H}{|}}{C}}-\overset{.}{\underset{H}{\overset{|}{C}}}-H$$

(b)

$$H-\underset{\underset{H}{|}}{C}-\overset{.}{C}-\underset{\underset{H}{|}}{C}-H$$
with C (H, H, H)

(c)

$$H-\underset{\underset{H}{|}}{\overset{\overset{H}{|}}{C}}-\overset{.}{\underset{H}{\overset{|}{C}}}-\underset{\underset{H}{|}}{\overset{\overset{H}{|}}{C}}-H$$

(d)

$$:\overset{..}{\underset{..}{I}}\cdot$$

4-2

(a)

(b) Free-radical halogenation substitutes a halogen atom for a hydrogen. Even if a molecule has only one type of hydrogen, substitution of the first of these hydrogens forms a new compound. Any remaining hydrogens in this product can compete with the initial reactant for the available halogen. Thus, chlorination of methane, $CH_4$, produces all possible substitution products: $CH_3Cl$, $CH_2Cl_2$, $CHCl_3$, and $CCl_4$.

If a molecule has different types of hydrogens, the reaction can generate a mixture of the possible substitution products.

(c) Production of $CCl_4$ or $CH_3Cl$ can be controlled by altering the ratio of $CH_4$ to $Cl_2$. To produce $CCl_4$, use an excess of $Cl_2$ and let the reaction proceed until all C—H bonds have been replaced with C—Cl bonds. Producing $CH_3Cl$ is more challenging because the reaction tends to proceed past the first substitution. By using a very large excess of $CH_4$ to $Cl_2$, perhaps 100 to 1 or even more, a chlorine atom is more likely to find a $CH_4$ molecule than it is to find a $CH_3Cl$, so only a small amount of $CH_4$ is transformed to $CH_3Cl$ by the time the $Cl_2$ runs out, with almost no $CH_2Cl_2$ being produced.

4-3

(a) This mechanism requires that one photon of light be added for each $CH_3Cl$ generated, a quantum yield of 1. The actual quantum yield is several hundred or thousand. The high quantum yield suggests a chain reaction, but this mechanism is not a chain; it has no propagation steps.

(b) This mechanism conflicts with at least two experimental observations. First, the energy of light required to break a H–$CH_3$ bond is 435 kJ/mole (104 kcal/mole, from Table 4-2); the energy of light determined by experiment to initiate the reaction is only 242 kJ/mole of photons (58 kcal/mole of photons, from text section 4-3A), much less than the energy needed to break this H–C bond. Second, as in (a), each $CH_3Cl$ produced would require one photon of light, a quantum yield of 1, instead of the actual number of several hundred or thousand. As in (a), there is no provision for a chain process, because all of the radicals generated are also consumed in the mechanism.

4-4

(a) The twelve hydrogens of cyclohexane are all on equivalent 2° carbons. Replacement of any one of the twelve will lead to the same product, chlorocyclohexane. Hexane, however, has hydrogens in three different positions: on carbon-1 (equivalent to carbon-6), carbon-2 (equivalent to carbon-5), and carbon-3 (equivalent to carbon-4). Monochlorination of hexane will produce a mixture of all three possible isomers: 1-, 2-, and 3-chlorohexane.

(b) The best conversion of cyclohexane to chlorocyclohexane would require the ratio of cyclohexane/chlorine to be a large number. If the ratio were small, as the concentration of chlorocyclohexane increased during the reaction, chlorine would begin to substitute for a second hydrogen of chlorocyclohexane, generating unwanted products. The goal is to have chlorine attack a molecule of cyclohexane before it ever encounters a molecule of chlorocyclohexane, so the concentration of cyclohexane should be kept high.

4-5

(a) $K_{eq} = e^{-\Delta G°/RT}$    ← $-2.1$ kJ/mole $= -2100$ J/mole

°K = temperature in the Kelvin scale
To be consistent with the textbook, and to avoid confusion with the symbol for an equilibrium constant, the degree symbol will be used with Kelvin temperatures.

$= e^{-(-2100 \text{ J/mole})/((8.314 \text{ J/°K-mole}) \cdot (298 \text{ °K}))}$

$= e^{2100/2478} = e^{0.847} = \boxed{2.3}$

(b) $K_{eq} = 2.3 = \dfrac{[CH_3SH]\,[HBr]}{[CH_3Br]\,[H_2S]}$

| | $[CH_3Br]$ | $[H_2S]$ | $[CH_3SH]$ | $[HBr]$ |
|---|---|---|---|---|
| initial concentrations: | 1 | 1 | 0 | 0 |
| final concentrations | $1-x$ | $1-x$ | x | x |

$K_{eq} = 2.3 = \dfrac{x \cdot x}{(1-x)(1-x)} = \dfrac{x^2}{1-2x+x^2}$ ⟹ $x^2 = 2.3\,x^2 - 4.6\,x + 2.3$

$0 = 1.3\,x^2 - 4.6\,x + 2.3$ ⟹ $x = 0.60$ , $1-x = 0.40$

(using quadratic equation)

$$\boxed{[CH_3SH] = [HBr] = 0.60 \text{ M}; \quad [CH_3Br] = [H_2S] = 0.40 \text{ M}}$$

4-6   2 acetone ⇌ diacetone alcohol

Assume that the initial concentration of acetone is 1 molar, and 5% of the acetone is converted to diacetone alcohol. NOTE THE MOLE RATIO. The coefficients in a chemical equation become the exponents in the equilibrium expression.

| | [acetone] | [diacetone alcohol] |
|---|---|---|
| initial concentrations: | 1 M | 0 |
| final concentrations: | 0.95 M | 0.025 M |

$K_{eq} = \dfrac{[\text{diacetone alcohol}]}{[\text{acetone}]^2} = \dfrac{0.025}{(0.95)^2} = \boxed{0.028}$

$\Delta G° = -2.303\ RT\ \log_{10} K_{eq} = -2.303\,((8.314 \text{ J/°K-mole}) \cdot (298 \text{ °K})) \cdot \log(0.028)$

$= \boxed{+8.9 \text{ kJ/mole} \quad (+2.1 \text{ kcal/mole})}$

4-7   $\Delta S°$ will be negative since two molecules are combined into one, a loss of freedom of motion. Since $\Delta S°$ is negative, $-T\Delta S°$ is positive; but $\Delta G°$ is a large negative number since the reaction goes to completion. Therefore, $\Delta H°$ must also be a large negative number, necessarily larger in absolute value than $\Delta G°$. We can explain this by formation of two strong C–H sigma bonds after breaking a strong H–H sigma bond and a WEAKER C=C pi bond.

4-8

(a) $\Delta S°$ is positive—one molecule became two smaller molecules with greater freedom of motion.
(b) $\Delta S°$ is negative—two smaller molecules combined into one larger molecule with less freedom of motion.
(c) $\Delta S°$ cannot be predicted since the number of molecules in reactants and products is the same.

4-9 (a)  initiation    (1)  Cl–Cl $\xrightarrow{h\nu}$  2 Cl •

propagation $\begin{cases} (2) & \text{Cl} \cdot \ + \ \text{H–CH}_2\text{CH}_3 \longrightarrow \text{H–Cl} \ + \ \cdot\text{CH}_2\text{CH}_3 \\ (3) & \text{Cl–Cl} \ + \ \cdot\text{CH}_2\text{CH}_3 \longrightarrow \text{Cl–CH}_2\text{CH}_3 \ + \ \text{Cl} \cdot \end{cases}$

termination $\begin{cases} \text{Cl} \cdot \ + \ \cdot\text{Cl} \longrightarrow \text{Cl–Cl} \\ \text{Cl} \cdot \ + \ \cdot\text{CH}_2\text{CH}_3 \longrightarrow \text{Cl–CH}_2\text{CH}_3 \\ \text{CH}_3\text{CH}_2 \cdot \ + \ \cdot\text{CH}_2\text{CH}_3 \longrightarrow \text{CH}_3\text{CH}_2\text{CH}_2\text{CH}_3 \end{cases}$

(b)    step (1)    break Cl–Cl          $\Delta H° = +242$ kJ/mole (+ 58 kcal/mole)

step (2)    break H–CH$_2$CH$_3$    $\Delta H° = +410$ kJ/mole (+ 98 kcal/mole)
            make H–Cl                $\underline{\Delta H° = -431 \text{ kJ/mole } (-103 \text{ kcal/mole})}$
                  step (2)           **$\Delta H° = -21$ kJ/mole (–5 kcal/mole)**

step (3)    break Cl–Cl              $\Delta H° = +242$ kJ/mole (+ 58 kcal/mole)
            make Cl–CH$_2$CH$_3$     $\underline{\Delta H° = -339 \text{ kJ/mole } (-81 \text{ kcal/mole})}$
                  step (3)           **$\Delta H° = -97$ kJ/mole (–23 kcal/mole)**

(c)  $\Delta H°$ for the reaction is the sum of the $\Delta H°$ values of the individual propagation steps:

$- 21$ kJ/mole $+ \ -97$ kJ/mole $= $ **$- 118$ kJ/mole**
$(- 5$ kcal/mole $+ \ -23$ kcal/mole $= $ **$- 28$ kcal/mole**)

4-10 (a)  initiation    (1)  Br–Br $\xrightarrow{h\nu}$  2 Br •

propagation $\begin{cases} (2) & \text{Br} \cdot \ + \ \text{H–CH}_3 \longrightarrow \text{H–Br} \ + \ \cdot\text{CH}_3 \\ (3) & \text{Br–Br} \ + \ \cdot\text{CH}_3 \longrightarrow \text{Br–CH}_3 \ + \ \text{Br} \cdot \end{cases}$

step (1)    break Br–Br          $\Delta H° = +192$ kJ/mole (+ 46 kcal/mole)

step (2)    break H–CH$_3$        $\Delta H° = +435$ kJ/mole (+ 104 kcal/mole)
            make H–Br            $\underline{\Delta H° = -368 \text{ kJ/mole } (-88 \text{ kcal/mole})}$
                  step (2)       **$\Delta H° = +67$ kJ/mole (+16 kcal/mole)**

step (3)    break Br–Br          $\Delta H° = +192$ kJ/mole (+ 46 kcal/mole)
            make Br–CH$_3$        $\underline{\Delta H° = -293 \text{ kJ/mole } (-70 \text{ kcal/mole})}$
                  step (3)       **$\Delta H° = -101$ kJ/mole (–24 kcal/mole)**

(b)  $\Delta H°$ for the reaction is the sum of the $\Delta H°$ values of the individual propagation steps:

$+ 67$ kJ/mole $+ \ -101$ kJ/mole $= $ **$- 34$ kJ/mole**
$(+ 16$ kcal/mole $+ \ -24$ kcal/mole $= $ **$- 8$ kcal/mole**)

4-11
(a) first order: the exponent of [ (CH$_3$)$_3$CCl ] in the rate law = 1
(b) zeroth order: [ CH$_3$OH ] does not appear in the rate law (its exponent is zero)
(c) first order: the sum of the exponents in the rate law = 1 + 0 = **1**

4-12
(a) first order: the exponent of [cyclohexene] in the rate law = 1
(b) second order: the exponent of [ Br$_2$ ] in the rate law = 2
(c) third order: the sum of the exponents in the rate law = 1 + 2 = **3**

4-13

(a) The reaction rate depends on neither [ethylene] nor [hydrogen], so it is zeroth order in both species. The overall reaction must be **zeroth order**.

(b) rate $= k_r$

(c) The rate law does not depend on the concentration of the reactants. It must depend, therefore, on the only other chemical present, the catalyst. Apparently, whatever is happening on the surface of the catalyst determines the rate, regardless of the concentrations of the two gases. Increasing the surface area of the catalyst, or simply adding more catalyst, would accelerate the reaction.

4-14

(a)

(b) $E_a = +13$ kJ/mole  ($+3$ kcal/mole)
(c) $\Delta H° = -4$ kJ/mole  ($-1$ kcal/mole)

4-15

(a)

(b) reverse:  $CH_3Cl + Cl• \longrightarrow Cl_2 + •CH_3$

(c) reverse: $E_a = +109$ kJ/mole $+ +4$ kJ/mole $= +113$ kJ/mole
       ($+26$ kcal/mole $+ +1$ kcal/mole $= +27$ kcal/mole)

4-16

(a)

Initiation steps are not included in energy calculations, nor in reaction energy diagrams. Because of the chain mechanism, one initiation photon can initiate several thousand reaction cycles.

4-16 continued

(b) The step leading to the highest energy transition state is rate-limiting. In this mechanism, the first propagation step is rate-limiting:

$$Br \bullet \ + \ CH_4 \longrightarrow HBr \ + \ \bullet CH_3$$

(c)  (1)  $\left[ \begin{array}{cc} \delta \bullet & \delta \bullet \\ Br \text{--------} Br \end{array} \right]^{\ddagger}$  ("$\delta \bullet$" means partial radical character on the atom)

(2)  $\left[ \begin{array}{c} H \quad \delta \bullet \\ H-C \text{--------} H \text{------} Br \\ H \end{array} \right]^{\ddagger}$  $\delta \bullet$

(3)  $\left[ \begin{array}{c} H \quad \delta \bullet \\ H-C \text{--------} Br \text{------} Br \\ H \end{array} \right]^{\ddagger}$  $\delta \bullet$

(d) $\Delta H°$ for the reaction is the sum of the $\Delta H°$ values of the individual propagation steps (refer to the solution to 4-10 (a) and (b)):

$$+ 67 \text{ kJ/mole } + \ - 101 \text{ kJ/mole } = \ - \textbf{34 kJ/mole}$$
$$( + 16 \text{ kcal/mole } + \ - 24 \text{ kcal/mole } = \ - \textbf{8 kcal/mole})$$

4-17

(a)  initiation  (1)  $I-I \xrightarrow{\ h\nu\ } 2 \ I \bullet$

propagation  (2)  $I \bullet \ + H-CH_3 \longrightarrow H-I \ + \ \bullet CH_3$

(3)  $I-I \ + \ \bullet CH_3 \longrightarrow I-CH_3 + \ I \bullet$

| step (1) | break I–I | $\Delta H° = + 151$ kJ/mole ( + 36 kcal/mole) |

| step (2) | break H–CH$_3$ | $\Delta H° = + 435$ kJ/mole  ( + 104 kcal/mole) |
|          | make H–I  | $\Delta H° = - \ 297$ kJ/mole  ( – 71 kcal/mole) |
|          | step (2)  | $\Delta H° = + \textbf{138 kJ/mole}$  ( + 33 kcal/mole) |

| step (3) | break I–I | $\Delta H° = + 151$ kJ/mole  ( + 36 kcal/mole) |
|          | make I–CH$_3$ | $\Delta H° = - 234$ kJ/mole  ( – 56 kcal/mole) |
|          | step (3)  | $\Delta H° = \textbf{–83 kJ/mole}$ ( –20 kcal/mole) |

(b) $\Delta H°$ for the reaction is the sum of the $\Delta H°$ values of the individual propagation steps:

$$+ 138 \text{ kJ/mole } + \ - 83 \text{ kJ/mole } = \ \textbf{+ 55 kJ/mole}$$
$$(+ 33 \text{ kcal/mole } + \ - 20 \text{ kcal/mole } = \ \textbf{+ 13 kcal/mole})$$

(c) Iodination of methane is unfavorable for both kinetic and thermodynamic reasons. Kinetically, the rate of the first propagation step must be very slow because it is very endothermic; the activation energy must be at least + 138 kJ/mole. Thermodynamically, the overall reaction is endothermic, so an equilibrium would favor reactants, not products; there is no energy decrease to drive the reaction to products.

4-18 Propane has six primary hydrogens and two secondary hydrogens, a ratio of 3 : 1. If primary and secondary hydrogens were replaced by chlorine at equal rates, the chloropropane isomers would reflect the same 3 : 1 ratio, that is, 75% 1-chloropropane and 25% 2-chloropropane.

4-19 (a)   $CH_3CH_2CH_2CH_3$   (b)   (c)

      ↑  ↑  ↑  ↑

      1°  2°  2°  1°

(d)

all are 2° H     OR     all are 2° H     (e)

all are 2° H except at bridgehead (labeled 3°)

4-20  $\Delta H°$ for abstraction of a 1° H from both ethane and propane are $+410$ kJ/mole ($+98$ kcal/mole). It is reasonable to use this same value for abstraction of the 1° H in isobutane.

3° H abstraction

$$Cl\bullet \;+\; CH_3-\overset{\overset{\displaystyle H}{|}}{\underset{\underset{\displaystyle CH_3}{|}}{C}}-CH_3 \;\longrightarrow\; H-Cl \;+\; CH_3-\overset{\bullet}{\underset{\underset{\displaystyle CH_3}{|}}{C}}-CH_3$$

| | |
|---|---|
| break 3° H–C(CH₃)₃ | $\Delta H° = +381$ kJ/mole ($+91$ kcal/mole) |
| make H–Cl | $\Delta H° = -431$ kJ/mole ($-103$ kcal/mole) |
| overall 3° H abstraction | $\boldsymbol{\Delta H° = }$ **–50 kJ/mole (–12 kcal/mole)** |

1° H abstraction

$$Cl\bullet \;+\; H-CH_2-\overset{\overset{\displaystyle H}{|}}{\underset{\underset{\displaystyle CH_3}{|}}{C}}-CH_3 \;\longrightarrow\; H-Cl \;+\; \bullet CH_2-\overset{\overset{\displaystyle H}{|}}{\underset{\underset{\displaystyle CH_3}{|}}{C}}-CH_3$$

| | |
|---|---|
| break 1° H–CH₂CH(CH₃)₂ | $\Delta H° = +410$ kJ/mole ($+98$ kcal/mole) |
| make H–Cl | $\Delta H° = -431$ kJ/mole ($-103$ kcal/mole) |
| overall 1° H abstraction | $\boldsymbol{\Delta H° = }$ **–21 kJ/mole (–5 kcal/mole)** |

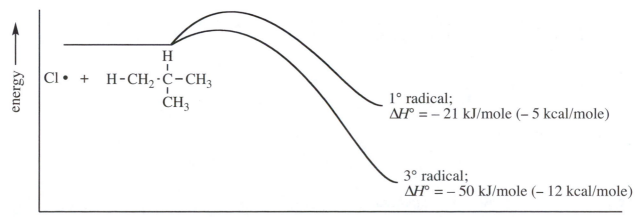

Since $\Delta H°$ for forming the 3° radical is more negative than $\Delta H°$ for forming the 1° radical, it is reasonable to infer that the activation energy leading to the 3° radical is lower than the activation energy leading to the 1° radical.

4-21

2-Methylbutane can produce four mono-chloro isomers. To calculate the relative amount of each in the product mixture, multiply the numbers of hydrogens that could lead to that product times the reactivity for that type of hydrogen. Each relative amount divided by the sum of all the amounts will provide the percent of each in the product mixture.

CICH$_2$—CHCH$_2$CH$_3$
(6 1° H) x (reactivity 1.0)
= 6.0 relative amount
$\frac{6.0}{23.5}$ x 100 %=
**26%**

(1 3° H) x (reactivity 5.5)
= 5.5 relative amount
$\frac{5.5}{23.5}$ x 100% = **23%**

(2 2° H) x (reactivity 4.5)
= 9.0 relative amount
$\frac{9.0}{23.5}$ x 100% = **38%**

CH$_3$—CHCH$_2$CH$_2$Cl
(3 1° H) x (reactivity 1.0)
= 3.0 relative amount
$\frac{3.0}{23.5}$ x 100% =
**13%**

4-22

(a) When heptane is burned, only 1° and 2° radicals can be formed (from either C—H or C—C bond cleavage). These are high-energy, unstable radicals that rapidly form other products. When isooctane (2,2,4-trimethylpentane, below) is burned, 3° radicals can be formed from either C—H or C—C bond cleavage. The 3° radicals are lower in energy than 1° or 2°, relatively stable, with lowered reactivity. Slower combustion translates to less "knocking."

isooctane (2,2,4-trimethylpentane)

Any indicated bond cleavage will produce a 3° radical.

(b)

$$H_3C—C—O—H \ + \ •R \longrightarrow H_3C—C—O• \ + \ R—H$$

an alkyl radical

When the alcohol hydrogen is abstracted from *tert*-butyl alcohol, a relatively stable *tert*-butoxy radical ( (CH$_3$)$_3$C—O • ) is produced. This low-energy radical is slower to react than alkyl radicals, moderating the reaction and producing less "knocking."

(c)

BDE +356 kJ/mol

benzylic

+ •R $\longrightarrow$

an alkyl radical

+ R—H

BDE –381 (3°) to –410 (1°) kJ/mol
EXOTHERMIC

toluene

*resonance stabilized*

When the benzylic hydrogen is abstracted from toluene, a relatively stable, *resonance-stabilized* benzylic radical is produced. This low-energy radical is energetically "easy" to produce, making toluene a reactive molecule with a high octane rating. (If you want to peek ahead for the resonance at benzylic positions, see the solution to 4-44(b).)

**4-23**

(a) 1° H abstraction

$$F\cdot \;+\; CH_3CH_2CH_3 \longrightarrow H-F \;+\; \cdot CH_2CH_2CH_3$$

break 1° H–CH$_2$CH$_2$CH$_3$    $\Delta H° = +410$ kJ/mole ( $+98$ kcal/mole)
make H–F    $\Delta H° = -569$ kJ/mole ( $-136$ kcal/mole)
overall 1° H abstraction    **$\Delta H° = -159$ kJ/mole ($-38$ kcal/mole)**

2° H abstraction

$$F\cdot \;+\; CH_3CH_2CH_3 \longrightarrow H-F \;+\; CH_3\overset{\bullet}{C}HCH_3$$

break 2° H–CH(CH$_3$)$_2$    $\Delta H° = +397$ kJ/mole ( $+95$ kcal/mole)
make H–F    $\Delta H° = -569$ kJ/mole ( $-136$ kcal/mole)
overall 2° H abstraction    **$\Delta H° = -172$ kJ/mole ($-41$ kcal/mole)**

(b) Fluorination is extremely exothermic and is likely to be indiscriminate in which hydrogens are abstracted. (In fact, C—C bonds are also broken during fluorination.)

(c) Free-radical fluorination is extremely exothermic. In exothermic reactions, the transition states resemble the starting materials more than the products, so while the 1° and 2° radicals differ by about 13 kJ/mole (3 kcal/mole), the transition states will differ by only a tiny amount: from Figure 4-11(b), only 4 kJ/mole (1 kcal/mole). For fluorination, then, the rate of abstraction for 1° and 2° hydrogens will be virtually identical. Product ratios will depend on statistical factors only.

Fluorination is difficult to control, but if propane were monofluorinated, the product mixture would reflect the ratio of the types of hydrogens: six 1° H to two 2° H, or 3 : 1 ratio, giving 75% 1-fluoropropane and 25% 2-fluoropropane.

**4-24**

Mechanism

initiation    $Br-Br \xrightarrow{h\nu} 2\ Br\cdot$

propagation

**4-25 (a)** Mechanism

initiation    $Br-Br \xrightarrow{h\nu} 2\ Br\cdot$

propagation

75

Copyright © 2013 Pearson Education, Inc.

4-25 continued

(b) Energy calculation uses the value for the allylic C—H bond from Table 4-2.

First propagation step
- break allylic H–CH[ring]    $\Delta H° = +364$ kJ/mole ($+87$ kcal/mole)
- make H–Br    $\Delta H° = -368$ kJ/mole ($-88$ kcal/mole)
- overall allylic H abstraction   **$\Delta H° = -4$ kJ/mole ($-1$ kcal/mole)**

Second propagation step
- break Br–Br    $\Delta H° = +192$ kJ/mole ($+46$ kcal/mole)
- make 2° C–Br    $\Delta H° = -285$ kJ/mole ($-68$ kcal/mole)
- overall C–Br formation   **$\Delta H° = -93$ kJ/mole ($-22$ kcal/mole)**

*The first propagation step is rate-limiting.*

structure of the transition state:

(c) The Hammond Postulate tells us that, in an exothermic reaction, the transition state is closer to the reactants in energy and in structure. Since the first propagation step is exothermic (although not by much), the transition state is closer to cyclohexene + bromine radical. This is indicated in the transition state structure by showing the H closer to the C than to the Br.

(d) A bromine radical will abstract the hydrogen with the lowest bond dissociation energy at the fastest rate. The allylic hydrogen of cyclohexene is more easily abstracted than a hydrogen of cyclohexane because the radical produced is stabilized by resonance. (Energy values below are per mole.)

aliphatic: 397 kJ (95 kcal) ⟶    *SLOWER*      ⟵ 364 kJ (87 kcal): allylic    *FASTER*

4-26

BHA radical

4-27

Vitamin E        stabilized by resonance— less reactive than RO•

4-28

The resonance forms on the next page do not include the simple benzene resonance forms as shown below; they are significant, but repetitive, so for simplicity, they are not drawn here. Braces are omitted.

many more      Ph = phenyl =

## 4-28 continued
The triphenylmethyl cation is so stable because of the delocalization of the charge. The more resonance forms a species has—especially equivalent resonance forms—the more stable it will be.

4-29  most stable  (d)  >  (c)  >  (b)  >  (a)  least stable

(d)
3° and resonance stabilized

(c) $CH_3 - \overset{\oplus}{\underset{\underset{CH_3}{|}}{C}}CH_2CH_3$
3°

(b) $CH_3 - \overset{\oplus}{\underset{\underset{CH_3}{|}}{C}HCHCH_3}$
2°

(a) $CH_3 - \underset{\underset{CH_3}{|}}{C}HCH_2\overset{\oplus}{C}H_2$
1°

4-30  most stable  (d)  >  (c)  >  (b)  >  (a)  least stable

(d)
3° and resonance stabilized

(c) $CH_3 - \overset{\bullet}{\underset{\underset{CH_3}{|}}{C}}CH_2CH_3$
3°

(b) $CH_3 - \overset{\bullet}{\underset{\underset{CH_3}{|}}{C}HCHCH_3}$
2°

(a) $CH_3 - \underset{\underset{CH_3}{|}}{C}HCH_2\overset{\bullet}{C}H_2$
1°

4-31

Negative charge delocalized on two oxygens lends a great deal of stability to this anion.

**77**

4-32

4-33

4-34  (a)  rate = k • [ CH$_3$O$^-$ ] [C$_4$H$_9$Br]   The reaction is first order in each, methoxide and 1-bromobutane.

(b)  If the solvent is reduced by half with the same amount of reactants, the concentration of each doubles. Doubling the concentration of each will increase the rate by a factor of 4, i.e., four times faster.

4-35
(a) and (c)

(b)  $\Delta H°$ is negative (decreases), so the reaction is exothermic.
(d)  The first transition state determines the rate since it is the highest energy point.  The *structure* of the first transition state resembles the *structure* of the intermediate since the *energy* of the transition state is closest to the *energy* of the intermediate.

4-36

4-37

4-38

The rate law is first order with respect to the concentrations of hydrogen ion and of *tert*-butyl alcohol, zeroth order with respect to the concentration of chloride ion, second order overall.
rate = k$_r$ [(CH$_3$)$_3$COH ] [ H$^+$ ]

4-39 (a) $CH_3CH_2CH(CH_3)_2$

$\uparrow$ $\uparrow$ $\uparrow$ $\uparrow$

1° 2° 3° 1°

(b) $(CH_3)_3CCH_2CH_3$

$\uparrow$ $\uparrow$ $\uparrow$

1° 2° 1°

(c)

(d)

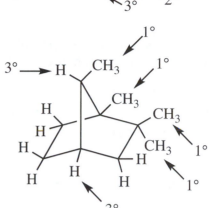

All are 2° H except for the
two types labeled.

(e)

All are 2° H except for the
two types labeled.

(f)

All are 2° H except as labeled.

4-40

(a) break H–$CH_2CH_3$ and I–I, make I–$CH_2CH_3$ and H–I

kJ/mole: ($+410$ + $+151$) + ($-222$ + $-297$) = **+ 42 kJ/mole**

kcal/mole: ($+98$ + $+36$) + ($-53$ + $-71$) = **+ 10 kcal/mole**

(b) break $CH_3CH_2$–Cl and H–I, make $CH_3CH_2$–I and H–Cl

kJ/mole: ($+339$ + $+297$) + ($-222$ + $-431$) = **−17 kJ/mole**

kcal/mole: ($+81$ + $+71$) + ($-53$ + $-103$) = **− 4 kcal/mole**

(c) break $(CH_3)_3C$–OH and H–Cl, make $(CH_3)_3C$–Cl and H–OH

kJ/mole: ($+381$ + $+431$) + ($-331$ + $-498$) = **− 17 kJ/mole**

kcal/mole: ($+91$ + $+103$) + ($-79$ + $-119$) = **− 4 kcal/mole**

(d) break $CH_3CH_2$–$CH_3$ and H–H, make $CH_3CH_2$–H and H–$CH_3$

kJ/mole: ($+356$ + $+435$) + ($-410$ + $-435$) = **− 54 kJ/mole**

kcal/mole: ($+85$ + $+104$) + ($-98$ + $-104$) = **− 13 kcal/mole**

(e) break $CH_3CH_2$–OH and H–Br, make $CH_3CH_2$–Br and H–OH

kJ/mole: ($+381$ + $+368$) + ($-285$ + $-498$) = **− 34 kJ/mole**

kcal/mole: ($+91$ + $+88$) + ($-68$ + $-119$) = **− 8 kcal/mole**

4-41 Numbers are bond dissociation energies in kcal/mole in the top line and kJ/mole in the bottom line.

$\langle\!\!\!\bigcirc\!\!\!\rangle$–$\overset{\bullet}{C}H_2$ > $CH_2{=}CH\overset{\bullet}{C}H_2$ > $(CH_3)_3\overset{\bullet}{C}$ > $(CH_3)_2\overset{\bullet}{C}H$ > $CH_3\overset{\bullet}{C}H_2$ > $\overset{\bullet}{C}H_3$

| | | | | | |
|---|---|---|---|---|---|
| 85 | 87 | 91 | 95 | 98 | 104 |
| 356 | 364 | 381 | 397 | 410 | 435 |
| benzyl | allyl | 3° | 2° | 1° | methyl |
| *most stable* | | | | | *least stable* |

4-42
(a)

Only one product; chlorination would work. Bromination on a 2° carbon would not be predicted to be a high-yielding process.

(b)

Chlorination would produce four constitutional isomers and would not be a good method to make only one of these. Monobromination at the 3° carbon would give a reasonable yield of the pure 3° bromide, in contrast to the results from chlorination.

4-42 continued

(c)

Chlorination would produce five constitutional isomers and would not be a good method to make only one of these. Monobromination would be selective for the 3° carbon and would give an excellent yield.

(d)

$$CH_3-\underset{\underset{CH_3}{|}}{\overset{\overset{CH_3}{|}}{C}}-\underset{\underset{CH_3}{|}}{\overset{\overset{CH_3}{|}}{C}}-CH_3 \longrightarrow CH_3-\underset{\underset{CH_3}{|}}{\overset{\overset{CH_3}{|}}{C}}-\underset{\underset{CH_3}{|}}{\overset{\overset{CH_3}{|}}{C}}-CH_2Cl$$

Only one product; chlorination would give a high yield. Monobromination would be very difficult since all hydrogens are on 1° carbons.

4-43
initiation    (1) Cl—Cl $\xrightarrow{h\nu}$ 2 Cl •

propagation

(2) Cl • + H—C(ring) $\longrightarrow$ H—Cl + •C(ring)

(3) Cl—Cl + •C(ring) $\longrightarrow$ Cl—C(ring) + Cl •

Termination steps are any two radicals combining.

4-44

(a) $\{CH_2=CH-\overset{\bullet}{C}H_2 \longleftrightarrow \overset{\bullet}{C}H_2-CH=CH_2\}$

(b)

(c) $\{CH_3-\overset{\overset{:O:}{||}}{C}-\overset{..}{\underset{..}{O}}• \longleftrightarrow CH_3-\overset{\overset{:O:}{|}}{C}=\overset{..}{\underset{..}{O}}\}$

(d)

(e)

(f)

**80**
Copyright © 2013 Pearson Education, Inc.

4-45

(a) <u>Mechanism</u>

initiation   Br — Br   $\xrightarrow{h\nu}$   2  Br •

The 3° allylic H is abstracted selectively (faster than any other type in the molecule), forming an intermediate represented by two *non-equivalent* resonance forms. Partial radical character on two different carbons of the allylic radical leads to two different products.

(b) There are two reasons why the H shown is the one that is abstracted by bromine radical: the H is 3° and it is allylic, that is, neighboring a double bond. Both of these factors stabilize the radical that is created by removing the H atom.

4-46 Where mixtures are possible, only the major product is shown.

(a) only one product possible

(b) 3° hydrogen abstracted selectively, faster than 2° or 1°

(c) 

decalin

3° hydrogen abstracted selectively, faster than 2°

(d) both 2° — formed in equal amounts

(e) CH₂CH₃ → CHCH₃ / Br   from resonance-stabilized benzylic radical

(f) All hydrogens at the starred positions are equivalent and *allylic*. The H from the lower right carbon has been removed to make an intermediate with two non-equivalent resonance forms, giving the two products shown. *Drawing the resonance forms is the key to answering this question correctly.*

**81**

4-47

(a) As $CH_3Cl$ is produced, it can compete with $CH_4$ for available $Cl \bullet$, generating $CH_2Cl_2$. This can generate $CHCl_3$, *etc*.

propagation steps

$$CH_4 + Cl\bullet \longrightarrow HCl + \bullet CH_3$$
$$\bullet CH_3 + Cl_2 \longrightarrow ClCH_3 + Cl\bullet$$
$$ClCH_3 + Cl\bullet \longrightarrow HCl + \bullet CH_2Cl$$
$$\bullet CH_2Cl + Cl_2 \longrightarrow CH_2Cl_2 + Cl\bullet$$
$$CH_2Cl_2 + Cl\bullet \longrightarrow HCl + \bullet CHCl_2$$
$$\bullet CHCl_2 + Cl_2 \longrightarrow CHCl_3 + Cl\bullet$$
$$CHCl_3 + Cl\bullet \longrightarrow HCl + \bullet CCl_3$$
$$\bullet CCl_3 + Cl_2 \longrightarrow CCl_4 + Cl\bullet$$

(b) To maximize $CH_3Cl$ and minimize formation of polychloromethanes, the ratio of methane to chlorine must be kept high (see solution to problem 4-2).

To guarantee that all hydrogens are replaced with chlorine to produce $CCl_4$, the ratio of chlorine to methane must be kept high.

4-48

(a) Pentane can produce three monochloro isomers. To calculate the relative amount of each in the product mixture, multiply the numbers of hydrogens which could lead to that product times the reactivity for that type of hydrogen. Each relative amount divided by the sum of all the amounts will provide the percent of each in the product mixture.

(b) $\dfrac{6.0}{33.0}$ x 100 = **18%**    $\dfrac{18.0}{33.0}$ x 100 = **55%**    $\dfrac{9.0}{33.0}$ x 100 = **27%**

4-49 (a) The second propagation step in the chlorination of methane is highly exothermic ($\Delta H° = -109$ kJ/mole ($-26$ kcal/mole)). The transition state resembles the reactants, that is, the Cl–Cl bond will be slightly stretched and the Cl–$CH_3$ bond will just be starting to form.

$$\left[ \begin{array}{c} \overset{\delta\bullet}{Cl} \underset{\text{stronger}}{------} \overset{}{Cl} \underset{\text{weaker}}{--------------} \overset{\delta\bullet}{CH_4} \end{array} \right]^{\ddagger}$$

(b) The second propagation step in the bromination of methane is highly exothermic ($\Delta H° = -101$ kJ/mole ($-24$ kcal/mole)). The transition state resembles the reactants, that is, the Br–Br bond will be slightly stretched and the Br–$CH_3$ bond will just be starting to form.

$$\left[ \begin{array}{c} \overset{\delta\bullet}{Br} \underset{\text{stronger}}{------} \overset{}{Br} \underset{\text{weaker}}{--------------} \overset{\delta\bullet}{CH_3} \end{array} \right]^{\ddagger}$$

**4-50** Two mechanisms are possible depending on whether HO • reacts with chlorine or with cyclopentane. It is reasonable to use the same bond dissociation energy for the 2° H of propane, 397 kJ/mole (from text Table 4-2), for the C—H bond in cyclopentane.

**Mechanism 1**                                                                                  all in kJ/mole

initiation
- (1)  $HO-OH \longrightarrow 2\ HO\bullet$                 $\Delta H^\circ = +213$ kJ/mole
- (2)  $HO\bullet + Cl-Cl \longrightarrow HO-Cl + Cl\bullet$     $\Delta H^\circ = +242 - 210 = +32$

propagation
- (3)  $Cl\bullet + $ (cyclopentane with 2 H's) $\longrightarrow H-Cl + $ (cyclopentyl radical with H)   $\Delta H^\circ = +397 - 431 = -34$
- (4)  $Cl-Cl + $ (cyclopentyl radical with H) $\longrightarrow$ (chlorocyclopentane with H) $+ Cl\bullet$   $\Delta H^\circ = +242 - 335 = -93$

**Mechanism 2**

initiation
- (1)  $HO-OH \longrightarrow 2\ HO\bullet$                 $\Delta H^\circ = +213$ kJ/mole
- (2)  $HO\bullet + $ (cyclopentane with 2 H's) $\longrightarrow H-OH + $ (cyclopentyl radical with H)   $\Delta H^\circ = +397 - 498 = -101$

propagation
- (3)  $Cl-Cl + $ (cyclopentyl radical with H) $\longrightarrow$ (chlorocyclopentane with H) $+ Cl\bullet$   $\Delta H^\circ = +242 - 335 = -93$
- (4)  $Cl\bullet + $ (cyclopentane with 2 H's) $\longrightarrow H-Cl + $ (cyclopentyl radical with H)   $\Delta H^\circ = +397 - 431 = -34$

In this case, the energies of initiation steps determine which mechanism is followed. The bond dissociation energy of HO—Cl is about 210 kJ/mole (about 50 kcal/mole), making initiation step (2) in mechanism 1 *endothermic* by about 30 kJ/mole (about 8 kcal/mole). In mechanism 2, initiation step (2) is *exothermic* by about 101 kJ/mole (24 kcal/mole); mechanism 2 is preferred. One strongly endothermic step can be enough to stop a mechanism.

**4-51**

$$Cl-\underset{\underset{H}{|}}{\overset{\overset{Cl}{|}}{C}}-H + {}^{\ominus}\!:\!\ddot{O}-H \longrightarrow H_2O + Cl-\underset{\underset{H}{|}}{\overset{\overset{Cl}{}}{C}}\!\!:^{\ominus} \longrightarrow \underset{\underset{H}{|}}{\overset{\overset{Cl}{\diagdown}}{C}}\!: + \ :\!\ddot{C}\!\ddot{l}\!:^{\ominus}$$

a carbene

**4-52** This critical equation is the key to this problem:  $\Delta G = \Delta H - T\,\Delta S$

At 1400 °K, the equilibrium constant is 1; therefore:

$$K_{eq} = 1 \implies \Delta G = -2.303\ RT(\log_{10}(1)) \implies \Delta G = 0 \implies \Delta H = T\,\Delta S$$

Assuming $\Delta H$ is about the same at 1400 °K as it is at calorimeter temperature:

$$\Delta S = \frac{\Delta H}{T} = \frac{-137\ \text{kJ/mole}}{1400\ °K} = \frac{-137{,}000}{1400}\ \text{J/°K-mole}$$

$$= -98\ \text{J/°K-mole}\ (-23\ \text{cal/°K-mole})$$

This is a large *decrease* in entropy, consistent with two molecules combining into one.

4-53

Assume that chlorine atoms (radicals) are still generated in the initiation reaction. Focus on the propagation steps. Bond dissociation energies are given below the bonds, in kJ/mole (kcal/mole).

$$Cl\bullet \;+\; H-CH_3 \longrightarrow H-Cl \;+\; \bullet CH_3 \qquad \Delta H = +4 \text{ kJ/mole } (+1 \text{ kcal/mole})$$
$$\phantom{Cl\bullet \;+\;}435\,(104)\phantom{H-CH_3 \longrightarrow}431\,(103)$$

$$Cl-Cl \;+\; \bullet CH_3 \longrightarrow Cl-CH_3 \;+\; Cl\bullet \qquad \Delta H = -109 \text{ kJ/mole } (-26 \text{ kcal/mole})$$
$$242\,(58)\phantom{\;+\; \bullet CH_3 \longrightarrow}351\,(84)$$

What happens when the different radical species react with iodine?

$$Cl\bullet \;+\; I-I \longrightarrow I-Cl \;+\; \bullet I \qquad \Delta H = -60 \text{ kJ/mole } (-14 \text{ kcal/mole})$$
$$\phantom{Cl\bullet \;+\;}151\,(36)\phantom{I-I \longrightarrow}211\,(50)$$

$$I-I \;+\; \bullet CH_3 \longrightarrow I-CH_3 \;+\; I\bullet \qquad \Delta H = -83 \text{ kJ/mole } (-20 \text{ kcal/mole})$$
$$151\,(36)\phantom{\;+\; \bullet CH_3 \longrightarrow}234\,(56)$$

Compare the second reaction in each pair: methyl radical reacting with chlorine is more exothermic than methyl radical reacting with iodine; this does not explain how iodine prevents the chlorination reaction. Compare the first reaction in each pair: chlorine atom reacting with iodine is very exothermic whereas chlorine atom reacting with methane is slightly endothermic. Here is the key: chlorine atoms will be scavenged by iodine before they have a chance to react with methane. Without chlorine atoms, the reaction comes to a dead stop.

4-54 (a)

(b) All energies are in kJ/mole. The abbreviation "c-Hx" stands for the cyclohexane ring.

Step 2: break H—Sn, make H—Br: +310 + −368 = −58 kJ/mole

Step 3: break c-Hx—Br, make Br—Sn: +285 + −552 = −267 kJ/mole WOW!

Step 4: break H—Sn, make c-Hx—H: +310 + −397 = −87 kJ/mole

The sum of the two propagation steps is: −267 + −87 = −354 kJ/mole —a hugely exothermic reaction.

4-55

**Mechanism 1:**

(1a) $Cl\bullet \; + \; O_3 \longrightarrow \; ClO\bullet \; + \; O_2$

(1b) $\mathbf{2}\;ClO\bullet \longrightarrow Cl-O-O-Cl$

(1c) $Cl-O-O-Cl \xrightarrow{h\nu} O_2 \; + \; \mathbf{2}\;Cl\bullet$

The biggest problem in Mechanism 1 lies in step (1b). The concentration of Cl atoms is very small, so at any given time, the concentration of ClO will be very small. The probability of two ClO radicals finding each other to form ClOOCl is virtually zero. Even though this mechanism shows a catalytic cycle with Cl• (starting the mechanism and being regenerated at the end), the middle step makes it highly unlikely.

**Mechanism 2:**

(2a) $O_3 \xrightarrow{h\nu} O_2 \; + \; O$

(2b) $Cl\bullet \; + \; O_3 \longrightarrow \; ClO\bullet \; + \; O_2$

(2c) $ClO\bullet \; + \; O \longrightarrow \; O_2 \; + \; Cl\bullet$

Step (2a) is the "light" reaction that occurs naturally in daylight. At night, the reaction reverses and regenerates ozone.

Step (2c) is the crucial step. A low concentration of ClO *will* find a relatively high concentration of O atoms because the "light" reaction is producing O atoms in relative abundance. Cl• is regenerated and begins propagation step (2b), continuing the catalytic cycle.

Mechanism 2 is believed to be the dominant mechanism in ozone depletion. Mechanism 1 can be discounted because of the low probability of step (1b) occurring, because two species in very low, catalytic concentration are required to find each other in order for the step to occur.

4-56

(a)

$$\left[ \begin{array}{c} H \\ | \\ H-C\text{--------}H\text{------}Cl \\ | \\ H \end{array} \right]^{\ddagger} \qquad \left[ \begin{array}{c} D \\ | \\ D-C\text{--------}D\text{------}Cl \\ | \\ D \end{array} \right]^{\ddagger}$$

In each case, the bond from carbon to H (D) is breaking and the bond from H (D) to Cl is forming.

(b) $C_2H_5-D \; + \; Cl_2 \longrightarrow \underbrace{C_2H_5-Cl \; + \; DCl}_{7\%} \; + \; \underbrace{C_2H_4DCl \; + \; HCl}_{93\%}$

D replacement: $7\% \div 1\,D = 7$ (reactivity factor)
H replacement: $93\% \div 5\,H = 18.6$ (reactivity factor)
relative reactivity of H : D abstraction $= 18.6 \div 7 = 2.7$

Each hydrogen is abstracted 2.7 times faster than deuterium.

(c) In both reactions of chlorine with either methane or ethane, the first propagation step is rate-limiting. The reaction of chlorine atom with methane is *endothermic* by 4 kJ/mole (1 kcal/mole), while for ethane this step is *exothermic* by 21 kJ/mole (5 kcal/mole). By the Hammond Postulate, differences in activation energy are most pronounced in *endothermic* reactions where the transition states most resemble the products. Therefore, a change in the methane molecule causes a greater change in its transition state energy than the same change in the ethane molecule causes in its transition state energy. Deuterium will be abstracted more slowly in both methane and ethane, but the rate effect will be more pronounced in methane than in ethane.

4-57   All energies are in kJ/mole.

initiation
{
(1)  HO⌢⌢OH  $\xrightarrow{\Delta}$  2  HO•                                    $\Delta H = + 213$ kJ/mole

(2)   HO•  +  I⌢CI$_3$  $\longrightarrow$  HO$-$I  +  •CI$_3$               $\Delta H = - 46$ kJ/mole
                 +188                              $-234$
}

propagation
{
(3)  [cyclopentane with H]  +397  +  •CI$_3$  $\longrightarrow$  [cyclopentyl radical]  +  H$-$CI$_3$   $\Delta H = - 21$ kJ/mole
                                                                                           $-418$

(4)  [cyclopentyl radical]  +  I⌢CI$_3$  $\longrightarrow$  [cyclopentyl-I]  +  •CI$_3$   $\Delta H = - 34$ kJ/mole
                                   +188                         $-222$
}

The sum of the two propagation steps is: $-21 + -34 = -55$ kJ/mole —a mildly exothermic reaction.

Note to the student: Stereochemistry is the study of molecular structure and reactions in three dimensions. Molecular models will be especially helpful in this chapter.

5-1   The best test of whether a household object is chiral is whether it would be used equally well by a left- or right-handed person. The chiral objects are the corkscrew, the writing desk, the can opener, the screw-cap bottle (only for refilling, however; in use, it would not be chiral), the rifle and the knotted rope. The corkscrew, the bottle top, and the rope each have a twist in one direction. The rifle, corkscrew, and desk are clearly made for right-handed users; the can opener is for a left-handed person. All the other objects are achiral and would feel equivalent to right- or left-handed users.

5-2
(a) *cis*

achiral—
identical mirror images

mirror

(b) *trans*

chiral—
enantiomers

mirror

(c) *cis* first, then *trans*

achiral—
identical mirror images

mirror

achiral—
identical mirror images

mirror

(d)

chiral—
enantiomers

mirror

(e)

chiral—
enantiomers

mirror

(f)

chiral—
enantiomers

mirror

87

5-3 Asymmetric carbon atoms are starred.

(a) enantiomers

(b) no asymmetric carbons— same structure

(c) enantiomers

(d) enantiomers

(e) no asymmetric carbon— same structure

(f) same structure— plane of symmetry (wavy line) explained in text Section 5-2C

(g) enantiomers

(h) enantiomers

(i) enantiomers

**5-4** You may have chosen to interchange two groups different from the ones shown here. The type of isomer produced will still be the same as listed here.

Interchanging any two groups around a chirality center (*) will create an enantiomer of the first structure.

**CH₂CH₃** Interchanging the Br and the H creates an enantiomer of the structure in Figure 5-5.

**CH₂CH₂CH₃** Interchanging the ethyl and the isopropyl creates an enantiomer of the structure in Figure 5-5.

On a double bond, interchanging the two groups on ONE of the stereocenters will create the other geometric (*cis-trans*) isomer. However, interchanging the two groups on BOTH of the stereocenters will give the original structure.

Original structure is *cis*.

Interchange H and CH₃ on top stereocenter to produce *trans* (interchanging bottom two groups will give the same structure).

**5-5**

(a) plane of symmetry

(b) plane of symmetry

(c) chiral—no plane of symmetry

(d) chiral—no plane of symmetry

(e) chiral—no plane of symmetry

(f) chiral—no plane of symmetry

(g) plane of symmetry

(h) plane of symmetry

This view is from the right side of the structure as drawn in the text.

**5-6** Place the 4th priority group away from you, where possible. Then determine if the sequence 1→2→3 is clockwise (*R*) or counter-clockwise (*S*). (There is a Problem-Solving Hint near the end of section 5-3 in the text that describes what to do when the 4th priority group is closest to you.)

(a)

*R*

This appears to be *S*, but group 4 is coming toward the viewer, so the opposite chirality must be assigned; it is actually *R*.

(b) *S*

Viewing from the bottom to put group 4 going away, the arrow is counterclockwise; the chirality is *S*.

(c)

*R*

Viewing from the front to put group 4 going away, the arrow is clockwise; the chirality is *R*.

5-6 continued

(d)

(e)

(f)

(g)

(h)

(i) $S$

Part (i) deserves some explanation. The difference between groups 1 and 2 hinges on what is on the "extra" oxygen.

$CH(OCH_3)_2$

$\Longrightarrow$  H—C—O—CH$_3$  with O—CH$_3$

higher priority

$O=C$ with H  $\Longrightarrow$  H—C—O  with O—C  *imaginary*

lower priority

5-7  There are no asymmetric carbons in 5-3 (b) and (e).

(a)  $S$  $C$  OH  H  CH$_3$  CH$_3$CH$_2$CH$_2$  $\vdots$  $R$  OH  H  H$_3$C  CH$_2$CH$_2$CH$_3$

(c)  $S$  COOH  C  H  H$_3$C  NH$_2$  $\vdots$  $R$  COOH  C  H  H$_2$N  CH$_3$

(d)  $S$  CH$_2$Br  C  H  CH$_3$CH$_2$  CH$_3$  $\vdots$  $R$  CH$_2$Br  C  H  H$_3$C  CH$_2$CH$_3$

(f)  $R$  H  H  $S$  Cl  Cl  $\vdots$  $R$  H  H  $S$  Cl  Cl

The identity of these two structures becomes more clear with the assignments of configuration.

(g)  H  C=C  R  H  H  CH$_3$  $\vdots$  $S$  H  C=C  H  H$_3$C  H

(h)  $S$  CH$_3$  $R$  $\vdots$  $R$  H$_3$C  $S$

What can you conclude about the designation of configuration between two enantiomers?

(i)  H  $S$  $S$  H  $\vdots$  R  H  R  H

90

5-8

2.0 g / 10.0 mL = 0.20 g/mL ; 100 mm = 1 dm

$$[\alpha]_D^{25} = \frac{+1.74°}{(0.20)(1)} = +8.7° \text{ for (+)-glyceraldehyde}$$

5-9

0.50 g / 10.0 mL = 0.050 g/mL ; 20 cm = 2 dm

$$[\alpha]_D^{25} = \frac{-5.1°}{(0.050)(2)} = -51° \text{ for (−)-epinephrine}$$

5-10

Measure using a solution of about one-fourth the concentration of the first. The value will be either + 45° or − 45°, which gives the sign of the rotation.

5-11

Whether a sample is dextrorotatory (abbreviated "(+)") or levorotatory (abbreviated "(−)") is determined experimentally by a polarimeter. Except for the molecule glyceraldehyde (see text Section 5-14), there is no direct, universal correlation between direction of optical rotation ((+) and (−)) and designation of configuration (R and S). In other words, one dextrorotatory compound might have R configuration while a different dextrorotatory compound might have S configuration.
(a) Yes, both of these are determined experimentally: the (+) or (−) by the polarimeter and the smell by the nose.
(b) No, R or S cannot be determined by either the polarimeter or the nose.
(c) The drawings show that (+)-carvone from caraway has the S configuration and (−)-carvone from spearmint has the R configuration.

(+)-carvone (caraway seed)          (−)-carvone (spearmint)

(For fun, ask your instructor if you can smell the two enantiomers of carvone. Some people are unable, presumably for genetic reasons, to distinguish the fragrance of the two enantiomers.)

5-12

(R)-2-bromobutane          (R)-butan-2-ol          (S)-butan-2-ol
                           one-third of mixture     two-thirds of mixture

Chapter 6 will explain how these mixtures come about. For this problem, the S enantiomer accounts for 66.7% of the butan-2-ol in the mixture and the rest, 33.3%, is the R enantiomer. Therefore, the excess of one enantiomer over the racemic mixture must be 33.3% of the S, the enantiomeric excess. (All of the R is "canceled" by an equal amount of the S, algebraically as well as in optical rotation.)

The optical rotation of pure (S)-butan-2-ol is + 13.5°. The optical rotation of this mixture is:
        33.3%  x  (+ 13.5°)  =  + 4.5°

*see next page for an alternative solution*          **91**

5-12 continued

(This algebraic approach has been suggested by Editorial Adviser Richard King.)
From the problem, the reaction produces twice as much (S)-(+)-butan-2-ol (the d isomer) as (R)-(−)-butan-2-ol (the l isomer): $d = 2l$

$$\text{e.e.} = \frac{d-l}{d+l} \times 100\% = \frac{2l-l}{2l+l} \times 100\% = \frac{l}{3l} \times 100\% = 33.3\%$$

The calculation of optical rotation of the mixture is the same as on the previous page.

5-13  The rotation of pure (+)-butan-2-ol is +13.5°.

$$\frac{\text{observed rotation}}{\text{rotation of pure enantiomer}} = \frac{+0.45°}{+13.5°} \times 100\% = 3.3\% \text{ optical purity}$$
$$= 3.3\% \text{ e.e.} = \text{excess of (+) over (−)}$$

To calculate percentages of (+) and (−):    (two equations in two unknowns)

$$(+) + (−) = 100\% \implies (−) = 100\% − (+)$$

$$(+) − (−) = 3.3\% \implies (+) − (100\% − (+)) = 3.3\%$$

$$\mathbf{2 (+) = 103.3\%}$$

> (+) = 51.6% (rounded)
> (−) = 48.4%

(This algebraic approach has been suggested by Editorial Adviser Richard King.)

$$\text{e.e.} = \frac{d-l}{d+l} \times 100\% = 3.33\% \implies \frac{d-l}{d+l} = \frac{3.33\%}{100\%} \implies$$

$$\left.\begin{array}{l} d-l = 3.33\% \\ \text{and } d+l = 100\% \end{array}\right\} \text{add these two equations}$$

$$2d = 103.33\% \implies d = 51.6\% \quad l = 100\% − d = 100\% − 51.6\% = 48.4\%$$

5-14  Drawing Newman projections is the clearest way to determine symmetry of conformations.

(a)

chiral —
optically active

(b)

plane includes Br and Cl

plane of symmetry containing
Br—C—C—Cl; not optically active

no asymmetric carbons

(c)

chiral —
optically active

(d)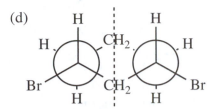

plane of symmetry—not optically active despite the presence of two asymmetric carbons

Br<sub>*</sub>   Br

H<sub>""</sub> *R*   *S* <sub>""</sub>H

(e)

Br     H

H   CH<sub>2</sub>   H

H   CH<sub>2</sub>   Br

H   H

no plane of symmetry— optically active (other chair form is equivalent—no plane of symmetry)

Br<sub>*</sub>   H

H<sub>""</sub> *R*   *R* <sub>""</sub> Br

(f)

Br ----- H

1   4

H   Br

plane of symmetry through C-1 and C-4—not optically active

Br """"""⟨   ⟩ Br

no asymmetric carbons

**Part (2)** Predictions of optical activity based on asymmetric centers give the same answers as predictions based on the most symmetric conformation. NOTE: The assignment of *R* and *S* above might be different from your answers if you happened to draw the enantiomer.

5-15
(a)

H     H

C=C=C<sub>""</sub>

Cl     Cl

No asymmetric carbons, but the molecule is chiral (an allene); the drawing below is a three-dimensional picture of the allene in (a) showing there is no plane of symmetry because the substituents of an allene are in different planes.

(b)

H     H

C=C=C<sub>""</sub>

Cl     CH<sub>3</sub>

No asymmetric carbons, but the molecule is chiral (an allene).

(c)

H     CH<sub>3</sub>

C=C=C<sub>""</sub>

Cl     CH<sub>3</sub>

No asymmetric carbons; this allene has a plane of symmetry between the two methyls (the plane of the paper), including all the other atoms because the two pi bonds of an allene are perpendicular, the Cl is in the plane of the paper and the plane of symmetry goes through it; not a chiral molecule.

(d)

H     H

C=C

Cl    C=C   H

H    H

planar molecule—no asymmetric carbons; not a chiral molecule

(e)

Cl

*

*

two asymmetric carbons and no plane of symmetry; a chiral compound

5-15 continued

(f)

No asymmetric carbons, but the molecule is chiral due to restricted rotation that precludes a plane of symmetry; the drawing below is a three-dimensional picture showing that the rings are perpendicular (hydrogens are not shown).

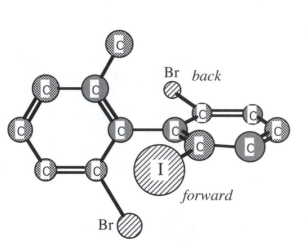

Br *back*

I *forward*

Br

(g)

No asymmetric carbons, and the groups are not large enough to restrict rotation; thus, it has a plane of symmetry and is not a chiral compound.

5-16

(a)

$$\begin{array}{c} COOH \\ H \underline{\quad\quad} OH \\ CH_3 \end{array}$$

$$\begin{array}{c} COOH \\ HO \underline{\quad\quad} H \\ CH_3 \end{array}$$
enantiomer

$$\begin{array}{c} H \\ CH_3 \underline{\quad\quad} COOH \\ OH \end{array}$$
enantiomer

$$\begin{array}{c} CH_3 \\ HO \underline{\quad\quad} H \\ COOH \end{array}$$
same

**Rules for Fischer projections:**
1. Interchanging any two groups an odd number of times (once, three times, etc.) makes an enantiomer. Interchanging any two groups an even number of times (e.g. twice) returns to the original stereoisomer.

(b)

$$\begin{array}{c} CH_2CH_3 \\ H \underline{\quad\quad} Br \\ CH_3 \end{array}$$

$$\begin{array}{c} CH_3 \\ Br \underline{\quad\quad} H \\ CH_2CH_3 \end{array}$$
same

$$\begin{array}{c} CH_2CH_3 \\ Br \underline{\quad\quad} H \\ CH_3 \end{array}$$
enantiomer

$$\begin{array}{c} CH_3 \\ H \underline{\quad\quad} Br \\ CH_2CH_3 \end{array}$$
enantiomer

2. Rotating the structure by 90° makes the enantiomer. Rotating by 180° returns to the original stereoisomer. (The second rule is an application of the first. Prove this to yourself.)

(c)

$$\begin{array}{c} CH_3 \\ HO \underline{\quad\quad} H \\ CH_2CH_3 \end{array}$$
(R)-butan-2-ol

$$\begin{array}{c} CH_3 \\ H \underline{\quad\quad} OH \\ CH_2CH_3 \end{array}$$
enantiomer

$$\begin{array}{c} CH_3 \\ HO \underline{\quad\quad} H \\ CH_2CH_3 \end{array}$$
same

$$\begin{array}{c} CH_2CH_3 \\ H \underline{\quad\quad} OH \\ CH_3 \end{array}$$
same

5-17

(a)
$$\begin{array}{c} CH_2OH \\ HO \underline{\quad\quad} H \\ CH_3 \end{array}$$

(b)
$$\begin{array}{c} CH_2OH \\ H \underline{\quad\quad} Br \\ CH_2CH_3 \end{array}$$

(c)
$$\begin{array}{c} CH_2Br \\ Br \underline{\quad\quad} H \\ CH_2CH_3 \end{array}$$

(d)
$$\begin{array}{c} CH_3 \\ HO \underline{\quad\quad} H \\ CH_2CH_3 \end{array}$$

(e)
$$\begin{array}{c} CHO \\ H \underline{\quad\quad} OH \\ CH_2OH \end{array}$$

5-18

mirror

(a)

CHO
H——OH
CH₂OH

CHO
HO——H
CH₂OH

180° rotation of the right structure does not give left structure; no plane of symmetry: chiral—**enantiomers**

(b)

CH₂OH
H——Br
CH₂OH

CH₂OH
---Br——H---
CH₂OH

180° rotation of the right structure gives same structure as on the left; also has plane of symmetry: **same structure**

(c)

CH₂Br
Br——Br
CH₃

CH₂Br
Br——Br
CH₃

plane of symmetry: **same structure**

(d)

CHO
H——OH
H——OH
CH₂OH

CHO
HO——H
HO——H
CH₂OH

180° rotation of the right structure does not give left structure; no plane of symmetry: chiral—**enantiomers**

(e)

CH₂OH
H——OH
H——OH
CH₂OH

CH₂OH
HO——H
HO——H
CH₂OH

180° rotation of the right structure gives same structure as on the left; also has plane of symmetry: **same structure**

(f)

CH₂OH
HO——H
H——OH
CH₂OH

CH₂OH
H——OH
HO——H
CH₂OH

180° rotation of the right structure does not give left structure; no plane of symmetry: chiral—**enantiomers**

5-19 If the Fischer projection is drawn correctly, the most oxidized carbon (most bonds to oxygen) will be at the top; this is the carbon with the greatest number of bonds to oxygen. Then the numbering goes from the top down.

(a) *R*

(b) no chiral center

(c) no chiral center

(d) *2R,3R*

(e) *2S,3R* (numbering down)

(f) *2R, 3R*

(g) *R*

(h) *S*

(i) *S*

**95**

5-20

(a) enantiomers—configurations at both asymmetric carbons inverted
(b) diastereomers—configuration at only one asymmetric carbon inverted
(c) diastereomers—configuration at only one asymmetric carbon inverted (the left carbon)
(d) constitutional isomers—C=C shifted position
(e) enantiomers—chiral, mirror images
(f) diastereomers—configuration at only one asymmetric carbon inverted (the top one)
(g) enantiomers—configuration at all asymmetric carbons inverted
(h) enantiomers—the front C is an asymmetric carbon atom; the two Newmans are mirror images
(i) diastereomers—configuration at only one chirality center (the nitrogen) inverted

5-21 It would be excellent practice to draw Newman projections of each structure! Symmetry is quickly revealed.

5-22

(a)

A     B     C     D

enantiomers: **A** and **B**; **C** and **D**
diastereomers: **A** and **C**; **A** and **D**; **B** and **C**; **B** and **D**

(b)

plane of symmetry
MESO

E

F     G

enantiomers: **F** and **G**
diastereomers: **E** and **F**; **E** and **G**

(c)

H     I     J     K

L     M     N     O

enantiomers: **H** and **I**; **J** and **K**; **L** and **M**; **N** and **O**
diastereomers: any pair that is not enantiomeric

(d)

P
*cis ring fusion*

plane of symmetry
MESO

Q     R
*trans ring fusion*

enantiomers: **Q** and **R**
diastereomers: **P** and **Q**; **P** and **R**

**97**

(e)

enantiomers:  **S** and **T**
diastereomers:  **S** and **U**; **S** and **V**;
**T** and **U**; **T** and **V**; **U** and **V**

5-23  Any diastereomeric pair could be separated by a physical process like distillation or crystallization. Diastereomers are found in parts (a), (b), and (d).  The structures in (c) are enantiomers; they could not be separated by normal physical means.

5-24

5-25  The asymmetric carbon atoms are indicated by asterisks.

**98**

5-26

(a)

chiral

(b)

```
        CH₂OH
        |
  H ——— * ——— OH
        |
        CH₃
        R
```

chiral

(c)

```
  H     H    H    Cl
   \    |    |   /
    C        C
   /|   *    *  |\
  H |   C       | Br
    H   |↗      ↖|
        H   OH
        S        S
```

chiral

(d)

```
  H     CH₃        H
   \    |          |
    C        *     C
   /|    C        ‖\
 CH₃|   /|↗    ↖  | CH₂
    |  H   CH₃
    S        
```

chiral

(e)

```
S         CH₂Br
 ↘        |
   H ——— * ——— Br
         |                plane of
  - - - - - - - - - - -   symmetry
   H ——— | ——— Br
         * 
 ↗       CH₂Br
R
```

meso; achiral

(f)

```
S         CH₂Br
 ↘        |
   H ——— * ——— Br
         |
  Br ——— | ——— H
         *
 ↗       CH₂Br
S
```

chiral

(g)

```
S         CH₃
 ↘        |
   H ——— * ——— Br
         |
   H ——— | ——— OH
         *
 ↗       CH₃
R
```

chiral

(h)

chiral

Both the front and back carbons
of this Newman projection are
asymmetric carbon atoms.

(i)

```
  Br              Br
    \            /
     C═══C═══C
    /            \
  Cl              Cl
```

chiral molecule, but no
asymmetric carbon atom

(j)

chiral

(k)

achiral

plane of symmetry

(l)

plane of symmetry
meso; achiral

5-27

(a)

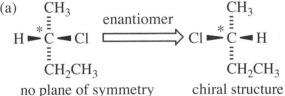

no plane of symmetry       chiral structure
no diastereomer
chiral structure

```
        CH₃                          CH₃
        |                            |
  H ▶ * C ◀ Cl    enantiomer   Cl ▶ * C ◀ H
        ⋮          ⟹                ⋮
        CH₂CH₃                       CH₂CH₃
```

(b)

```
  CH₃  CH₃                        H₃C    CH₃
    \   |  CH₃                  H₃C  \   |
     \  | ⁄                        ⋯⋯   |
      \ |⁄                              |
   *   ⟍⟍ H      enantiomer     H    * ⟋
    ⬡                ⟹            ⬡
```

no plane of symmetry            chiral structure
no diastereomer
chiral structure
```

**99**
Copyright © 2013 Pearson Education, Inc.
```

5-27 continued

(c)

no plane of symmetry
chiral structure

enantiomer

chiral structure

diastereomer

chiral structure
(Inverting two groups on the
bottom asymmetric carbon
instead of the top one would
also give a diastereomer.)

(d)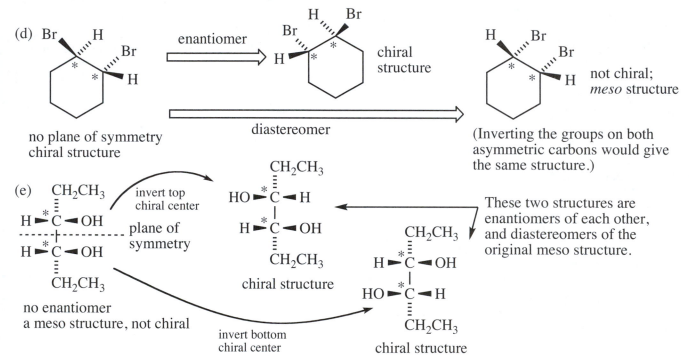

enantiomer

chiral
structure

diastereomer

not chiral;
*meso* structure

no plane of symmetry
chiral structure

(Inverting the groups on both
asymmetric carbons would give
the same structure.)

(e)

invert top
chiral center

plane of
symmetry

no enantiomer
a meso structure, not chiral

invert bottom
chiral center

chiral structure

chiral structure

These two structures are
enantiomers of each other,
and diastereomers of the
original meso structure.

racemic mixture of enantiomers; each is chiral with no plane of symmetry

(f)

diastereomer

+

diastereomer

plane of
symmetry

This meso structure is a diastereomer of each of the
enantiomers; it is not chiral.

**100**

**5-28**

(a)
CH₂OH
H———OH
CH₃

(b)
CHO
H———Br
CH₃

(c)
CH₂OH
H———Br
HO———H
CH₃

(d)
CH₂OH
HO———H
H———OH
CH₃

**5-29** Your drawings may look different from these and still be correct. Check configuration by assigning *R* and *S* to be sure.

(a)
COOH
CH₃———H, NH₂

(b)
CHO
HOCH₂———OH, H

(c)
CH₂OH
CH₃———Cl, Br

(d)
Cl—H           H—Br
CH₃           CH₂OH

**5-30**
(a) same (meso)—plane of symmetry, superimposable
(b) enantiomers—configuration inverted at both asymmetric carbon atoms
(c) enantiomers—configuration inverted at both asymmetric carbon atoms
(d) enantiomers—Solve this problem by switching two groups at a time to put the groups in the same positions as in the first structure; it takes three switches to make the identical compound, so they are enantiomers; an even number of switches would prove they are the same structure.
(e) diastereomers—Front carbon has same configuration (rotate Br down), back carbon is mirror image.
(f) diastereomers—configuration inverted at only one asymmetric carbon
(g) enantiomers—configuration inverted at both asymmetric carbon atoms
(h) same compound—Rotate the right structure 180° around a horizontal axis and it becomes the left structure; it might help to assign *R* and *S* to the C—Br groups: one is *R* and the other is *S*, and this makes it easier to see how they are superimposable.
(i) enantiomers—nonsuperimposable mirror images

Enantiomers have identical chemical and physical properties and cannot be separated by normal methods like distillation or crystallization. Diastereomers, however, have different properties and are therefore separable: the structures in parts (e) and (f) could be separated by normal physical methods.

**5-31** Drawing the enantiomer of a chiral structure is as easy as drawing its mirror image.

(a)
CH₃
H———Br, Cl

(b)
CHO
Br———H
CH₂OH

(c)
CHO
HO———H
HO———H
HO———H
CH₂OH

(d)
[aromatic biphenyl structure with CH₃, CH₃, I, I]

(e)
CH₃
C=C=C
H          H, Br

(f)
H CH₃
H directly behind CH₃
plane of symmetry—no enantiomer

(g)
CH₃
H
H
H₃C

(h)
OH
H———Br
H———CH₃
CH₃

**101**

**5-32**

(a)  1.00 g / 20.0 mL = 0.0500 g/mL ;   20.0 cm = 2.00 dm

$$[\alpha]_D^{25} = \frac{-1.25°}{(0.0500)\,(2.00)} = -12.5°$$

(b)  0.050 g / 2.0 mL = 0.025 g/mL ;   2.0 cm = 0.20 dm

$$[\alpha]_D^{25} = \frac{+0.043°}{(0.025)\,(0.20)} = +8.6°$$

**5-33**  The 32% of the mixture that is (–)-tartaric acid will cancel the optical rotation of the 32% of the mixture that is (+)-tartaric acid, leaving only (68 – 32) = 36% of the mixture as excess (+)-tartaric acid to give measurable optical rotation.  The specific rotation will therefore be only 36% of the rotation of pure (+)-tartaric acid:  (+ 12.0°) x 36% = + 4.3°

(This algebraic approach has been suggested by Editorial Adviser Richard King. )

$$\text{e.e.} = \frac{d-l}{d+l} \text{ x } 100\% = \frac{68-32}{68+32} \text{ x } 100\% = 36\%$$

The optical rotation of this mixture is:   36%  x  (+ 12.0°)  =  + 4.3°

**5-34**

(a)

(b) Rotation of the enantiomer will be equal in magnitude, opposite in sign: – 15.90°.

(c)  The rotation – 7.95° is what percent of – 15.90°?

$$\frac{-7.95°}{-15.90°} \text{ x } 100\% = 50\% \text{ e.e.}$$

There is 50% excess of (R)-2-iodobutane over the racemic mixture; that is, another 25% must be R and 25% must be S.  The total composition is 75% (R)-(–)-2-iodobutane and 25% (S)-(+)-2-iodobutane.

**5-35**  All structures in parts (a) and (b) of this problem are chiral.

(a)

enantiomers:  **A** and **B; C** and **D**
diastereomers:  **A** and **C; A** and **D;**
**B** and **C; B** and **D**

(b)

enantiomers:  **E** and **F; G** and **H**
diastereomers:  **E** and **G; E** and **H; F** and **G; F** and **H**

**5-35 continued**

(c) This structure is a challenge to visualize. A model helps. One way to approach this problem is to assign *R* and *S* configurations. Each arrow shows a change at one asymmetric carbon.

Summary
• *RSSS* (**K**) is the enantiomer of *RRRS* (**M**)

• *SSSS* (**L**) is the enantiomer of *RRRR* (**P**)

• **J** is a meso structure; it has chirality centers and is superimposable on its mirror image.

• **N** and **O** are enantiomers, and are diastereomers of all of the other structures.

**Give yourself a gold star if you got this correct!**

**5-36**
(a) Benzylic bromination follows a free-radical chain mechanism.

This benzylic radical is stabilized by resonance; see the solution to 4-44(b) for benzylic resonance forms.

Abstraction of only the benzylic H gives a resonance-stabilized intermediate.

5-36 continued

(b) and (c)

(d) These are diastereomers. *R,R* and *S,R* have one asymmetric carbon with the same configuration and one of opposite configuration.

(e) It is possible but unlikely that they will be produced in a 50:50 mixture. Unlike racemic mixtures of enantiomers that must be 50:50, diastereomers can be and usually are unequal mixtures.

(f) Diastereomers have different physical properties like melting point and boiling point, so in theory, they could be separated by a physical method like distillation or crystallization.

5-37

(a)

(b)
| 2S,4R | 2S,4R | 2R,4R | 2S,4S |
| equivalent | equivalent | | |
| to 2R,4S | to 2R,4S | | |

(c) According to the IUPAC designation described in text Section 5-2B, a chirality center is "any atom holding a set of ligands in a spatial arrangement which is not superimposable on its mirror image." An asymmetric carbon must have four different groups on it, but in **A** and **B**, C-3 has two groups that are identical (except for their stereochemistry). C-3 holds its groups in a spatial arrangement that is superimposable on its mirror image, so it is not a chirality center. But it is a stereocenter: in structure **A**, interchanging the H and Br at C-3 gives structure **B**, a diastereomer of A; therefore, C-3 is a stereocenter.

(d) In structure **C** or **D**, C-3 is not a stereocenter. Inverting the H and the Br, then rotating the structure 180°, shows that the same structure is formed. Therefore, interchanging two atoms at C-3 does *not* give a stereoisomer, so C-3 does not fit the definition of a stereocenter.

5-38 The Cahn-Ingold-Prelog priorities of the groups are the circled numbers in (a).

(a)

5-38 continued

(b) The reaction did not occur at the asymmetric carbon atom, so the configuration has not changed—the reaction went with retention of configuration at the asymmetric carbon.

(c) The *name* changed because the *priority* of groups in the Cahn-Ingold-Prelog system of nomenclature changed. When the alkene became an ethyl group, its priority changed from the highest priority group to priority 2. (We will revisit this anomaly in problem 6-21(c).)

(d) There is no general correlation between *R* and *S* designation and the physical property of optical rotation. Professor Wade's poetic couplet makes an important point: do not confuse an object and its properties with the *name* for that object. (Scholars of Shakespeare have come to believe that this quote from Juliet is a veiled reference to designation of *R,S* configuration versus optical rotation of a chiral molecule. Shakespeare was *way* ahead of his time.)

5-39

(a) The product has no asymmetric carbon atoms but it has three stereocenters: the carbon with the OH, plus both carbons of the double bond. Interchange of two bonds on any of these makes the enantiomer.

(b) The product is an example of a chiral compound with no asymmetric carbons. Like the allenes, it is classified as an "extended tetrahedron"; that is, it has four groups that extend from the rigid molecule in four different directions. (A model will help.) In this structure, the plane containing the COOH and carbons of the double bond is perpendicular to the plane bisecting the OH and H and carbon that they are on. Since the compound is chiral, it is capable of being optically active.

(c) As shown in text Figure 5-16, Section 5-6, catalytic hydrogenation that creates a new chirality center creates a racemic mixture (both enantiomers in a 1:1 ratio). A racemic mixture is not optically active. In contrast, by using a chiral enzyme to reduce the ketone to the alcohol (as in part (b)), an excess of one enantiomer was produced, so the product was optically active.

5-40

$$
\begin{array}{c}
\text{CHO} \\
\text{H} \rule[0.5ex]{1.5em}{0.4pt}\!\!\rule{0.4pt}{2.5ex}\!\!\rule[0.5ex]{1.5em}{0.4pt}\ \text{OH} \\
\text{CH}_2\text{OH}
\end{array}
$$

D-(+)-glyceraldehyde has *R* configuration.

AND

$$
\begin{array}{c}
\text{CH}_2\text{OH} \\
\text{H} \rule[0.5ex]{1em}{0.4pt}_{\text{S}}\!\!\rule[0.5ex]{1em}{0.4pt}\ \text{OH} \\
\text{H} \rule[0.5ex]{1em}{0.4pt}_{\text{R}}\!\!\rule[0.5ex]{1em}{0.4pt}\ \text{OH} \\
\text{CH}_2\text{OH}
\end{array}
$$

Reduced form is optically inactive because it is MESO.

$$\Longrightarrow$$

$$
\begin{array}{c}
\text{CHO} \\
\text{H} \rule[0.5ex]{1em}{0.4pt}_{\text{R}}\!\!\rule[0.5ex]{1em}{0.4pt}\ \text{OH} \\
\text{H} \rule[0.5ex]{1em}{0.4pt}_{\text{R}}\!\!\rule[0.5ex]{1em}{0.4pt}\ \text{OH} \\
\text{CH}_2\text{OH}
\end{array}
$$

The Cahn-Ingold-Prelog priorities change! The OH = #1, but the CHO = #2, so C-2 is now *R*!

D-(−)-erythrose must have the (2R,3R) configuration.

5-41 (a) The key to this problem is shown in Figure 5-18: *trans*-cyclooctene is chiral because of the twist created by the strained ring (called a chirality helix). Therefore, *cis*-cyclooctene has a chiral diastereomer (*cis* and *trans* are diastereomers) and it fits the original definition, but neither *cis* nor *trans* has chirality centers, so the *cis* isomer does not fit the working definition. (Without the information in Figure 5-18, or a model, this could not have been deduced from principles.)

*cis*
MESO

*trans*
CHIRAL
see Fig 5-18

(b) Let's try nona-2,3,6,7-tetraene:

plane of symmetry

MESO

a chiral diastereomer

Consider the "twist" of an allene: in the top structure, viewing down C2 to C3, the $CH_3$ to $CH_2$ twist is clockwise; from the other end, the $CH_3$ to $CH_2$ has counterclockwise twist.

However, in the bottom diastereomer, both ends have a clockwise twist; this is a chiral compound although it has no chirality centers.

105

6-1 In problems like part (a), draw out the whole structure to detect double bonds:
(a) vinyl halide     (b) alkyl halide     (c) alkyl halide
(d) alkyl halide     (e) vinyl halide     (f) aryl halide

structure
in part (a)

6-2 (a)     (b)     (c)     (d)     (e)

(f)    (g)     (h) 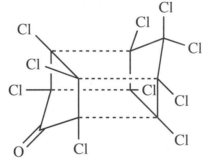 Cl   or   $(CH_3)_3CCl$

You may have drawn the other
enantiomer. Either is correct.

6-3 IUPAC name; common name; degree of halogen-bearing carbon

(a) 1-chloro-2-methylpropane; isobutyl chloride; 1° halide
(b) 2-bromo-2-methylpropane; *tert*-butyl bromide; 3° halide
(c) 1-chloro-2-methylbutane; no common name; 1° halide
(d) 4-fluoro-1,1-dimethylcyclohexane; no common name; 2° halide
(e) 4-bromo-3-methylheptane; no common name; 2° halide
(f) *cis*-1-bromo-2-chlorocyclobutane; no common name; both 2° halides;
also correct is (1*R*,2*S*)-1-bromo-2-chlorocyclobutane

6-4

Kepone®

Chlordane®

**106**

6-5

(a) Table 6-1 shows that in all cases, the iodide has the lowest dipole moment of all four halides. Even though the C—I bond length is longer than C—Cl, the larger electronegativity of Cl makes a more significant contribution to the dipole moment. Ethyl chloride has a larger dipole moment.

(b) 1-Bromopropane has a polar C—Br bond and has a large dipole moment. Cyclopropane has no electronegative atom and has essentially zero dipole moment.

(c) The isomer with two bromine atoms *cis* to each other will have a large dipole moment. The *trans* isomer has the individual bond dipole moments pointing in opposite directions, so *trans*-2,3-dibromobut-2-ene has essentially zero dipole moment.

(d) Two chlorine atoms *cis* to each other on a ring like *cis*-1,2-dichlorocyclobutane will have a large molecular dipole moment. When two chlorine atoms point in opposite directions in a molecule as in *trans*-1,3-dichlorocyclobutane, they effectively cancel and the molecular dipole moment is essentially zero.

6-6

(a) *n*-Butyl bromide (1-bromobutane) has a higher molecular weight and less branching, and boils at a higher temperature than isopropyl bromide.

(b) *tert*-Butyl bromide has a higher molecular weight and a larger halogen, and despite its greater branching, boils at a higher temperature than isopropyl chloride.

(c) 1-Bromobutane has a higher molecular weight and a larger halogen, and boils at a higher temperature than 1-chlorobutane.

6-7 From Table 3-2, the density of hexane is 0.66; it will float on the water layer (d 1.00). From Table 6-2, the density of chloroform is 1.50; water will float on the chloroform. Water is immiscible with many organic compounds; whether water is the top layer or bottom layer depends on whether the other material is less dense or more dense than water. (This is an important consideration to remember in lab procedures.) Water and ethanol are miscible, so only one phase would appear after shaking these two together.

6-8

(a) Step (1) is initiation; steps (2) and (3) are propagation.

(1) $\text{Br}-\text{Br} \xrightarrow{\ h\nu\ } 2\ \text{Br}\bullet$

(2) $\text{H}_2\text{C}=\underset{\text{H}}{\text{C}}-\text{CH}_2\text{H} \ + \ \text{Br}\bullet \ \longrightarrow \ \text{HBr} \ + \ \left\{ \text{H}_2\text{C}=\underset{\text{H}}{\text{C}}-\dot{\text{C}}\text{H}_2 \longleftrightarrow \dot{\text{H}}_2\text{C}-\underset{\text{H}}{\text{C}}=\text{CH}_2 \right\}$

(3) $\text{H}_2\text{C}=\underset{\text{H}}{\text{C}}-\dot{\text{C}}\text{H}_2 \ + \ \text{Br}-\text{Br} \ \longrightarrow \ \text{H}_2\text{C}=\underset{\text{H}}{\text{C}}-\underset{\underset{\text{Br}}{|}}{\text{CH}_2} \ + \ \text{Br}\bullet$

(b) Step (2): break allylic C—H, make H—Br: kJ/mole: $+364 - (+368) = -4$ kJ/mole

kcal/mole: $+87 - (+88) = -1$ kcal/mole

Step (3): break Br—Br, make allylic C—Br: kJ/mole: $+192 - (+280) = -88$ kJ/mole

kcal/mole: $+46 - (+67) = -21$ kcal/mole

$\Delta H^0$ overall $= -4 + -88 = -92$ kJ/mole    $(-1 + -21 = -22$ kcal/mole $)$

This is a very exothermic reaction; it is reasonable to expect a small activation energy in step (1), so this reaction should be very rapid.

6-9   (a) propagation steps

NBS produces low concentrations of $Br_2$, so the mechanism shown here uses $Br_2$ as the source of bromine, even though it originally came from NBS.

repeats chain mechanism

The resonance-stabilized allylic radical intermediate has radical character on both the 1° and 3° carbons, so bromine can bond to either of these carbons producing two isomeric products.

(b) Allylic bromination of cyclohexene gives 3-bromocyclohex-1-ene regardless of whether there is an allylic shift. Either pathway leads to the same product. If one of the ring carbons were somehow marked or labeled, then the two products could be distinguished. (We will see in following chapters how labeling is done experimentally.)

This second structure from the allylic shift is *identical* to the first structure— only one compound is produced here.

6-10
(a)

This compound has only one type of hydrogen—only one monochlorine isomer can be produced.

(b)

Bromination has a strong preference for abstracting hydrogens that give the most stable radical intermediates, like 3° in this case.

(c)

(NBS can also be used for benzylic bromination)

Bromine atom will abstract the hydrogen giving the most stable radical; in this case, the radical intermediate will be stabilized by resonance with the benzene ring.

(d)

This second structure from the allylic shift is *identical* to the first structure— only one compound is produced here.

(NBS can also be used for benzylic bromination)

Bromine atom will abstract the hydrogen giving the most stable radical; in this case, the radical intermediate will be stabilized by resonance with the benzene ring.

108

6-11
(a) Substitution—Br⁻ is the leaving group; $CH_3O^-$ is the nucleophile.
(b) Elimination—when OH is protonated, $H_2O$ is the leaving group.
(c) Elimination—both Br atoms are lost; iodide ion is a nucleophile that reacts at Br.

6-12 (a) $CH_3(CH_2)_4CH_2-OCH_2CH_3$     (b) $CH_3(CH_2)_4CH_2-CN$     (c) $CH_3(CH_2)_4CH_2-OH$

6-13 (a) The rate law is first order in both 1-bromobutane, $C_4H_9Br$, and methoxide ion. If the concentration of $C_4H_9Br$ is lowered to one-fifth the original value, the rate must decrease to one-fifth; if the concentration of methoxide is doubled, the rate must also double. Thus, the rate must decrease to two-fifths of the original rate, 0.02 mole/L per second:

$$\text{rate} = \underbrace{(\,0.05 \text{ mole/L per second}\,)}_{\text{original rate}} \times \underbrace{\frac{(\,0.1 \text{ M}\,)}{(\,0.5 \text{ M}\,)}}_{\substack{\text{change in}\\ C_4H_9Br}} \times \underbrace{\frac{(\,2.0 \text{ M}\,)}{(\,1.0 \text{ M}\,)}}_{\substack{\text{change in}\\ NaOCH_3}} = \underbrace{0.02 \text{ mole/L per second}}_{\text{new rate}}$$

A completely different way to answer this problem is to solve for the rate constant $k$, then put in new values for the concentrations.

rate $= k$ [$C_4H_9Br$] [$NaOCH_3$] ⟹ 0.05 mole $L^{-1}$ sec$^{-1} = k$ (0.5 mol $L^{-1}$) (1.0 mol $L^{-1}$) ⟹
rate constant $k = 0.1$ L mol$^{-1}$ sec$^{-1}$

rate $= k$ [$C_4H_9Br$] [$NaOCH_3$] $= (0.1$ L mol$^{-1}$ sec$^{-1}) (0.1$ mol $L^{-1}) (2.0$ mol $L^{-1}) = 0.02$ mole $L^{-1}$ sec$^{-1}$

(b)

nucleophile          electrophile                    transition state                    initial product          leaving group

(c)     $NaOCH_2CH_2CH_2CH_3$  +  $CH_3Br$  ⟶
         sodium butoxide          bromomethane

         $CH_3CH_2CH_2CH_2O-CH_3$   +   NaBr
         1-methoxybutane

:$NH_3$
base

final product

6-14  Organic and inorganic products are shown here for completeness.

(a) $(CH_3)_3C-O-CH_2CH_3$     +   KBr

(b) $HC\equiv C-CH_2CH_2CH_2CH_3$     +   NaCl

(c) $(CH_3)_2CHCH_2-\overset{\oplus}{N}H_3$ Br⁻ $\xrightarrow{NH_3}$ $(CH_3)_2CHCH_2-NH_2$   +   $NH_4^+$ Br⁻

The first $NH_3$ replaces the Br. The second $NH_3$ removes H⁺ from the N, leaving R-$NH_2$.

(d) $CH_3CH_2CH_2-CN$   +   NaI

(e) I   +   NaCl

(f) F   +   KCl     (18-Crown-6 is the catalyst and does not change; $CH_3CN$ is the solvent.)

**6-15** All reactions in this problem follow the same pattern; the only difference is the nucleophile ($^-$:Nuc). Only the nucleophile is listed below. (Cations like $Na^+$ or $K^+$ accompany the nucleophile but are simply spectator ions and do not take part in the reaction; they are not shown here.)

1-chlorobutane

(a) $HO^-$     (b) $F^-$ from KF/18-crown-6     (c) $I^-$    (d) $^-CN$     (e) $HC\equiv C:^-$

(f) $^-OCH_2CH_3$     (g) excess $NH_3$ (or $^-NH_2$)

**6-16**
(a) $(CH_3CH_2)_2NH$ is a better nucleophile—less hindered.
(b) $(CH_3)_2S$ is a better nucleophile—S is larger, more polarizable than O.
(c) $PH_3$ is a better nucleophile—P is larger, more polarizable than N.
(d) $CH_3S^-$ is a better nucleophile—anions are better than neutral atoms of the same element.
(e) $(CH_3)_3N$ is a better nucleophile—less electronegative than oxygen, better able to donate an electron pair.
(f) $CH_3COO^-$ is a better nucleophile—more basic, electrons less delocalized than in $CF_3COO^-$ because of inductive effect of F substituents.
(g) $CH_3CH_2CH_2O^-$ is a better nucleophile—less branching, less steric hindrance.
(h) $I^-$ is a better nucleophile—larger, more polarizable than Cl.

**6-17** A mechanism uses arrows to show *electron movement*. An arrow must begin at either a bond or an unshared electron pair (or a single electron in radical reactions). "Et" is the abbreviation for an ethyl group.

Protonation converts $OCH_2CH_3$ to a good leaving group so that bromide can effect substitution.

**6-18** The type of carbon with the halide, and relative leaving group ability of the halide, determine the reactivity.

methyl iodide > methyl chloride > ethyl chloride > isopropyl bromide >> neopentyl bromide,

*most reactive*                   1°            2°       *tert*-butyl iodide } *least reactive*

                                                3°

Predicting the relative order of neopentyl bromide and *tert*-butyl iodide would be difficult because both would be extremely slow.

**6-19** In all cases, the less hindered structure is the better $S_N2$ substrate.

(a) 2-methyl-1-iodopropane (1° versus 3°)

(b) cyclohexyl bromide (2° versus 3°)

(c) isopropyl bromide (no substituent on neighboring carbon)

(d) 2-chlorobutane (Even though this is a 2° halide, it is easier to attack than the 1° neopentyl type in 1-chloro-2,2-dimethylbutane—see structure and the solution to Problem 6-18.)

(e) 1-iodobutane (1° versus 2°)

a neopentyl halide—
hindered to backside attack
by neighboring methyl groups

6-20 All S$_N$2 reactions occur with inversion of configuration at carbon.

(a)

transition state

*trans* ———— *inversion* ————→ *cis*

(b)

(c)

(d)

Fluoride is a bad leaving group; see the solution to 6-21(a).

(e)

(f)

6-21

(a) The best leaving groups are the weakest bases. Bromide ion is so weak it is not considered at all basic; it is an excellent leaving group. Fluoride is moderately basic, by far the most basic of the halides. It is a terrible leaving group. Bromide is many orders of magnitude better than fluoride in leaving group ability.

## 6-21 continued

(b)

transition state

(c) As noted on the structure above, the configuration is inverted even though the designations of the configuration for both the starting material and the product are $S$; the oxygen of the product has a lower priority than the bromine it replaces. Refer to the solution to problem 5-38 for the caution about confusing absolute configuration with the *designation* of configuration.

(d) The result is perfectly consistent with the $S_N2$ mechanism. Even though both the reactant and the product have the $S$ designation, the configuration has been inverted: the nomenclature priority of fluorine changes from second (after bromine) in the reactant to first (before oxygen) in the product. While the designation may be misleading, the structure shows with certainty that an inversion has occurred.

## 6-22

6-23 The structure that can form the more stable carbocation will undergo $S_N1$ faster.

(a) 2-Bromopropane: will form a 2° carbocation.
(b) 2-Bromo-2-methylbutane: will form a 3° carbocation.
(c) Allyl bromide is faster than propyl bromide: allyl bromide can form a resonance-stabilized intermediate.
(d) 2-Bromopropane: will form a 2° carbocation.
(e) 2-Iodo-2-methylbutane is faster than *tert*-butyl chloride (iodide is a better leaving group than chloride).
(f) 2-Bromo-2-methylbutane (3°) is faster than ethyl iodide (1°); although iodide is a somewhat better leaving group, the difference between 3° and 1° carbocation stability dominates.

6-24 Ionization is the rate-determining step in $S_N1$. Anything that stabilizes the carbocation intermediate will speed the reaction. Both of these compounds form resonance-stabilized intermediates.

allylic

benzylic

6-25

acting as nucleophile

acting as base

6-26 It is important to analyze the structure of carbocations to consider if migration of any groups from adjacent carbons will lead to a more stable carbocation. As a general rule, if rearrangement would lead to a more stable carbocation, a carbocation will rearrange. (Beginning with this problem, only those unshared electron pairs involved in a particular step will be shown.)

(a)

$2°$ carbocation

nucleophilic attack on unrearranged carbocation

unrearranged product

continued on next page

6-26 (a) continued
nucleophilic attack after carbocation rearrangement

Methyl shift to the
2° carbocation forms a
more stable 3° carbocation.

rearranged product

(b)

$$\Delta$$

2° carbocation

nucleophilic attack on unrearranged carbocation

unrearranged
product

nucleophilic attack after carbocation rearrangement

hydride
shift

Hydride shift to the
2° carbocation forms a
more stable 3° carbocation.

rearranged
product

6-26 continued

(c)

2° carbocation

Note: braces are used to indicate the ONE chemical species represented by multiple resonance forms.

nucleophilic attack on unrearranged carbocation

unrearranged product

The most basic species in a mixture is the most likely to remove a proton. In this reaction, acetic acid is more basic than iodide ion.

nucleophilic attack after carbocation rearrangement

allylic—resonance-stabilized

rearranged product

(removes H⁺ as above)

plus two other resonance forms as shown above

*continued on next page*

**115**
Copyright © 2013 Pearson Education, Inc.

6-26(c) continued

<u>Comments on 6-26(c)</u>
(1)  The hydride shift to a 2° carbocation generates an allylic, resonance-stabilized 2° carbocation.

(2)  The double-bonded oxygen of acetic acid is more nucleophilic because of the resonance forms it can have after attack.  (See Solved Problem 1-5 and Problem 1-16 in the text.)

(3)  Attack on only one carbon of the allylic carbocation is shown.  In reality, both positive carbons would be attacked in equal amounts, but they would give the identical product *in this case*.  In other compounds, however, attack on the different carbons might give different products.  ALWAYS CONSIDER ALL POSSIBILITIES.

(d)

The 1° carbocation shown here would be very unstable.  Rearrangement usually happens at the same time as the leaving group leaves.

<u>hydride shift followed by nucleophilic attack</u>

acting as base

<u>alkyl migration (ring expansion) followed by nucleophilic attack</u>

acting as base

6-27

(a)

$S_N1$, 3°,
weak nucleophile

(b) $CH_3-CHCH_2OCH_3$ with $CH_3$ substituent

$S_N2$, 1°,
strong nucleophile

(c)

$S_N1$, 3°,
weak nucleophile

(d)

$S_N1$ on 2° C,
weak nucleophile

(e)

$S_N2$ on 2° C, strong nucleophile

**116**

**6-28**

(R)-2-bromobutane

$H_2O, \Delta$ / $S_N2$— inversion → (S)-butan-2-ol

$H_2O, \Delta$ / $S_N1$— racemization → (R)-butan-2-ol + (S)-butan-2-ol

50 : 50 mixture — racemic

If $S_N1$, which gives racemization, occurs exactly twice as fast as $S_N2$, which gives inversion, then the racemic mixture (50 : 50 R + S) is 66.7% of the mixture and the rest, 33.3%, is the S enantiomer from $S_N2$. Therefore, the excess of one enantiomer over the racemic mixture must be 33.3%, the enantiomeric excess. (In the racemic mixture, the R and S "cancel" each other algebraically as well as in optical rotation.)

The optical rotation of pure (S)-butan-2-ol is + 13.5°. The optical rotation of this mixture is:

33.3%  x  + 13.5°  =  + 4.5°

**6-29**

(a) Methyl shift may occur simultaneously with ionization.

This 1° carbocation would be very unstable.

(b) Alkyl shift may occur simultaneously with ionization.

This 1° carbocation would be very unstable.

alkyl shift — ring expansion

also produced from the solvent as nucleophile

**117**

## 6-30
(a)

The $S_N1$ mechanism begins with ionization to form a carbocation, attack of a nucleophile, and in the case of ROH nucleophiles, removal of a proton by a base to form a neutral product.

(b) In the E1 reaction, the solvent (ethanol, in this case) serves two functions: it aids the ionization process by solvating both the leaving group (bromide) and the carbocation; and second, it serves as a base to remove the proton from a carbon adjacent to the carbocation in order to form the carbon–carbon double bond. The $S_N1$ mechanism adds a third function to the solvent: the first step is the same as in E1, ionization to form the carbocation; the second step has the solvent acting as a nucleophile—this step is different from E1; third, the solvent acts like a base and removes a proton, although from an oxygen ($S_N1$) and not a carbon (E1). Solvents are versatile!

6-31 This solution does not show complete mechanisms. Rather, it shows the general sequence of all four pathways leading to five different products.

2-bromo-3-methylbutane

ionization

hydride shift

unrearranged 2° carbocation

rearranged 3° carbocation

$S_N1$    E1    $S_N1$    E1

Five unique products shown in the boxes. One alkene can be produced from either the 2° or the 3° carbocation.

same compound

two possible E1 products

two possible E1 products

6-32

E1

E1

$S_N1$

$S_N1$

6-33 Substitution products are shown first, then elimination products.

(a)

OCH₃

+

(b) OCH₂CH₃

+

+

(c)

OCH₂CH₃

+

+

OCH₂CH₃

+

+

without rearrangement

with rearrangement

(d)

OH, CH₃

+

OH, CH₃

+

CH₃

CH₃

CH₂

$S_N1$ product without rearrangement

$S_N1$ product with rearrangement

major E1 product with or without rearrangement

minor

6-34 Et is the abbreviation for ethyl, so EtOH is ethanol.

CH₃

Br

EtOH

Δ

Br⁻

CH₃

C, CH₃

HOEt

CH₃

H

O⁺

Et

HOEt

$S_N1$ product:

CH₃

O

Et

+ EtOH₂⁺

continued on next page

**119**

**6-34 continued**

E1—MAJOR product—
more substituted alkene

E1—MINOR product—
less substituted alkene

+ EtOH$_2$ $\oplus$

+ EtOH$_2$ $\oplus$

**6-35**

without rearrangement

$$CH_3-\overset{\overset{\displaystyle CH_3}{|}}{\underset{\underset{\displaystyle CH_3}{|}}{C}}-\overset{}{\underset{\underset{\displaystyle OCH_2CH_3}{|}}{CH}}-CH_3$$

with rearrangement

$$CH_3-\overset{\overset{\displaystyle OCH_2CH_3}{|}}{\underset{\underset{\displaystyle CH_3}{|}}{C}}-\overset{}{\underset{\underset{\displaystyle CH_3}{|}}{CH}}-CH_3$$

**6-36**

NaOCH$_2$CH$_3$

E2—minor product—less substituted alkene (monosubstituted)

+ E2—major product—more substituted alkene (trisubstituted)

+ OCH$_2$CH$_3$ S$_N$2—probably very small amount because of steric hindrance

**6-37**

(a) NaOCH$_3$

minor alkene—monosubstituted

+ major alkene—disubstituted (*cis + trans*)

+ OCH$_3$ S$_N$2 product—some will be formed on a 2° carbon

(b) NaOCH$_3$

major alkene—trisubstituted (*cis + trans*)

+ minor alkene—disubstituted

No S$_N$2 product will be formed; S$_N$2 cannot occur at a 3° carbon.

(c) NaOH

minor alkene—monosubstituted

+ major alkene—trisubstituted

+ OH S$_N$2 product—very small amount because 2° carbon is hindered

(d) NaOEt

H atoms that are removed are shown

minor alkene—disubstituted

+ major alkene—trisubstituted

+ OEt S$_N$2 product showing inversion at the carbon—some of this product will be formed on a 2° carbon

*see top of next page for another view of this reaction*

**120**

6-37(d) continued

6-38 In systems where free rotation is possible, the H to be abstracted by the base and the leaving group (Br here) must be anti-coplanar. The E2 mechanism is a concerted, one-step mechanism, so the arrangement of the other groups around the carbons in the starting material is retained in the product; there is no intermediate to allow time for rotation of groups. (Models will help.)

H and Br anti-coplanar      transition state      only geometric isomer possible from a concerted mechanism

- - - - - - = stretched bond being broken or formed

orbital picture of this reaction

The other diastereomer has two groups interchanged on the back carbon of the Newman projection, where it must give the *cis*-alkene.

6-39 In this problem, all new internal alkenes form *cis* isomers as well as the *trans* isomers shown.

Section 1

Section 2

Section 3

Section 4

Section 5

**122**

6-40

(a) Ethoxide is a strong base/nucleophile—second-order conditions. The 1° bromide favors substitution over elimination, so $S_N2$ will predominate over E2.

substitution—major                elimination—minor

(b) Methoxide is a strong base/nucleophile—second-order conditions. The 2° chloride will undergo $S_N2$ by backside attack as well as E2 to make a mixture of alkenes.

substitution                elimination—                elimination—
                            minor alkene                major alkene
                                                        (cis + trans)

(c) Ethoxide is a strong base/nucleophile—second-order conditions. The 3° chloride is hindered and cannot undergo $S_N2$ by backside attack. E2 is the only route possible.

$$CH_3—C=CHCH_3 \atop \underset{CH_3}{|} \quad major \qquad CH_2=C—CH_2CH_3 \atop \underset{CH_3}{|} \quad minor$$

(d) Heating in ethanol are conditions for solvolysis, an $S_N1$ reaction. Heat also promotes elimination.

$$\overset{OCH_2CH_3}{\underset{|}{}} \atop CH_3—\overset{|}{C}—CH_2CH_3 \atop \underset{CH_3}{|} \quad S_N1 \qquad CH_3—C=CHCH_3 \atop \underset{CH_3}{|} \quad E1$$

Substitution and elimination are possible.

major elimination product

(e) Hydroxide is a strong base/nucleophile—second-order conditions. The 1° iodide is more likely to undergo $S_N2$ than E2, but both products will be observed.

$$CH_3–CHCH_2OH \atop \underset{CH_3}{|} \qquad CH_3–C=CH_2 \atop \underset{CH_3}{|}$$

substitution—major                elimination—minor

(f) Silver nitrate in ethanol/water are ionizing conditions for 1° alkyl halides that will lead to rearrangement followed by substitution on the 3° carbocation. (No heat, so E1 is unlikely.)

$$\overset{OCH_2CH_3}{\underset{|}{}} \atop CH_3—\overset{|}{C}—CH_3 \atop \underset{CH_3}{|}$$ from ethanol as nucleophile

$$\overset{OH}{\underset{|}{}} \atop CH_3—\overset{|}{C}—CH_3 \atop \underset{CH_3}{|}$$ from water as nucleophile

(g) Ethoxide in ethanol on a 3° halide will lead to E2 elimination; there will be no substitution.

major alkene trisubstituted                minor alkene disubstituted

(h) Heating a 3° halide in methanol is quintessential first-order conditions, either E1 or $S_N1$ (solvolysis).

substitution ($S_N1$)                elimination (E1):                (trace)

**6-41** "Ph" is the abbreviation for the phenyl substituent.

(NBS can also be used for benzylic bromination)

Benzylic bromination is very selective; dibromination might occur if an excess of Br$_2$ is used.

solvolysis conditions—S$_N$1
If the reaction mixture got too warm, some E1 elimination might occur.

Using a hindered base like *tert*-butoxide will give exclusively the E2 product, with no competing S$_N$2 as KOH or NaOCH$_3$ might give.

**6-42**

(a) [structure with Cl]

(b) [structure with Br]

(c) Br [structure with Br]

(d) Cl−C−CH$_2$OH with Cl above and Cl below

(e) [cyclohexane with Cl and CH$_3$]

(f) Cl
H−C−H
Cl

(g) Cl
Cl−C−H
Cl

(h) [cyclopentane with Cl]

(i) CH$_2$CH$_3$
I−C−CH$_3$
CH$_3$

**6-43**

(a) 2-bromo-2-methylpentane
(b) 1-chloro-1-methylcyclohexane
(c) 1,1-dichloro-3-fluorocycloheptane
(d) 4-(2-bromoethyl)-3-(fluoromethyl)-2-methylheptane
(e) 4,4-dichloro-5-cyclopropyl-1-iodoheptane
(f) *cis*-1,2-dichloro-1-methylcyclohexane

**6-44** Ease of backside attack (less steric hindrance) decides which undergoes S$_N$2 faster in all these examples except (b).

(a) [structure with Cl]  faster than  [structure with Cl]  Primary R-X reacts faster than 2° R-X.

(b) [structure with I]  faster than  [structure with Cl]  Iodide is a better leaving group than chloride.

(c) [structure with Cl]  faster than  [structure with Cl]  less branching on a neighboring carbon

(d) [structure with Br]  faster than  [structure with Br]  Same neighboring branching, so 1° is faster than 2°.

(e) [cyclohexyl]−CH$_2$Cl  faster than  [cyclohexyl]−Cl  Primary R-X reacts faster than 2° R-X.

(f) [structure with Br]  faster than  [structure with Br]  less branching on a neighboring carbon

**6-45** Formation of the more stable carbocation decides which undergoes $S_N1$ faster in all these examples except (d).

(a)    3°    [structure: tert-butyl chloride with Cl]    faster than    [structure: 3-chloropentane]    2°

(b)    [structure: 2-chlorobutane]    faster than    [structure: 1-chlorohexane with Cl]
       2°  Cl                                                                1°

(c)    [cyclohexyl–Br]    faster than    [cyclohexyl–CH₂Br]
            2°                                  1°

(d)    [cyclohexyl–I]    faster than    [cyclohexyl–Cl]    (leaving group ability)

(e)    [structure: 1-phenylethyl bromide, Br]    faster than    [structure: 2-bromo-3-methylpentane]
       2° benzylic! (resonance!)                              Br   2°

(f)    [structure: 3-bromocyclohexene, Br]    faster than    [structure: bromocyclohexane, Br]
       2° allylic!  (resonance!)                              2°

**6-46**   For $S_N2$, reactions should be designed such that the nucleophile attacks the least highly substituted alkyl halide.  ("X" stands for a halide:  Cl, Br, or I.)

(a)    [cyclohexyl]—CH₂X  +  HO⁻    ⟶    [cyclohexyl]—CH₂OH

(b)    [cyclohexyl]—S⁻    +  X–CH₂CH₃    ⟶    [cyclohexyl]—SCH₂CH₃

(c)    [structure with OH and X]  —HO⁻→  [cyclic intermediate with O⁻ and X]  ⟶  [tetrahydropyran ring with O]    some  [structure with OH]  also produced

(d)    [cyclohexyl]—CH₂X  +  NH₃    ⟶    [cyclohexyl]—CH₂NH₂
                              excess

(e)    H₂C=CHCH₂X    +  ⁻CN    ⟶    H₂C=CHCH₂CN

(f)    HC≡C⁻    +  X–CH₂CH₂CH₃    ⟶    HC≡C–CH₂CH₂CH₃

6-47

(a) (1)

(2)

Synthesis (1) would give a better yield of the desired ether product. (1) uses $S_N2$ attack of a nucleophile on a 1° carbon, while (2) requires attack on a more hindered 2° carbon. Reaction (2) would give a lower yield of substitution, with more elimination.

(b)     CANNOT DO $S_N2$ ON A 3° CARBON!

Better to do $S_N2$ on a methyl carbon:

6-48

(a) $S_N2$—second order: reaction rate doubles: rate = $k$ [EtBr] [KO-$t$-Bu]
(b) $S_N2$—second order: reaction rate increases six times
(c) Virtually all reaction rates, including this one, increase with a temperature increase.

6-49 This is an $S_N1$ reaction; the rate law depends only on the substrate concentration, not on the nucleophile concentration:    rate = $k$ [$C_4H_9Br$]
(a) no change in rate
(b) the rate triples, dependent only on [$tert$-butyl bromide]
(c) Virtually all reaction rates, including this one, increase with a temperature increase.

6-50 The key to this problem is that iodide ion is both an excellent nucleophile AND leaving group. Substitution on chlorocyclohexane is faster with iodide than with cyanide (see Table 6-3 for relative nucleophilicities). Once iodocyclohexane is formed, substitution by cyanide is much faster on iodocyclohexane than on chlorocyclohexane because iodide is a better leaving group than chloride. So two fast reactions involving iodide replace a slower single reaction, resulting in an overall rate increase.

6-51 Only the solvolysis ($S_N1$) products are shown. Elimination (E1) products are possible too.

(a)

(b)

(c)

$S_N1$ without rearrangement

$S_N1$ with rearrangement

(d)        rearrangement— methyl shift        rearrangement— ethyl shift

6-52

(a)

(b)

(i)        equivalent resonance forms

(ii)

(iii)        equivalent resonance forms

(iv)        Loss of bromide gives unstable 1° carbocation that quickly rearranges to a 2° allylic carbocation.

equivalent resonance forms—same as from part (iii)

(c) Only one substitution product arises from equivalent resonance forms.

(i)        $OCH_2CH_3$

(ii)        $CH_2OCH_2CH_3$        $CH_2$

        +

        H        $OCH_2CH_3$

(iii)        $OCH_2CH_3$

        cis + trans

(iv)        $OCH_2CH_3$

        cis + trans        same as part (iii)

6-53

6-55 Reactions would also give some elimination products; only the substitution products are shown here.

(a)

$N \equiv C \blacktriangleright \overset{\displaystyle CH_3}{\underset{\displaystyle CH_2CH_3}{\vdots}} \blacktriangleleft H$

$S_N2$ gives inversion:
only product

(b)

$H \blacktriangleright \overset{\displaystyle CH_3}{\underset{\displaystyle }{\vdots}} \blacktriangleleft OH$
$H \blacktriangleright \underset{\displaystyle CH_2CH_3}{\overset{\displaystyle }{C}} \blacktriangleleft CH_3$

$S_N2$ only
with inversion

(c)

$CH_3CH_2O \blacktriangleright \overset{\displaystyle CH_2CH_3}{\underset{\displaystyle CH(CH_3)_2}{\vdots}} \blacktriangleleft CH_3$ + $CH_3 \blacktriangleright \overset{\displaystyle CH_2CH_3}{\underset{\displaystyle CH(CH_3)_2}{\vdots}} \blacktriangleleft OCH_2CH_3$

solvolysis, $S_N1$, racemization

6-56
(a)  $CH_3CH_2OCH_2CH_3$

(b)  ⬡—$CH_2CH_2CN$

(c)  ⬡—$SCH_2CH_3$

(d)  $CH_3(CH_2)_8\overset{\displaystyle C\equiv CH}{\underset{\displaystyle }{CH_2}}$

(e)  ⬡$\overset{\oplus}{N}$—$CH_3$   $I^-$

(f)  $(CH_3)_3C-CH_2CH_2NH_2$

(g)  [furan ring with O]   [ring with OH and OH]
      also possible

(h)  $HO\text{\tiny\textbar\textbar\textbar}$⬡$\text{\tiny\textbar\textbar\textbar}CH_3$

**128**

6-57    substitution

CH₃ structures (Fischer projections):

H—Br, Br—H (top), CH₃ bottom — 2S,3S

Br—H, H—Br (top), CH₃ bottom — 2R,3R

+

racemic mixture

Regardless of which bromine is substituted on each molecule, the same mixture of products results.

Each of the substitution products has one chiral center inverted from the starting material. The mechanism that accounts for inversion is $S_N2$. If an $S_N1$ process were occurring, the product mixture would also contain 2R,3R and 2S,3S diastereomers. Their absence argues against an $S_N1$ process occurring here.

↓ KOH

HO—H, Br—H (CH₃ top, CH₃ bottom) — 2R,3S

+

H—OH, H—Br (CH₃ top, CH₃ bottom) — 2S,3R

racemic mixture

elimination

H and Br
anti-coplanar

trans

The other enantiomer gives the same product (you should prove this to yourself).

The absence of *cis* product is evidence that only the E2 elimination is occurring in one step through an anti-coplanar transition state with no chance of rotation. If E1 had been occurring, rotation around the carbocation intermediate would have been possible, leading to both the *cis* and *trans* products.

6-58

(a)    $\dfrac{+15.58°}{+15.90°}$ x 100% = 98% of original optical activity = 98% e.e.

Thus, 98% of the *S* enantiomer and 2% racemic mixture gives an overall composition of 99% *S* and 1% *R*.

(b)  The 1% of radioactive iodide has produced exactly 1% of the *R* enantiomer. Each substitution must occur with inversion, a classic $S_N2$ mechanism.

6-59

(a)  An $S_N2$ mechanism with inversion will convert *R* to its enantiomer, *S*. An accumulation of excess *S* does not occur because it can also react with bromide, regenerating *R*. The system approaches a racemic mixture at equilibrium.

$R$ + Br⁻  →($S_N2$ inversion)→  $S$ + Br⁻

**129**

Copyright © 2013 Pearson Education, Inc.

**6-59 continued**

(b) In order to undergo substitution and therefore inversion, $HO^-$ would have to be the leaving group, but $HO^-$ is never a leaving group in $S_N2$. No reaction can occur.

(c) Once the OH is protonated, it can leave as $H_2O$. Racemization occurs in the $S_N1$ mechanism because of the planar, achiral carbocation intermediate which "erases" all stereochemistry of the starting material. Racemization occurs in the $S_N2$ mechanism by establishing an equilibrium of *R* and *S* enantiomers, as explained in 6-59(a).

The symbol H—A is used for a generic acid, usually a catalyst in a mechanism. The conjugate base is $A^-$.

**6-60**

(a)

major      minor      trace amount

The carbocation produced in this E1 elimination is 3° and will not rearrange. The product ratio follows the Zaitsev rule.

(b)

trace—from unrearranged 2° carbocation

from either unrearranged 2° carbocation or rearranged 3° carbocation

from rearranged 3° carbocation

The carbocation produced in this E1 elimination is 2° and can either eliminate to give the first two alkenes, or can rearrange by a hydride shift to a 3° carbocation which would produce the last two products. The amounts of the last two products are not predictable as they are both trisubstituted, but the first product will certainly be the least.

(c)

trace—from unrearranged 2° carbocation

major from rearranged 3° carbocation

minor from rearranged 3° carbocation

The carbocation produced in this E1 elimination is 2° and can either eliminate to give the first alkene, or can rearrange by a methyl shift to a 3° carbocation which would produce the last two products. The middle product is major as it is tetrasubstituted versus disubstituted for the last structure and monosubstituted for the first structure.

6-61
(a)

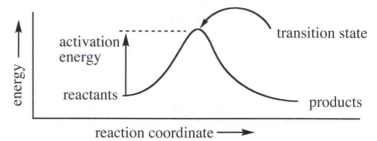

Methoxide is a strong base and nucleophile so the reaction is second order. The Br is on a 3° carbon so $S_N2$ is not possible, only E2. The transition state shows bond-forming and bond-breaking with dashed bonds.

The reaction-energy diagram for a one-step mechanism like E2 has one peak.

(b)

The reaction-energy diagram for a two-step mechanism like E1 has two peaks, with the first step as the slow step with higher activation energy.

3° carbocation intermediate

intermediate carbocation

6-62
(a)

(b)

$S_N1$

$S_N2$

**131**

**6-63** NBS generates bromine which produces bromine radical. Bromine radical abstracts an allylic hydrogen, resulting in a resonance-stabilized allylic radical. The allylic radical can bond to bromine at either of the two carbons with radical character.

**6-64** The bromine radical from NBS will abstract whichever hydrogen produces the most stable intermediate; in this structure, that is a benzylic hydrogen, giving the resonance-stabilized benzylic radical.

(Even though three carbons of the ring have some radical character, these are minor resonance contributors. The product is most stable when the ring has all three double bonds intact, necessitating that the bromine bond to the benzylic carbon.)

**6-65** Two related factors could explain this observation. First, as carbocation stability increases, the leaving group will be less tightly held by the carbocation for stabilization; the more stable carbocations are more "free" in solution, meaning more exposed. Second, more stable carbocations will have longer lifetimes, allowing the leaving group to drift off in the solvent, leading to more possibility for the incoming nucleophile to attack from the side that the leaving group just left.

The less stable carbocations hold tightly to their leaving groups, preventing nucleophiles from attacking this side. Backside attack with inversion is the preferred stereochemical route in this case.

**6-66**

2° carbocation          3° carbocation

mechanisms continued
on next page

**132**

6-66 continued
<u>substitution on 2° carbocation</u>

<u>substitution on rearranged 3° carbocation</u>

<u>elimination from rearranged 3° carbocation—two possible products</u>

major alkene product

<u>elimination from 2° carbocation</u>

rotate

This is the minor alkene product; it can come only from the carbocation before rearrangement.

6-67

Nuc⁻ :  $S_N2$  CH₂CH₃

H₃CH₂C   O⁺   CH₂CH₃

BF₄⁻ ⟶  Nuc—CH₂CH₃

+

H₃CH₂C   O   CH₂CH₃  } The leaving group is a molecule of diethyl ether.

From Table 6-2, all of the ethyl halides are liquids. Triethyloxonium tetrafluoroborate is a solid which is easier to handle and often safer.

6-68

$S_N1$

formate nucleophile

Ionization of the 1° chloride gives a cation stabilized by resonance, leading to an $S_N1$ mechanism. Elimination cannot occur because no neighboring carbon has a hydrogen atom.

**133**

**6-69** The energy, and therefore the structure, of the transition state determines the rate of a reaction. Any factor that lowers the energy of the transition state will speed the reaction.

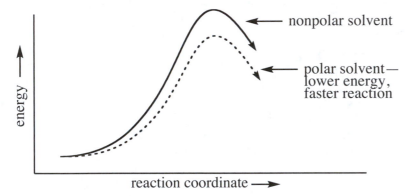

This example of $S_N2$ is unusual in that the nucleophile is a neutral molecule—it is not negatively charged. The transition state is beginning to show the positive and negative charges of the products (ions), so the transition state is more charged than the reactants. The polar transition state will be stabilized in a more polar solvent through dipole-dipole interactions, so the rate of reaction will be enhanced in a polar solvent.

**6-70** The problem is how to explain this reaction:

facts
1) second order, but several thousand times faster than similar second-order reactions without the NEt$_2$ group
2) NEt$_2$ group migrates

Solution

Clearly, the NEt$_2$ group is involved. The nitrogen is a nucleophile and can do an internal nucleophilic substitution ($S_Ni$), a very fast reaction for entropy reasons because two different molecules do not have to come together.

The slower step is attack of HO$^-$ on intermediate **3**; the N is a good leaving group because it has a positive charge. Where will HO$^-$ attack **3**? On the less substituted carbon, in typical $S_N2$ fashion.

This overall reaction is fast because of the *neighboring group assistance* in forming **3**. It is second order because the HO$^-$ group and **3** collide in the slow step (not the *only* step, however). And the NEt$_2$ group "migrates", although in two steps.

**6-71** The symmetry of this molecule is crucial.

(a)

$$CH_3 - \overset{\overset{\displaystyle Ph}{|}}{\underset{\underset{\displaystyle H}{|}}{C}} - \overset{\overset{\displaystyle H}{|}}{\underset{\underset{\displaystyle Br}{|}}{C}} - \overset{\overset{\displaystyle Ph}{|}}{\underset{\underset{\displaystyle H}{|}}{C}} - CH_3$$

Regardless of which adjacent H is removed by *tert*-butoxide, the product will be 2,4-diphenylpent-2-ene.

(b) Here are a Newman projection, a three-dimensional representation, and a Fischer projection of the required diastereomer. On both carbons 2 and 4, the H has to be anti-coplanar with the bromine while leaving the other groups to give the same product. Not coincidentally, the correct diastereomer is a meso structure.

H and Br are anti-coplanar

**6-72** All five products (boxed) come from rearranged carbocations. Rearrangement, which may occur simultaneously with ionization, can occur by hydride shift to the 3° methylcyclopentyl cation, or by ring expansion to the cyclohexyl cation.

This 1° carbocation would be very unstable.

hydride shift

alkyl shift shown on next page

## 6-72 continued
### alkyl shift (ring expansion)

6-73 Begin with a structure of (S)-2-bromo-2-fluorobutane. Since there is no H on C-2, the lowest priority group must be the $CH_3$. The Br has highest priority, then F, then $CH_2CH_3$, and $CH_3$ is fourth. Sodium methoxide is a strong base and nucleophile, so the reaction must be second order, E2 or $S_N2$.

(a)

Br · · · CH₃  NaOCH₃  →  H—⟨π bond⟩—CH₃   H₃C · · · Br  NaOCH₃  H₃C—⟨π bond⟩—H

rotate back carbon 120°     π bond

π bond

In regular structural formulas, the reaction would give three products including the stereoisomers shown above.

NaOCH₃ →

$H_3C$ ... $H$ / $F$ ... $CH_3$  *trans*
+
$H_3C$ ... $CH_3$ / $F$ ... $H$  *cis*
+
$H_2C$ ... $H$ / $F$ ... $CH_3$  minor product from E2

(b)

⁴  ³      ⁴  ³
NaOCH₃ →
¹Br  F² $S_N2$   ¹F  OCH₃²
(S)            (S)

In these structures, the numbers 1 to 4 indicate the group's priority in the Cahn-Ingold-Prelog system.

A cursory analysis of the *designation* of configuration would suggest to the uncritical mind that this reaction proceeded with retention of configuration—but that would be wrong! You know by now that a careful analysis is required. In the Cahn-Ingold-Prelog system, the F in the starting material was priority group 2, but in the product, because Br has left, F is now the first priority group. So even though the *designation* of configuration suggests retention of configuration, the molecule has actually undergone inversion as would be expected with an $S_N2$ reaction. (See the solution to problem 6-21 for a similar example.)

6-74

(a) Only the propagation steps are shown. NBS provides a low concentration of $Br_2$ which generates bromine radical in ultraviolet light. In the starting material, all 8 allylic hydrogens are equivalent.

(b) The allylic carbocation has two resonance forms showing that two carbons share the positive charge. The ethanol nucleophile can attack either of these carbons, giving the $S_N1$ products; or loss of an adjacent H will give the E1 product.

$S_N1$

E1

**137**

(c) The first step in this first-order solvolysis is ionization, followed by rearrangement.

hydride shift

allylic!

E1

$S_N1$

(d) The first step in this first-order solvolysis is ionization, followed by rearrangement.

$S_N1$ product from unrearranged carbocation

rearrangement by alkyl shift

E1 on rearranged carbocation

mechanisms continued on next page

6-74(d)  continued

S$_N$1 on rearranged carbocation

6-75

(a)  E2—
one step

$$H_2C=CH-CH_3 \;+\; H_2O \;+\; Br^-$$

S$_N$2—
one step

$$CH_3-CH-CH_3 \;+\; Br^-$$
with OH above the central carbon

(b)  In the E2 reaction, a C—H bond is broken.  When D is substituted for H, a C—D bond is broken, slowing the reaction.  In the S$_N$2 reaction, no C—H (C—D) bond is broken, so the rate is unchanged.

(c)  These are first-order reactions.  The slow, rate-determining step is the first step in each mechanism.

The only mechanism of these two involving C—H bond cleavage is the E1, but the C—H cleavage does NOT occur in the slow, rate-determining step.  Kinetic isotope effects are observed only when C—H (C—D) bond cleavage occurs in the rate-determining step.  Thus, we would expect to observe *no change in rate* for the deuterium-substituted molecules in the E1 or S$_N$1 mechanisms.  (In fact, this technique of measuring isotope effects is one of the most useful tools chemists have for determining what mechanism a reaction follows.)

6-76  Both products are formed through E2 reactions.  The difference is whether a D or an H is removed by the base.  As explained in Problem 6-75, C—D cleavage can be up to 7 times slower than C—H cleavage, so the product from C—H cleavage should be formed about 7 times as fast.  This rate preference is reflected in the 7 : 1 product mixture.   ("Ph" is the abbreviation for a benzene ring.)

requires C—D bond cleavage;
slow; minor product

requires C—H bond cleavage;
7 times faster; major product

Note to the student: The most common question from students studying Chapter 6 material is: How do I know whether a reaction will follow an $S_N1$, $S_N2$, E1, or E2 mechanism? The answer is not always simple even for experienced chemists, and mixtures of substitution and elimination products are typical. Here are some generalizations that can provide guidelines about how to make reasonable predictions. Most instructors have their own version of this "concept map"; this one was modified from one of my colleagues, Dr. Hima Joshi, who has permitted its use here.

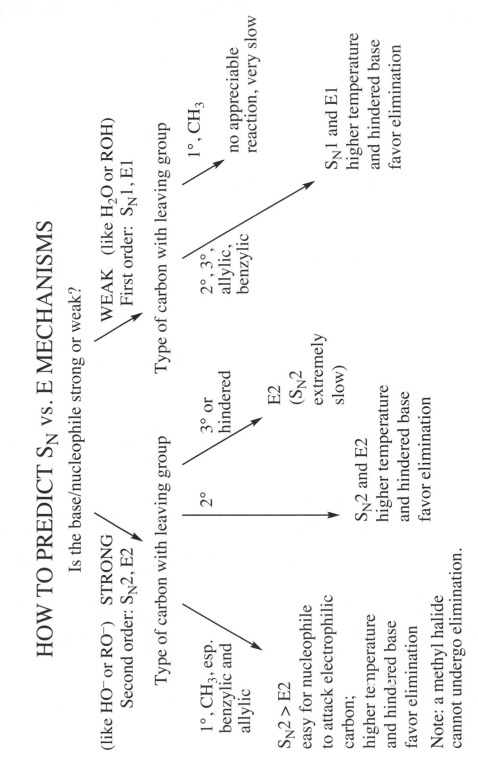

# HOW TO PREDICT $S_N$ vs. E MECHANISMS

Is the base/nucleophile strong or weak?

(like HO⁻ or RO⁻) **STRONG**
Second order: $S_N2$, E2

**WEAK** (like $H_2O$ or ROH)
First order: $S_N1$, E1

Type of carbon with leaving group

2°

3° or hindered

E2
($S_N2$ extremely slow)

$S_N2$ and E2
higher temperature and hindered base favor elimination

1°, $CH_3$, esp. benzylic and allylic

$S_N2 > E2$
easy for nucleophile to attack electrophilic carbon;
higher temperature and hindered base favor elimination

Note: a methyl halide cannot undergo elimination.

Type of carbon with leaving group

1°, $CH_3$

no appreciable reaction, very slow

2°, 3°, allylic, benzylic

$S_N1$ and E1
higher temperature and hindered base favor elimination

7-1  The number of elements of unsaturation in a hydrocarbon formula is given by:

$$\frac{2(\#C) + 2 - (\#H)}{2}$$

(a)  Solve this equation for #H.  Three double bonds plus one ring make 4 elements of unsaturation.

$$4 = \frac{2(9) + 2 - (\#H)}{2} \implies \boxed{\#H = 12}$$

$C_9H_{12}$ has 4 elements of unsaturation.

(b)  $C_6H_{12} \implies \dfrac{2(6) + 2 - (12)}{2}$ = 1 element of unsaturation

(c)  Many examples are possible.  Yours may not match these structures, but all possible answers must have either one double bond or one ring, that is, one element of unsaturation.

7-2    $C_4H_6 \implies \dfrac{2(4) + 2 - (6)}{2}$ = 2 elements of unsaturation

Of the nine structures, two have 1 triple bond, two have 2 double bonds, four have 1 ring plus 1 double bond, and one has 2 rings.

7-3  Hundreds of examples of $C_4H_6NOCl$ are possible.  Yours may not match these, but all must contain two elements of unsaturation.

7-4  Many examples of these formulas are possible.  Yours may not match these, but correct answers must have the same number of elements of unsaturation.  *Check your answers in your study group.*

(a)  $C_4H_4Cl_2 \implies C_4H_6 = 2$

(b)  $C_4H_8O \implies C_4H_8 = 1$

(c)  $C_6H_8O_2 \implies C_6H_8 = 3$

(d)  $C_5H_5NO_2 \Rightarrow C_{5.5}H_5 = 4$

$C \equiv C - NH_2$ (with dioxole ring)

$N \equiv C$ ... $O$ ... $O$ (lactone structure)

(e)  $C_6H_3NClBr \Rightarrow C_{6.5}H_5 = 5$   $HC \equiv C - C = C - C \equiv CH$   with substituents $Br$, $NH$, $Cl$

$N = C = CH_2$ with $Br$, $Cl$ on cyclobutene

$C \equiv N$ on cyclopentadiene with $Br$, $Cl$

---

Note to the student: The IUPAC system of nomenclature is undergoing many changes, most notably in the placement of position numbers. The new system places the position number close to the functional group designation, which is what this Solutions Manual will attempt to follow; however, you should be able to use and recognize names in either the old or the new style. Ask your instructor which system to use.

---

7-5  The *E/Z* system is unambiguous and is generally preferred for designating stereochemistry around a double bond. However, *cis/trans* are still used for geometric isomers from substituents on a ring.

(a)  4-methylpent-1-ene
(b)  2-ethylhex-1-ene (Number the longest chain *containing the double bond*.)
(c)  penta-1,4-diene
(d)  penta-1,2,4-triene
(e)  2,5-dimethylcyclopenta-1,3-diene
(f)  4-vinylcyclohex-1-ene ("1" is optional)
(g)  3-phenylprop-1-ene ("1" is optional)
(h)  *trans*-3,4-dimethylcyclopent-1-ene ("1" is optional)
(Note that *trans* applies to the methyl groups, not the C=C.)
(i)  7-methylenecyclohepta-1,3,5-triene
(j)  (2E,4Z)-5,6-dimethylhepta-2,4-diene

For comparison, these are the old IUPAC names, which will not be used in this Solutions Manual.

(a)  4-methyl-1-pentene
(b)  2-ethyl-1-hexene
(c)  1,4-pentadiene
(d)  1,2,4-pentatriene
(e)  2,5-dimethyl-1,3-cyclopentadiene
(f)  4-vinyl-1-cyclohexene ("1" is optional)
(g)  3-phenyl-1-propene ("1" is optional)
(h)  *trans*-3,4-dimethyl-1-cyclopentene ("1" is optional)
(Note that *trans* applies to the methyl groups, not the C=C.)
(i)  7-methylene-1,3,5-cycloheptatriene
(j)  (2E,4Z)-5,6-dimethyl-2,4-heptadiene

---

7-6   Parts (b) and (e) do not show *cis,trans* isomerism.

(a)

(Z)-hex-3-ene
(*cis*-hex-3-ene)

(E)-hex-3-ene
(*trans*-hex-3-ene)

(c)

(2Z,4Z)-hexa-2,4-diene
*cis,cis*-hexa-2,4-diene

(2Z,4E)-hexa-2,4-diene
*cis,trans*-hexa-2,4-diene

(2E,4E)-hexa-2,4-diene
*trans,trans*-hexa-2,4-diene

(d)

(Z)-3-methylpent-2-ene

(E)-3-methylpent-2-ene

"*Cis*" and "*trans*" are not clear for this example; "*E*" and "*Z*" are unambiguous.

**7-6 continued**

(f)

cis-3,4-dibromocyclopent-1-ene
(Both Br atoms could also be
up, the enantiomer.)

trans-3,4-dibromocyclopent-1-ene
(Both Br atoms could also
be inverted, the enantiomer.)

In this example, *cis* and *trans* refer
to the relative positions of the Br
atoms, not the double bond. All
double bonds in rings of 7 or fewer
atoms must be *cis*.

**7-7**

(a)

2,3-dimethylpent-2-ene
(neither *cis* nor *trans*)

(b)

trans-3-ethylhexa-1,4-diene
(Whether *cis* or *trans* is not
specified in the problem; the vinyl
group is part of the main chain.
This could also be named as the *E*
isomer.)

(c)

1-methylcyclopentene

(d)

5-chlorocyclohexa-1,3-diene
(Positions of double bonds need
to be specified.)

(e)

1,4-dimethylcyclohexene

(f)

(*E*)-2,5-dibromo-3-ethylpent-2-ene
(The *cis* designation does not apply.)

**7-8**

(a)

(*E*)-3-bromo-2-chloropent-2-ene

(*Z*)-3-bromo-2-chloropent-2-ene

(b)

(2*E*,4*E*)-3-ethylhexa-2,4-diene

(2*Z*,4*E*)-

(2*E*,4*Z*)-

(2*Z*,4*Z*)-

(c)

(*Z*)-3-bromo-2-methylhex-3-ene

(*E*)-3-bromo-2-methylhex-3-ene

(d)

(*Z*)-penta-1,3-diene

(*E*)-penta-1,3-diene

(e)

no geometric isomer

7-8 continued

(f)

(2Z,5E)-3,7-dichloroocta-2,5-diene     (2E,5E)-3,7-dichloroocta-2,5-diene

(2Z,5Z)-3,7-dichloroocta-2,5-diene     (2E,5Z)-3,7-dichloroocta-2,5-diene

(g) no geometric isomers (An *E* double bond would be too highly strained.)

(h)

(Z)-cyclodecene       (E)-cyclodecene

(i)

(1E,5E)-cyclodeca-1,5-diene    (1Z,5E)-cyclodeca-1,5-diene    (1Z,5Z)-cyclodeca-1,5-diene

7-9 From Table 7-1, approximate heats of hydrogenation can be determined for similarly substituted alkenes. The energy difference is approximately 6 kJ/mole (1.4 kcal/mole), the more substituted alkene being more stable.

*gem*-disubstituted
117 kJ/mole
(28.0 kcal/mole)

tetrasubstituted
111 kJ/mole
(26.6 kcal/mole)

7-10 Use the values in Table 7-1. Alternatively, the relative values in Figure 7-8 could be used to reach the same conclusions. Minor discrepancies between kcal and kJ are due to rounding.

(a)   2 x (*cis*-disubstituted − *trans*-disubstituted) =

2 x (120 − 116) = 8 kJ/mole more stable for *trans, trans*

(2 x (28.6 − 27.6) = 2 kcal/mole)

*trans,trans*
more stable

*cis,cis*

--------------------------------------------------

(b)   monosubstituted − *gem*-disubstituted = 127 − 119 = 8 kJ/mole

(30.3 − 28.0 = 2.3 kcal/mole)

2-methylbut-1-ene is more stable

disubst.
more stable

monosubst.

--------------------------------------------------

(c)   *gem*-disubstituted − trisubstituted = 117 − 113 = 4 kJ/mole

(28.0 − 26.9 = 1.1 kcal/mole)

2-methylbut-2-ene is more stable

trisubst.
more stable

disubst.

--------------------------------------------------

**144**

7-10 continued

(d)    *gem*-disubstituted − tetrasubstituted = 117 − 111 = 6 kJ/mole
(28.0 − 26.6 = 1.4 kcal/mole)

2,3-dimethylbut-2-ene is more stable

tetrasubst.    disubst.
more stable

7-11

(a)  Slightly strained but stable with C=C in a 5-membered ring.    (b) Could not exist as the ring size must be 8 atoms or greater to include *trans* double bond—see answer to part (e).

(c)

The *trans* refers to the two methyls since this compound can exist, rather than the *trans* referring to the alkene, a molecule that could not exist because the ring would have too much strain. A ring must have 8 or more atoms before a *trans* double bond can exist in the ring.

(d)

stable—*trans* in 10-membered ring

(e) Unstable at room temperature—it cannot have *trans* alkene in 7-membered ring (possibly isolable at very low temperature—this type of experiment is one of the challenges chemists attack with gusto).
(f) Stable—the alkene is not at a bridgehead.
(g) Unstable—violates Bredt's Rule (alkene at bridgehead in 6-membered ring).
(h) Stable—alkene is at bridgehead in 8-membered ring.
(i) Unstable—violates Bredt's Rule (alkene at bridgehead in 7-membered ring).

7-12
(a) The dibromo compound should boil at a higher temperature because of its much larger molecular weight.
(b) The *cis* should boil at a higher temperature than the *trans* as the *trans* has a zero dipole moment and therefore no dipole-dipole interactions.
(c) 1,2-Dichlorocyclohexene should boil at a higher temperature because of its much larger molecular weight and larger dipole moment than cyclohexene.

7-13

(a) 3°  H₃C  Br  →  NaOH / acetone →    major E2  +  minor E2    Strong base (hydroxide) will give second order reactions; a 3° alkyl halide precludes $S_N2$, so E2 is the only option.

(b) 3°  H₃C  Br  →  (CH₃CH₂)₃N: →    minor E2  +  major E2    E2 is only option; hindered base gives less substituted double bond as major isomer.

(c) Cl  2°  →  NaOCH₃ / CH₃OH →    OCH₃ $S_N2$  +  E2    Major product is difficult to predict for 2° halides; both $S_N2$ and E2.

145

7-13 continued

(d)  2°

$\xrightarrow[\text{(CH}_3)_3\text{COH}]{\text{NaOC(CH}_3)_3}$

E2

NaOC(CH$_3$)$_3$ is a bulky base and a poor nucleophile, minimizing S$_N$2.

7-14

7-15 (a)

Since hydroxide is a strong base, the reaction will be second order: on a 2° alkyl bromide, both substitution and elimination will occur.

(b)

meso

*hindered base*
(CH$_3$CH$_2$)$_3$N:

only alkene isomer
E2

Note that these products are geometric isomers. If a reaction starts with stereoisomers and produces stereoisomers, that is the definition of a *stereospecific* reaction.

(c)

one enantiomer
of the *d,l* pair

*hindered base*
(CH$_3$CH$_2$)$_3$N:

only alkene isomer
E2

(d)

$\xrightarrow[\text{acetone}]{\text{NaOH}}$

S$_N$2 is the only mechanism possible because no H can get coplanar with Cl; an elimination product would violate Bredt's Rule.

## 7-15 continued

(e)

Models show that the H on C-3 cannot be anti-coplanar with the Cl on C-2. Thus, this E2 elimination must occur with a *syn*-coplanar orientation: the D must be removed as the Cl leaves.

(f)

H and Br must be 180°.

*hindered base:*
*E2 only*

π bond

only E2 product

**7-16** As shown in Solved Problem 7-3, the H and the Br must have *trans*-diaxial orientation for the E2 reaction to occur. In part (a), the *cis* isomer has the methyl in the equatorial position, the $NaOCH_3$ can remove a hydrogen from either C-2 or C-6, giving a mixture of alkenes where the most highly substituted isomer is the major product (Zaitsev). In part (b), the *trans* isomer has the methyl in the axial position at C-2, so no elimination can occur at C-2. The only possible elimination orientation is toward C-6.

(a)

Either H can be removed.

major    +    minor     Zaitsev orientation

(b)

Only this H can be removed.

Only alkene formed; stereochemistry of E2 precludes other isomer from forming.

**7-17** E2 elimination requires that the H and the leaving group be anti-coplanar; in a chair cyclohexane, this requires that the two groups be *trans* diaxial. However, when the bromine atom is in an axial position, there are no hydrogens in axial positions on adjacent carbons, so no elimination can occur.

no hydrogens *trans* diaxial to the Br

Br ← axial

**147**

7-18 (a)

Showing the chair form of the decalins makes the answer clear. The top isomer locks the H and the Br that eliminate into a *trans*-diaxial conformation—optimum for E2 elimination. The bottom isomer has Br equatorial where it is exceedingly slow to eliminate.

(b)

Br must be axial to eliminate by E2, forcing the large isopropyl group axial, a very unstable conformer; this *trans* isomer will be exceedingly slow to react as the preferred conformer will be diequatorial.

In this *cis* isomer, the Br is in the optimum axial position for elimination while the large isopropyl group is in the energetically preferred equatorial position; this will be the reactive conformer.

7-19   Models are a big help for this problem.

(a)

NaOCH₃

the only H
*trans* diaxial

the only elimination
product

(b)

NaOCH₃

both *trans* diaxial—
gives two products

(from elimination of H and Br)

+

(from elimination of D and Br)

**148**

7-20

both Br anti-coplanar

transition state

····· = stretched bond being broken or formed

only geometric isomer possible from a concerted mechanism

transition state

pi bond

cis

7-21

(S,S)

both Br atoms anti-coplanar

transition state

····· = stretched bond being broken or formed

only geometric isomer possible from a concerted mechanism

Br atoms at 180°

transition state

pi bond

cis

7-22
(a) NaI / acetone → cyclohexene

(b) (R,R) both Br atoms anti-coplanar → NaI / acetone → CH₃CH₂...C=C...CH₂CH₂CH₃ with H below

*cis*-hept-3-ene
(Z)-hept-3-ene

(c) rotate → Br atoms at 180° → I⁻ → *trans*-pent-2-ene

(d)
NaI / acetone →

(S)-3-bromo-1-ethyl-3-methylcyclohexene

The two bromine atoms must be anti-coplanar in order to eliminate. In a 6-membered ring, these groups must be *trans* to each other, in diaxial positions. Only the bromines at C-1 and C-2 fit that requirement; the bromines at C-2 and C-3 cannot eliminate.

(e)
rotation → NaI, acetone →

*trans*-cyclodecene

In the first conformation shown, the bromine atoms are not coplanar and cannot eliminate. Rotation in this large ring can place the two bromines anti-coplanar, generating a *trans* alkene in the ring. (Use models!)

7-23 The stereochemical requirement of E2 elimination is anti-coplanar; in cyclohexanes, this translates to *trans*-diaxial. Both dibromides are *trans*, but because the *tert*-butyl group must be in an equatorial position, only the left molecule can have the bromines diaxial. The one on the right has both bromines locked into equatorial positions, from which they cannot undergo E2 elimination.

*trans*-diaxial — can do E2

*trans*-diequatorial — cannot do E2

7-24
(a)

as nucleophile

same

as base

as base

(b)

same

as nucleophile

as base

as base

carbocation rearrangement

2°

3°

same product mixture as in part (a)

**151**

7-25 In these mechanisms, the base removing the final proton is shown as water. Other alcohols are just as likely to serve as the base.

(a)

(b)

*cis + trans*
major isomer

minor isomer

(c)

from unrearranged 2° carbocation

greatest amount

from rearranged 3° carbocation

greatest amount

least amount

## 7-25 continued

(d)

When proposing mechanisms, always look for what bonds are broken and what new bonds must be formed. In this problem, the new ring must come from the pi electrons forming a new sigma bond at a 3° carbon, which must have been a carbocation.

## 7-26

(a) $\Delta G°$ $= \Delta H° - T\Delta S°$

$= +116,000 \text{ J/mol} - 298\ °K\ (117\ \text{J/°K•mol}) = +116,000 \text{ J/mol} - 34,866 \text{ J/mol}$

$= +81,134 \text{ J/mol} = +81.1 \text{ kJ/mol}\ (+19.3 \text{ kcal/mole})$

$\Delta G°$ is positive, the reaction is **disfavored** at 25°C.

(b) $\Delta G_{1000} = +116,000 \text{ J/mol} - 1273\ °K\ (117\ \text{J/°K•mol}) = +116,000 \text{ J/mol} - 148,941 \text{ J/mol}$

$= -32,941 \text{ J/mol} = -32.9 \text{ kJ/mol}\ (-8.0 \text{ kcal/mole})$

$\Delta G$ is negative, the reaction is **favored** at 1000°C.

## 7-27

(a) Basic and nucleophilic mechanism: $Ba(OH)_2$ is a strong base.
(b) Acidic and electrophilic mechanism: the catalyst is $H^+$.
(c) Free radical chain reaction: the catalyst is a peroxide that initiates free radical reactions.
(d) Acidic and electrophilic mechanism: the catalyst $BF_3$ is a strong Lewis acid.

## 7-28

(a)

## 7-28 continued

(b)

SN2, Na+ ⊖:Ö–CH₃ as a nucleophile → + NaBr

E2, Na+ ⊖:Ö–CH₃ as a base → + NaBr + CH₃OH

(c)

H—OSO₃H →

$\Delta$ → H₂O +

Pi electrons form sigma bond at carbocation. "B"

"B"

"A"

H₂O: "A"

new bond

## 7-29

(a)

H—OPO₃H₂ →

$\Delta$ →

H₂O: →

+ H₂O

(b)

$CH_2$—ÖH

H—OSO₃H →

$\Delta$ / – H₂O →

$CH_2$⊕

This 1° carbocation would be very unstable.

rearrangements shown on the next page

**154**

7-29 (b)  continued

Two possible rearrangements

1.  Hydride shift

2.  Alkyl shift—ring expansion

(c)  carbocation formation

without rearrangement

$E + Z$

with hydride shift

**155**

## 7-29 continued
(d)

E1 product from
unrearranged
carbocation

rearrangement by alkyl shift

E1 on rearranged carbocation

## 7-30

(a) all positions equivalent → $\xrightarrow[hv]{Br_2}$ Br

(b) from (a) Br $\xrightarrow[\Delta]{KOH}$ or KO-t-Bu

(c) from (a) Br $\xrightarrow[\text{cold to avoid E2}]{\substack{S_N2 \\ NaOEt}}$ OEt  or $S_N1$: EtOH, $\Delta$

(d) from (b) $\xrightarrow[hv]{NBS}$ Br  or low conc. of $Br_2$, hv

(e) from (d) Br $\xrightarrow[\Delta]{KOH}$ or KO-t-Bu

(f) from (a) Br $\xrightarrow[\text{cold to avoid E2}]{\substack{S_N2 \\ NaOH}}$ OH  or $S_N1$: $H_2O$, $\Delta$

## 7-31

(a)   (b)   (c) Br, Br   (d)   (e)

(f) Z   (g)   (h) Br, Z   (i) Z, E

**7-32** These names follow the modern IUPAC system of placement of position numbers.
(a) 2-ethylpent-1-ene (Number the longest chain *containing the double bond*.)
(b) 3-ethylpent-2-ene
(c) (3E,5E)-2,6-dimethylocta-1,3,5-triene
(d) (E)-4-ethylhept-3-ene
(e) 1-cyclohexylcyclohexa-1,3-diene
(f) (3Z,5Z)-6-chloro-3-(chloromethyl)octa-1,3,5-triene

**7-33** (a) *E*  (b) neither—two methyl groups on one carbon  (c) *Z*  (d) *Z*

**7-34**
(a)

(Z)-1-fluoro-      (E)-1-fluoro-      2-fluoro-      3-fluoro-      fluorocyclopropane
prop-1-ene         prop-1-ene         prop-1-ene     prop-1-ene

(b) C₄H₇Br has one element of unsaturation, but no rings are permitted in the problem, so all the isomers must have one double bond. Only four isomers are possible with four carbons and one double bond, so the 12 answers must have these four skeletons with a Br substituted in all possible positions.

(c) Cholesterol, $C_{27}H_{46}O$, has five elements of unsaturation. If only one of those is a pi bond, the other four must be rings. (The structure of cholesterol can be found in text section 25-6.)

**7-35** (a) Neither *cis* nor *trans* is defined for the double bond beginning at C-2 because none of the four groups are the same.

(b)    2Z,4E         2E,4E         2Z,4Z         2E,4Z

All are 3-chlorohepta-2,4-diene.

The *E/Z* nomenclature is unambiguous and is preferred for all four of these isomers.

**7-36** Parts (a) and (d) have no geometric isomers.

(b)  *trans*-pent-2-ene   *cis*-pent-2-ene       (c)  *trans*-hex-3-ene   *cis*-hex-3-ene
     (E)-pent-2-ene       (Z)-pent-2-ene              (E)-hex-3-ene       (Z)-hex-3-ene

(e)  *trans*-1,2-dibromopropene    *cis*-1,2-dibromopropene
     (E)-1,2-dibromopropene        (Z)-1,2-dibromopropene

**157**

**7-36 continued**

(f) *Cis* and *trans* do not apply to the C=C at C-1, only the one at C-3. The *E/Z* system is unambiguous.

**7-37**

(a)

F, F    H, F
C=C      C=C
H   H    F   H
         dipole moment = 0

(b)

Br, Br    Br, CH₃
C=C       C=C
H   H     CH₃  Br
          dipole moment = 0

(c)

Cl, Cl    Cl, Cl
C=C       C=C
Br   Br   H   H
          larger dipole moment
          (no bromines opposing
          the dipole of the chlorines)

**7-38**

(a)  (b) major    minor   (c) major   minor   (d) minor   major   minor

**7-39** Major alkene isomers are shown. Minor alkene isomers would also be produced in parts (a), (b) and (d).

(a)

$$CH_3-\overset{\overset{\displaystyle H}{|}}{C}-\overset{\overset{\displaystyle H}{|}}{\underset{\underset{\displaystyle OH}{|}}{C}}-CH_3 \xrightarrow[\Delta]{H_2SO_4} CH_3-\overset{\overset{\displaystyle H}{|}}{C}=\overset{\overset{\displaystyle H}{|}}{C}-CH_3 + H_2O$$

$E + Z$

minor

(b)

+ NaOC(CH₃)₃ ⟶      + NaBr + HOC(CH₃)₃

hindered base

minor

(c)

$$CH_3-\overset{\overset{\displaystyle H}{|}}{\underset{\underset{\displaystyle Br}{|}}{C}}-\overset{\overset{\displaystyle H}{|}}{\underset{\underset{\displaystyle Br}{|}}{C}}-CH_3 + NaI \xrightarrow{acetone} CH_3-\overset{\overset{\displaystyle H}{|}}{C}=\overset{\overset{\displaystyle H}{|}}{C}-CH_3 + NaBr + IBr$$

$E + Z$

(d)

$$CH_3-\overset{\overset{\displaystyle CH_3}{|}}{\underset{\underset{\displaystyle H}{|}}{C}}-\overset{\overset{\displaystyle CH_3}{|}}{\underset{\underset{\displaystyle Br}{|}}{C}}-CH_3 \xrightarrow{NaOH, \Delta} \overset{H_3C}{\underset{H_3C}{}}C=C\overset{CH_3}{\underset{CH_3}{}} + NaBr + H_2O$$

minor

**7-40**

(a) [cyclopentane with two Br] NaI, acetone → cyclopentene

(b) [cyclopentanol] $H_2SO_4$ , Δ → cyclopentene

(c) [bromocyclopentane] $KOC(CH_3)_3$ → cyclopentene    Other bases could also be used. The hindered bases would minimize the amount of substitution that would occur.

(d) [cyclopentane] $Br_2$ , hv → [bromocyclopentane] $KOC(CH_3)_3$ → cyclopentene

**7-41**

(a) [alkene]    only product

(b) [alkene] + [alkene] (major) + [alkene] (minor)

(c) [alkene] + [alkene]

(d) [methylcyclohexene] (major) + [methylcyclohexene] (minor)

(e)  NaOH →   The E2 mechanism requires anti-coplanar orientation of H and Br.

**7-42**  The bromides are shown here.  Chlorides or iodides would also work.

(a)  [hexyl bromide]

(b) [Br structures] or [structure]

(c) [structure with two Br]

(d) [bromomethylcyclohexane]

(e) [dimethyl dibromocyclohexane structure]

**7-43**

(a)  There are two reasons why alcohols do not dehydrate with strong base.  The potential leaving group, hydroxide, is itself a strong base and therefore a terrible leaving group.  Second, the strong base deprotonates the —OH faster than any other reaction can occur, consuming the base and making the leaving group anionic and therefore even worse.

$$ROH + {}^-OC(CH_3)_3 \rightleftharpoons RO^- + HOC(CH_3)_3$$

(b)  A halide is already a decent leaving group.  Since halides are extremely weak bases, the halogen atom is not easily protonated, and even if it were, the leaving group ability is not significantly enhanced.  The hard step is to remove the adjacent H, something only a strong base can do—and strong bases will not be present under strong acid conditions.

**7-44**

(a) [alkene]   (b) [methylenecyclopentane]   (c) [methylcyclohexene]   (d) [alkene]   rearrangement

7-45
(a)

(b)

This 1° carbocation would be very unstable.

$+\ H_2O$

hydride shift

alkyl shift
ring expansion

This alkene could be produced from the unrearranged carbocation, but the 1° carbocation has such a short lifetime that it is more reasonable to propose the pathway through the rearranged 3° carbocation.

(c)

7-46
(a)

major          minor

(b)

major          minor

(c)

major          minor

(d)

Only product from E2—anti-coplanar is possible only from carbon without $CH_3$ by removing H and Cl.

D

**7-46 continued**

(e)

The conformation must be considered because the leaving group must be in an axial position. The *tert*-butyl group is so large that it must be equatorial, locking the conformation into the chair shown. As a result, only the Cl at position 4 is axial, so that is the one that must leave, not the one at position 3. Two isomers are possible but it is difficult to say which would be formed in greater amount.

(f)
Only one of the two Br atoms is in the required axial position for E2 elimination. There are axial H atoms at the two adjacent carbons, so two alkene products can be formed. Predicting major and minor would be difficult.

**7-47**

E1 works well because only one carbocation and only one alkene are possible. Substitution is not a problem here. The only nucleophiles are water, which would simply form starting material by a reverse of the dehydration, and bisulfate anion. Bisulfate anion is an extremely weak base and poor nucleophile; if it did attack the carbocation, the unstable product would quickly re-ionize, with no net change, back to the carbocation.

**7-48**

The driving force for this rearrangement is the great stability of the resonance-stabilized, protonated carbonyl group.

**7-49** NBS contains traces of bromine; when combined with HBr, NBS produces more bromine. Bromine contains small amounts of bromine radical. Bromine radical abstracts an allylic hydrogen, resulting in a resonance-stabilized allylic radical. The allylic radical can bond to bromine at either of the two carbons with radical character. See the solution to problem 6-63.

The major product is usually the one with the double bond in the more stable position.

**7-50**

E2 dehydrohalogenation requires anti-coplanar arrangement of H and Br, so specific *cis-trans* isomers (**B** or **C**) are generated depending on the stereochemistry of the starting material. Removing a hydrogen from C-4 (achiral) will give about the same mixture of *E* and *Z* (**A**) from either diastereomer.

**7-51**

more stable than    by 4 kJ/mole

more stable than    by 16 kJ/mole

Steric crowding by the *tert*-butyl group is responsible for the energy difference. In *cis*-but-2-ene, the two methyl groups have only slight interaction. However, in the 4,4-dimethylpent-2-enes, the larger size of the *tert*-butyl group crowds the methyl group in the *cis* isomer, increasing its strain and therefore its energy.

7-52

| endocyclic trisubstituted | exocyclic disubstituted | endocyclic trisubstituted | exocyclic trisubstituted |

9 kJ/mole         5 kJ/mole

A standard principle of science is to compare experiments that differ by only one variable. Changing more than one variable clouds the interpretation, possibly to the point of invalidating the experiment.

The first set of structures compares endo and exocyclic double bonds, but the degree of substitution on the alkene is also different, so this comparison is not valid—we are not isolating simply the exo or endocyclic effect. The second pair is a much better measure of endo versus exocyclic stability because both alkenes are trisubstituted, so the degree of substitution plays no part in the energy values. Thus, 5 kJ/mole is a better value.

7-53 (a)          (b) No reaction—the two bromines are not *trans* and therefore cannot be *trans*-diaxial.      (c)

*trans*-diaxial

(d) No reaction—the bromines are *trans*, but they are diequatorial because of the locked conformation of the *trans*-decalin system. E2 can occur only when the bromines are *trans* AND diaxial.

7-54 In E2, the two groups to be eliminated must be coplanar. In conformationally mobile systems like acyclic molecules, or in cyclohexanes, anti-coplanar is the preferred orientation where the H and leaving group are 180° apart. In rigid systems like norbornanes, however, SYN-coplanar (angle 0°) is the only possible orientation and E2 will occur, although at a slower rate than anti-coplanar.

     The structure having the H and the Cl syn-coplanar is the *trans*, which undergoes the E2 elimination. (It is possible that the *other* H and Cl eliminate from the *trans* isomer; the results from this reaction cannot distinguish between these two possibilities.)

*cis*
extremely slow to eliminate—
H and Cl not coplanar

*trans*

syn-coplanar

7-55 It is interesting to note that even though three-membered rings are more strained than four-membered rings, three-membered rings are far more common in nature than four-membered rings. Rearrangement from a four-membered ring to something else, especially a larger ring, will happen quickly.

This 1° carbocation would be very unstable.

mechanism continued on next page

**163**

7-55 continued
<u>without rearrangement</u>

This alkene could be produced from the unrearranged carbocation, but the 1° carbocation has such a short lifetime that it is more reasonable to propose the pathway through the rearranged 3° carbocation.

<u>with rearrangement—hydride shift</u>

3° carbocation, but
still in a strained
4-membered ring

<u>with rearrangement—alkyl shift—ring expansion</u>

2° carbocation on
5-membered ring—
HOORAY!

MAJOR

7-56

1° carbocation—terrible!—will rearrange; can't do hydride shift, must do alkyl shift = ring expansion.

This is an unstable carbocation even though it is 3°; bridgehead carbons cannot be sp² (planar—try to make a model), so this carbocation does a hydride shift to a 2°, *more stable* carbocation.

Abstraction of adjacent
H gives bridgehead
alkene—violates
Bredt's Rule.

↓ hydride shift

continued on next page

**164**

7-56 continued

↓ hydride shift

This 2° carbocation loses an adjacent H to form an alkene; can't form at bridgehead (Bredt's Rule)— only one other choice.

7-57 "Ph" = phenyl

(a)

$$CH_3CHCHCH_3 \xrightarrow[\text{E2}]{NaOCH_3} CH_3-C=CHCH_3 + CH_3CHCH=CH_2$$

(with Ph and Br substituents)

Ph | major    Ph | minor

(b) H and Br must be anti-coplanar in the transition state.

2R,3R

↓ NaOCH₃

methyls *cis*

transition state

(c)

2S,3R

methyls *trans*

(d) The 2S,3S is the mirror image of 2R,3R; it would give the mirror image of the alkene that 2R,3R produced (with two methyl groups *cis*). The alkene product is planar, not chiral, so its mirror image is the same: the 2S,3S and the 2R,3R give the same alkene.

7-58

(a)

remove H from CH₃ → Radical character on 1° and 2° carbons— least stable of these three. (Will be slowest to form.)

remove H from C-3 → Radical character on 2° and 3° carbons— most stable of these three. (Will be fastest to form.)

remove H from C-6 → Radical character on two 2° carbons, equivalent resonance forms—almost as stable as the 2°/3° radical.

(b)

165

## 7-58 continued

(c)

two products from the
1°/2° radical; very minor

Two products from the 2°/3°
radical; the first has a
trisubstituted C=C—MAJOR.

**SAME COMPOUND**

Only ONE compound is
produced from the 2°/2°
radical, with a trisubstituted
C=C—MAJOR.

(d) The two compounds in the boxes are the major products found in the reaction mixture.

## 7-59

A pi bond is formed from two parallel p orbitals. Rotating around a double bond breaks the pi bond by forcing the p orbitals to be perpendicular. If anything can stabilize the transition state, the rotational energy barrier will be lower.

Each radical is on a 2° carbon, not particularly
stable. Two radicals on adjacent carbons are
particularly unstable.

Each radical is on a *benzylic* carbon; each is
stabilized by resonance. Review the
solution to 4-44(b) or 6-64 for benzylic
resonance forms.

An *unequal* distribution of electrons between the
two sp$^2$ carbons gives resonance forms
maximizing the number of pi bonds, all atoms
with full octets, and negative charge on the more
electronegative atoms! This large degree of
stabilization explains the lowest rotational
energy barrier: this transition state is the easiest
to form.

8-1  *Major* products are produced in greatest amount; they are not necessarily the *only* products produced.

(a)

(b)        (c)

(d)        Produced in about
equal amounts;
each is a mixture
of *cis* and *trans*.

Both 2° carbocations will form.

8-2
(a)

Because the allylic carbocation
has partial positive charge at two
carbons, the bromide nucleophile
can bond at either electrophilic
carbon, giving two products.

3-bromobut-
1-ene

1-bromobut-2-ene

(b)

This orientation produces a carbocation
intermediate that is stabilized by
resonance with the Cl atom.

This possible intermediate is not produced; it is a 2° carbocation without any stabilization.

**8-3**

**(a) initiation steps**

propagation steps

The 3° radical is more stable than the 2° radical.

1-bromo-2-methylcyclopentane (*cis* + *trans*)

+ • Br  recycles in first propagation step

**(b) initiation steps**

$$\text{(CH}_3)_3\text{C} - \overset{..}{\underset{..}{\text{O}}} \cdot + \text{ H} - \text{Br} \longrightarrow (\text{CH}_3)_3\text{C} - \overset{..}{\underset{..}{\text{O}}}\text{H} + \cdot \text{Br}$$

propagation steps

(Recall that "Ph" is the abbreviation for "phenyl".)

The benzylic radical is stabilized by resonance, and more stable than the aliphatic radical.

+ • Br  recycles in first propagation step

2-bromo-1-phenylpropane

**8-4**

**(a)** ⟶ HBr / ROOR ⟶ Br

**(b)** ⟶ HBr ⟶ Br

**(c)** OH ⟶ H₂SO₄ / Δ ⟶ ⟶ HBr ⟶ Br

Note: A good synthesis uses major products as intermediates, not minor products. Knowing orientation of addition and elimination is critical to using reactions correctly.

**(d)** OH ⟶ H₂SO₄ / Δ ⟶ ⟶ HBr / ROOR ⟶ Br

8-5

2,3-dimethylbutan-2-ol

+ H₃O⁺

2,3-dimethylbut-2-ene

+ H₃O⁺

8-6
(a) from 3° carbocation

(b)

(c) from 3° benzylic carbocation

8-7

mercurinium ion

(a)

"Ac" = acetyl

"OAc" or "AcO" = acetate

(b)

NaBH₄
demercuration

8-8
(a) plus the enantiomer

(b)

(c) plus the enantiomer of each

Note: When new chiral centers are generated from achiral or racemic reactants, the products are racemic mixtures. This book will indicate a racemic mixture by adding "plus the enantiomer".

**169**

**8-8 continued**

**(d)**

The *cis* and *trans* stereoisomers of each positional isomer would also be produced.

**8-9**

**(a)**

**(b)**

Heat favors elimination.

**(c)**

Using an acid-catalyzed hydration in part (c) would initially form a 2° carbocation that would quickly rearrange to 3° on carbon-3. The desired product would not be synthesized.

**8-10**

**(a), (b)**

**(c), (d)**

**(e), (f)**

plus the enantiomer

**8-11**

**(a)**

**(b)**

**(c)**

**8-12**  The attack of borane on 1-methylcyclopentene is equally likely from the **top face or the** bottom face, leading to a racemic mixture of the *trans* isomer.

racemic mixture

**8-13** The products are racemic.

(a)

*trans* product
plus the enantiomer

(b)

major

minor—steric
hindrance to
attack of $BH_3$

(c)

$BH_3$ goes on the less
hindered top (exo) face
of the C=C.

**8-14**

(a)

$BH_3 \cdot THF$   $H_2O_2$   $HO^-$

enantiomers

(b)

$BH_3 \cdot THF$   $H_2O_2$   $HO^-$

enantiomers

The enantiomeric pair produced from the Z-alkene is diastereomeric with the other enantiomeric pair produced from the E-alkene. Hydroboration-oxidation is stereospecific, that is, each alkene gives a specific set of stereoisomers, not a random mixture.

**8-15**

(a)

$Hg(OAc)_2$   $NaBH_4$
$H_2O$

(b)

$BH_3 \cdot THF$   $H_2O_2$   $HO^-$

plus the enantiomer

(c)

$H_2SO_4$   $\Delta$   $BH_3 \cdot THF$   $H_2O_2$   $HO^-$

plus the enantiomer

8-16 (a)

The *planar* carbocation is responsible for non-stereoselectivity. The bromide nucleophile can attack from the top or bottom, leading to a mixture of stereoisomers. The addition is therefore a mixture of syn and anti addition.

(b)

Two methyl groups are always *cis* to each other.

In contrast to part (a), hydroboration has no planar intermediate. Borane adds in a molecular addition with *syn* stereochemistry, and replacement of B with O proceeds with retention of stereochemistry. All of the steps in the process are stereospecific, so the product will be one diastereomer (although a racemic mixture).

8-17 During bromine addition to either the *cis*- or *trans*-alkene, two new chiral centers are being formed. Neither alkene (nor bromine) is optically active, so the product cannot be optically active.

The *cis*-but-2-ene gives two chiral products, a racemic mixture. However, *trans*-but-2-ene (shown on the next page), because of its symmetry, gives only one *meso* product, that can never be chiral. The "optical *in*activity" is built into this symmetric molecule.

This can be seen by following what happens to the configuration of the chiral centers from the intermediates to products, below. (The key lies in the symmetry of the intermediate and *inversion* of configuration when bromide attacks.)

IDENTICAL—SYMMETRIC

A Br⁻ nucleophile could attack at either carbon.

2S,3S          ENANTIOMERS    2R,3R

See the **TRANS** case on the next page.

**172**

**TRANS**

A Br⁻ nucleophile could attack at either carbon.

Bromide attack on this bromonium ion gives the same results.

2R,3S    IDENTICAL—MESO    2R,3S

CONCLUSION: Anti addition of a symmetric reagent to a symmetric *cis*-alkene gives racemic product, while anti addition to a *trans*-alkene gives meso product. (We will see shortly that syn addition to a *cis*-alkene gives meso product, and syn addition to a *trans*-alkene gives racemic product. Stay tuned.)

8-18 Enantiomers of chiral products are also produced but not shown.

(a) [reaction scheme] + the enantiomer

(b) Three new asymmetric carbons are produced in this reaction. All stereoisomers will be produced with the restriction that the two adjacent chlorines on the ring must be *trans*.

(c) [reaction scheme] *E* plus the enantiomer

The bromonium ion could form on either face of the C=C, and bromide will attack the other carbon of the bromonium ion as well. Either leads to equal amounts of the two enantiomers.

**173**

8-18 continued

(d)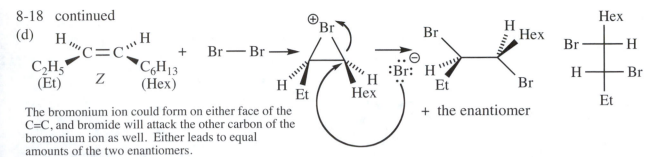

The bromonium ion could form on either face of the C=C, and bromide will attack the other carbon of the bromonium ion as well. Either leads to equal amounts of the two enantiomers.

8-19   The *trans* product results from water attacking the bromonium ion from the face opposite the bromine. Equal amounts of the two enantiomers result from the equal probability that water will attack either C-1 or C-2.

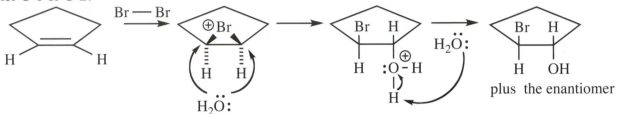

Water will do nucleophilic attack at either carbon.

equal amounts of enantiomers = racemic mixture

8-20

from Solved Problem 8-5

The bromonium ion shown here is the enantiomer of the one shown in Solved Problem 8-5.

enantiomer of product in Solved Problem 8-5

from Solved Problem 8-6:  The bromonium ion shown is meso as it has a plane of symmetry; attack by the nucleophile at the other carbon from what is shown in the text will create the enantiomer.

Show products from both nucleophiles attacking at this carbon.

These are the enantiomers of the structures shown in the text.

8-21   The chiral products shown here will be racemic mixtures.

(a) $H_3C$  OH
Cl

plus the enantiomer

(b)
HO
Br
plus the enantiomer

(c) $H_3C$   H   Cl
HO      H
$CH_3$
plus the enantiomer

(d) $H_3C$   H   Cl
HO   H   $CH_3$
plus the enantiomer

(e) $CH_3$  OH      $CH_3$  Cl
Br                Br
+

plus the enantiomers

**8-22 (a)**

**(b)**

plus the enantiomer

**(c)** plus the enantiomer

**8-23**

**(a)**   **(b)**   **(c)** *cis*   **(d)**

**8-24**  Limonene, $C_{10}H_{16}$, has three elements of unsaturation. Upon catalytic hydrogenation, the product, $C_{10}H_{20}$, has one element of unsaturation. Two elements of unsaturation have been removed by hydrogenation —these must have been pi bonds, either two double bonds or one triple bond. The one remaining unsaturation must be a ring. Thus, limonene must have one ring and either two double bonds or one triple bond. (The structure of limonene is shown in the text in Problem 8-23(d), and the hydrogenation product is shown above in the solution to 8-23(d).)

**8-25**  The BINAP ligand is an example of a conformationally hindered biphenyl as described in text section 5-9A and Figure 5-17. The groups are too large to permit rotation around the single bond connecting the rings, so the molecules are locked into one chiral twist or its mirror image.

This *simplified* three-dimensional drawing of one of the enantiomers shows that the two naphthalene rings are twisted almost perpendicular to each other. Molecular models will help visualize this concept.

**8-26**  Stereochemistry of the starting material is retained in the product. Products are racemic.

**(a)**   **(b)**

**8-27**

**(a)**   **(b)**   Probably minor—the OH crowds the top face of the C=C.

probably major

**(c)** *The original 6-membered ring is shown in bold bonds; arrows point to new bonds.*

new bonds

*side view*

BOTTOM VIEW

site of original double bond

**175**

8-28

(a)

(b)

(c)

8-29 All chiral products in this problem are racemic mixtures.

(a)        (b)        (c)        (d)

8-30 HA is a generic acid catalyst.

(a)

ENANTIOMERS

(H₂O removes
H⁺ from oxygen.)

−H⁺

(b)

mechanism continued on next page

**8-30 continued** IDENTICAL—MESO

Remember the lesson from Problem 8-17: anti addition of a symmetric reagent to a symmetric *cis*-alkene gives racemic product, while anti addition to a *trans*-alkene gives meso product. This fits the definition of a *stereospecific* reaction, where different stereoisomers of the starting material (*cis* and *trans*) are converted into different stereoisomers of product (a *dl-pair* and *meso form*).

**8-31**

**8-32** All chiral products in this problem are racemic mixtures.

**8-33**

8-34  All these reactions begin with achiral reagents; therefore, all the chiral products are racemic.

Refer to the observation in the solution to Problem 8-35.

8-35

(a) [alkene] $\xrightarrow[\text{H}_2\text{O}_2]{\text{OsO}_4}$ [product] HO  OH    meso

(b) [alkene] $\xrightarrow[\text{H}_2\text{O}]{\text{CH}_3\text{CO}_3\text{H}}$ [product] HO  OH    racemic (d,l)

(c) [alkene] $\xrightarrow[\text{H}_2\text{O}]{\text{CH}_3\text{CO}_3\text{H}}$ rotate [product] HO  OH    meso

(d) [alkene] $\xrightarrow[\text{H}_2\text{O}_2]{\text{OsO}_4}$ rotate [product] HO  OH    racemic (d,l)

Have you noticed yet?  For symmetric alkenes and symmetric reagents (addition of two identical X groups):

| | |
|---|---|
| cis-alkene + **syn** addition → meso | Assume that cis/syn/meso    +1 x +1 = +1 |
| cis-alkene + **anti** addition → racemic | are "same", and trans/anti/    +1 x −1 = −1 |
| trans-alkene + **syn** addition → racemic | racemic are "opposite".    −1 x +1 = −1 |
| trans-alkene + **anti** addition → meso | Then any combination can    −1 x −1 = +1 |
| | be predicted, just like math! |

**8-36** Solve these ozonolysis problems by working backwards, that is, by "reattaching" the two carbons of the new carbonyl groups into alkenes. Here's a hint. When you cut a circular piece of string, you still have only one piece. When you cut a linear piece of string, you have two pieces. Same with molecules. If ozonolysis forms only one product with two carbonyls, the C=C had to have been in a ring. If ozonolysis gives two molecules, the C=C had to have been in a chain.

(a) Two carbonyls from ozonolysis are in a chain, so alkene had to have been in a ring.

(b) Two carbonyls from ozonolysis are in two different products, so alkene had to have been in a chain, not a ring.

(c) Two carbonyls from ozonolysis are in two different products, so alkene had to have been in a chain, not a ring.

*E* or *Z* of alkene cannot be determined from products.

**8-37**

(a) $O_3$   $Me_2S$

(b) $KMnO_4$   $\Delta$

(c) $O_3$   $Me_2S$

(d) $O_3$   $Me_2S$

(e) $KMnO_4$   $\Delta$

(f) $KMnO_4$   cold

OH groups are *cis* ; product is racemic.

**8-38** The representation for a generic acid is H—A, where A is the conjugate base.

(a) H—A

8-38 continued

(b) Catalytic $BF_3$ reacts with trace amounts of water to form the probable catalyst:

catalyst

dimer

etc.

tetramer

trimer

8-39  H—A symbolizes a generic acid.

alkenes

polymers
(colored)

Note: The wavy bond symbol is used in the following
solutions to indicate the continuation of a polymer chain.

8-40

1° radical, and *not* resonance-stabilized—
**this orientation is not observed**

Orientation of addition always generates the more stable intermediate; the energy difference between a 1°
radical (shown above) and a benzylic radical is huge.  The phenyl substituents must necessarily be on
alternating carbons because the orientation of attack is always the same—not a random process.

**8-41** Each monomer has two carbons in the backbone, so the substituents on the monomer will repeat every two carbons in the polymer. Wavy bonds indicate continuation of the polymer chain.

polyvinyl chloride, PVC

polytetrafluoroethylene, PTFE, Teflon®

polyacrylonitrile, Orlon®

**8-42**

Wavy bonds mean that the chain continues.

**8-43** The accepted mechanism of olefin metathesis includes an intermediate with **a four-membered ring** where one of the atoms in the ring is the metal catalyst, abbreviated [M]. All steps are **equilibria**.

Step 1:

*cis* and *trans*

Step 2:

catalyst is regenerated

NEW PRODUCTS

**8-44** An excellent discussion of the olefin metathesis reaction is available at the Nobel Institute website:

http://nobelprize.org/nobel_prizes/chemistry/laureates/2005/chemadv05.pdf

The catalyst needs to have the short end of the molecule attached to the metal:

$[M] + [M] \longrightarrow [M]=CHCH_2CH_3$
catalyst

*liquid*

*gas*     *E + Z*

Unnumbered problem in the Problem-Solving Strategy section:

plus the enantiomer

plus the enantiomer

**182**

**8-45** In the spirit of this problem, all starting materials will have six carbons or fewer and only one carbon-carbon double bond. Reagents may have other atoms.

(a)

(b)

anti-Markovnikov
syn stereochemistry

major product from resonance
stabilized radical on 1° and 3° C

plus the
enantiomer

(c)

Only substitution
is possible.

**8-46**

(a)

Even though this product comes from an allylic radical intermediate with the possibility of two products, the major product is this one, with the double bond in the more stable tetrasubstituted position.

(b)

from (a)

$H_2O$, Δ

$S_N1$ conditions

(Avoid KOH that would do E2.)

(c)

$Br_2$

anti
addition

plus the
enantiomer

(d)

KMnO4

$HO^-$

cold, dilute

or $OsO_4$, $H_2O_2$

*syn* addition

(e)

$HCO_3H$

$H_3O^+$

anti addition

plus the
enantiomer

## 8-46 continued

(f) $\xrightarrow[\text{E2 conditions}]{\text{KO-}t\text{-Bu, }\Delta}$

from (c)   or Et$_3$N, a bulky, hindered base

(g) $\xrightarrow{\text{Br}_2, \text{H}_2\text{O}}$ plus the enantiomer

a bromohydrin

(h) $\xrightarrow[\text{conditions}]{\text{KOH, }\Delta}$

from (g)   E2 conditions

(i) $\xrightarrow[\text{HO}^-, \Delta]{\text{KMnO4}}$

or (1) O$_3$, (2) Me$_2$S

## 8-47  All chiral products in this problem are racemic mixtures.

(a)

(b)

(c) $\longrightarrow$

intermediate

(d) +

(e)

(f)

Peroxides do not affect HCl addition.

(g)

trans

(h)

(i)

(j)

(k)

(l)

(m)

(n)

(o) +

(p) $\xrightarrow{\text{NaBH}_4}$

HgOAc

intermediate

(q)

## 8-48  (a)

### initiation

RO—OR $\longrightarrow$ 2 RO •

RO • + H—Br $\longrightarrow$ ROH + Br •

propagation

(b)

(c)

2° carbocation

3° carbocation

from unrearranged carbocation

from rearranged carbocation

(d)

(e)

OR

# 8-48 continued

(f)

Water could attack either carbon with the same results.

(g)

Methanol could attack either carbon of the epoxide but it must attack from the opposite side of the ring.

racemic mixture

(h)

Recall that "Ph" is the abbreviation for phenyl.

# 8-49

(a) Hg(OAc)₂, H₂O → NaBH₄

(b) HBr, ROOR

(c) 1) BH₃ • THF  2) H₂O₂ , HO⁻

(d) 1) O₃ , then 2) Me₂S or KMnO₄, Δ

(e) Hg(OAc)₂, CH₃OH → NaBH₄

(f) OsO₄, H₂O₂ or cold, dilute KMnO₄

(g) CH₂I₂, Zn(Cu)

(h) Cl₂, H₂O

(i) HBr → KOH, Δ → MCPBA

Alternatively, hydration followed by dehydration to methylcyclohexene would give the same product.

**186**

8-50

(a)

8-51 For clarity, the new bonds formed in this mechanism are shown in bold.

8-52 Without divinylbenzene, individual chains of polystyrene are able to slide past one another. Addition of divinylbenzene during polymerization forms bridges, or *crosslinks*, between chains, adding strength and rigidity to the polymer. Divinylbenzene and similar molecules are called crosslinking agents.

Two polystyrene chains crosslinked by a divinylbenzene monomer shown in the dashed oval.

**187**

**8-53** A peroxy radical is shown as the initiator. Newly formed bonds are shown in bold. ▬

**8-54**
(a) new bonds   + H₂C=CH₂

(b) new bonds   + H₂C=CHCH₂OH

(c) new bonds   + H₂C=CH₂

**8-55**
(a) 2 [M]=CHR → Z + E + H₂C=CH₂

(b) + [M]=CHR → Z + E   Z + E

Ring-opening metathesis: the 5 carbons from cyclopentene become the middle 5 carbons of the product.

**8-56** Once the bromonium ion is formed, it can be attacked by either nucleophile, bromide or chloride, leading to the mixture of products.

8-57  Two orientations of attack of bromine radical are possible:

(A)  anti-Markovnikov

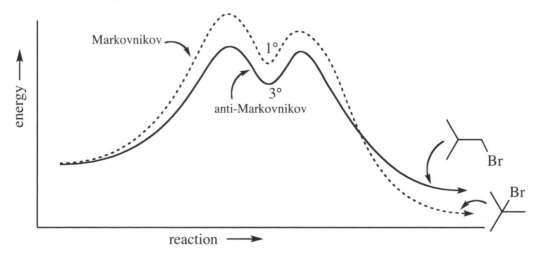

3° radical

(B)  Markovnikov

1° radical

The first step in the mechanism is endothermic and rate determining.  The 3° radical produced in anti-Markovnikov attack (A) of bromine radical is several kJ/mole more stable than the 1° radical generated by Markovnikov attack (B).  The Hammond Postulate tells us that it is reasonable to assume that the activation energy for anti-Markovnikov addition is lower than for Markovnikov addition.  This defines the first half of the energy diagram.

The relative stabilities of the final products are somewhat difficult to predict.  (Remember that stability of final products does not necessarily reflect relative stabilities of intermediates; this is why a thermodynamic product can be different from a kinetic product.)  From bond dissociation energies (kJ/mole) in Table 4-2:

| anti-Markovnikov | | Markovnikov | |
|---|---|---|---|
| H to 3° C | 381 | H to 1° C | 410 |
| Br to 1° C | 285 | Br to 3° C | 272 |
| | 666 kJ/mole | | 682 kJ/mole |

If it takes more energy to break bonds in the Markovnikov product, it must be lower in energy, therefore, more stable—OPPOSITE OF STABILITY OF THE INTERMEDIATES!  Now we are ready to construct the energy diagram.

It is the anti-Markovnikov product that is the kinetic product, not the thermodynamic product; the anti-Markovnikov product is obtained since its rate-determining step has the lower activation energy.

**8-58**  Recall these facts about ozonolysis:  each alkene cleaved by ozone produces two carbonyl groups; an alkene in a chain produces two separate products; an alkene in a ring produces one product in which the two carbonyls are connected.

(a)

+ $CH_2=O$

(b)

+

(c)

=

(d)

**8-59**  Chiral products in this problem are racemic.

(a)

$$\xrightarrow[\text{H}_2\text{O}]{\text{CH}_3\text{CO}_3\text{H}}$$

(b)

$$\xrightarrow[\text{H}_2\text{O}_2]{\text{OsO}_4}$$

(or cold, dilute $KMnO_4$)

*cis* + **syn**
(*cis* double bond plus
**syn** stereochemistry of addition)

$$\xleftarrow[\text{H}_2\text{O}]{\text{CH}_3\text{CO}_3\text{H}}$$

*trans* + **anti**
(*trans* double bond plus
**anti** stereochemistry of addition)

(c)

$$\xrightarrow{\text{Br}_2}$$

**Anti** addition of Br$_2$ *requires trans* alkene to give meso product.

This structure shows
a *trans* alkene in a
10-membered ring,
just the rotated view
of the structure to
the right.

$$\xrightarrow{\text{Br}_2}$$

rotate around C-2

*trans*-cyclodecene

(d)

$$\xrightarrow[\text{H}_2\text{O}]{\text{Cl}_2}$$

(e)

$$\xrightarrow[\text{HO}^-]{\text{BH}_3 \cdot \text{THF} \quad \text{H}_2\text{O}_2}$$

(f)

   or   or

$$\xrightarrow[\text{CH}_3\text{OH}]{\text{Hg(OAc)}_2 \quad \text{NaBH}_4}$$

**190**
Copyright © 2013 Pearson Education, Inc.

8-60

A) Unknown X, $C_5H_9Br$, has one element of unsaturation. **X** reacts with neither bromine nor $KMnO_4$, so the unsaturation in **X** cannot be an alkene; it must be a ring.

B) Upon treatment with strong base (*tert*-butoxide), **X** loses H and Br to give **Y**, $C_5H_8$, which does react with bromine and $KMnO_4$; it must have an alkene and a ring. Only one isomer is formed.

C) Catalytic hydrogenation of **Y** gives methylcyclobutane. This is a BIG clue because it gives the carbon skeleton of the unknown. **Y** must have a double bond in the methylcyclobutane skeleton, and **X** must have a Br on the methylcyclobutane skeleton.

D) Ozonolysis of **Y** gives a dialdehyde **Z**, $C_5H_8O_2$, which contains all the original carbons, so the alkene cleaved in the ozonolysis had to be in the ring.

Let's consider the possible answers for **X** and see if each fits the information.

1)
if this is **X**  →  KO-*t*-Bu  →  this could be **Y**; only one product  →  ozonolysis  →  [product] + $O=CH_2$  DOES NOT FIT OZONOLYSIS RESULTS so this cannot be **X** and **Y**.

2)
if this is **X**  →  KO-*t*-Bu  →  major + minor  **Y** would be a mixture of alkenes, but the elimination gives only one product. We already saw in example 1 that the exocyclic double bond does not fit the ozonolysis results so this structure cannot be **X**.

3)
if this is **X**  →  KO-*t*-Bu  →  major + minor  **Y** would be a mixture of alkenes, but the elimination gives only one product, so this structure of **X** is not consistent with the information provided.

4)
if this is **X**  →  KO-*t*-Bu  →  SAME COMPOUND; this could be **Y**; only one product  →  ozonolysis  →  **Z**, a dialdehyde

The correct structures for **X**, **Y**, and **Z** are given in the fourth possibility. The only structural feature of **X** that remains undetermined is whether it is the *cis* or *trans* isomer.

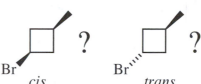

cis                trans

8-61 The clue to the structure of α-pinene is the ozonolysis. Working backwards shows the alkene position.

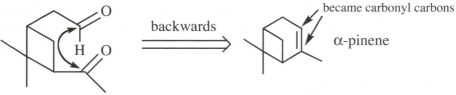

backwards                became carbonyl carbons

α-pinene

continued on next page

## 8-61 continued

After ozonolysis, the two carbonyls are still connected; the alkene must have been in a ring, so reconnect the two carbonyl carbons with a double bond.

Bredt's Rule: no double bond at bridgehead

**8-62** The two products from permanganate oxidation must have been connected by a double bond at the carbonyl carbons. Whether the alkene was $E$ or $Z$ cannot be determined by this experiment.

$CH_3(CH_2)_{12}CH=CH(CH_2)_7CH_3$

shown here as $Z$, the naturally occurring isomer

## 8-63

must have three alkenes in this skeleton

There are several ways to attack a problem like this. One is the trial-and-error method, that is, put double bonds in all possible positions until the ozonolysis products match. There are times when the trial-and-error method is useful (as in simple problems where the number of possibilities is few), but this is not one of them.

Let's try logic. Analyze the ozonolysis products carefully—what do you see? There are only two methyl groups, so one of the three terminal carbons in the skeleton (C-8, C-9, or C-10) has to be a $=CH_2$. Do we know which terminal carbon has the double bond? Yes, we can deduce that. If C-10 were double-bonded to C-4, then after ozonolysis, C-8 and C-9 must still be attached to C-7. However, in the ozonolysis products, there is no branched chain, that is, no combination of C-8 + C-9 + C-7 + C-1. What if C-7 had a double bond to C-1? Then we would have acetone, $CH_3COCH_3$, as an ozonolysis product—we don't. Thus, we can't have a double bond from C-4 to C-10. One of the other terminal carbons (C-8 or C-9) must have a double bond to C-7.

continued on next page

The other two double bonds have to be in the ring, but where? The products do not have branched chains, so double bonds must appear at both C-1 and C-4. There are only two possibilities for this requirement.

Ozonolysis of **I** would give fragments containing one carbon, two carbons, and seven carbons. Ozonolysis of **II** would give fragments containing one carbon, four carbons, and five carbons. Aha! Our mystery structure must be **II**.

(Editorial comment: Science is more than a collection of facts. The application of observation and logic to solve problems by *deduction* and *inference* is a critical scientific skill, one that distinguishes humans from algae.)

**8-64** In this type of problem, begin by determining which bonds are broken and which are formed. These will always give clues as to what is happening.

formed

broken

$H^+$ goes to most electronegative atom.

Protonated epoxide opens to give the most stable carbocation (3°).

3° carbocation looks for electrons, finds them at nearby alkene, forming a 6-membered ring (yes!)—leaves a 3° carbocation.

**8-65** See the solution to Problem 8-35 for simplified examples of these reactions.

(a) *trans* + **syn** → racemic

(b) *trans* + **anti** → meso

8-65 continued

(c)

$cis$ + **anti** → racemic

(d)

$cis$ + **syn** → meso

8-66

8-67 By now, these rearrangements should not be so "unexpected".

Alkyl migration with ring expansion gives 3° carbocation in 6-membered ring—carbocation nirvana!

You must be asking yourself, "Why didn't the methyl group migrate?" To which you answered by drawing the carbocation that would have been formed:

The new carbocation is indeed 3°, but it is only in a 5-membered ring, not quite as stable as in a 6-membered ring. In all probability, some of the product from methyl migration would be formed, but the 6-membered ring would be the major product.

**8-68**   Each alkene will produce two carbonyls upon ozonolysis or permanganate oxidation.  Oxidation of the unknown generated four carbonyls, so the unknown must have had two alkenes.  There is only one possibility for their positions.

**8-69**

(a)  Fumarase catalyzes the addition of H and OH, a hydration reaction.
(b)  Fumaric acid is planar and cannot be chiral.  Malic acid does have a chiral center and is chiral.  The enzyme-catalyzed reaction produces only the *S* enantiomer, so the product must be optically active.
(c)  One of the fundamental rules of stereochemistry is that optically inactive starting materials produce optically inactive products.  Sulfuric-acid-catalyzed hydration would produce a racemic mixture of malic acid, that is, equal amounts of *R* and *S*.
(d)  If the product is optically active, then either the starting materials or the catalyst were chiral.  We know that water and fumaric acid are not chiral, so we must infer that fumarase is chiral.
(e)  The D and the OD are on the "same side" of the Fischer projection (sometimes called the "erythro" stereoisomer).  These are produced from either:  (1) syn addition to *cis* alkenes, or (2) anti addition to *trans* alkenes.  We know that fumaric acid is *trans*, so the addition of D and OD must necessarily be anti.
(f)  Hydroboration is a syn addition.

As expected, *trans* alkene plus syn addition puts the two groups on the "opposite" side of the Fischer projection (sometimes called "threo").

**8-70**

(a)

Copyright © 2013 Pearson Education, Inc.

(b)

bromonium ion

$C_7H_{13}BrO$

8-71  The addition of $BH_3$ to an alkene is reversible.  Given heat and time, the borane will eventually "walk" its way to the end of the chain through a series of addition-elimination cycles.  The most stable alkylborane has the boron on the end carbon; eventually, the series of equilibria leads to the ultimate borane product that is oxidized to the primary alcohol.

dec-5-ene

+ $BH_3$

+ $BH_3$

+ $BH_3$

most stable

$H_2O_2$, $HO^-$

decan-1-ol

8-72 First, we explain *how* the mixture of stereoisomers results, then *why*.

We have seen many times that the bridged halonium ion permits attack of the nucleophile only from the opposite side.

expected:

*trans* only
(plus the enantiomer)

A mixture of *cis* and *trans* could result only if attack of chloride were possible from both top and bottom, something possible only if a *carbocation* existed at this carbon.

actual:

*trans*          *cis*

This picture of the p orbitals of benzene shows resonance overlap with the p orbital of the carbocation. The chloride nucleophile can form a bond to the positive carbon from either the top or the bottom.

*Why* does a carbocation exist here? Not only is it 3°, *it is also next to a benzene ring (benzylic) and therefore resonance-stabilized.* This resonance stabilization would be forfeited in a halonium ion intermediate.

plus the enantiomers of each

Refer to Appendix 3 in this manual for a format to organize reactions for studying.

9-1 Each pi bond = 1 element of unsaturation; each ring = 1 e.u.
(a) parsalmide: 4 double bonds + 1 triple bond + 1 ring = **7** unsaturations
ethynyl estradiol: 3 double bonds + 1 triple bond + 4 rings = **9** unsaturations
dynemicin A: 11 double bonds + 2 triple bonds + 7 rings = **22** unsaturations

(b) Many other structures are possible in each case; the number of unsaturations must be consistent.

(1) $C_6H_{10}$ ⟹ **2**  (2) $C_8H_{12}$ ⟹ **3**  (3) $C_7H_8$ ⟹ **4**

$CH_3CH_2C\equiv CCH_2CH_3$

$CH_3CH_2CH_2CH_2C\equiv CH$

$HC\equiv C$—⟨cyclohexane⟩

$HC\equiv C$—$CH_2CH=CHCH_2CH_2CH_3$

$HC\equiv C$—⟨cyclopentene⟩

$HC\equiv C$—$C\equiv C$—$CH_2CH_2CH_3$

9-2 New IUPAC names are given. The asterisk (*) denotes acetylenic hydrogens of terminal alkynes.

(a)  $CH_3CH_2CH_2$—$C\equiv C$—$\overset{*}{H}$   $CH_3CH_2$—$C\equiv C$—$CH_3$   $(CH_3)_2CH$—$C\equiv C$—$\overset{*}{H}$
     pent-1-yne                     pent-2-yne              3-methylbut-1-yne

(b) Do these isomer problems systematically: draw the 6-carbon chains first, then the 5-carbon chains, *etc*.

$\overset{*}{H}C\equiv C$—⟨chain⟩
hex-1-yne

$H_3C$-$C\equiv C$—⟨chain⟩
hex-2-yne

⟨chain⟩—$C\equiv C$—⟨chain⟩
hex-3-yne

$\overset{*}{H}C\equiv C$—⟨chain⟩
3-methylpent-1-yne

$H_3C$-$C\equiv C$—⟨chain⟩
4-methylpent-2-yne

$\overset{*}{H}C\equiv C$—⟨chain⟩
4-methylpent-1-yne

$\overset{*}{H}C\equiv C$—⟨chain⟩
3,3-dimethylbut-1-yne

9-3 Acetylene would likely decompose into its elements. The decomposition reaction below is exothermic ($\Delta H° = -234$ kJ/mole) as well as having an increase in entropy. Thermodynamically, at 1500°C, an increase in entropy will have a large effect on $\Delta G$ (remember $\Delta G = \Delta H - T\Delta S$). Kinetically, almost any activation energy barrier will be overcome at 1500°C.

$$HC\equiv CH \xrightarrow{1500\ °C} 2\ C\ +\ H_2$$

9-4 Adding sodium amide to the mixture will produce the sodium salt of hex-1-yne, leaving hex-1-ene untouched. Distillation will remove the hex-1-ene, leaving the non-volatile acetylide salt behind.

$\left.\begin{array}{l} CH_2=CHCH_2CH_2CH_2CH_3 \\[2mm] H-C\equiv C-CH_2CH_2CH_2CH_3 \end{array}\right\}$ $\xrightarrow{NaNH_2}$ $\left\{\begin{array}{l} CH_2=CHCH_2CH_2CH_2CH_3 \\[2mm] Na^+\ {}^{\ominus}\!:\!C\equiv C-CH_2CH_2CH_2CH_3 \quad \text{non-volatile salt} \end{array}\right.$

9-5 The key to this problem is to understand that *a proton donor will react only with the conjugate base of a weaker acid*. See Appendix 2 at the end of this Solutions Manual for an in-depth discussion of acidity.

(a)  $H-C\equiv C-H$ + $NaNH_2$ ⟶ $H-C\equiv C\!:^{\ominus}$ $Na^+$ + $NH_3$

(b)  $H-C\equiv C-H$ + $CH_3Li$ ⟶ $H-C\equiv C\!:^{\ominus}$ $Li^+$ + $CH_4$

(c)  no reaction: $NaOCH_3$ is not a strong enough base

(d)  no reaction: $NaOH$ is not a strong enough base

(e)  $H-C\equiv C\!:^{\ominus}$ $Na^+$ + $CH_3OH$ ⟶ $H-C\equiv C-H$ + $NaOCH_3$  (opposite of (c))

(f)  $H-C\equiv C\!:^{\ominus}$ $Na^+$ + $H_2O$ ⟶ $H-C\equiv C-H$ + $NaOH$  (opposite of (d))

(g)  no reaction : $H-C\equiv C\!:^{\ominus}$ $Na^+$ is not a strong enough base

(h)  no reaction: $NaNH_2$ is not a strong enough base

(i)  $CH_3OH$ + $NaNH_2$ ⟶ $NaOCH_3$ + $NH_3$

**9-6**

$$H-C\equiv C-H \xrightarrow{NaNH_2} H-C\equiv C:^{\ominus} Na^+ \xrightarrow{CH_3CH_2Br} H-C\equiv C-CH_2CH_3$$

$$\downarrow NaNH_2$$

$$CH_3(CH_2)_5-C\equiv C-CH_2CH_3 \xleftarrow{CH_3(CH_2)_5Br} Na^+ \ ^{\ominus}:C\equiv C-CH_2CH_3$$

**9-7** To be a feasible synthesis, the desired product must have the sp carbons from acetylene bonded to $CH_2$ groups that came from 1° halides in the precursor.

(a) $H-C\equiv C-H \xrightarrow[\text{2) } CH_3CH_2CH_2CH_2Br]{\text{1) } NaNH_2} H-C\equiv C-CH_2CH_2CH_2CH_3$

(b) $H-C\equiv C-H \xrightarrow[\text{2) } CH_3CH_2CH_2Br]{\text{1) } NaNH_2} H-C\equiv C-CH_2CH_2CH_3 \xrightarrow[\text{2) } CH_3I]{\text{1) } NaNH_2} CH_3-C\equiv C-CH_2CH_2CH_3$

(c) $H-C\equiv C-H \xrightarrow[\text{2) } CH_3CH_2Br]{\text{1) } NaNH_2} H-C\equiv C-CH_2CH_3 \xrightarrow[\text{2) } CH_3CH_2Br]{\text{1) } NaNH_2} CH_3CH_2-C\equiv C-CH_2CH_3$

(d) The desired product cannot be synthesized by an $S_N2$ reaction—would require attack on a 2° alkyl halide.

$$CH_3-C\equiv C:^{\ominus} Na^+ \ + \ \underset{\underset{CH_3}{|}}{\overset{\overset{Br \nearrow 2°}{|}}{CHCH_2CH_3}} \ \times\!\!\!\longrightarrow \ CH_3-C\equiv C-\underset{\underset{CH_3}{|}}{CHCH_2CH_3}$$

strong nucleophile; must be second order

low yields; not practical

(e) $H-C\equiv C-H \xrightarrow[\text{2) } CH_3I]{\text{1) } NaNH_2} CH_3-C\equiv C-H \xrightarrow[\text{2) } BrCH_2\underset{\underset{CH_3}{|}}{CHCH_3}]{\text{1) } NaNH_2} CH_3-C\equiv C-CH_2\underset{\underset{CH_3}{|}}{CHCH_3}$

(f) $H-C\equiv C-H \xrightarrow[\text{2) } Br(CH_2)_8Br]{\text{1) } NaNH_2}$

$H-C\equiv C-CH_2$ (chain with Br)

$$\downarrow NaNH_2$$

$Na^+ \ ^{\ominus}:C\equiv C-CH_2$ (chain with Br)

$H_2C-C\equiv C-CH_2$ (ring) $\longleftarrow$

Intramolecular cyclization of large rings must be carried out in dilute solution so the last $S_N2$ displacement will be *intra*molecular (within one molecule) and not *inter*molecular (between two molecules).

**9-8**

(a) $H-C\equiv C-H \xrightarrow[\text{2) } H_2C=O]{\text{1) } NaNH_2} \xrightarrow{H_2O} H-C\equiv C-CH_2OH$

(b) $H-C\equiv C-H \xrightarrow[\text{2) } CH_3I]{\text{1) } NaNH_2} CH_3-C\equiv C-H \xrightarrow[\underset{O}{\overset{||}{\underset{\text{3) } H_2O}{\text{2) } HCCH_2CH_2CH_3}}}]{\text{1) } NaNH_2} CH_3-C\equiv C-\overset{\overset{OH}{|}}{CH}CH_2CH_2CH_3$

**199**

9-8 continued

(c) $H-C\equiv C-H$ $\xrightarrow[\text{2) Ph} \overset{O}{\underset{}{\underset{}{\parallel}} CH_3]{\text{1) NaNH}_2}$ $\xrightarrow{H_2O}$ $H_3C-\overset{\overset{OH}{|}}{\underset{\underset{Ph}{|}}{C}}-C\equiv C-H$

(d) $H-C\equiv C-H$ $\xrightarrow[\text{2) CH}_3I]{\text{1) NaNH}_2}$ $CH_3-C\equiv C-H$ $\xrightarrow[\text{2)} \overset{O}{\parallel}]{\text{1) NaNH}_2}$ $\xrightarrow{H_2O}$ $CH_3-C\equiv C-\overset{\overset{OH}{|}}{\underset{\underset{CH_3}{|}}{C}}CH_2CH_3$

9-9  In a synthesis problem, draw the target to see what new bonds need to be made during the synthesis.

$H-C\equiv C-CH_2CH_3$ $\xrightarrow[\text{2) Ph} \overset{O}{\parallel} CH_3]{\text{1) NaNH}_2}$ $\xrightarrow{H_2O}$ $H_3C-\overset{\overset{OH}{|}}{\underset{\underset{Ph}{|}}{\underset{2}{C}}}\underset{3}{C}\equiv\underset{4}{C}-\underset{5}{C}H_2\underset{6}{C}H_3$   2-phenylhex-3-yn-2-ol

⋯ new C—C bond

9-10  The mechanism is likely to be two sequential E2 reactions.  KOH usually makes internal alkynes.

1-phenylprop-1-yne

9-11  The mechanism is likely to be two sequential E2 reactions.  Where possible, $NaNH_2$ makes terminal alkynes because the terminal proton is removed by $NaNH_2$ giving a stable acetylide ion as the initial product. (This is called a *thermodynamic sink*, or a *potential energy well*, that is, a valley on the energy diagram.)

3-phenylprop-1-yne

This acetylide ion remains in the reaction mixture until a proton source is added.

9-12

(a)  $CH_3CH_2-C\equiv C-CH_2CH_2CH_2CH_3$ $\xrightarrow[\text{Lindlar's catalyst}]{H_2}$ 

(b)  $H_3C-C\equiv C-CH_2CH_3$ $\xrightarrow[\text{NH}_3]{\text{Na}}$

9-12 continued

(c)

One of the useful "tricks" of organic chemistry is the ability to convert one stereoisomer into another. Having a pair of reactions like the two reductions shown in parts (a) and (b) that give opposite stereochemistry is very useful, because a common intermediate (the alkyne) can be transformed into either stereoisomer.

(d)  $CH_3CH_2-C\equiv C-H$ $\xrightarrow[\text{2) CH}_3\text{CH}_2\text{Br}]{\text{1) NaNH}_2}$ $CH_3CH_2-C\equiv C-CH_2CH_3$ $\xrightarrow[\substack{\text{Lindlar's}\\\text{catalyst}}]{\text{H}_2}$ 

9-13

$CH_3CH_2CH_2-C\equiv C-H$ $\xrightarrow[\substack{\text{2)}\\\text{3) H}_2\text{O}}]{\text{1) NaNH}_2}$ $CH_3CH_2CH_2-C\equiv C-\underset{\underset{Ph}{|}}{\overset{\overset{OH}{|}}{CH}}$ $\xrightarrow[\substack{\text{Lindlar's}\\\text{catalyst}}]{\text{H}_2}$ 

(Z)-1-phenylhex-2-en-1-ol

9-14   The goal is to add only one equivalent of bromine, always avoiding an excess of bromine, because two molecules of bromine could add to the triple bond if bromine were in excess. If the alkyne is added to the bromine, the first drops of alkyne will encounter a large excess of bromine. Instead, adding bromine to the alkyne will always ensure an excess of alkyne and should give a good yield of dibromo product.

9-15

2° better than
1° carbocation

2° carbocation and
resonance-stabilized

## 9-16

(a) PhC≡CH + 2 HBr → Ph–C(Br)(Br)–CH₃

$$\text{PhC}\equiv\text{CH} + 2\ \text{HBr} \longrightarrow \text{Ph}-\overset{\overset{\displaystyle Br}{|}}{\underset{\underset{\displaystyle Br}{|}}{C}}-CH_3$$

(b) 
$$\text{CH}_3\text{CH}_2\text{CH}_2-\text{C}\equiv\text{CH} + 2\ \text{HCl} \longrightarrow \text{CH}_3\text{CH}_2\text{CH}_2-\overset{\overset{\displaystyle Cl}{|}}{\underset{\underset{\displaystyle Cl}{|}}{C}}-CH_3$$

Electrophilic addition to terminal alkynes follows Markovnikov orientation.

(c) cyclooctyne + 2 HBr → 1,1-dibromocyclooctane

$$\text{(d)}\quad \overset{3\quad 2\quad 1}{\text{C}\equiv\text{C}-\text{CH}_3} + 2\ \text{HCl} \longrightarrow \underset{\underset{\displaystyle Cl\ \ Cl}{}}{\overset{2\quad 1}{3}} + \underset{\underset{\displaystyle }{}}{\overset{Cl\ \ Cl}{\underset{3}{\overset{2\ \ 1}{}}}}$$

Electrophilic addition to unsymmetric internal alkynes will give mixtures of isomers.

## 9-17

initiation:

$$\text{RO}-\text{OR} \xrightarrow{h\nu} 2\ \text{RO}\bullet$$

$$\text{RO}\bullet + \text{H}-\text{Br} \longrightarrow \text{RO}-\text{H} + \text{Br}\bullet$$

propagation:

$$\text{Br}\bullet + \text{H}-\text{C}\equiv\text{C}-\text{CH}_2\text{CH}_2\text{CH}_3 \longrightarrow \text{H}-\text{C}=\overset{\overset{\displaystyle \bullet}{}}{\underset{\underset{\displaystyle Br}{|}}{C}}-\text{CH}_2\text{CH}_2\text{CH}_3$$

$$\text{H}-\overset{}{\text{C}}=\overset{\bullet}{\text{C}}-\text{CH}_2\text{CHCH}_3 + \text{H}-\text{Br} \longrightarrow \text{H}-\text{C}=\text{C}-\text{CH}_2\text{CH}_2\text{CH}_3 + \text{Br}\bullet$$
(with Br on first C, Br H on second C)

The 2° radical is more stable than 1°. The anti-Markovnikov orientation occurs because the bromine radical attacks first to make the most stable radical, in contrast to electrophilic addition where the H⁺ bonds first (see the solution to 9-15).

## 9-18

(a) $\text{H}-\text{C}\equiv\text{C}-(\text{CH}_2)_3\text{CH}_3 \xrightarrow{\text{1 eq. Cl}_2} \text{H}-\overset{}{\text{C}}=\overset{}{\text{C}}-(\text{CH}_2)_3\text{CH}_3$ (Cl, Cl)  $E + Z$

(b) $\text{H}-\text{C}\equiv\text{C}-(\text{CH}_2)_3\text{CH}_3 \xrightarrow[\text{ROOR}]{\text{HBr}} \text{H}-\overset{}{\text{C}}=\overset{}{\text{C}}-(\text{CH}_2)_3\text{CH}_3$ (Br, H)  $E + Z$

(c) $\text{H}-\text{C}\equiv\text{C}-(\text{CH}_2)_3\text{CH}_3 \xrightarrow{\text{HBr}} \text{H}-\overset{}{\text{C}}=\overset{}{\text{C}}-(\text{CH}_2)_3\text{CH}_3$ (H, Br)

(d) $\text{H}-\text{C}\equiv\text{C}-(\text{CH}_2)_3\text{CH}_3 \xrightarrow[\text{CCl}_4]{\text{2 Br}_2} \text{H}-\overset{\overset{\displaystyle Br}{|}}{\underset{\underset{\displaystyle Br}{|}}{C}}-\overset{\overset{\displaystyle Br}{|}}{\underset{\underset{\displaystyle Br}{|}}{C}}-(\text{CH}_2)_3\text{CH}_3$

9-18 continued

(e) $H-C\equiv C-(CH_2)_3CH_3$ $\xrightarrow[\substack{\text{Lindlar's}\\ \text{catalyst}\\ \text{(or Na, NH}_3\text{)}}]{H_2}$ $\underset{\substack{|\\H}}{\overset{}{H}}-\underset{\substack{|\\H}}{\overset{}{C}}=C-(CH_2)_3CH_3$ $\xrightarrow{HBr}$ $CH_3-\underset{\substack{|\\Br}}{CH}-(CH_2)_3CH_3$

(f) $H-C\equiv C-(CH_2)_3CH_3$ $\xrightarrow{\textbf{2 HBr}}$ $H-\underset{\substack{|\\H}}{\overset{\substack{H\\|}}{C}}-\underset{\substack{|\\Br}}{\overset{\substack{Br\\|}}{C}}-(CH_2)_3CH_3$

9-19 Dilute $H_2SO_4$ (aq) is shown here as $H_3O^+$.

The role of the mercury catalyst is not shown in this mechanism. As a Lewis acid, it may act like the proton in the first step, helping to form vinyl cations; the mercury is replaced when acid is added. See further explanation on the next page.

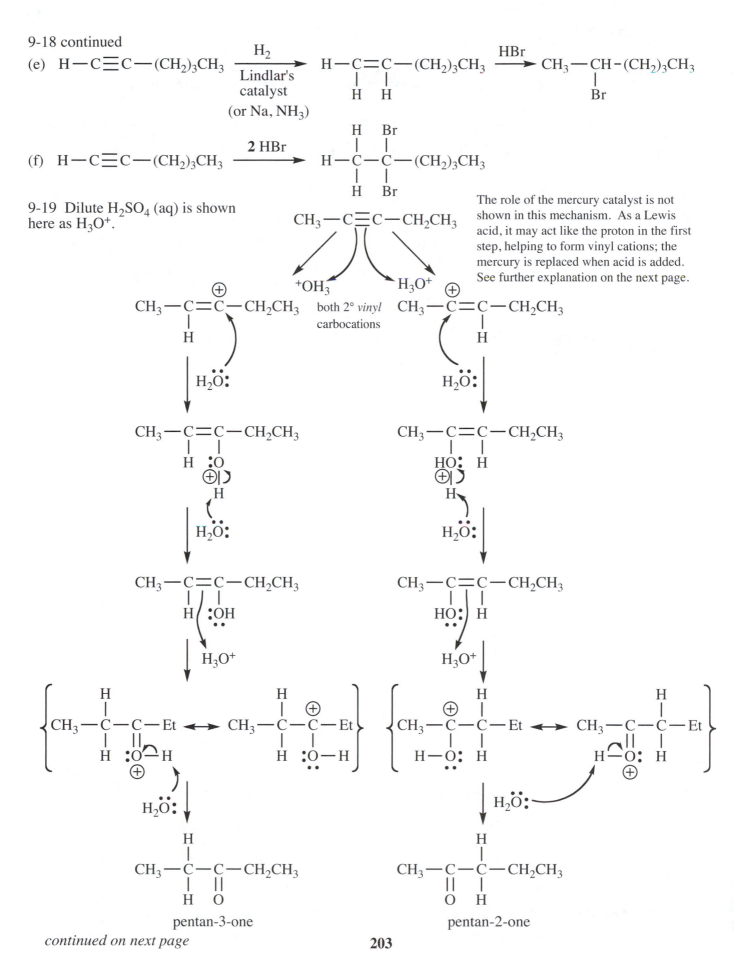

pentan-3-one                    pentan-2-one

*continued on next page*

**203**

**9-19 continued**

The role of the mercury catalyst is not shown in this mechanism. As a Lewis acid, it may act like the proton in the first step, helping to form vinyl cations; the mercury is replaced when acid is added.

$$CH_3-\overset{\oplus}{C}=C-CH_2CH_3 \quad\xrightarrow[\substack{H_2O \\ \text{removes} \\ H^+}]{H_2O \quad -H^+}\quad CH_3-\underset{\underset{\oplus Hg}{|}}{C}=\underset{\underset{OH}{|}}{C}-CH_2CH_3 \quad\xrightarrow{H_3O^+}\quad CH_3-\underset{\underset{H}{|}}{C}=\underset{\underset{OH}{|}}{C}-CH_2CH_3$$

**9-20**

(a) But-2-yne is symmetric. Either orientation produces the same product.

$$CH_3-C\equiv C-CH_3 \xrightarrow{Sia_2BH} \overset{CH_3\quad CH_3}{\underset{H\quad\; BSia_2}{C=C}} \xrightarrow[HO^-]{H_2O_2} \overset{CH_3\quad CH_3}{\underset{H\quad\; OH}{C=C}} \xrightarrow{HO^-} CH_3CH_2-\overset{O}{\overset{||}{C}}-CH_3$$

(b) Pent-2-yne is not symmetric. Different orientations of attack will lead to different products on any unsymmetrical internal alkyne.

$$CH_3-C\equiv C-CH_2CH_3$$

Sia$_2$BH (left)    Sia$_2$BH (right)

Left path:

$$\overset{CH_3\quad CH_2CH_3}{\underset{H\quad\;\; BSia_2}{C=C}} \xrightarrow{H_2O_2,\ HO^-} \overset{CH_3\quad CH_2CH_3}{\underset{H\quad\;\; OH}{C=C}} \xrightarrow{HO^-} \underset{CH_3CH_2-\overset{O}{\overset{||}{C}}-CH_2CH_3}{}$$

Right path:

$$\overset{CH_3\quad CH_2CH_3}{\underset{Sia_2B\quad\;\; H}{C=C}} \xrightarrow{H_2O_2,\ HO^-} \overset{CH_3\quad CH_2CH_3}{\underset{HO\quad\;\; H}{C=C}} \xrightarrow{HO^-} CH_3-\overset{O}{\overset{||}{C}}-CH_2CH_2CH_3$$

**9-21**

(a) (1) ![ketone structure]    (2) H—[aldehyde structure]

(b) (1) [ketone structure] + [ketone structure]

    (2) same mixture as in (b) (1)

(c) (1) [ketone structure]

    (2) same as in (c) (1)

(d) (1) [cyclic ketone structure]

    (2) same as in (d) (1)

9-22

(a)

disiamylborane, Sia$_2$BH

(b) There is too much steric hindrance in Sia$_2$BH for the third B—H to add across another alkene. The reagent can add to alkynes because alkynes are linear and attack is not hindered by bulky substituents.

9-23

(a) (1)
$$HO-\overset{\overset{O}{\|}}{C}-\overset{\overset{O}{\|}}{C}-(CH_2)_3CH_3$$
Oxidation of a terminal alkyne with neutral KMnO$_4$ produces the ketone and carboxylic acid without cleaving the carbon-carbon bond.

(2)
$$CO_2 + HO-\overset{\overset{O}{\|}}{C}-(CH_2)_3CH_3$$
Oxidation of a terminal alkyne with warm, basic KMnO$_4$ cleaves the carbon-carbon bond, producing the carboxylic acid after acid work-up, and carbon dioxide.

(b) (1)
$$CH_3-\overset{\overset{O}{\|}}{C}-\overset{\overset{O}{\|}}{C}-CH_2CH_2CH_3$$

(2)
$$CH_3-\overset{\overset{O}{\|}}{C}-OH + HO-\overset{\overset{O}{\|}}{C}-CH_2CH_2CH_3$$

(c) (1)
$$CH_3CH_2-\overset{\overset{O}{\|}}{C}-\overset{\overset{O}{\|}}{C}-CH_2CH_3$$

(2)
$$CH_3CH_2-\overset{\overset{O}{\|}}{C}-OH \quad (2 \text{ equivalents})$$

(d) (1)
$$CH_3\underset{\underset{CH_3}{|}}{CH}-\overset{\overset{O}{\|}}{C}-\overset{\overset{O}{\|}}{C}-CH_2CH_3$$

(2)
$$CH_3\underset{\underset{CH_3}{|}}{CH}-\overset{\overset{O}{\|}}{C}-OH + HO-\overset{\overset{O}{\|}}{C}-CH_2CH_3$$

(e) (1)

(2)

9-24

(a) $CH_3-C\equiv C-(CH_2)_4-C\equiv C-CH_3$

(b)

9-25 When proposing syntheses, begin by analyzing the target molecule, looking for smaller pieces that can be combined to make the desired compound. This is especially true for targets that have more carbons than the starting materials; immediately, you will know that a carbon-carbon bond forming reaction will be necessary.

People who succeed at synthesis *know the reactions*—there is no shortcut. Practice the reactions for each functional group until they become automatic. See Appendix 3 in this Manual for a suggestion on how to organize reactions.

(a) analysis of target

from acetylene

3° acetylenic alcohols made from acetylide plus ketones

from alkylation of acetylide

forward direction:

Put on less reactive group first.

$$H-C\equiv C-H \; + \; NaNH_2 \longrightarrow H-C\equiv C:^{-} \; Na^{+} \xrightarrow{\quad Br\diagup\diagdown\diagup \quad} H-C\equiv C-\diagup\diagdown\diagup$$

$$\downarrow NaNH_2$$

$$\xleftarrow{H_2O} \xleftarrow{\quad} Na^{+} \; :^{-}C\equiv C-\diagup\diagdown\diagup$$

(left product: 3° acetylenic alcohol with OH group)

(b)

Analysis of target: cyclopropanes are made by carbene insertion into alkenes. To get *cis* substitution around cyclopropane, stereochemistry of the alkene precursor must be *cis*. *Cis* alkenes come from catalytic hydrogenation of an alkyne using Lindlar's catalyst.

$$H-C\equiv C-H \xrightarrow{NaNH_2 \;\; CH_3I} H_3C-C\equiv C-H \xrightarrow{NaNH_2 \;\; CH_3CH_2Br} H_3C-C\equiv C-CH_2CH_3$$

$$\downarrow H_2 \;\; \text{Lindlar's catalyst}$$

cyclopropane product $\xleftarrow{CH_2I_2 \;\; Zn, CuCl}$ cis-alkene ($H_3C$ / $CH_2CH_3$)

(c)

Analysis of target: epoxides are made by direct epoxidation of alkenes. To get *trans* substitution around the epoxide, stereochemistry of the alkene precursor must be *trans*. *Trans* alkenes come from sodium/ammonia reduction of an alkyne.

$$H-C\equiv C-H \xrightarrow{NaNH_2 \;\; CH_3CH_2Br} CH_3CH_2-C\equiv C-H \xrightarrow{NaNH_2 \;\; CH_3CH_2CH_2Br}$$

$$CH_3CH_2-C\equiv C-CH_2CH_2CH_3$$

epoxide product $\xleftarrow{MCPBA}$ trans-alkene $\xleftarrow{Na \; / \; NH_3}$ $CH_3CH_2-C\equiv C-CH_2CH_2CH_3$

9-25 continued
(d) *meso*-hexane-3,4-diol

Analysis of target: diols are prepared from the C=C by two methods: cold, dilute $KMnO_4$ (or $OsO_4$), a syn addition; or with a peroxy acid that makes the epoxide, followed by hydrolysis, an anti addition of the two OH groups. Recall from the solution to 8-35 that in order to produce a meso structure, a syn addition requires a *cis* alkene, and an anti addition requires a *trans* alkene.

Both syntheses begin with the preparation of hex-3-yne:

$$H-C\equiv C-H \xrightarrow{NaNH_2} \xrightarrow{CH_3CH_2Br} CH_3CH_2-C\equiv CH \xrightarrow{NaNH_2} \xrightarrow{CH_3CH_2Br} CH_3CH_2-C\equiv C-CH_2CH_3$$

Route A: via an epoxide on a *trans* alkene

$$CH_3CH_2-C\equiv C-CH_2CH_3 \xrightarrow[NH_3]{Na} \begin{array}{c} H \quad CH_2CH_3 \\ \diagup \diagdown \\ CH_3CH_2 \quad H \end{array} \xrightarrow{MCPBA} \begin{array}{c} H \quad CH_2CH_3 \\ \diagup \diagdown \\ CH_3CH_2 \quad O \quad H \end{array} \xrightarrow{H_3O^+} \text{meso product}$$

Route B: via *syn* hydroxylation on a *cis* alkene

$$CH_3CH_2-C\equiv C-CH_2CH_3 \xrightarrow[\substack{Lindlar's \\ catalyst}]{H_2} \begin{array}{c} H \quad H \\ \diagup \diagdown \\ CH_3CH_2 \quad CH_2CH_3 \end{array} \xrightarrow[\substack{or\ OsO_4,\ H_2O_2}]{cold,\ dilute\ KMnO_4} \text{meso product}$$

9-26
(a) [cyclooctyne] $\xrightarrow[\substack{Lindlar's \\ catalyst}]{H_2}$ [cyclooctene]

(b) [cyclooctyne] $\xrightarrow[\substack{Pt\ or\ Pd\ or \\ Ni\ catalyst}]{H_2}$ [cyclooctane]

(c) [cyclooctene] $\xrightarrow[\substack{anti \\ addition}]{Br_2}$ [dibromocyclooctane, Br wedge, Br dash]
from (a)     two Br *trans*

(d) [cyclooctyne] $\xrightarrow[HgSO_4]{H_2O,\ H_2SO_4}$ [cyclooctanone]
OR 1) $Sia_2BH$, 2) $H_2O_2$, $HO^-$

(e) [cyclooctyne] $\xrightarrow{2\ HBr}$ [2,2-dibromocyclooctane]

(f) [cyclooctene] $\xrightarrow[OR\ Br_2,\ hv]{NBS,\ hv}$ [bromocyclooctene] allylic bromination
from (a)

(g) [cyclooctyne] $\xrightarrow[\substack{H_2O \\ neutral}]{KMnO_4}$ [cyclooctanedione]

(h) [cyclooctene with H's] $\xrightarrow[2)\ Me_2S]{1)\ O_3}$ [dialdehyde]
from (a)

(i) [cyclooctyne] $\xrightarrow[2)\ H^+]{1)\ KMnO_4,\ NaOH}$ OR $\xrightarrow[2)\ H_2O]{1)\ O_3}$ [HO-C(=O)-...-C(=O)-OH with ketone]

**207**

**9-27**

(a) $CH_3-C\equiv C-(CH_2)_4CH_3$    (b) $CH_3CH_2-C\equiv C-CH_2CH_2CH(CH_3)_2$    (c)

(d)    (e) $CH_3CH_2-C\equiv C-\underset{\underset{\displaystyle CH_3}{|}}{C}HCH_2CH_2CH_3$

(f)    (g)    (h)

(i) $H-C\equiv C-CH_2-C\equiv C-CH_2CH_3$    (j) $H-C\equiv C-CH=CH_2$    (k)

**9-28**

(a) ethylmethylacetylene    (b) phenylacetylene    (c) *sec*-butylpropylacetylene
(d) *sec*-butyl-*tert*-butylacetylene

**9-29**

(a) 4-phenylpent-2-yne    (b) (*E*)-3-methylhept-2-en-4-yne    (c) 2,2,5-trimethylhept-3-yne
(d) 4,4-dibromopent-2-yne    (e) 3-methylhex-4-yn-3-ol    (f) 1-cyclopentylbut-2-yne

**9-30**

(a)

(b)

(c)

(d)

(e)

(f)

(g)

(h)

(i)

undec-6-yn-5-ol

**9-31**

$CH_3CH_2-\overset{\displaystyle O}{\overset{\|}{C}}$ ... $:\overset{\ominus}{C}\equiv CH$ $\xrightarrow{H_2O}$ $CH_3CH_2-\overset{\displaystyle OH}{\underset{\underset{\displaystyle CHCl}{\overset{|}{HC}}}{\overset{|}{\underset{|}{C}}}-C\equiv CH$   ethchlorvynol

with $HC\equiv CH$ and $NaNH_2$

**9-32**

$H-C\equiv C-H$ $\xrightarrow{NaNH_2}$ $\xrightarrow{CH_3(CH_2)_7Br}$ $CH_3(CH_2)_7-C\equiv C-H$ $\xrightarrow{NaNH_2}$ $\xrightarrow{CH_3(CH_2)_{12}Br}$ $CH_3(CH_2)_7-C\equiv C-(CH_2)_{12}CH_3$

$\xrightarrow[\text{Lindlar's catalyst}]{H_2}$ 

$\underset{\underset{\displaystyle muscalure}{}}{\overset{\displaystyle CH_3(CH_2)_7 \quad (CH_2)_{12}CH_3}{\underset{\displaystyle H \qquad H}{C=C}}}$

**9-33**

(a) $CH_2=\overset{\overset{\displaystyle Cl}{|}}{C}CH_2CH_2CH_3$

(b) $H_3C-\overset{\overset{\displaystyle Cl}{|}}{\underset{\underset{\displaystyle Cl}{|}}{C}}-CH_2CH_2CH_3$

(c) $CH_3CH_2CH_2CH_2CH_3$

(d) $CH_2=CHCH_2CH_2CH_3$

(e) $\underset{\displaystyle Br \qquad CH_2CH_2CH_3}{\overset{\displaystyle H \qquad Br}{C=C}}$

(f) $H-\overset{\overset{\displaystyle Br}{|}}{\underset{\underset{\displaystyle Br}{|}}{C}}-\overset{\overset{\displaystyle Br}{|}}{\underset{\underset{\displaystyle Br}{|}}{C}}-CH_2CH_2CH_3$

(g) $HO-\overset{\overset{\displaystyle O}{\|}}{C}-\overset{\overset{\displaystyle O}{\|}}{C}-CH_2CH_2CH_3$

(h) $CO_2 + HO-\overset{\overset{\displaystyle O}{\|}}{C}-CH_2CH_2CH_3$

(i) $H_2C=CHCH_2CH_2CH_3$

(j) $Na^+ \;:\overset{\ominus}{C}\equiv C-CH_2CH_2CH_3$

(k) $H_3C-\overset{\overset{\displaystyle O}{\|}}{C}-CH_2CH_2CH_3$

(l) $H-\overset{\overset{\displaystyle O}{\|}}{C}-CH_2CH_2CH_2CH_3$

**9-34**

(a) $H_3C-\overset{\overset{\displaystyle Br}{|}}{\underset{\underset{\displaystyle Br}{|}}{C}}-CH_2CH_3$ $\xrightarrow[150°]{NaNH_2}$ $\xrightarrow{H_2O}$ $H-C\equiv C-CH_2CH_3$

(b) $H_3C-\overset{\overset{\displaystyle Br}{|}}{\underset{\underset{\displaystyle Br}{|}}{C}}-CH_2CH_3$ $\xrightarrow[200°]{KOH}$ $H_3C-C\equiv C-CH_3$

(c) $CH_3CH_2-C\equiv C-H$ $\xrightarrow{NaNH_2}$ $CH_3CH_2-C\equiv \overset{\ominus}{C}:$ $Na^+$

$\downarrow CH_3CH_2CH_2CH_2Br$

$CH_3CH_2-C\equiv C-CH_2CH_2CH_2CH_3$

(d) $\underset{\displaystyle H \qquad CH_2CH_2CH_3}{\overset{\displaystyle H_3C \qquad H}{C=C}}$ $\xrightarrow[CCl_4]{Br_2}$ $\underset{\displaystyle H \;\; H}{\overset{\displaystyle Br \;\; Br}{CH_3C-CCH_2CH_2CH_3}}$ $\xrightarrow[200°]{KOH}$ $H_3C-C\equiv C-CH_2CH_2CH_3$

(e) $H_3C-\overset{\overset{\displaystyle Br}{|}}{\underset{\underset{\displaystyle Br}{|}}{C}}-CH_2CH_2CH_2CH_3$ $\xrightarrow[\substack{150° \\ 2)\,H_2O}]{1)\,NaNH_2}$ $H-C\equiv C-CH_2CH_2CH_2CH_3$

**209**

**9-34 continued**

(f)

$\xrightarrow[\text{catalyst}]{\text{Lindlar's}}^{H_2}$ cis

(g)

$\xrightarrow[\text{NH}_3]{\text{Na}}$ trans

(h) $HC\equiv C-CH_2CH_2CH_2CH_3 \xrightarrow[\substack{H_2SO_4 \\ HgSO_4}]{H_2O} \left[ CH_2=\overset{\overset{\displaystyle OH}{|}}{C}CH_2CH_2CH_2CH_3 \right] \longrightarrow H_3C-\overset{\overset{\displaystyle O}{\|}}{C}CH_2CH_2CH_2CH_3$

unstable enol

(i) $HC\equiv C-CH_2CH_2CH_2CH_3 \xrightarrow[\substack{2) H_2O_2, \\ HO^-}]{1) Sia_2BH} \left[ \overset{\overset{\displaystyle OH}{|}}{CH}=CHCH_2CH_2CH_2CH_3 \right] \longrightarrow H-\overset{\overset{\displaystyle O}{\|}}{C}CH_2CH_2CH_2CH_2CH_3$

unstable enol

(j) 
$\xrightarrow[\text{CCl}_4]{\text{Br}_2}$ $H_3C-\overset{\overset{\displaystyle Br}{|}}{\underset{\underset{\displaystyle H}{|}}{C}}-\overset{\overset{\displaystyle Br}{|}}{\underset{\underset{\displaystyle H}{|}}{C}}CH_2CH_2CH_3$ $\xrightarrow[200°]{\text{KOH}}$ $H_3C-C\equiv C-CH_2CH_2CH_3$

major

$\xrightarrow[\text{catalyst}]{H_2 \,|\, \text{Lindlar's}}$

**9-35** All four syntheses in this problem begin with the same reaction of benzyl bromide with acetylide ion:

(a) $PhCH_2-C\equiv C:^{\ominus}$ + $Br$ $\longrightarrow$ $PhCH_2-C\equiv C$ 6-phenylhex-1-en-4-yne

allyl bromide

(b) $PhCH_2-C\equiv C:^{\ominus}$ + $Br$ $\longrightarrow$ $PhCH_2-C\equiv C$ $\xrightarrow[\substack{\text{quinoline} \\ \text{Lindlar's catalyst}}]{\substack{H_2 \\ Pd/BaSO_4}}$

ethyl bromide

*cis*-1-phenylpent-2-ene

(c) $PhCH_2-C\equiv C:^{\ominus}$ + $Br$ $\longrightarrow$ $PhCH_2-C\equiv C$ $\xrightarrow[\text{NH}_3]{\text{Na}}$

ethyl bromide

*trans*-1-phenylpent-2-ene

**9-35 continued**

(d) The diol with the two OH groups on the same side in the Fischer projection is the equivalent of a meso structure, although this one is not meso because the top and bottom group are different. Still, it gives a clue as to its synthesis. The "meso" diol can be formed by either a syn addition to a *cis* double bond, or by an anti addition to a *trans* double bond. We saw the same thing in the solution to 9-25 (d).

*cis* — syn addition → "meso" diol ← anti addition — *trans*

product from part (b)

product from part (c)

racemic

**9-36**

(a) $CH_3CH_2-C\equiv C-CH_2CH_3$

(b) $CH_3CH_2-C\equiv C-H$ + $CH_2=C(CH_3)CH_3$  elimination on 3° halide

(c) $CH_3CH_2-C\equiv C-CH_2OH$ (after $H_2O$ workup)

(d) $CH_3CH_2-C\equiv C-$ (cyclohexyl with OH) (after $H_2O$ workup)

(e) $CH_3CH_2-C\equiv C-CH(OH)CH_2CH_2CH_3$ (after $H_2O$ workup)

(f) $CH_3CH_2-C\equiv C-H$ + $Na^+$ $^{\ominus}O-$(cyclohexyl)

(g) $CH_3CH_2-C\equiv C-C(OH)(CH_3)-CH_2CH_3$ (after $H_2O$ workup)

**9-37**

(a) $HC\equiv C-H \xrightarrow{NaNH_2} HC\equiv C:^{\ominus} Na^+ \xrightarrow{CH_3CH_2CH_2CH_2Br} HC\equiv C-CH_2CH_2CH_2CH_3$

(b) $HC\equiv C-H \xrightarrow{NaNH_2} HC\equiv C:^{\ominus} Na^+ \xrightarrow{CH_3CH_2CH_2Br} HC\equiv C-CH_2CH_2CH_3$

$\xrightarrow[CH_3I]{NaNH_2} H_3C-C\equiv C-CH_2CH_2CH_3$

(c) $H_3C-C\equiv C-CH_2CH_2CH_3 \xrightarrow[\text{Lindlar's catalyst}]{H_2}$ 

synthesized in part (b)

(d) $H_3C-C\equiv C-CH_2CH_2CH_3 \xrightarrow[NH_3]{Na}$

synthesized in part (b)

**211**

9-37 continued

(e) $HC\equiv C-CH_2CH_2CH_2CH_3$    $\xrightarrow[\text{HOOH}]{\substack{\text{2 equiv.}\\ \text{HBr}}}$    $\underset{\displaystyle \underset{Br}{|}}{\overset{\displaystyle \overset{Br}{|}}{HC}}-CH_2CH_2CH_2CH_2CH_3$    anti-Markovnikov orientation

synthesized in part (a)

(f) $HC\equiv C-CH_2CH_2CH_2CH_3$    $\xrightarrow{\textbf{2 HBr}}$    $H_3C-\underset{\displaystyle \underset{Br}{|}}{\overset{\displaystyle \overset{Br}{|}}{C}}-CH_2CH_2CH_2CH_3$    Markovnikov orientation

synthesized in part (a)

(g) $H-C\equiv C-CH_2CH_2CH_3$    $\xrightarrow[\text{2) }H_2O_2,\ HO^-]{\text{1) Sia}_2BH}$    $H-\overset{\displaystyle \overset{O}{\|}}{C}CH_2CH_2CH_2CH_3$

from (b)

(h) $H-C\equiv C-CH_2CH_2CH_3$    $\xrightarrow[\substack{H_2SO_4\\ HgSO_4}]{H_2O}$    $H_3C-\overset{\displaystyle \overset{O}{\|}}{C}CH_2CH_2CH_3$

from (b)

(i) $HC\equiv C-H$ $\xrightarrow{NaNH_2}$ $HC\equiv C\overset{\ominus}{:}$ $Na^+$ $\xrightarrow{CH_3CH_2Br}$ $HC\equiv C-CH_2CH_3$ $\xrightarrow{NaNH_2}$

$\Big\downarrow CH_3CH_2Br$

Alkene must be *cis* to produce the (±) product from anti addition.

*Review the stereochemistry in the solution to Problem 8-35 of this Solutions Manual.*

$CH_3CH_2-C\equiv C-CH_2CH_3$

(j) $HC\equiv C-H$ $\xrightarrow{NaNH_2}$ $HC\equiv C\overset{\ominus}{:}$ $Na^+$ $\xrightarrow{CH_3I}$ $HC\equiv C-CH_3$ $\xrightarrow[\text{2) }CH_3I]{\text{1) }NaNH_2}$ $H_3C-C\equiv C-CH_3$

Alternatively, *trans*-but-2-ene could be dihydroxylated with anti stereochemistry using aqueous peracetic acid.

*Review the stereochemistry in the solution to Problem 8-35 of this Solutions Manual.*

meso

$\xrightarrow[H_2O_2]{OsO_4}$

$H_2 \Big\downarrow$ Lindlar's catalyst

Alkene must be *cis* to produce the meso product from syn addition.

(k) $HC\equiv C-H$ $\xrightarrow{NaNH_2}$ $HC\equiv C\overset{\ominus}{:}$ $Na^+$ $\xrightarrow{CH_3CH_2Br}$ $HC\equiv C-CH_2CH_3$

$\Big\downarrow NaNH_2$

$H_3C-\underset{\displaystyle \underset{CH_3}{|}}{\overset{\displaystyle \overset{OH}{|}}{C}}-C\equiv C-CH_2CH_3$ $\xleftarrow{H_2O}$

212

Copyright © 2013 Pearson Education, Inc.

9-38

Compound **X**
$$\xrightarrow[\text{Pt}]{\textbf{5 H}_2}$$
cyclohexyl—$CH_2CH_2CH_2CH_3$

⟹ The fact that five equivalents of hydrogen are consumed says that **X** must have five pi bonds in the above carbon skeleton.

1) $O_3$
2) $Me_2S, H_2O$

$$\underset{\text{O}}{\overset{\text{O}}{H-C-CH_2CH_2-C-C-H}} + \underset{\text{O}}{\overset{\text{O}}{H-C-C-H}} + \underset{\text{O}}{\overset{\text{O}}{H-C-C-OH}} + \underset{\text{O}}{\overset{\text{O}}{H-C-OH}}$$

from C≡C

6 carbonyls ⟹ 3 alkenes    2 carboxylic acids ⟹ 1 alkyne

Ph—CH=CH—C≡CH

Compound **X**

Whether the alkene is *cis* or *trans* cannot be determined from these results.

9-39  Compound **Z**

ozonolysis ⟹ $CH_3(CH_2)_4-\overset{O}{\overset{||}{C}}-H$    $CH_3-\overset{O}{\overset{||}{C}}-CH_2-\overset{O}{\overset{||}{C}}-OH$    $HO-\overset{O}{\overset{||}{C}}-H$

from alkene          from alkyne

Compound **Z**:

$$CH_3(CH_2)_4CH=\underset{\overset{|}{CH_3}}{C}-CH_2-C≡C-H$$

Whether the alkene is *E* or *Z* cannot be determined from this information.

9-40  This synthesis begins the same as the solution to problem 9-35:

$$HC≡CH \xrightarrow{NaNH_2} HC≡C:^{\ominus} + CH_2Br \longrightarrow PhCH_2-C≡CH \xrightarrow{NaNH_2} PhCH_2-C≡C:^{\ominus}$$

benzyl bromide

The anion will add across the carbonyl group of the aldehyde:

$$\text{cyclopentyl}-\overset{O}{\overset{||}{C}}-H + PhCH_2-C≡C:^{\ominus} \longrightarrow \xrightarrow{H_3O^+}$$

acid-catalyzed dehydration    $H_2SO_4, \Delta$

MCPBA

**213**
Copyright © 2013 Pearson Education, Inc.

9-41

(a)   $CH_3CH_2-C\equiv C-H$  $\xrightarrow{\text{1) Sia}_2\text{BH}}_{\text{2) H}_2\text{O}_2,\ \text{HO}^-}$  $CH_3CH_2CH_2-\overset{\overset{\displaystyle O}{\|}}{C}-H$

(b)

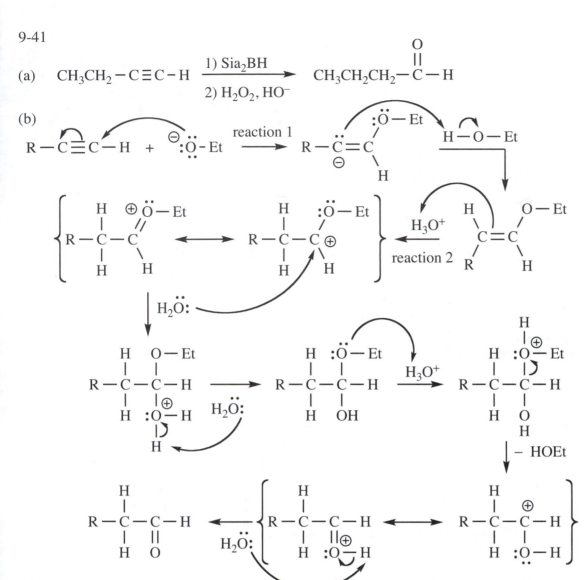

(c)   <u>alkyne</u>

$R-C\equiv CH$  $\xrightarrow{RO^-}$  $R-\overset{..}{\underset{\ominus}{C}}=CH$
                                              |
                                              OR

$sp^2$ carbanion

<u>alkene</u>

$R-\underset{\overset{|}{H}}{C}=CH_2$  $\xrightarrow{RO^-}$  $R-\overset{..}{\underset{\overset{|}{\ominus}}{C}}-\underset{\overset{|}{OR}}{CH_2}$
                                                                      |
                                                                      H

$sp^3$ carbanion

The closer that electrons are to the nucleus, the more stable. An s orbital is closer to a nucleus than a p orbital is, as p orbitals are elongated away from the nucleus. An $sp^2$ carbanion is more stable than an $sp^3$ carbanion because the $sp^2$ carbanion has 33% s character and the electron pair is closer to the positive nucleus than in an $sp^3$ carbanion which is only 25% s character. The $sp^2$ carbanion is easier to form because of its relative stability.

9-42 Diols are made by two reactions from Chapter 8: either *syn*-dihydroxylation with $OsO_4$ or cold $KMnO_4$, or *anti*-dihydroxylation via an epoxide using a peroxyacid and water. As this problem says to use inorganic reagents, the solution shown here will use $OsO_4$.

Recall the stereochemical requirements of syn addition as outlined in this Solutions Manual, Problem 8-35:

*cis*-alkene + **syn** addition → meso                  *trans*-alkene + **syn** addition → racemic (±)

*cis*-alkene + **anti** addition → racemic (±)          *trans*-alkene + **anti** addition → meso

*continued on next page*

**214**

9-42 continued

Part (a) asks for the synthesis of the meso isomer, so syn addition will have to occur on the *cis*-alkene. Part (b) will require syn addition to the *trans*-alkene to give the (±) product.

(a)

$$HC \equiv C-H \xrightarrow{NaNH_2} HC \equiv C: \ominus \ Na^+$$

Alkene must be *cis* to produce the meso product from syn addition, so reduction is done with Lindlar's catalyst to produce the *cis* alkene.

(b)

$$HC \equiv C-H \xrightarrow{NaNH_2} HC \equiv C: \ominus \ Na^+$$

Alkene must be *trans* to produce the (±) product from syn addition, so reduction is done with Na/NH$_3$.

9-43  Each unknown has molecular formula C$_8$H$_{12}$ with 3 elements of unsaturation.

(a)

W

Addition of H$_2$ to the triple bond gives cyclooctane. Cleavage of the triple bond with ozone will produce an 8-carbon chain with a carboxylic acid at each end, as shown in the solution to 9-26(i).

(b)

X

Addition of H$_2$ to the double bonds gives cyclooctane. Cleavage of the double bonds with ozone will produce two 4-carbon chains with an aldehyde at each end.

(c)

Y

Addition of H$_2$ to the double bonds gives cyclooctane. Cleavage of the double bonds with ozone will produce one 3-carbon chain and one 5-carbon chain with an aldehyde at each end.

(d)

Z

bicyclo[4.2.0]octane

Addition of H$_2$ to the double bond gives bicyclo[4.2.0]octane. Cleavage of the double bond with ozone opens the larger ring but retaining the cyclobutane with 1-carbon and 3-carbon chains with an aldehyde at each end.

**215**

# CHAPTER 10—STRUCTURE AND SYNTHESIS OF ALCOHOLS

10-1 The 1993 IUPAC recommendations put the position number before the group it describes.

(a) 2-phenylbutan-2-ol

(b) (E)-5-bromohept-3-en-2-ol

(c) 4-methylcyclohex-3-en-1-ol ("1" is optional)

(d) *trans*-2-methylcyclohexan-1-ol  ("1" is optional)
    or (1R,2R)-2-methylcyclohexan-1-ol

(e) (E)-2-chloro-3-methylpent-2-en-1-ol

(f) (2R,3S)-2-bromohexan-3-ol

10-2  IUPAC name first, then common name.

(a) butan-2-ol; *sec*-butyl alcohol

(b) cyclopropanol; cyclopropyl alcohol

(c) 1-cyclobutylpropan-2-ol; no common name

(d) 3-methylbutan-1-ol; isopentyl alcohol
    (also isoamyl alcohol)

10-3 Only constitutional isomers are requested, not stereoisomers, and only structures with an alcohol group.

(a) $C_3H_8O$    OH  propan-1-ol

 propan-2-ol

(b) $C_4H_{10}O$

butan-1-ol       butan-2-ol      2-methylpropan-1-ol    2-methylpropan-2-ol

(c)  $C_3H_6O$ has one element of unsaturation, either a double bond or a ring.

cyclopropanol    prop-2-en-1-ol   prop-1-en-2-ol *      prop-1-en-1-ol (*E* or *Z*)

(d) $C_3H_4O$ has two elements of unsaturation, so each structure must have either a triple bond, or two double bonds, or a three-membered ring and a double bond.  All structures must contain an OH.  (In the name, the "e" is dropped from "yne" because it is followed by a vowel in "ol".)

HO–C≡C–CH₃   HC≡C–CH₂   H₂C=C=CH *        —OH               
*                                                                
prop-1-yn-1-ol                                               
              prop-2-yn-1-ol   propa-1,2-dien-1-ol  cycloprop-1-en-1-ol   cycloprop-2-en-1-ol

*The structures with the OH bonded directly to the carbon-carbon double bond are called *enols* or *vinyl alcohols*.  The structure with OH on a carbon-carbon triple bond is called an *ynol*.  These compounds are unstable, although the structures are legitimate.

10-4    (a) 8,8-dimethylnonane-2,7-diol

(b) octane-1,8-diol

(c) *cis*-cyclohex-2-ene-1,4-diol

(d) 3-cyclopentylheptane-2,4-diol

(e) *trans*-cyclobutane-1,3-diol

10-5 There are four structural features to consider when determining solubility in water:  1) molecules with fewer carbons will be more soluble in water (assuming other things being equal); 2) branched or otherwise compact structures are more soluble than linear structures; 3) more hydrogen-bonding groups will increase solubility; 4) an ionic form of a compound will be more soluble in water than the nonionic form.

(a) Cyclohexanol is more soluble because its alkyl group is more compact than in hexan-1-ol.

(b) 4-Methylphenol is more soluble because its hydrocarbon portion is more compact than in heptan-1-ol, and phenols form particularly strong hydrogen bonds with water.

(c) 3-Ethylhexan-3-ol is more soluble because its alkyl portion is more spherical than in octan-2-ol.

(d) Cyclooctane-1,4-diol is more soluble because it has two OH groups which can hydrogen bond with water, whereas hexan-2-ol has only one OH group.  (The ratio of carbons to OH is 4 to 1 in the former compound and 6 to 1 in the latter; the smaller this ratio, the more soluble.)

(e) These are enantiomers and will have identical solubility.

10-6  Dimethylamine molecules can hydrogen bond among themselves so it takes more energy (higher temperature) to separate them from each other.  Trimethylamine has no N-H and cannot hydrogen bond, so it takes less energy to separate these molecules from each other, despite its higher molecular weight.

10-7  See Appendix 2 at the back of this Solutions Manual for a review of acidity and basicity.
(a)  Methanol is more acidic than *tert*-butyl alcohol.  The greater the substitution, the lower the acidity.
(b)  2-Chloropropan-1-ol is more acidic because the electron-withdrawing chlorine atom is closer to the OH group than in 3-chloropropan-1-ol.
(c)  2,2-Dichloroethanol is more acidic because two electron-withdrawing chlorine atoms increase acidity more than just the one chlorine in 2-chloroethanol.
(d)  2,2-Difluoropropan-1-ol is more acidic because fluorine is more electronegative than chlorine; the stronger the electron-withdrawing group, the more acidic the alcohol.

10-8
sulfuric acid  >>  acetic acid  >  2-chloroethanol  >  water  >  ethanol  >

$H_2SO_4$      $CH_3COOH$      $ClCH_2CH_2OH$      $H_2O$      $CH_3CH_2OH$              *least*
*most*                                                          *acidic*
*acidic*                            *tert*-butyl alcohol  >  hex-1-yne  >  ammonia  >  hexane
                                 $(CH_3)_3COH$    $C_4H_9-C\equiv C-H$    $NH_3$    $C_6H_{14}$

Sulfuric acid is one of the strongest acids known, and acetic acid is the strongest type of organic acid.  On the other extreme, alkanes like hexane are the least acidic compounds.  Alkynes are more acidic than ammonia.  The N–H bond in ammonia is less acidic than any O–H bond.  Among the four compounds with O–H bonds, the tertiary alcohols are the least acidic.  Water is more acidic than most alcohols including ethanol.  However, if a strong electron-withdrawing substituent like chlorine is near the alcohol group, the acidity increases enough so that it is more acidic than water.  (Determining exactly where water appears in this list is the most difficult part.)

10-9  Resonance forms of phenoxide anion show the negative charge delocalized onto the ring only at carbons 2, 4, and 6:

Nitro group at position 2

Nitro at position 2 delocalizes negative charge.

Nitro group at position 3

Nitro at position 3 cannot delocalize negative charge at position 2 or 4 — no resonance stabilization.

continued on next page

**217**

10-9 continued
Nitro group at position 4

Nitro at position 4 delocalizes negative charge.

Only when the nitro group is at one of the negative carbons will the nitro have a stabilizing effect (via resonance). Thus, 2-nitrophenol and 4-nitrophenol are substantially more acidic than phenol itself, but 3-nitrophenol is only slightly more acidic than phenol (due to the inductive effect).

10-10

(a) Structure **A** is a phenol because the OH is bonded to a benzene ring. As a phenol, it will be acidic enough to react with sodium hydroxide to generate a phenoxide ion that will be fairly soluble in water. Structure **B** is a 2° benzylic alcohol, not a phenol, not acidic enough to react with NaOH.
(b) Both of these organic compounds will be soluble in an organic solvent like dichloromethane. Shaking this organic solution with aqueous sodium hydroxide will ionize the phenol **A**, making it more polar and water soluble; it will be extracted from the organic layer into the water layer, while the alcohol will remain in the organic solvent. Separating these immiscible solvents will separate the original compounds. The alcohol can be retrieved by evaporating the organic solvent. The phenol can be isolated by acidifying the basic aqueous solution and filtering if the phenol is a solid, or separating the layers if the phenol is a liquid.

10-11 The Grignard reaction needs a solvent containing an ether functional group: (b), (f), (g), and (h) are possible solvents. Dimethyl ether, (b), is a gas at room temperature, however, so it would have to be liquefied at low temperature for it to be a useful solvent.

10-12 (a) $CH_3CH_2MgBr$    (b) Li + LiI    (c) F—⬡—MgBr    (d) + LiCl

10-13 Any of three halides—chloride, bromide, iodide, but not fluoride—can be used. Ether is the typical solvent for Grignard reactions. The new C—C bond is shown in bold. ▬

(a)

(b)

(c)

----

Note: the alternative arrow symbolism could also be used, where the two steps are numbered around one arrow:

1) ether
2) $H_3O^+$   OK

ether
$H_3O^+$

NO! My BAD!
This means that water is present with ether during the Grignard reaction!

**10-14** Any of three halides—chloride, bromide, iodide, but not fluoride—can be used. Grignard reactions are always performed in ether solvent; ether is not shown here.

(a) two methods

The newly-formed C—C bond is shown in bold. ▬

(b) two methods

(c) two methods

**10-15** Grignard reactions are always performed in ether. Here, the ether is not shown.

(a) Any of the three bonds shown in bold can be formed by adding a Grignard reagent across a ketone, followed by aqueous acid workup.

(i) $CH_3CH_2MgBr$
+

(ii)

+ PhMgBr

(iii) $CH_3CH_2CH_2MgBr$
+

(b) Ph–MgBr +

(c) $CH_3CH_2$–MgI +

## 10-15 continued
(d) three methods, all with $H_3O^+$ workup

(i)     (ii)     (iii)

## 10-16

(This is just a nucleophilic substitution where Cl is the leaving group. The unusual feature is that it occurs at a carbonyl carbon.)

ketone intermediate

10-17   Acid chlorides or esters will work as starting materials in these reactions. The typical solvent for Grignard reactions is ether; it is not shown here.

(a)   $Ph-\overset{\overset{O}{\|}}{C}-Cl$   + 2 PhMgBr   $\xrightarrow{H_3O^+}$   $Ph-\overset{\overset{Ph}{|}}{\underset{Ph}{C}}-OH$

(b)

(c)   $Ph-\overset{\overset{O}{\|}}{C}-Cl$   + 2   (cyclohexyl)—MgCl   $\xrightarrow{H_3O^+}$

## 10-18

(a)   $H-\overset{\overset{O}{\|}}{C}-OEt$ + (allyl)MgBr  →  aldehyde intermediate

10-18 continued

(b) (i) 
$$\underset{\text{HC-OEt}}{\overset{O}{\|}} + 2\ CH_3CH_2MgBr \xrightarrow{\ } \xrightarrow{H_3O^+} $$ 

(ii) 
$$\underset{\text{HC-OEt}}{\overset{O}{\|}} + 2 \ \bigg\langle\!\!\!\bigcirc\!\!\!\bigg\rangle\!\!-MgBr \xrightarrow{\ } \xrightarrow{H_3O^+} $$ 

(iii) 
$$\underset{\text{HC-OEt}}{\overset{O}{\|}} + 2 \ \diagup\!\!\!\diagdown\!\!\!\diagup MgBr \xrightarrow{\ } \xrightarrow{H_3O^+} $$ 

10-19 Ether is the typical solvent in Grignard reactions. The new C—C bond is shown in bold. ▬

(a) 
$$\bigg\langle\!\!\!\bigcirc\!\!\!\bigg\rangle\!\!-MgBr + \triangle\!\!\!\!O \xrightarrow{H_3O^+} $$ 

(b) 
$$\diagup\!\!\diagdown\!\!\diagup MgCl + \triangle\!\!\!\!O \xrightarrow{H_3O^+} $$ 

(c) 
$$\bigg\langle MgI + \triangle\!\!\!\!O \xrightarrow{H_3O^+} $$ 

10-20

(a) 
$$HC\!\equiv\!\overset{\ominus}{C}\!:\ + \ \overset{O}{\triangle} \xrightarrow{\ } \xrightarrow{H_3O^+} HC\!\equiv\!C\!-\!CH_2CH_2OH $$ 

(b) 
$$CH_3CH_2C\!\equiv\!\overset{\ominus}{C}\!:\ + \ \overset{O}{\triangle} \xrightarrow{\ } \xrightarrow{H_3O^+} CH_3CH_2C\!\equiv\!C\!-\!CH_2CH_2OH $$ 

10-21 Often, there are several synthetic routes to each structure; the ones shown here are representative. The new bonds formed are shown here in bold. Your answers may be different and still be correct.

(a) 
$$\diagup\!\!\diagdown\!\!\diagup Br \xrightarrow{Li} \xrightarrow{CuI} \left(\diagup\!\!\diagdown\!\!\diagup\right)_2 CuLi \xrightarrow{\ \diagup\!\!\diagdown\!\!\diagup Br\ } $$ 

(b) 
$$\diagup\!\!=\!\!Br \xrightarrow{Li} \xrightarrow{CuI} \left(\diagup\!\!=\right)_2 CuLi \xrightarrow{\ } $$ 

**221**
Copyright © 2013 Pearson Education, Inc.

10-21 continued

(c)

Alternatively, coupling lithium dicyclohexylcuprate with 1-bromobutane would also work. As this mechanism is not a typical $S_N2$, it is not as susceptible to steric hindrance like acetylide ion substitution or a similar $S_N2$ reaction.

(d)

10-22 These reactions are acid-base reactions in which an acidic proton (or deuteron) is transferred to a basic carbon in either a Grignard reagent or an alkyllithium.

(a)

+ Mg(OD)I

(b)  $CH_3CH_2CH_2CH_3$  +  $LiOCH_2CH_3$

(c)

H  +  $BrMg^{\oplus}$  $^{\ominus}$:C≡C–CH$_2$CH$_3$

(d)

+

(e)

+  Mg(OD)Br

10-23 Grignard reagents are incompatible with acidic hydrogens and with electrophilic, polarized multiple bonds like C=O, $NO_2$, etc.
(a)  As Grignard reagent is formed, it would instantaneously be protonated by the N—H present in other molecules of the same substance.
(b)  As Grignard reagent is formed, it would immediately attack the ester functional group present in other molecules of the same substance.
(c)  Care must be taken in how reagents are written above and below arrows. If reagents are numbered "1. ... 2. ... etc.", it means they are added in separate steps, the same as writing reagents over separate arrows. If reagents written around an arrow are not numbered, it means they are added all at once in the same mixture. In this problem, the ketone is added in the presence of aqueous acid. The acid will immediately protonate and destroy the Grignard reagent before reaction with the ketone can occur.
(d)  The ethyl Grignard reagent will be immediately protonated and consumed by the OH. This reaction *could* be made to work, however, by adding two equivalents of ethyl Grignard reagent: the first to consume the OH proton, the second to add across the ketone. Aqueous acid will then protonate both oxygens.

10-24 Sodium borohydride does not reduce carboxylic acids or esters.

(a)  $CH_3(CH_2)_8CH_2OH$      (b)  no reaction      (c)  no reaction ($PhCOO^-$ before acid work-up)      (d)

(e)

(f)  ester  $\left\{ \rule{0pt}{30pt}\right.$

10-25 Lithium aluminum hydride reduces carboxylic acids and esters as well as other carbonyl groups.

(a)  $CH_3(CH_2)_8CH_2OH$      (b)  $CH_3CH_2CH_2OH$  +  $HOCH_3$      (c)  $PhCH_2OH$      (d)

(e)

+  $HOCH_3$

(f)

10-26
(a)

OR

OR

(b)

(c)

(d)

LiAlH$_4$ will NOT give the desired product. LiAlH$_4$ will also reduce the ester in addition to the ketone.

10-27  Approximate pKa values are shown below each compound. Refer to text Tables 1-5, 9-2, and 10-3, and Appendix 4 at the back of the text.

$$CH_3SO_3H > CH_3COOH > CH_3SH > CH_3OH > CH_3C\equiv CH > CH_3NH_2 > CH_3CH_3$$

| $< 0$ | 4.74 | $\approx 10.5$ | 15.5 | 25 | $\approx 35$ | 50 |

most acidic                                                                                          least acidic

10-28
(a)  4-methylpentane-2-thiol
(b)  (Z)-2,3-dimethylpent-2-ene-1-thiol
(c)  cyclohex-2-ene-1-thiol ("1" is optional)

10-29

3-methylbutane-1-thiol

but-2-ene-1-thiol

OR

HBr, 40° C
(see Problem 8-2)

10-30 Bromobenzene is abbreviated PhBr. The phenyl Grignard reagent is abbreviated PhMgBr.

(a)

PhBr $\xrightarrow[\text{ether}]{\text{Mg}}$ PhMgBr + (aldehyde) $\xrightarrow{H_3O^+}$ (alcohol, Ph)

OR

PhMgBr from (a)
+
(epoxide)
with $H_3O^+$ workup

(b)

(alcohol, Ph) $\xrightarrow[\Delta]{H_2SO_4}$ (Ph alkene) + $H_2O$

from (a)

(c) two methods

(Ph alkene) $\xrightarrow[\text{THF}]{BH_3} \xrightarrow[HO^-]{H_2O_2}$ (Ph, OH)

from (b)

(d)

(Ph alkene) $\xrightarrow[h\nu]{\text{NBS}}$ (Ph, Br) $\xrightarrow[\Delta]{H_2O}$ (Ph, OH) Either $S_N1$ or $S_N2$ with NaOH will work.

from (b)                        *Trans* is more stable.

(e)

PhMgBr + (ketone) $\xrightarrow{H_3O^+}$ (Ph, OH)

from (a)

(f) (Ph, OH) $\xrightarrow[\Delta]{H_2SO_4}$ (Ph alkene)

major

10-31
(a) 5-methyl-4-propylheptan-2-ol; 2°
(b) 4-(1-bromoethyl)heptan-3-ol; 2°
(c) 6-chloro-3-phenyloctan-3-ol; 3°
(d) 3-bromocyclohex-3-en-1-ol; 2° ("1" is optional)
(e) *cis*-4-chlorocyclohex-2-en-1-ol; 2° ("1" is optional)
also possible is (1R,4S)-
(f) (*E*)-4,5-dimethylhex-3-en-1-ol; 1°
(g) (1-cyclopentenyl)methanol; 1°

10-32
(a) 4-chloro-1-phenylhexane-1,5-diol
(b) *trans*-cyclohexane-1,2-diol
(c) 3-nitrophenol
(d) 4-bromo-2-chlorophenol

10-33

(a) ... (b) ... (c) ... (d) ... (e) ...

(f) ... (g) ... (h) ... (i) ... (j) $CH_3S-SCH_3$

(k) ...

10-34
(a) Hexan-1-ol will boil at a higher temperature as it is less branched than 3,3-dimethylbutan-1-ol.
(b) Hexan-2-ol will boil at a higher temperature because its molecules hydrogen bond with each other, whereas molecules of hexan-2-one have no intermolecular hydrogen bonding.
(c) Hexane-1,5-diol will boil at a higher temperature as it has two OH groups for hydrogen bonding. Hexan-2-ol has only one group for hydrogen bonding.
(d) Hexan-2-ol will boil at a higher temperature because it has a higher molecular weight than pentan-2-ol. All other structural features of the two molecules are the same, so they should have the same intermolecular forces.

10-35 Refer to Table 10-4 to compare acidities of different functional groups. The strength of an acid is determined by the stability of its conjugate base.

(a) 3-Chlorophenol is more acidic than cyclopentanol. In general, phenols are many orders of magnitude more acidic than alcohols because phenoxide anions are stabilized by resonance.

(b) Cyclohexanethiol is more acidic than cyclohexanol. S is beneath O on the periodic table, and acidity increases down the periodic table. Larger anions are more stable because a negative charge on a larger atom is distributed over a larger volume, with lower electron density and greater delocalization of the negative charge.

(c) Cyclohexanecarboxylic acid is more acidic than cyclohexanol. In general, carboxylic acids are many orders of magnitude more acidic than alcohols because carboxylate anions are stabilized by resonance.

(d) 2,2-Dichlorobutan-1-ol is more acidic than butan-1-ol because of the two electron-withdrawing substituents near the acidic functional group.

10-36
(a) Propan-2-ol is the most soluble in water as it has the fewest carbons and the most branching.
(b) Cyclohexane-1,2-diol is the most soluble as it has two OH groups for hydrogen bonding. Cyclohexanol has only one OH group; chlorocyclohexane cannot hydrogen bond and is the least soluble.
(c) Phenol is the most soluble as it can hydrogen bond better than cyclohexanol. Phenol is a stronger acid and will be ionized to a greater extent than cyclohexanol. Cyclohexanol has low solubility, and 4-methylcyclohexanol has the added hydrophobic methyl group, decreasing its water solubility.

10-37

(a)

$$\xrightarrow[\text{H}_2\text{O}]{\text{Hg(OAc)}_2 \quad \text{NaBH}_4}$$

(b)

$$\xrightarrow[\text{2) H}_2\text{O}_2, \text{HO}^-]{\text{1) BH}_3 \bullet \text{THF}}$$

(c)

$$\xrightarrow[\text{HO}^-]{\text{BH}_3 \bullet \text{THF} \quad \text{H}_2\text{O}_2}$$

(d)

$$\xrightarrow[\text{H}_2\text{O}]{\text{Hg(OAc)}_2 \quad \text{NaBH}_4}$$

10-38 Products after aqueous acid workup:

(a)

(b)

(c)

(d) This problem confuses a lot of people. When a Grignard reagent is added to a compound that has an OH group, the first thing that happens is that the Grignard reacts by removing the H⁺ to give O⁻.

+ 1 CH₃MgI
(only one equivalent of Grignard reagent added)

+ CH₄

If no more Grignard reagent is added before acid hydrolysis, then the starting material is recovered.

If a second equivalent (or excess) of Grignard reagent is added before hydrolysis, then it will add at the ketone. Acid hydrolysis will give the diol.

225
Copyright © 2013 Pearson Education, Inc.

# 10-38 continued

(e) [cyclopentane-CH₂-OH structure]

(f) Ph₂C(Ph)(OH) — Ph Ph / Ph OH structure

(g) [structure with OH, Ph, Ph and OH]

(h) [dicyclopentyl structure with Ph and OH]

(i) HO-[cyclohexane]-CH₂-C(=O)-OCH₃

(j) HO-[cyclohexane]-CH₂CH₂-OH

(k) [decalin with OH]

(l) [decalin with H and OH]
plus the enantiomer

(m) HO...CH₃...H...OH...H structure
plus the enantiomer

(n) HO...CH₃...H...H...OH structure
plus the enantiomer

(o) [diene structure]

10-39 All Grignard reactions are run in ether solvent. Two arrows are shown indicating that the Grignard reaction is allowed to proceed, and then in a second step, dilute aqueous acid is added. The new C—C bonds are shown in bold. ▬

(a) [aldehyde] + BrMg[CH₂CH₃] → $H_3O^+$ → [alcohol product with OH]

(b) [alkyl bromide] $\xrightarrow[\text{ether}]{\text{Mg}}$ $CH_2O$ $H_3O^+$ → [alcohol product with OH]

(c) [acetaldehyde] + BrMg—[cyclohexane] → $H_3O^+$ → [alcohol product]

(d) [cyclohexyl Br] $\xrightarrow[\text{ether}]{\text{Mg}}$ [epoxide] $H_3O^+$ → [product with OH]

(e) [phenyl Br] $\xrightarrow[\text{ether}]{\text{Mg}}$ $CH_2O$ $H_3O^+$ → [benzyl alcohol]

(f) [cyclohexyl]—$\overset{\text{O}}{\overset{||}{\text{C}}}$OCH₂CH₃ + 2 CH₃MgI → $H_3O^+$ → [cyclohexyl]—$\overset{\text{CH}_3}{\underset{\text{CH}_3}{\text{C}}}$—OH

(g) [benzaldehyde] + BrMg—[cyclopentane] → $H_3O^+$ → [product with OH, phenyl and cyclopentyl]

10-40

(a) $\xrightarrow[\text{}]{\text{BH}_3 \bullet \text{THF}} \xrightarrow[\text{HO}^-]{\text{H}_2\text{O}_2}$ plus the enantiomer

(b) $\xrightarrow[\Delta]{\text{KMnO}_4 \text{ or HNO}_3 \text{ or NaOCl}}$ Many strong oxidizing agents will transform thiols to sulfonic acids.

(c) $\xrightarrow[\text{CH}_3\text{OH}]{\text{NaBH}_4}$

(d) $\xrightarrow[\text{Pt catalyst}]{\text{1 eq. H}_2}$

$\text{H}_2/\text{Pt reduces C=C faster than C=O}$

(e) $\xrightarrow[\text{CH}_3\text{OH}]{\text{NaBH}_4}$

(f) $\xrightarrow{\text{LiAlH}_4} \xrightarrow{\text{H}_3\text{O}^+}$

10-41

(a) $\text{MgBr} + \text{CH}_2\text{O} \xrightarrow{\text{ether}} \xrightarrow{\text{H}_3\text{O}^+}$

(b) $\xrightarrow[\text{}]{\text{BH}_3 \bullet \text{THF}} \xrightarrow[\text{HO}^-]{\text{H}_2\text{O}_2}$

(c) $\xrightarrow{\text{NaOH}}$

(d) $\xrightarrow[\text{ether}]{\text{Mg}} \triangle \xrightarrow{\text{H}_3\text{O}^+}$

(e) $+ \text{NaSH} \longrightarrow$

(f) $\xrightarrow{\text{Li}} \xrightarrow{\text{CuI}}$ $\text{CuLi} + $ $\longrightarrow$

10-42   The position of the equilibrium can be determined by the strength of the acids or the bases.  The stronger acid and stronger base will always react to give the weaker acid and base, so the side of the equation with the weaker acid and base will be favored at equilibrium.  See Appendix 2 in this Solutions Manual for a review of acidity.

(a)   $CH_3CH_2O^{\ominus}$ +  —OH  $\rightleftharpoons$  $CH_3CH_2OH$  +  —O$^{\ominus}$

         stronger        stronger                 weaker      weaker
         base            acid                       acid      base

**products favored**

(b)   $KOH$  +  $Cl$——OH  $\rightleftharpoons$  $H_2O$  +  $Cl$——O$^{\ominus}$ $K^+$

         stronger     stronger            weaker      weaker
         base         acid              acid      base

**products favored**

(c)   +  $CH_3O^{\ominus}$  $\rightleftharpoons$  +  $CH_3OH$

         stronger     stronger           weaker     weaker
         acid       base           base      acid

**products favored**

(d)   +  $KOH$  $\rightleftharpoons$  $H_2O$  +  —O$^{\ominus}$ $K^+$

                  weaker       stronger
 weaker        base          acid       stronger
 acid                              base

**reactants favored**

(e)   $(CH_3)_3CO^{\ominus}$ +  $CH_3CH_2OH$  $\rightleftharpoons$  $(CH_3)_3COH$  +  $CH_3CH_2O^{\ominus}$

         stronger      stronger           weaker      weaker
         base        acid                acid      base

**products favored**

(f)   $(CH_3)_3CO^{\ominus}$ +  $H_2O$  $\rightleftharpoons$  $(CH_3)_3COH$  +  $HO^-$

         stronger     stronger          weaker     weaker
         base        acid           acid      base

**products favored**

(g)   $KOH$  +  $CH_3CH_2OH$  $\rightleftharpoons$  $H_2O$  +  $CH_3CH_2O^{\ominus}$ $K^+$

 weaker    weaker           stronger   stronger
 base      acid            acid       base

**reactants favored**

10-43

(a)

NaBH$_4$ / CH$_3$OH

OR

1) LiAlH$_4$
2) H$_3$O$^+$

OR

1) LiAlH$_4$
2) H$_3$O$^+$

OR

1) LiAlH$_4$
2) H$_3$O$^+$

(b)

NaBH$_4$ / CH$_3$OH

OR

1) LiAlH$_4$
2) H$_3$O$^+$

(c)

NaBH$_4$ / CH$_3$OH

OR

1) LiAlH$_4$
2) H$_3$O$^+$

(d)

NaBH$_4$ / CH$_3$OH

OR

1) LiAlH$_4$
2) H$_3$O$^+$

(e)

NaBH$_4$ / CH$_3$OH

OR

1) LiAlH$_4$
2) H$_3$O$^+$

OR

or ester

1) LiAlH$_4$
2) H$_3$O$^+$

(f)

NaBH$_4$ / CH$_3$OH

The milder NaBH$_4$ reduces aldehydes and ketones but not carboxylic acids and esters; the stronger LiAlH$_4$ is needed to reduce acids and esters.

**10-44** (a) The goal is to synthesize the target compound (boxed) from starting materials of six carbons or fewer. The product has 11 carbons, so the logical "disconnection" in working backwards is one six-carbon fragment and the cyclopentane ring that could be joined in a Grignard reaction. The tetrasubstituted C=C will come from dehydration of a tertiary alcohol produced in two possible Grignard reactions.

Route 1:

Route 2:

(b) The goal is to synthesize the target compound (boxed) from starting materials of six carbons or fewer. The product has 12 carbons, so the logical "disconnection" in working backwards is two six-carbon fragments which could be joined in a Grignard reaction. The best way to make epoxides is from the double bond, and double bonds are made from alcohols which are the products of Grignard reactions.

(c) The goal is to synthesize the target compound (boxed) from starting materials of six carbons or fewer. The product has 14 carbons, so the logical "disconnection" in working backwards is two six-carbon fragments and the ethyl group put on with a Williamson ether synthesis at the end.

**10-45** All steps are reversible.

**10-46** The symbol H—A represents a generic acid, where A$^-$ is the conjugate base.

(a)

+ H—A

(b)

+ H—A

(c)

+ :Ö-H

**10-47**

A → (Mg, ether) → B → (D$_2$O) → C

1) (acetone) 2) H$_3$O$^+$

isobutylcyclohexane

D → (H$_2$SO$_4$, Δ) → E → (2 Br$_2$) → F

excess H$_2$ / Pt

**10-48** This mechanism is similar to cleavage of the epoxide in ethylene oxide by Grignard reagents. The driving force for the reaction is relief of ring strain in the 4-membered cyclic ether, which is why it will undergo a Grignard reaction whereas most other ethers will not.

R-MgX + □-Ö: → R-⊓-Ö:$^-$ MgX$^+$

10-49 When mixtures of isomers can result, only the major product is shown.

10-50 The most important component in the deskunking mixture is hydrogen peroxide. Thiols are oxidized to structures having one, two, or three oxygens on the sulfur; all of these functional groups are acidic, less volatile so they don't reach the nose, and less stinky. The sodium bicarbonate is basic enough to ionize these acids, making them water soluble where the soap can wash them away.

10-51

Note about the reaction from **G** to non-1-yne: This reaction is a rearrangement of internal alkynes to terminal alkynes similar to what is described in text Section 9-8. Problem 10-51 does not rely on your knowing this reaction because the product of the reaction, non-1-yne, was given to you in the problem.

10-52 This problem is reminiscent of problem 8-49.

(a) Hg(OAc)₂ / H₂O, NaBH₄

(b) 1) BH₃ • THF 2) H₂O₂, HO⁻

(c) OsO₄ / H₂O₂ or cold, dilute KMnO₄

catalytic H₂SO₄

HBr

KOH, Δ

Parts (d) and (e) require moving the double bond into the ring, to form methylcyclohexene. Adding a catalytic amount of strong acid would do this, and addition-elimination will accomplish the same thing.

Alternatively, the tertiary alcohol product in part (a) could be dehydrated to give methylcyclohexene as the major product.

10-52 continued—See the note at the bottom of the previous page regarding (d) and (e).

(d)

$$\xrightarrow[\text{2) H}_2\text{O}_2 \text{, HO}^-]{\text{1) BH}_3 \bullet \text{THF}}$$

syn addition of H and OH with anti-Markovnikov orientation

plus the enantiomer

(e)

$$\xrightarrow[\text{H}_2\text{O}]{\text{Cl}_2}$$

chlorohydrin formation with anti addition of Cl and OH

plus the enantiomer

(f)

$$\xrightarrow{\text{MCPBA}} \xrightarrow[\text{ether}]{\text{PhMgBr}} \xrightarrow{\text{H}_3\text{O}^+}$$

New C—C bond requires Grignard; OH is not on same carbon where new C—C is formed suggesting the epoxide intermediate.

10-53

I
mol. wt. 60 g/mole
boiling point 82 °C
dipole moment 1.66 D
pK$_a$ 16.5

II
mol. wt. 168 g/mole
boiling point 58 °C
dipole moment 0.32 D
pK$_a$ 9.3

(a) Boiling point is a rough indicator of the strength of intermolecular forces, of which three common ones apply to organic compounds: van der Waal's forces (or London forces), the weakest; dipole-dipole interactions for molecules with a permanent dipole moment; and hydrogen-bonding in compounds with OH or NH bonds, the strongest. Both **I** and **II** have the alcohol functional group and are likely to form hydrogen bonds with their neighbors, although an argument could be made that the slightly larger F substituents with the greater electron clouds would make hydrogen bonding more difficult in **II**. Rather than stretch a possibility, let's focus on something obvious: there is a huge difference in the dipole moments of these two compounds. Even though the molecular weight of **II** is much greater, thereby increasing van der Waal's forces, the dipole moment of **II** is very small. We must conclude that the significantly decreased dipole-dipole interaction of **II** is more important in boiling point than the increased van der Waal's interaction, a weaker force.

(b) **I** has a large dipole moment because of the bond polarizations due to the electronegative oxygen. In **II**, however, the bond polarization in the alcohol is counteracted by six F atoms; recall that F is the most electronegative element in the periodic table. The oxygen pulling in one direction is balanced against the six F atoms pulling partially away from the oxygen. (If you have studied vectors in physics, consider each polarized C—F bond as a vector with a "down" component and a "left" or "right" component; you will see that some portion of the left CF$_3$ group cancels the same portion of the right CF$_3$, leaving only a portion of the C—F polarization to cancel the C—O polarization. Make a model!)

(c) **The strength of an acid is determined by the stability of its conjugate base**. The anion of **I** has no particular stabilization. The anion of **II**, however, has six F atoms pulling electron density by the inductive effect; that is, the negative charge on the oxygen is partially shared by the six electronegative F atoms pulling electron density through sigma bonds. The anion of **II** is much more stable than the anion of **I**, making **II** the stronger acid.

11-1

(a) Both reactions are oxidations.

(b) oxidation, oxidation, reduction, oxidation

(c) One carbon is oxidized and one carbon is reduced—no net change (elimination of H and OH).

(d) reduction: C—O is replaced by C—H

(e) oxidation (addition of $X_2$)

(f) Neither oxidation nor reduction—the C still has two bonds to O.

(g) neither oxidation nor reduction (addition of HX)

(h) first step: neither oxidation nor reduction (elimination of $H_2O$); second step: reduction (addition of $H_2$)

(i) oxidation: adding an O to each carbon of the double bond

(j) The first reaction is oxidation as a new C—O bond is formed to each carbon of the alkene; the second reaction is neither oxidation nor reduction, as $H_2O$ is added to the epoxide, and each carbon still has one bond to oxygen.

(k) oxidation: adding a Cl to one carbon and an O to the other

(l) Neither oxidation nor reduction: overall, only H and OH are added, so there is no net oxidation nor reduction.

11-2

**11-3**

(a) HO−Cr(=O)(=O)−OH ⟶ HO−Cr(=O)−OH

Cr begins with bonds to four oxygen atoms and ends with bonds to three oxygen atoms. Whether the bonds between the metals and the oxygen are single or double is not important—note this is NOT true of carbon! What matters here is the number of oxygen atoms bonded to the oxidizing atom.

(b) ⊖O−Cl ⟶ Cl ⊖   Cl begins with one bond to oxygen and ends with no bonds to oxygen.

(c) H₃C−S(=O)−CH₃ ⟶ H₃C−S−CH₃   S begins bonded to one oxygen and ends with no bonds to oxygen.

(d)   Iodine begins with four bonds to oxygen and ends with two bonds to oxygen.

(e)   C begins with one bond to O and (at least) one bond to H. It ends with two bonds to O and one fewer bond to H. This is one definition of oxidation: replacing a C−H bond with a C−O bond.

**11-4** Note that PCC and DMP (and Swern) oxidation stop at the aldehyde when oxidizing a primary alcohol. Chromic acid and NaOCl take a primary alcohol to the carboxylic acid.

(a) CH₃(CH₂)₆⌒OH  $\xrightarrow[\text{H}_2\text{SO}_4]{\text{Na}_2\text{Cr}_2\text{O}_7}$  AND  $\xrightarrow{\text{NaOCl}}$  CH₃(CH₂)₆−C(=O)−OH

CH₃(CH₂)₆⌒OH  $\xrightarrow[\text{These do not oxidize aldehydes.}]{\text{PCC}}$ AND $\xrightarrow{\text{DMP}}$  CH₃(CH₂)₆−C(=O)−H

(b) All four reagents give the same ketone product with a secondary alcohol.

[O] is the general abbreviation for an oxidizing agent.

(c)

$\xrightarrow[\text{H}_2\text{SO}_4]{\text{Na}_2\text{Cr}_2\text{O}_7}$  AND  $\xrightarrow{\text{NaOCl}}$

$\xrightarrow[\text{These do not oxidize aldehydes.}]{\text{PCC}}$ AND $\xrightarrow{\text{DMP}}$

(d) All four reagents give no reaction with a tertiary alcohol.

$\xrightarrow{\text{[O]}}$  NR (no reaction)

**236**
Copyright © 2013 Pearson Education, Inc.

To the student: For simplicity, this Solutions Manual will use these laboratory methods of oxidation:

—PCC (pyridinium chlorochromate) to oxidize 1° alcohols to aldehydes
—$H_2CrO_4$ (chromic acid) to oxidize 1° alcohols to carboxylic acids, and 2° alcohols to ketones

Understand that other choices are legitimate; for example, DMP and Swern oxidation works as well as PCC in the preparation of aldehydes, and chlorine bleach (NaOCl) will oxidize a 1° alcohol to a carboxylic acid as well as chromic acid does. All of these five oxidizing agents will convert a 2° alcohol to a ketone. If you have a question about the appropriateness of a reagent you choose, consult the table in the text before Problem 11-3.

**11-5** None of the five oxidation reagents affects the 3° alcohol. All five oxidize the 2° alcohol to a ketone. Chromic acid and NaOCl oxidize the 1° alcohol to COOH, whereas PCC, DMSO/oxalyl chloride (Swern), and DMP oxidize the 1° alcohol to an aldehyde.

(b) PCC OR
(d) DMSO/ClCOCOCl OR
(e) DMP

(a) $H_2CrO_4$ OR
(c) NaOCl

**11-6** Cr reagents shown over the arrow. Non-Cr reagents added in text after the reaction.

(a)
PCC
non-Cr: DMP or DMSO/ClCOCOCl (Swern)

(b)
$H_2CrO_4$
(Z or E not specified)
non-Cr: NaOCl

(c)
$H_2CrO_4$
or PCC
non-Cr: DMP or DMSO/ClCOCOCl (Swern) or NaOCl

(d)
$H_2CrO_4$
or PCC
EtMgBr
ether
$H_3O^+$
non-Cr: DMP or DMSO/ClCOCOCl (Swern) or NaOCl

(e)
PCC
EtMgBr
ether
$H_3O^+$
non-Cr: DMP or DMSO/ClCOCOCl (Swern)

(f)
$H_2SO_4$
Δ
– $H_2O$
1) $BH_3 • THF$
2) $H_2O_2$, $HO^-$
$H_2CrO_4$
or PCC
non-Cr: DMP or DMSO/ClCOCOCl (Swern) or NaOCl

237

**11-7** A chronic alcoholic has induced more ADH enzyme to be present to handle large amounts of imbibed ethanol, so requires more ethanol "antidote" molecules to act as a competitive inhibitor to "tie up" the extra enzyme molecules.

**11-8**

$$CH_3-\overset{\overset{OH}{|}}{CH}-\overset{\overset{OH}{|}}{CH_2} \xrightarrow{[O]} CH_3-\overset{\overset{O}{||}}{C}-\overset{\overset{O}{||}}{CH} \xrightarrow{[O]} CH_3-\overset{\overset{O}{||}}{C}-\overset{\overset{O}{||}}{COH}$$

$$\text{pyruvaldehyde} \qquad\qquad \text{pyruvic acid}$$

Pyruvic acid is a normal metabolite
in the breakdown of glucose ("blood sugar").

**11-9** From this problem on, "Ts" will refer to the "tosyl" or "p-toluenesulfonyl" group:

$$Ts \Longrightarrow -\overset{\overset{O}{||}}{\underset{\underset{O}{||}}{S}}-\underset{}{\bigcirc}-CH_3$$

(a) $CH_3CH_2-OTs$ + $KO-\overset{\overset{CH_3}{|}}{\underset{\underset{CH_3}{|}}{C}}-CH_3$ ⟶ $CH_3CH_2O-\overset{\overset{CH_3}{|}}{\underset{\underset{CH_3}{|}}{C}}-CH_3$ + KOTs

Lower temperature favors substitution. Higher temperature favors elimination.

(E2 is possible with this hindered base; the product would be ethylene, $CH_2{=}CH_2$.)

(b) + NaI ⟶ + NaOTs

(c) + NaCN ⟶ + NaOTs    inversion$-S_N2$

(d)

(e) + Na$^+$ $:\overset{\ominus}{C}{\equiv}CH$ ⟶ $C{\equiv}CH$ + NaOTs

**11-10** All parts begin with forming the tosylate.

**11-11**

(a)

**11-11 continued**

(b)

(c)

major + minor

(d)

**11-12**

(a) $S_N1$ on 3° alcohol

carbocation intermediate

(b) $S_N2$ on 1° alcohol

**11-13**

**11-14** The two standard qualitative tests are:

1) <u>chromic acid</u>—distinguishes 3° alcohol from either 1° or 2°

3°                                                          1°, 2°

2) <u>Lucas test</u>—distinguishes 1° from 2° from 3° alcohol by the rate of reaction

**3°** insoluble—"cloudy" in < 1 minute

**2°** insoluble—"cloudy" in 1-5 minutes

**1°** insoluble—"cloudy" in > 6 minutes
(No observable reaction at room temp.)

**239**

11-14 continued

(a)

Lucas:        cloudy in 1-5 min.
$H_2CrO_4$:   immediate blue-green

              cloudy in < 1 min.
              no reaction—stays orange

(b)

Lucas:        cloudy in 1-5 min.
$H_2CrO_4$:   immediate blue-green

              no reaction
              no reaction—stays orange

(c)

Lucas:        no reaction
$H_2CrO_4$:   DOES NOT DISTINGUISH—immediate blue-green for both

              cloudy in 1-5 min.

(d)

Lucas:        cloudy in < 1 min. **
$H_2CrO_4$:   DOES NOT DISTINGUISH—
              immediate blue-green for both

              no reaction

(**Remember that allylic cations are resonance-stabilized and are about as stable as 3° cations. Thus, they will react as fast as 3° in the Lucas test, even though they may be 1°. Be careful to notice subtle but important structural features!)

(e)

Lucas:        no reaction
$H_2CrO_4$:   DOES NOT DISTINGUISH—stays orange for both

              cloudy in < 1 min.

11-15

Even though 1°, the neopentyl carbon is hindered to backside attack, so $S_N2$ cannot occur easily. Instead, an $S_N1$ mechanism occurs, with rearrangement.

11-16

This 3° carbocation is planar at the $C^+$ so that the $Cl^-$ can approach from the top or bottom giving both the *cis* and *trans* isomers.

approach from above

approach from below

methyls *trans*

methyls *cis*

**240**
Copyright © 2013 Pearson Education, Inc.

**11-17**

ZnCl₂ ⊖ ... CH₃ ... OH ... H ... ZnCl₂ → H CH₃ O⊕ H H → CH₃ ⊕C 2° H → hydride shift *rearrangement* → 3° CH₃ ⊕C → (from HCl) :Cl:⊖ → CH₃ Cl

**11-18**

**3** ⟨(CH₃)₂C—CH₂OH⟩ + PBr₃ ⟶ **3** ⟨(CH₃)₂C—CH₂Br⟩ + P(OH)₃

**6** $CH_3(CH_2)_{14}CH_2OH$ + **2** P + **3** $I_2$ ⟶ **6** $CH_3(CH_2)_{14}CH_2I$ + **2** P(OH)₃

**3** (cyclopentyl)OH + PBr₃ ⟶ **3** (cyclopentyl)Br + P(OH)₃

**11-19**

(a) OH ⟨cyclohexane ring⟩ CH₃ →[SOCl₂] Cl ⟨cyclohexane ring⟩ CH₃ *retention*

(b) OH ⟨cyclohexane ring⟩ CH₃ →[TsCl / pyridine] OTs ⟨cyclohexane ring⟩ CH₃ →[NaCl] Cl ⟨cyclohexane ring⟩ CH₃

S_N2—*inversion*

Another possible answer would be to use PCl₃ or PCl₅.

**11-20**

(a) ⟨reaction mechanism with SOCl₂⟩

D O—H + Cl—S(=O)—Cl → D O⊕(H)—S—O⊖ with Cl, Cl → D O⊕(H)—S=O with Cl + :Cl:⊖

↓ − HCl

D O—S=O with Cl →

⟨allylic!⟩ resonance structures: { D ⊕CH ... ↔ ⊕C D ... }

+ :O⊖—S=O with Cl

↓

SO₂ + :Cl:⊖

D ⟨cyclohexene ring⟩ Cl + D Cl ⟨cyclohexene ring⟩

**241**

11-20 continued

(b) The key is that the intermediate carbocation is allylic, very stable, and relatively long-lived. It can therefore escape the ion pair and become a "free" carbocation. The nucleophilic chloride can attack any carbon with positive charge, not just the one closest. Since two carbons have partial positive charge, two products result.

11-21

**11-21 continued**

(d)

S_N1 carbocation intermediate can be attacked from either side by chloride.

$S_N1$ at 2° carbon—carbocation intermediate can be attacked from either side by bromide.

S_N2 with inversion of configuration

S_N2 with inversion of configuration

retention of configuration   *See the Problem Solving Hint next to Problem 11-20 of the text.*

**11-22** Water is also produced in each of these dehydration reactions.

(a) $H_2SO_4$, $\Delta$
major     +     minor

(b) $H_2SO_4$, $\Delta$
major (*cis* + *trans*)     +     minor     Rearrangement is likely.

(c) $H_2SO_4$, $\Delta$
major (*cis* + *trans*)     +     minor

(d) $H_2SO_4$, $\Delta$
major     +     minor

(e) $H_2SO_4$, $\Delta$
major     +     minor     +     trace     Rearrangement is likely.

11-23

good leaving group

Cyclohexene was formed without a carbocation intermediate.

This product can react with two more alcohols to become leaving groups in the E2 elimination.

11-24

Both mechanisms begin with protonation of the oxygen.

One mechanism involves another molecule of ethanol acting as a base, giving elimination.

The other mechanism involves another molecule of ethanol acting as a nucleophile, giving substitution.

11-25  An equimolar mixture of methanol and ethanol would produce all three possible ethers.  The difficulty in separating these compounds would preclude this method from being a practical route to any one of them.  This method is practical only for symmetric ethers, that is, where both alkyl groups are identical.

$$CH_3CH_2OH + HOCH_3 \xrightarrow[\Delta]{H_2SO_4} H_2O + CH_3CH_2OCH_3 + CH_3OCH_3 + CH_3CH_2OCH_2CH_3$$

11-26

(a)

(b)

(c)

hydride shift

This 1° carbocation would be very unstable.

ring expansion

(d)

starting material redrawn to show relationship to product

**245**

11-27

11-28

(a)

Methyl shift

Alkyl shift—ring *contraction*

(b)

11-29

Similar to the pinacol rearrangement, this mechanism involves a carbocation next to an alcohol, with rearrangement to a protonated carbonyl. Relief of some ring strain in the cyclopropane is an added advantage of the rearrangement.

Another oxygen in the reaction mixture is the likely base, removing this proton.

11-30

(a) **2** $H_3C-\overset{\overset{\displaystyle O}{||}}{C}-H$   (b) cyclopentanone + $O=CH_2$   (c) acetophenone + $H-\overset{\overset{\displaystyle O}{||}}{C}-CH_2CH_3$   (d) bicyclic dialdehyde

11-31   (a) $CH_3CH_2CH_2-\overset{\overset{\displaystyle O}{||}}{C}-Cl$ + $HOCH_2CH_2CH_3$     (b) $CH_3(CH_2)_3OH$ + $Cl-\overset{\overset{\displaystyle O}{||}}{C}CH_2CH_3$

(c) $H_3C-\!\!\!\bigcirc\!\!\!-OH$ + $Cl-\overset{\overset{\displaystyle O}{||}}{C}CH(CH_3)_2$     (d) cyclopropyl$-OH$ + $Cl-\overset{\overset{\displaystyle O}{||}}{C}-\bigcirc$

11-32 *The strength of an acid is determined by the stability of its conjugate base.* The more stable the conjugate base, the stronger the acid.

The methanesulfonate anion is stabilized by resonance and by induction: It has three equivalent resonance forms, plus the sulfur atom is more electronegative than carbon and plays a small role in stabilizing the negative charge on oxygen. The acetate ion has two equivalent resonance forms, but no inductive effect to stabilize the anion. Acetate is good, but the methanesulfonate ion is even better.

11-33 Proton transfer (acid-base) reactions are much faster than almost any other reaction. Methoxide will act as a base and remove a proton from the oxygen much faster than methoxide will act as a nucleophile and displace water.

11-34

(a) $CH_3CH_2OH$ $\xrightarrow[\text{pyridine}]{\text{TsCl}}$ $CH_3CH_2OTs$

⬡—OH $\xrightarrow{\text{Na}}$ ⬡—$\overset{\ominus}{\text{O}:}$ $Na^+$ } $\longrightarrow$ ⬡—$OCH_2CH_3$

(b) There are two problems with this attempted bimolecular dehydration. First, all three possible ether combinations of cyclohexanol and ethanol would be produced. Second, heat and sulfuric acid are the conditions for dehydrating secondary alcohols like cyclohexanol, so elimination would compete with substitution.

11-35

(a) What the student did:

$Na^+$ $\overset{\ominus}{O}$ ... H / $H_3C$ — $CH_2CH_3$  + TsO—$CH_2CH_3$  $\longrightarrow$  $CH_3CH_2$—O ... H / $H_3C$ — $CH_2CH_3$

sodium (S)-but-2-oxide                                  (S)-2-ethoxybutane

The product also has the S configuration, not the R. Why? The substitution is indeed an $S_N2$ reaction, but the *substitution did not take place at the chiral center*, so the configuration of the starting material is retained, not inverted.

(b) There are two ways to make (R)-2-ethoxybutane. Start with (R)-butan-2-ol, make the anion, and substitute on ethyl tosylate similar to part (a), or do an $S_N2$ inversion at the chiral center of (S)-butan-2-ol. $S_N2$ works better at 1° carbons so the former method would be preferred to the latter.

(c) This is not the optimum method because it requires $S_N2$ at a 2° carbon, as discussed in part (b).

HO ... H / $H_3C$ — $CH_2CH_3$ $\xrightarrow[\text{pyridine}]{\text{TsCl}}$ Ts—O ... H / $H_3C$ — $CH_2CH_3$ $\xrightarrow[\substack{\text{low temperature}\\\text{(high temp.}\\\text{favors elimination)}}]{Na^+ \,\, {}^-OCH_2CH_3}$ H ... $OCH_2CH_3$ / $H_3C$ — $CH_2CH_3$

(S)-butan-2-ol          (S)-2-butyl tosylate          (R)-2-ethoxybutane
                        no inversion yet              INVERSION!

11-36

⬡—OH $\xrightarrow{\text{NaOH}}$ ⬡—$\overset{\ominus}{O}:$ $Na^+$ + $CH_3$—O—$SO_3CH_3$ $\longrightarrow$ ⬡—$OCH_3$ + $\overset{\ominus}{:O}$—$SO_3CH_3$

11-37

(a) $CH_3CH_2CH_2OH$ $\xrightarrow{\text{PCC}}$ $CH_3CH_2-\overset{O}{\overset{\|}{C}}-H$ $\xrightarrow[\text{2) } H_3O^+]{\text{1) } CH_3CH_2MgBr}$ [2-pentanol structure with OH] $\xrightarrow{H_2CrO_4}$ [3-pentanone structure]

(b) $CH_3CH_2-\overset{O}{\overset{\|}{C}}-Cl$ $\xrightarrow[\text{2) } H_3O^+]{\text{1) } 2 \,\, CH_3CH_2MgBr}$ $CH_3CH_2-\overset{\overset{\displaystyle CH_2CH_3}{|}}{\underset{\underset{\displaystyle OH}{|}}{C}}-CH_2CH_3$ $\xrightarrow{H_2SO_4}$ $CH_3CH=\overset{\overset{\displaystyle CH_2CH_3}{|}}{C}-CH_2CH_3$

1) $BH_3 \bullet$ THF
2) $H_2O_2$, $HO^-$

$CH_3\overset{\overset{\displaystyle CH_2CH_3}{|}}{\underset{\underset{\displaystyle O}{\|}}{C}}CHCH_2CH_3$ $\xleftarrow{H_2CrO_4}$ $CH_3\overset{\overset{\displaystyle CH_2CH_3}{|}}{CH}\underset{\underset{\displaystyle OH}{|}}{CH}CH_2CH_3$

11-38

(a)

Any peroxy acid can be used to form the epoxide, which is cleaved to the *trans*-diol in aqueous acid.

(b)

(c)

OR:

(d)

(e) There are several possible combinations of Grignard reactions on aldehydes or ketones. This is one example. Your example may be different and still be correct.

## 11-38 continued

(f)

In a complex synthesis, it is worth the time to analyze what pieces need to be put together to create the carbon skeleton; this is called "retrosynthesis" (reverse synthesis), or the "disconnection" approach. This target has a cyclopentane ring and two four-carbon fragments, so the new C—C bonds should be made using Grignard reactions.

## 11-39 Stereochemistry is not specified in this problem.

(a)

(b)

(c)

or PCC or DMP or Swern or NaOCl

(d)

from (a)

(e)

from (c)

(f)

OR

from (b)

(g)

(h)

from (a)                from (d)

**11-40**

(a) [structure: propyl... OH, labeled 1°] —SOCl₂→ [structure with Cl]

—PBr₃→ [structure with Br]

—P / I₂→ [structure with I]

(b) [cyclopentanol, 2° OH] —SOCl₂→ [cyclopentyl Cl]

—PBr₃→ [cyclopentyl Br]

—P / I₂→ [cyclopentyl I]

(c) [3° OH cyclohexane] —HCl→ [Cl structure]

—HBr→ [Br structure]

—HI→ [I structure]

(d) [2° OH methylcyclohexane] —SOCl₂→ [Cl structure]

—PBr₃→ [Br structure]

—P / I₂→ [I structure]

**11-41**

(a) H,,,OTs [structure] R

(b) H,,,Br [structure] R (from inversion)

(c) [cyclooctanone, O]

(d) [cyclopentane carbaldehyde O, H]

(e) [cyclopentane COOH]

(f) [cyclopentyl Cl]

(g) [chain Br]

(h) [cyclooctane CH₂OMgBr] + CH₃CH₃

(i) [structure OCH₃]

(j) [alkene] + CH₃OH

(k) [cyclopentene]

(l) H [structure] O, O, H

(m) [chain O chain]

(n) [alkene] major + [alkene] minor + EtOH

(o) [chain aldehyde O, H]

(p) [cyclopentane chain O, H with ethyl]

**11-42**

(a) [cyclohexanol OH] —Na→ [cyclohexyl O⁻ Na⁺] —CH₃CH₂Br→ [cyclohexyl OCH₂CH₃]

Williamson ether synthesis

11-42 continued

(b)

(c)

(d)

11-43 Major product for each reaction is shown.

(a) *cis + trans*—rearranged

(b) *cis + trans*

(c) *cis + trans*

(d)

(e)

(f) rearranged

Note that (d), (e), and (f) produce the same alkene.

11-44

(a)

(b) $CH_3ONO_2$

(c) $CH_3CH_2O-\underset{\underset{OCH_2CH_3}{|}}{\overset{\overset{O}{||}}{P}}-OH$

(d)

(e)

11-45

cis  $\xrightarrow[\Delta]{H_2SO_4}$  Zaitsev product  $\xleftarrow[\Delta]{H_2SO_4}$  trans

only product

These two dehydrations follow the E1 mechanism with a common carbocation intermediate. The stereochemistry plays no role in the E1 mechanism.

Elimination of the tosylate with base follows the E2 mechanism with the stereochemical requirement that the H and the OTs must be anti-coplanar; see text section 6-20.

252

## 11-46

(a)

$S$ → SOCl$_2$ → $S$—retention

(b)

$S$ → TsCl / pyridine → $S$ → KBr → $R$—inversion

Alternatively, PBr$_3$ could be used.

(c)

$S$ → TsCl / pyridine → $S$ (Keep cold to avoid elimination.) → NaOH → $R$—inversion

## 11-47

(a)

H—Br → $-H_2O$ → 2° hydride shift → 3° → :Br:⁻ → 

(b) PBr$_3$ converts alcohols to bromides without rearrangement because no carbocation intermediate is produced. Alternatively, making the tosylate and displacing with bromide would also work.

OH → PBr$_3$ → Br

## 11-48

(a)

OH → PCC → 

(b)

OH → PBr$_3$ → Br

(c)

OH → Na → O⁻ Na⁺ → CH$_3$I → OCH$_3$

(d)

SOCl$_2$ → *retention*

OR

TsCl / pyridine → Cl⁻ / S$_N$2 → *inversion*

**11-48 continued**

(e)

(f)

(g)

(h)

*cis*                *cis*

**11-49**

(a)   *inversion*

PBr₃

1) TsCl, pyridine
2) NaBr

*inversion, S_N2*   (e)

(b)   *retention*

SOCl₂

HCl
ZnCl₂

*cis* and *trans*

*S_N1*

(c)

HBr

*cis* and *trans*

*S_N1*

(d)

**11-50**

(a)

Lucas:      no reaction

cloudy in 1-5 min.

(b)

Lucas:      cloudy in 1-5 min.
H₂CrO₄:   immediate blue-green

cloudy in < 1 min.
no reaction—stays orange

(c)

Lucas:      cloudy in 1-5 min.
H₂CrO₄:   immediate blue green

no reaction
no reaction—stays orange

(d)

Lucas:      cloudy in 1-5 min.
H₂CrO₄:   immediate blue green

no reaction
no reaction—stays orange

(e)

Lucas:      no reaction

cloudy in < 1 min.

**254**

11-51

(a)

(b)

The last three resonance forms are similar to the first three; the change is that the electrons are shown in alternate positions in the benzene ring. To be rigorously correct, these three resonance forms should be included, but most chemists would not write them since they do not reveal extra charge delocalization; understand that they would still be significant, even if not written with the others.

(c)

11-52

11-53

Alternatively, the last step could dehydrate the 3° alcohol, making this alkene **Z**.

Keep cold to avoid elimination.

**255**

**11-54**

The migration directly above does NOT occur
as the cation produced is not resonance-stabilized.

*An alternative mechanism could be proposed: protonate the ring oxygen, open the ring to a 2°*
*carbocation followed by a hydride shift to a resonance-stabilized cation, ring closure, and dehydration.*

hydride shift to resonance-
stabilized cation

2° carbocation

H⁺ off one O, H⁺
on the other O

**11-55**

(a)

In this presentation of the mechanism, the
rearrangement is shown concurrently with
cleavage of the C-O bond with no 1°
carbocation intermediate.

$+ \ H_2O - \overset{-}{Z}nCl_2$ ⎫
⎬ $Cl^- + H_2O + ZnCl_2$
$H - Cl$ ⎭ $H^+$

alkyl shift—
ring expansion

Recall that 1° carbocations
probably do not exist; this could
be considered a transition state.

ring expansion from 1°
carbocation to 2°, *resonance-*
*stabilized* carbocation

NOTE:

(b)

Once this 3° carbocation is formed, removal of adjacent protons produces the compounds shown.

(c)  All three of the products go through a common carbocation intermediate.

rearrangement

− H₂O

This is the product from a pinacol rearrangement.

**11-56**

(a)

CH₃OH → PBr₃ → CH₃Br

OR:  alternative ending

after hydrolysis of
Grignard product

SN1

11-56 continued

(b)

(c)

or use the S$_N$1 method
shown in part (a)

(d)

(e)

from (b)

(f)

(g)

(h)

**258**

**11-57** For a complicated synthesis like this, begin by working backwards. Try to figure out where the carbon framework came from; in this problem we are restricted to alcohols containing five or fewer carbons. The dashed boxes show the fragments that must be assembled. The most practical way of forming carbon-carbon bonds is by Grignard reactions. The epoxide must be formed from an alkene, and the alkene must have come from dehydration of an alcohol produced in a Grignard reaction.

**11-58**

(a) Both of these pseudo-syntheses suffer from the misconception that incompatible reagents or conditions can co-exist. In the first example, the $S_N1$ conditions of ionization cannot exist with the $S_N2$ conditions of sodium methoxide. The tertiary carbocation in the first step would not wait around long enough for the sodium methoxide to be added in the second step. (The irony is that the first step by itself, the solvolysis of *tert*-butyl bromide in methanol, would give the desired product without the sodium methoxide.)

In the second reaction, the acidic conditions of the first step in which the alcohol is protonated are incompatible with the basic conditions of the second step. If basic sodium methoxide were added to the sulfuric acid solution, the instantaneous acid-base neutralization would give methanol, sodium sulfate, and the starting alcohol. No reaction on the alcohol would occur.

(b)

Several synthetic sequences are possible for the second synthesis.

11-59

Compound X : — must be a 1° or 2° alcohol with an alkene; no reaction with Lucas leads to a 1° alcohol; can't be allylic as this would give a positive Lucas test.

Compound Y : — must be a cyclic ether, not an alcohol and not an alkene; other isomers of cyclic ethers are possible.

11-60

TsCl
pyridine

this reaction
works fine,
but wait.......

NaOCH₃

NO REACTION!
cannot do an S_N2 reaction

The Williamson ether synthesis is an S_N2 displacement of a leaving group by an alkoxide ion. In addition to being a 3° substrate, this tosylate cannot undergo an S_N2 reaction for two reasons. First, backside attack cannot occur because the back side of the bridgehead carbon is blocked by the other bridgehead. Second, the bridgehead carbon cannot undergo inversion because of the constraints of the bridged ring system.

Backside
attack is
blocked.

side view

This carbon cannot invert
which is required in the S_N2
mechanism.

side view

alternative synthesis: $R-OH \xrightarrow{Na} R-O^- Na^+ \xrightarrow{CH_3I} R-OCH_3$

11-61  Let's begin by considering the facts.

The axial alcohol is oxidized ten times as fast as the equatorial alcohol. (In the olden days, this observation was used as evidence suggesting the stereochemistry of a ring alcohol.)

$H_2CrO_4$
faster

$H_2CrO_4$
slower

Second, it is known that the oxidation occurs in two steps: 1) formation of the chromate ester; and 2) loss of H and chromate to form the C=O. Let's look at each mechanism.

AXIAL

$H_2CrO_4$
Step 1

Base
(H₂O)

Step 2

FASTER

*continued on next page*

## 11-61 continued

### EQUATORIAL

So what do we know about these systems? We know that substituents are more stable in the equatorial position than in the axial position because any group at the axial position has 1,3-diaxial interactions. So what if Step 1 were the rate-limiting step? We would expect that the equatorial chromate ester would form faster than the axial chromate ester; since this is contrary to what the data show, Step 1 is not likely to be rate-limiting. How about Step 2? If the elimination is rate limiting, we would expect the approach of the base (probably water) to the equatorial hydrogen (axial chromate ester) would be faster than the approach of the base to the axial hydrogen (equatorial chromate ester). Moreover, the axial ester is more motivated to leave due to steric congestion associated with such a large group. This is consistent with the relative rates of reaction from experiment. Thus, it is reasonable to conclude that the second step of the mechanism is rate-limiting.

## 11-62

(a)

Protonation of the OH makes a good leaving group. There are two reaction paths possible, $S_N1$ and $S_N2$.

$S_N1$ would give a racemic mixture.

$S_N2$ would give 100% S.

The data show that the product has "racemization with excess inversion", that is, more S than R. The best explanation is that a mixture of $S_N1$ and $S_N2$ is happening. This is not surprising as secondary halides are on the fence between the two mechanisms.

There is another explanation: when the water leaves, it might not leave all the way so it blocks the incoming bromide ion from giving the product with retention of configuration, and the $S_N2$ type backside attack is the dominant pathway.

(b)

This mechanism is the reverse of bromohydrin formation, passing through the same bromonium ion intermediate, leading to a racemic mixture of *trans*-1,2-dibromocyclopentane. The participation of the Br is called *neighboring group assistance* and explains the difference in results between (a) and (b).

(a)

It is equally likely for protonation to occur first on the ring oxygen, followed by ring opening, then replacement of $OCH_3$ by water.

My colleague Dr. Kantorowski suggests this alternative. He and I will arm wrestle to determine which mechanism is correct.

(b)

11-64 One of the keys to solving these "roadmap" or "structure proof" problems is making inferences from each piece of information given. Ask the question: "What is this fact telling me?"

Q has molecular formula $C_6H_{12}O$ ⟹ Q has one element of unsaturation ⟹ Q has a ring OR
$\qquad\qquad\qquad\qquad\qquad\qquad\qquad\qquad\qquad\qquad\qquad\qquad\qquad\qquad\qquad\qquad$ C=C or C=O

Q cannot be separated into enantiomers ⟹ Q does not have an asymmetric carbon atom

Q does not react with $Br_2$, $KMnO_4$, $H_2$ ⟹ Q has no C=C

Q reacts with $H_2SO_4$ and loses $H_2O$ ⟹ Q is an alcohol, not a C=O ⟹ Q has a ring

R has C=C; ozonolysis gives one acyclic product, S ⟹ R has a C=C in its ring

R has enantiomers ⟹ the dehydration that produced the C=C also created an asymmetric carbon atom

S is a ketoaldehyde ⟹ one C of the C=C has an H, the other has an R

Putting all this together gives a 4-membered ring: 3 is too small to fit the optical activity results, 5 is too large.

# CHAPTER 12—INFRARED SPECTROSCOPY AND MASS SPECTROMETRY

*See p. 276 for some useful web sites with infrared and mass spectra.*

12-1   The table is completed by recognizing that: $(\bar{\nu})(\lambda) = 10{,}000$

| $\bar{\nu}$ ( cm$^{-1}$ ) | 4000 | **3300** | **3003** | **2198** | 1700 | 1640 | 1600 | 400 |
|---|---|---|---|---|---|---|---|---|
| $\lambda$ ( $\mu$m ) | 2.50 | 3.03 | 3.33 | 4.55 | **5.88** | **6.10** | **6.25** | 25.0 |

12-2   In general, only bonds with dipole moments will have an IR absorption.  Bonds in a completely symmetric environment will have at most a weak absorption.

12-3

(a)  Alkene:  C=C at 1640 cm$^{-1}$ , =C—H at 3080 cm$^{-1}$ , saturated C—H below 3000 cm$^{-1}$ ; 1820 is overtone of 910 cm$^{-1}$.

(b)  Alkane:  no peaks indicating sp or sp$^2$ carbons present; only saturated C—H below 3000 cm$^{-1}$.

(c)  This IR shows more than one group.  There is a terminal alkyne shown by:  C≡C at 2100 cm$^{-1}$ , and ≡C—H at 3300 cm$^{-1}$ .  These signals indicate an aromatic hydrocarbon as well:  =C—H at 3050 cm$^{-1}$, and C=C at ≈1600 cm$^{-1}$ .  It is unusual to see no saturated C—H below 3000 cm$^{-1}$ .

12-4

(a)  2° amine, R—NH—R:  one peak at 3300 cm$^{-1}$ indicates an N—H bond;  this spectrum also shows a C=C at 1640 cm$^{-1}$ , unsaturated =C—H at 3070 cm$^{-1}$, saturated C—H below 3000 cm$^{-1}$ ; 1840 cm$^{-1}$  is overtone of 920 cm$^{-1}$.

(b)  Carboxylic acid:  the extremely broad absorption in the 2500-3500 cm$^{-1}$  range, with a "shoulder" around 2500-2700 cm$^{-1}$ , and a C=O at 1710 cm$^{-1}$ , are compelling evidence for a carboxylic acid; saturated C—H below 3000 cm$^{-1}$.

(c)  Alcohol:  strong, broad O—H at 3330 cm$^{-1}$ ; saturated C—H below 3000 cm$^{-1}$.

12-5

(a)  Conjugated ketone:  the small peak at 3030 cm$^{-1}$ suggests =C—H, and the strong peak at 1685 cm$^{-1}$  is consistent with a ketone conjugated with the alkene.  The C=C is indicated by a very small peak around 1620 cm$^{-1}$ consistent with conjugation; saturated C—H below 3000 cm$^{-1}$ ; 3440 is overtone of 1685 cm$^{-1}$ .

(b)  Ester:  the C=O absorption at 1738 cm$^{-1}$ (higher than the ketone's 1710 cm$^{-1}$ ), in conjunction with the strong C—O at 1200 cm$^{-1}$, points to an ester; saturated C—H below 3000 cm$^{-1}$ .

(c)  Amide:  the two peaks at 3160-3360 cm$^{-1}$  are likely to be an NH$_2$ group; the strong peak at 1640 cm$^{-1}$ is too strong for an alkene, so it must be a different type of C=X, in this case a C=O, so low because it is part of an amide; saturated C—H below 3000 cm$^{-1}$.

12-6

(a)  The moderate peak at 1642 cm$^{-1}$ indicates a C=C, consistent with the =C—H at 3080 cm$^{-1}$ .  This appears to be a simple alkene.  Saturated C—H below 3000 cm$^{-1}$; 1825 cm$^{-1}$ is overtone of peak at 915 cm$^{-1}$.

12-6 continued

(b) The strong absorption at 1691 cm$^{-1}$ is unmistakably a conjugated C=O. The smaller peak at 1626 cm$^{-1}$ indicates a C=C, probably conjugated with the C=O. The two peaks at 2712 cm$^{-1}$ and at 2814 cm$^{-1}$ represent H—C=O confirming that this is an aldehyde. Unsaturated =C—H above 3000 cm$^{-1}$ ; saturated C—H below 3000 cm$^{-1}$ .

(c) The strong peak at 1650 cm$^{-1}$ is C=C, probably conjugated with C=O as it is unusually strong. The 1703 cm$^{-1}$ peak appears to be a conjugated C=O, undeniably a carboxylic acid because of the strong, broad O—H absorption from 2400-3400 cm$^{-1}$ . The unsaturated =C—H is obscured by the strong O—H; saturated C—H below 3000 cm$^{-1}$ .

(d) The C=O absorption at 1742 cm$^{-1}$ coupled with C—O at 1220 cm$^{-1}$ suggest an ester but 1742 cm$^{-1}$ is too high to be conjugated. The small peak at 1604 cm$^{-1}$ , peaks above 3000 cm$^{-1}$, and peaks in the 600-800 cm$^{-1}$ region indicate a benzene ring. It has both unsaturated C—H above 3000 cm$^{-1}$ and saturated C—H below 3000 cm$^{-1}$ .

12-7

(a) The M and M+2 peaks of equal intensity identify the presence of bromine. The mass of M (156) minus the weight of the lighter isotope of bromine (79) gives the mass of the rest of the molecule: $156 - 79 = 77$. The $C_6H_5$ (phenyl) group weighs 77; this compound is bromobenzene, $C_6H_5Br$ .

(b) The m/z 127 peak shows that iodine is present. The molecular ion minus iodine gives the remainder of the molecule: $156 - 127 = 29$ . The $C_2H_5$ (ethyl) group weighs 29; this compound is iodoethane, $C_2H_5I$ .

(c) The M and M+2 peaks have relative intensities of about 3:1, a sure sign of chlorine. The mass of M minus the mass of the lighter isotope of chlorine gives the mass of the remainder of the molecule: $90 - 35 = 55$. A fragment of mass 55 is not one of the common alkyl groups (15, 29, 43, 57, *etc.*, increasing in increments of 14 mass units ($CH_2$)), so the presence of unsaturation and/or an atom like oxygen must be considered. In addition to the chlorine atom, mass 55 could be $C_4H_7$ or $C_3H_3O$. Possible molecular formulas are $C_4H_7Cl$ or $C_3H_3ClO$.

(d) The odd-mass molecular ion indicates the presence of an odd number of nitrogen atoms (always begin by assuming *one* nitrogen). The rest of the molecule must be: $115 - 14 = 101$; this is most likely $C_7H_{17}$. $C_7H_{17}N$ is the correct formula of a molecule with no elements of unsaturation. The seven carbons probably include alkyl groups like ethyl or propyl or isopropyl. A hint about spectra with odd-mass molecular ions: Look for fragment peaks with even masses to confirm that the odd-mass peak is the parent and not just a fragment from an invisible molecular ion.

12-8  Recall that radicals are not detected in mass spectrometry; only positively charged ions are detected.

The fragment giving m/z 57 is a primary carbocation, less stable than the more abundant secondary carbocations of m/z 43 and 71.

**12-9**

m/z 100 → CH₃—CH(CH₃)⁺ (m/z 43) + •CH₂—CH(CH₃)—CH₃ (mass 57)

→ CH₃—CH(CH₃)• (mass 43) + ⁺CH₂—CH(CH₃)—CH₃ (m/z 57)

→ CH₃—CH(CH₃)—CH₂—CH(CH₃)⁺ (m/z 85) + •CH₃ (mass 15)

**12-10**  The molecular weight of each isomer is 116 g/mole, so the molecular ion appears at m/z 116.  The left half of each structure is the same; loss of a three-carbon radical gives a stabilized cation, each with m/z 73:

m/z 73          m/z 73

Where the two structures differ is in the alpha-cleavage on the right side of the oxygen.  Alpha-cleavage on the left structure loses two carbons, whereas alpha-cleavage on the right structure loses only one carbon.

alpha-cleavage    loss of •CH₂CH₃          alpha-cleavage    loss of •CH₃

m/z 87          m/z 101

**12-11**  2,6-Dimethylheptan-4-ol, C₉H₂₀O, has molecular weight 144.  The highest mass peak at 126 is *not* the molecular ion, but rather is the loss of water (18) from the molecular ion.

m/z 144 → H₂O + (m/z 126)

The peak at m/z 111 is loss of another 15 (CH₃) from the fragment of m/z 126.  This is called allylic cleavage; it generates a 2°, allylic, resonance-stabilized carbocation.

m/z 126 → (m/z 111) + •CH₃ (mass 15)

The peak at m/z 87 results from fragmentation on one side of the alcohol:

m/z 144 → alpha-cleavage → m/z 87  resonance-stabilized + (mass 57)

**266**

**12-12**  Table 12-2 and text Appendix 2 are particularly helpful in this problem.  All values have units of $cm^{-1}$.

(a)  C=C stretch between 1640 and 1680, weak or absent because it is symmetrically substituted, so there is essentially no dipole change; the =C—H stretch appears just above 3000.

(b)  broad, strong O—H stretch centered around 3300

(c)  strong C=O stretch about 1710

(d)  C≡C stretch below 2200; ≡C—H stretch , sharp, around 3300

(e)  broad N—H stretch of medium strength centered around 3300, with one spike

(f)  broad, strong O—H stretch centered around 3000, covering the C—H stretch, often with a shoulder around 2700; strong C=O stretch around 1710

(g)  sharp, strong C≡N stretch around 2200

(h)  strong C=O stretch about 1735; often a strong C—O stretch around 1200

(i)  strong C=O stretch around 1650—sometimes amides show two peaks; strong N—H stretch around 3300 with two spikes

**12-13**   Divide the numbers into 10,000 to arrive at the answer.

(a)  1603 $cm^{-1}$     (b)  2959 $cm^{-1}$     (c)  1709 $cm^{-1}$     (d)  1739 $cm^{-1}$     (e)  2212 $cm^{-1}$     (f)  3300 $cm^{-1}$

**12-14**

(a)

(b)

(c)

(d)

**12-15**

(a)

**267**

12-15 continued

(b) 1645 cm⁻¹

1620 cm⁻¹
conjugated           not conjugated

(c) CH₃(CH₂)₃ – C – H ← 1725 cm⁻¹     and     1710 cm⁻¹ → CH₃(CH₂)₂ – C – CH₃
        O                                                              O

2700-2800 cm⁻¹
two small peaks

(d) CH₃CH₂CH₂ – C – N – H, ← 1650 cm⁻¹     and     CH₃CH₂ – C – CH₂CH₃ ← 1710 cm⁻¹
                    O                                                    O
                    H } 3300 cm⁻¹
                       two peaks

(e) CH₃(CH₂)₅ – C≡C – H     and     CH₃(CH₂)₆ – C≡N
         3300 cm⁻¹

2100-2200 cm⁻¹          2200-2300 cm⁻¹
weak to moderate       moderate to
intensity              strong intensity

(f) CH₃CH₂CH₂ – C – OH ← 1710 cm⁻¹     and     CH₃ – CH – CH₂ – C – H ← 1725 cm⁻¹
                O                                       O – H  3300 cm⁻¹  broad, strong
                                                                O
2500-3500 cm⁻¹                                  2700-2800 cm⁻¹
very broad                                      two small peaks

(g) O ← 3300 cm⁻¹     and     O
       H broad, strong
                              1710 cm⁻¹
1200 cm⁻¹                     strong

(h) O ← 1685 cm⁻¹ , strong     and     O ← 1710 cm⁻¹ , strong
      ← 1620 cm⁻¹                         ← 1645 cm⁻¹ , weak to moderate
      absent or weak
conjugated  relative to C=O    not conjugated

12-16

(a) O ← 1690 cm⁻¹
    C – OH
H
 C=C   2400-3400 cm⁻¹
H    CH₃  2900-3000 cm⁻¹
  1640 cm⁻¹

(b) H  O ← 1715 cm⁻¹
       (3420 cm⁻¹ overtone)
   CH₃ – C – C – CH₃
2900-     CH₃
3000 cm⁻¹

(c) H ← 3000-3100 cm⁻¹
    CH₂ – C≡N
            2250 cm⁻¹
1600 cm⁻¹   2900-3000 cm⁻¹

(d) H ← 3000-3100 cm⁻¹
    H₂  ← 2900-3000 cm⁻¹
    C
    C – O
    H₂   H  3200-3500 cm⁻¹
            1050 cm⁻¹
1600 cm⁻¹

12-17

(a)

$$\left[ CH_3 - CH - CH_2CH_2CH_3 \right]^{\cdot +}$$

with $CH_3$ branch (71), 43 marked — m/z 86

$\longrightarrow$

$\overset{+}{C}H - CH_2CH_2CH_3$ with $CH_3$ — m/z 71

$CH_3 - \overset{+}{C}H$ with $CH_3$ — m/z 43

(b)

$$\left[ CH_3 - CH = C - CH_2 - CH_2CH_3 \right]^{\cdot +}$$ with $CH_3$ branch, 69

m/z 98—allylic cleavage

$\longrightarrow$

$\left\{ CH_3 - CH = C - \overset{+}{C}H_2 \longleftrightarrow CH_3 - \overset{+}{C}H - C = CH_2 \right\}$ with $CH_3$ branches

m/z 69

(c)

$$\left[ CH_3 - CH - CH_2 - CH \cdot CH_3 \right]^{+}$$ with OH (87) and $CH_3$ branches, 45

m/z 102

alpha-cleavage $\longrightarrow$

$\left\{ :\!\ddot{O}H \; CH - CH_2 - \overset{+}{C}H \cdot CH_3 \longleftrightarrow \overset{+}{\ddot{O}H} \; CH - CH_2 - CH \cdot CH_3 \right\}$ with $CH_3$

m/z 87

$\downarrow$ $- H_2O$

$$\left[ CH_3 - CH = CH - CH \cdot CH_3 \right]^{\cdot +}$$ with $CH_3$

m/z 84

alpha-cleavage $\longrightarrow$

$\left\{ CH_3 - \overset{+}{C}H \;:\!\ddot{O}H \longleftrightarrow CH_3 - CH \; \overset{+}{\ddot{O}H} \right\}$

m/z 45

(d)

$$\left[ \text{phenyl—CH}_2\text{—CH(CH}_3)_2 \right]^{\cdot +}$$ 77, 43, 57, 91 marked

m/z 134

$\longrightarrow$ benzyl cation ($\overset{+}{C}H_2$ on phenyl)

rearrange $\longrightarrow$ tropylium ion, m/z 91

The *tropylium ion* is a characteristic fragment from phenylalkanes. It has six more equivalent resonance forms—particularly stable.

$C^{+}$ on benzene ring — m/z 77

$H\overset{+}{C}$ with two $CH_3$ — m/z 43

$H_2\overset{+}{C}$ — rearrange $\longrightarrow$ $H_3C - \overset{+}{C} - CH_3$ with $CH_3$ — m/z 57

(e)

$$\left[ \text{cyclohexyl—O—C(CH}_3)_3 \right]^{\cdot +}$$ 83, 43, 127 marked

m/z 142

$\longrightarrow$ $\overset{+}{C}H$ on cyclohexyl — m/z 83

$H\overset{+}{C}$ with $CH_3$, $CH_3$ — m/z 43

$\downarrow$ alpha-cleavage

$\left\{ \text{cyclohexyl} - \ddot{O} - \overset{+}{C}H \text{ with } CH_3 \longleftrightarrow \text{cyclohexyl} - \overset{+}{O} = CH \text{ with } CH_3 \right\}$

m/z 127

**269**

12-17 continued

(f) CH₃CH₂—C(H₂)—N(H)—C(CH₃)(CH₃)CH₃  $m/z\ 115$  (57, 86, 100)

→ { $H_2C$—N(H)—C(CH₃)(CH₃)CH₃ ↔ $H_2C$=N(H)—C(CH₃)(CH₃)CH₃ }  $m/z\ 86$

→ C(CH₃)(CH₃)—CH₃  $m/z\ 57$

→ { CH₃CH₂—C(H₂)—N(H)—C(CH₃)CH₃ ↔ CH₃CH₂—C(H₂)—N(H)=C(CH₃)CH₃ }  $m/z\ 100$

(g) [ Ph—C(=O)—CH₃ ]⁺  $m/z\ 120$  (77, 105, 43)

→ { Ph—C(=O)⁺ ↔ Ph—C≡O⁺ ↔ plus other resonance forms with (+) on ring }  $m/z\ 105$

→ Ph—C⁺  $m/z\ 77$

→ { C(=O)—CH₃ ↔ ⁺O≡C—CH₃ }  $m/z\ 43$

These are three alpha-cleavages.

(h) [ H₃C—C(H)(CH₃)—C(H)(Br)—C(H₂)—CH₃ ]⁺  $m/z\ 164,\ 166$  (43, 85, 121,123)

→ { HC⁺—C(H₂)—CH₃ with Br ↔ HC=C(H₂)—CH₃ with ⁺Br }  $m/z\ 121,\ 123$

→ H₃C—C(H)(CH₃)⁺  $m/z\ 43$

→ H₃C—C(H)(CH₃)—C(H)⁺—C(H₂)—CH₃  $m/z\ 85$

Any fragment containing Br will reflect the two isotopes of Br with approximately 50:50 natural abundance.

12-18

(a) [ n-octane chain ]⁺  $m/z\ 114$  (57, 71, 85)

→ ⁺CH₂ (chain)  $m/z\ 85$  +  H₂C•—CH₃  mass 29

→ CH₂⁺ (chain)  $m/z\ 71$  +  H₂C• (ethyl)  mass 43

→ ⁺CH₂ (chain)  $m/z\ 57$  +  H₂C• (propyl)  mass 57

12-18 continued

(b)

$$\left[\overset{CH_3}{\underset{\text{(cyclohexane ring)}}{\bigcirc}}\right]^{+\cdot} \longrightarrow \overset{\oplus}{\underset{\text{(ring)}}{C}}\overset{H}{} \quad + \quad \cdot CH_3$$

m/z 98       m/z 83     mass 15

(c)

$$\left[CH_3-\overset{CH_3}{\overset{|}{C}}=CH-CH_2\overset{69}{\vert}CH_3\right]^{+\cdot} \longrightarrow \left\{CH_3-\overset{CH_3}{\overset{|}{C}}=CH-\overset{\oplus}{C}H_2 \longleftrightarrow CH_3-\overset{CH_3}{\overset{|}{C}}-CH=CH_2\right\}$$

m/z 84             m/z 69 (with $\oplus$ on central C in second form)

$$+ \quad \cdot CH_3$$
mass 15

(d)

$$\left[\overset{OH}{\underset{\underset{31}{CH_2}}{\overset{|}{}}}\!\!\vert\, CH_2\text{-}CH_2CH_2CH_3\right]^{+\cdot} \longrightarrow \left\{\overset{:\overset{\cdot\cdot}{O}H}{\underset{\overset{\oplus}{C}H_2}{\overset{|}{}}} \longleftrightarrow \overset{\oplus\overset{\cdot\cdot}{O}H}{\underset{CH_2}{\overset{\|}{}}}\right\} + \cdot CH_2CH_2CH_2CH_3$$

m/z 88            m/z 31        mass 57

$\downarrow -H_2O$

$$\left[CH_2=CH-CH_2\overset{41}{\vert}CH_2\overset{55}{\vert}CH_3\right]^{+\cdot} \longrightarrow CH_2=CH-CH_2-\overset{\oplus}{C}H_2 \quad + \quad \cdot CH_3$$

m/z 70           m/z 55      mass 15

$\downarrow$

$$\left\{CH_2=CH-\overset{\oplus}{C}H_2 \longleftrightarrow \overset{\oplus}{C}H_2-CH=CH_2\right\} \quad + \quad \cdot CH_2CH_3$$

m/z 41           mass 29

(e)

$$\left[\overset{77}{\underset{}{\bigcirc}}\vert\overset{H}{\underset{}{N}}\vert\underset{106}{}CH_2CH_3\right]^{+\cdot}$$

m/z 121

$\longrightarrow \overset{\oplus}{\underset{}{\bigcirc}}\overset{}{C} \quad$ m/z 77    $+ \quad \cdot NHCH_2CH_3$
                                 mass 44

alpha-cleavage

$$\left\{\overset{H}{\underset{}{\bigcirc}}\!\!\overset{\overset{\cdot\cdot}{N}}{}\overset{}{\underset{\oplus}{C}H_2} \longleftrightarrow \overset{\oplus H}{\underset{}{\bigcirc}}\!\!\overset{N}{}=CH_2\right\} \quad + \quad \cdot CH_3$$

m/z 106                   mass 15

12-18 continued

(f)

Any fragment containing Br will reflect the two isotopes of Br with approximately 50:50 natural abundance.

12-19

(a) The characteristic frequencies of the OH absorption and the C=C absorption will indicate the presence or absence of the groups. A spectrum with an absorption around 3300 cm$^{-1}$ will have some cyclohexanol in it; if that same spectrum also has a peak at 1645 cm$^{-1}$, then the sample will also contain some cyclohexene. Pure samples will have peaks representative of only one of the compounds and not the other. Note that *quantitation* of the two compounds would be very difficult by IR because the strength of absorptions are very different. Usually, other methods are used in preference to IR for quantitative measurements.

(b) Mass spectrometry can be misleading with alcohols. Usually, alcohols dehydrate in the inlet system of a mass spectrometer, and the characteristic peaks observed in the mass spectrum are those of the alkene, not of the parent alcohol. For this particular analysis, mass spectrometry would be unreliable and perhaps misleading.

12-20

(a) The "student prep" compound must be 1-bromobutane. The most obvious feature of the mass spectrum is the pair of peaks at M and M+2 of approximately equal heights, characteristic of a bromine atom. Loss of bromine (79) from the molecular ion at 136 gives a mass of 57, $C_4H_9$, a butyl group. Which of the four possible butyl groups is it? The peaks at 107 (loss of 29, $C_2H_5$) and 93 (loss of 43, $C_3H_7$) are consistent with a linear chain, not a branched chain. A branched chain is more likely to lose $CH_3$ (loss of 15).

(b) The base peak at 57 is so strong because the carbon-halogen bond is the weakest in the molecule. Typically, loss of a halogen is the dominant fragmentation in alkyl halides.

**12-21**

(a) Deuterium has twice the mass of hydrogen, but similar spring constant, k. Compare the frequency of C—D vibration to C—H vibration by setting up a ratio, changing only the mass (substitute 2m for m).

$$\frac{\nu_D}{\nu_H} = \frac{\sqrt{k/2m}}{\sqrt{k/m}} = \frac{\sqrt{1/2}\ \sqrt{k/m}}{\sqrt{k/m}} = \sqrt{1/2} = 0.707$$

$$\nu_D = 0.707\ \nu_H = 0.707\ (3000\ \text{cm}^{-1}) \approx \textbf{2100 cm}^{-1}$$

(b) The functional group most likely to be confused with a C—D stretch is the alkyne (carbon-carbon triple bond), which appears in the same region and is often very weak.

(If you have had physics, you may recognize this version of Hooke's Law that describes the motion of weights on a spring.)

**12-22**

The most likely fragmentation of 2,2,3,3-tetramethylbutane will give a 3° carbocation, the most stable of the common alkyl cations. The molecular ion should be small or non-existent while m/z 57 is likely to be the base peak, whereas the molecular ion peaks will be more prominent for octane and for 3,4-dimethylhexane.

**12-23**

(a) The information that this mystery compound is a hydrocarbon makes interpreting the mass spectrum much easier. (It is relatively simple to tell if a compound has chlorine, bromine, or nitrogen by a mass spectrum, but oxygen is difficult to determine by mass spectrometry alone.) A hydrocarbon with molecular ion of 110 can have only 8 carbons (8 x 12 = 96) and 14 hydrogens. The formula $C_8H_{14}$ has two elements of unsaturation.
(b) The IR will be useful in determining what the elements of unsaturation are. Cycloalkanes are generally not distinguishable in the IR. An alkene should have an absorption around 1600-1650 cm$^{-1}$; none is present in this IR. An alkyne should have a small, sharp peak around 2200 cm$^{-1}$—PRESENT AT 2120 cm$^{-1}$! Also, a sharp peak around 3300 cm$^{-1}$ indicates a hydrogen on an alkyne, so the alkyne is at one end of the molecule. Both elements of unsaturation are accounted for by the alkyne. (Also, alkynes smell bad, perhaps because of that linear geometry poking the nose like a sharp stick!)
(c) The only question is how are the other carbons arranged. The mass spectrum shows a progression of peaks from the molecular ion at 110 to 95 (loss of $CH_3$), to 81 (loss of $C_2H_5$), to 67 (loss of $C_3H_7$). The mass spectrum suggests it is a linear chain. The extra evidence that hydrogenation of the mystery compound gives *n*-octane verifies that the chain is linear. The original compound must be oct-1-yne.

12-23 continued

(d) The base peak is so strong because the ion produced is stabilized by resonance.

12-24

(a) and (b) The mass spec is consistent with the formula of the alkyne, $C_8H_{14}$, mass 110. The IR is not consistent with the alkyne, however. Often, symmetrically substituted alkynes have a miniscule C≡C peak, so the fact that the IR does not show this peak does not prove that the alkyne is absent. The important evidence in the IR is the significant peak at 1620 cm$^{-1}$ and the =C—H absorption above 3000 cm$^{-1}$; this absorption is characteristic of a conjugated diene. Instead of the alkyne being formed, the reaction must have been a double elimination to the diene.

12-25

(a)

(b)

(c)

12-26

(a)

2-methylhexan-2-ol
$C_7H_{16}O$   mol. wt. 116

## 12-26 continued

(b) The molecular ion is not visible in the spectrum. Alcohols typically dehydrate in the hot inlet system of the mass spectrometer, especially true for 3° alcohols that are the easiest type to dehydrate. The two fragmentations that produce a resonance-stabilized carbocation give the major peaks in the spectrum at m/z 59 and 101.

This is a good example of how the peak at m/z 101 was probably NOT the molecular ion, because there are no significant even mass peaks that come from it. The peak at 101 must have been a fragment peak.

12-27 The unknown compound has peaks in the mass spec at m/z 198, 155, 127, 71, and 43. It is helpful that the masses of two of the fragments, 155 and 43, sum to 198, as do the other two fragment masses, 127 and 71. We can say with certainty that the molecular ion is at m/z 198, and that the unknown is a relatively simple molecule with two main fragmentations.

This is a fairly high mass for a simple compound; some heavy group must be present. What is NOT present is N because of the even molecular ion mass, nor Cl nor Br because of the lack of isotope peaks, nor a phenyl group because of the absence of a peak at 77. The progression of alkyl group masses: 15, 29, **43**, 57, **71**, 85, 99—includes two of the peaks, so it appears that the unknown contains a propyl group and a pentyl group (the propyl could be part of the pentyl group). The 127 fragment is key; the fragment $C_9H_{19}$ has this mass, but we would expect much more fragmentation from a nine carbon piece. There must be some other explanation for this 127 peak.

And there is! There is one piece—more specifically, one atom—that has mass 127: iodine! In all probability, the iodine atom is attached to a fragment of mass 71 which is $C_5H_{11}$, a pentyl group. We cannot tell with certainty what isomer it is, so unless there is some other evidence, let's propose a branched chain isomer, 1-iodo-2-methylbutane.

The 155 fragment probably has this bridged structure because iodine is so big:

## 12-28

(a)

broad, strong OH peak around 3300 cm$^{-1}$

strong peak around 1715 cm$^{-1}$

**12-28 continued**

**(b)**

$1620 \text{ cm}^{-1}$

strong peak around $1690 \text{ cm}^{-1}$

two peaks at $2700$ and $2800 \text{ cm}^{-1}$

$1600 \text{ cm}^{-1}$

OH broad, strong OH peak around $3300 \text{ cm}^{-1}$

$1640 \text{ cm}^{-1}$

$1610\text{-}1620 \text{ cm}^{-1}$ — less conjugation

**(c)**

OH very broad $2500\text{-}3500 \text{ cm}^{-1}$ with a "shoulder" around $2700 \text{ cm}^{-1}$

broad COOH band missing

still has a strong OH at $3300 \text{ cm}^{-1}$ but different from COOH

**12-29**

Use the MS to determine how large a molecule the unknown is. With highest peak at m/z 121, an odd molecular ion suggests the presence of N. If one N, then $121 - 14 = 107$ for C and H: the only plausible formula is $C_8H_{11}N$, unless oxygen or some other element is present. If $C_8H_{11}N$, this formula has 4 elements of unsaturation, the minimum needed for a benzene ring. The mass spectrum shows a peak at 77, the phenyl group, so a monosubstituted benzene ring is part of the structure. The other significant peak in the MS is at 106, M − 15, so a methyl group is present.

Pieces from the MS:

$- CH_3$     $+ \ N \ + \ C \ + \ 3 \ H$

From the IR: The most important peak is the single peak at $3400 \text{ cm}^{-1}$ showing the presence of N—H. The only two possible structures are:

OR

This structure would have lost $CH_3$ as its major fragment, predicting that m/z 106 should be the base peak—it is! This must be the correct structure.

This structure would have a m/z 91 for the benzyl group much higher than the m/z 106. This structure does not fit the data.

---

If you wish to find IR spectra and mass spectra of common compounds, there are two websites that are very helpful. Entering a name or molecular formula will give isomers from which to choose the desired structure and the IR spectrum or MS if available in their database.

http://webbook.nist.gov/
"NIST" is the U.S. National Institute of Standards and Technology.

http://riodb01.ibase.aist.go.jp/sdbs/cgi-bin/cre_index.cgi?lang=eng
This is from the National Institute of Advanced Industrial Science and Technology of Japan.

A benzene ring can be written with three alternating double bonds or with a circle in the ring. All of the carbons and hydrogens in an unsubstituted benzene ring are equivalent, regardless of the symbolism used.

          equivalent to

*The Japanese website listed at the bottom of p. 276 also gives proton and carbon NMR spectra.*

**Reminder: The word "spectrum" is singular; the word "spectra" is plural.**

13-1

(a) $\dfrac{650 \text{ Hz}}{300 \times 10^6 \text{ Hz}} = 2.17 \times 10^{-6} = 2.17$ ppm downfield from TMS

(b) The chemical shift does not change with field strength: $\delta$ 2.17 at both 60 MHz and 300 MHz.

(c) $(2.17 \text{ ppm}) \times (60 \text{ MHz}) = (2.17 \times 10^{-6}) \times (60 \times 10^6 \text{ Hz}) = 130$ Hz

13-2 Numbers are chemical shift values, in ppm, derived from Table 13-3 and Appendix 1 in the text. *Your predictions should be in the given range, or within 0.5-1.0 ppm of the given value.*

(a)

a = δ 5-6
b = δ 0.9

(b)

a = δ 7.2
b = δ 2.3

(c)

a = δ 7.2
b = δ 3.6

(d)

a = δ 2-5
b = δ 2.5
c = δ 1-2

(e)

a = δ 10-12
b ≈ δ 2.5 (CH₃ next to C=O is around 2.1)
c ≈ δ 2.7 (CH₃ next to benzene is around 2.3)
d = δ 7.2

a CH₂ is about 0.4 greater than CH₃

(f)

a = δ 3-4
b = δ 1-2

(The hydrogens labeled "d" are not equivalent. They appear at roughly the same chemical shift because the substituent is neither strongly electron-donating nor withdrawing.)

13-3

(a) CH₃CH₂CH₂Cl
　　　c　b　a

three types of H

(b) CH₃CHCH₃
　　　b　a　b
　　　　|
　　　　Cl

two types of H

(c)
$$\begin{array}{c}a\ CH_3\ b\ c\\CH_3-\overset{|}{\underset{\underset{a}{|}}{C}}-CH_2CH_3\\CH_3\end{array}$$

three types of H

(d)
$$\begin{array}{cc}a & a\\H_3C & CH_3\\ \diagdown & \diagup\\ b\ HC-CH\ b\\ \diagup & \diagdown\\ H_3C & CH_3\\ a & a\end{array}$$

two types of H

13-4
(a)

four types of H

(b) The three types of aromatic hydrogens appear in a relatively small space around δ 7.2. The signal is complex because all the peaks from the three types of hydrogens overlap.

three types of H (for (e))

five types of H (for (f))

Note: NMR spectra drawn in this Solutions Manual will represent peaks as single lines. These lines may not look like "real" peaks, but this avoids the problem of variation among spectrometers and printers. Individual spectra may look different from the ones presented here, but all of the important information will be contained in these representational spectra.

13-5

$$\begin{array}{c}c\\CH_3\ O\ \ \ \ O\\H_3C-\overset{|}{\underset{\underset{c}{|}}{C}}-O-\overset{\|}{C}-CH_2-\overset{\|}{C}-CH_3\\CH_3\end{array}$$
　　a　　b
　　c

c 9H

b 3H

a 2H

TMS

δ (ppm)

13-6  The three spectra are identified with their structures. Data are given as chemical shift values, with the integration ratios of each peak given in parentheses.

Spectrum (a)

$$\begin{array}{c}b\\OH\\c\ \ \ |\ \ \ a\\CH_3-\overset{|}{\underset{\underset{c}{|}}{C}}-C\equiv C-H\\CH_3\end{array}$$

a = δ 2.4 (1) (1H)
b = δ 2.6 (1) (1H)
c = δ 1.5 (6) (6H)

Spectrum (b)

CH₃O—⟨ring⟩—OCH₃
(aromatic H's labeled a, b)

a = δ 6.8 (2) (4H)
b = δ 3.7 (3) (6H)

**278**

Copyright © 2013 Pearson Education, Inc.

13-6 continued

Spectrum (c)

$$
\begin{array}{c}
\text{b} \\
\text{CH}_3 \quad \text{a} \\
\text{b} \quad | \\
\text{CH}_3-\overset{|}{\underset{|}{\text{C}}}-\text{CH}_2\text{Br} \\
\text{Br}
\end{array}
$$

a = δ 3.9 (1) (2H)   (This is the compound in Problem 13-2(f). You
b = δ 1.9 (3) (6H)   may wish to check your answer to that question
against the spectrum.)

13-7 Chemical shift values are approximate and may vary slightly from yours. The splitting and integration values should match exactly, however.

(c)

Note: the three types of aromatic protons above are *accidentally equivalent* because the isopropyl group has little effect on their chemical shifts.

(To show the pattern of peaks, this septet is larger than it would appear on a real spectrum.)

(d)

All four aromatic protons are equivalent.

(e)

No splitting is observed when *chemically equivalent* hydrogens are adjacent to each other, as with hydrogens "b" above.

**280**

13-8

(a)

H δ 5.8
δ 2.4 H₂C
δ 1.8 H₂C
H₃C    CH₃    H δ 6.7

δ 1.2

The signals at 1.8 and 2.4 are triplets because each set of protons has two neighboring hydrogens. The N+1 rule correctly predicts each to be a triplet.

(b)

δ 1.1
H₃C   CH₃   δ 7.3 H
δ 1.5 H₂C
δ 1.65 H₂C
C    CH₃
H₂    δ 1.7
δ 2.1    allylic
allylic

CH₃ δ 2.3
H δ 6.2

The three CH₂ groups can be distinguished by chemical shift and by splitting. The allylic CH₂ will be the farthest downfield of the three, at 2.1. The signal at 1.5 is a triplet, so that must be the one with only two neighboring protons. The CH₂ showing the multiplet at 1.65 must be the one between the other two CH₂ groups.

13-9

(a)

The three types of aromatic protons are *accidentally equivalent*.

a 5H

b 1H J = 15 Hz

c 1H J = 15 Hz

d 9H

a H      H a
a H—          —H a
H—          —H
H a  C=C  c H
H b      C(CH₃)₃ d

TMS

(b)

b CH₃O          c CH₃
C=C
Cl          H a

a 1H

b 3H

c 3H

TMS

13-9 continued

(c)

e
$(CH_3)_3C$

$\overset{O}{\overset{\|}{C}}$–$OCH_2CH_3$
c    d

$C=C$

H    H
a    b

e
9H

d
3H

c
2H

a
1H

b
1H

TMS

δ (ppm)

(d)

a
1H

11
The chemical shift
of —COOH is
variable. It is
usually a broad
peak.

b
2H

c
2H

The electron-withdrawing
carbonyl group deshields the
protons adjacent to it, moving
them downfield from their
usual position at δ 7.2.

d
CH₃

c
H

c
H

b
H

b
H

$O=\overset{}{C}$–OH
a

d
3H

TMS

δ (ppm)

13-10  The formula $C_3H_2NCl$ has three elements of unsaturation.  The IR peak at 1650 cm$^{-1}$ indicates an
alkene, while the absorption at 2200 cm$^{-1}$ must be from a nitrile (not enough carbons left for an alkyne).
These two groups account for the three elements of unsaturation.  So far, we have:

$C≡N$

$C=C$    + 2 H  +  Cl

H    $C≡N$

$C=C$

Cl    H

The NMR gives the coupling constant for the two protons as 14 Hz.
This large *J* value shows the two protons as *trans* (*cis*, *J* = 10 Hz;
geminal, *J* = 2 Hz).  The structure must be the one in the box.

13-11

(a)  $C_3H_7Cl$—no elements of unsaturation;  three types of protons in the ratio of 2 : 2 : 3 .

a    b    c
Cl—$CH_2CH_2CH_3$

a = δ 3.8 (triplet, 2H); b = δ 2.1 (multiplet, 2H);
c = δ 1.3 (triplet, 3H)

(The only other isomer is 2-
chloropropane with two types of
protons in the ratio of 6 : 1.  This is
why you did all of those isomer
problems early in the course!)

**282**

13-11 continued

(b) $C_9H_{10}O_2$—five elements of unsaturation; four protons in the aromatic region of the NMR indicate a disubstituted benzene; the pair of doublets with $J = 8$ Hz indicate the substituents are on opposite sides of the ring (*para*).

The other NMR signals are two 3H singlets, two $CH_3$ groups. One at $\delta$ 3.9 must be a $CH_3O$ group. The other at $\delta$ 2.4 is most likely a $CH_3$ group on the benzene ring.

$+ \ 3\,C \ + \ 6\,H \ + \ 2\,O \ +$
1 element of unsaturation

$+ \ OCH_3 \ + \ C \ + \ O \ +$
1 element of unsaturation

One way to assemble these pieces consistent with the NMR is:

a = $\delta$ 7.9 (doublet, 2H)
b = $\delta$ 7.2 (doublet, 2H)
c = $\delta$ 3.9 (singlet, 3H)
d = $\delta$ 2.4 (singlet, 3H)

(Another plausible structure is to have the methoxy group directly on the ring and to put the carbonyl between the ring and the methyl. This does not fit the chemical shift values quite as well as the above structure, as the methyl would appear around $\delta$ 2.1 or 2.2 instead of 2.4. It would also put the $OCH_3$ on the ring, moving the ring H upfield to about 6.8.)

13-12   $H_c$ , $\delta$ 5.16

13-13

(a)

a = $\delta$ 12.1 (broad singlet, 1H)
b = $\delta$ 5.8 (doublet, 1H)
c = $\delta$ 7.1 (multiplet, 1H)
d = $\delta$ 2.2 (quartet, 2H)
e = $\delta$ 1.5 (sextet, 2H)
f = $\delta$ 0.9 (triplet, 3H)

(b) The vinyl proton at $\delta$ 7.1 is $H_c$; it is coupled with $H_b$ and $H_d$, with two different coupling constants, $J_{bc}$ and $J_{cd}$, respectively. The value of $J_{bc}$ can be measured most precisely from the signal for $H_b$ at $\delta$ 5.8; the two peaks are separated by about 15 Hz, corresponding to 0.05 ppm in a 300 MHz spectrum. The value of $J_{cd}$ appears to be about the standard value 8 Hz, judging from the signal at $\delta$ 7.1. The splitting tree would appear as shown on the next page.

13-13 (b) continued

13-14

(a)

a = δ 9.7 (doublet, 1H)
b = δ 6.7 (doublet of doublets, 1H)
c = δ 7.5 (doublet, 1H)
d = δ 7.5 (doublet, 2H ortho to substituent)
e = δ 7.4 (broad singlet, 3H)

The doublet for $H_c$ at δ 7.4 OVERLAPS the 5 benzene protons.

(b) $J_{ab}$ can be determined most accurately from $H_a$ at δ 9.7: $J_{ab} \approx 8$ Hz, about the same as "normal" alkyl coupling.

$J_{bc}$ can be measured from $H_b$ at δ 6.7, as the distance between either the first and third peaks or the second and fourth peaks (see diagram below): $J_{bc} \approx 18$ Hz, about double the "normal" alkyl coupling.

(c)

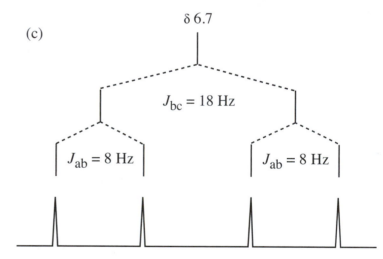

13-15

(a) a = δ 1.7       (b) a = doublet
    b ≈ δ 6.8            b = multiplet (two overlapping quartets—see part (c))
    c = δ 5-6           c = doublet
    d = δ 2.1           d = singlet

13-15 continued

(c)  using $J_{bc} = 15$ Hz and $J_{ab} = 7$ Hz:

13-16

replace $H_c$ → cis diastereomer

replace $H_d$ → trans diastereomer

13-17
(a)

replace $H_a$

replace $H_b$

Non-superimposable mirror images = enantiomers; therefore, $H_a$ and $H_b$ are *enantiotopic* and are not distinguishable by NMR.

(b)

replace $H_c$ on the left

replace $H_c$ on the right

enantiomers

(c)  The $H_d$ protons are also enantiotopic.

13-18

(a)

Br H b
| |
CH₃(a)—C—C—CH₃(e)
| |
H H
c d

This compound has five types of protons. $H_c$ and $H_d$ are diastereotopic. a = δ 1.5; b = δ 3.6; c,d = δ 1.7; e = δ 1.0

(b)

This compound has six types of protons. $H_c$ and $H_d$ are diastereotopic, as are $H_e$ and $H_f$. a = δ 2-5; b = δ 3.9; c,d = δ 1.6; e,f = δ 1.3

(c)

This compound has six types of protons. $H_e$ and $H_f$ are diastereotopic. a,b,c = δ 7.2; d ≈ δ 5.0; e,f = δ 3.6

(d)

a        Cl
H        /
 \      /
  C = C
 /      \
H        H
b        c

This compound has three types of protons. $H_a$ and $H_b$ are diastereotopic. a,b = δ 5-6; c = δ 7-8

13-19 (a)

$CH_3CH_2—\overset{..}{\underset{..}{O}}—H$ + H–A ⟶ $CH_3CH_2—\overset{\oplus}{\underset{..}{O}}—H$ + :A⊖ ⟶ $CH_3CH_2—\overset{H}{\underset{..}{O}}:$ + H–A

(b)

$CH_3CH_2—\overset{..}{\underset{..}{O}}—H$ + :A⊖ ⟶ $CH_3CH_2—\overset{..}{\underset{..}{O}}:⊖$ + H–A ⟶ $CH_3CH_2—\overset{H}{\underset{..}{O}}:$ + :A⊖

13-20 The protons from the OH in ethanol exchange with the deuteriums in D₂O. Thus, the OH in ethanol is replaced with OD, which does not absorb in the NMR. What happens to the H? It becomes HOD, which can usually be seen as a broad singlet around δ 5.25. (If the solvent is CDCl₃, the immiscible HOD will float on top of the solvent, out of the spectrometer beam, and its signal will be missing.)

$$ROH + D_2O \longrightarrow ROD + HOD$$

CH₃CH₂OH + D₂O

3H

2H

HOD

Peaks from OH are usually broad because of rapid proton transfer. This is especially true in water, or HOD.

TMS

10  9  8  7  6  5  4  3  2  1  0
δ (ppm)

**286**

13-21

(a) The formula $C_4H_{10}O_2$ has no elements of unsaturation, so the oxygens must be alcohol or ether functional groups. The doublet at $\delta$ 1.2 represents 3H and must be a $CH_3$ next to a CH. The peaks centered at $\delta$ 1.65 integrating to 2H appear to be an uneven quartet and signify a $CH_2$ between two sets of non-equivalent protons; apparently the coupling constants between the non-equivalent protons are not equal, leading to a complicated pattern of overlapping peaks. The remaining five hydrogens appear in four groups of 1H, 1H, 1H, and 2H, between $\delta$ 3.7 and 4.2. The 2H multiplet at $\delta$ 3.78 is a $CH_2$ next to O, with complex splitting (doublet of doublets) due to diastereotopic neighbors. The 1H multiplet at $\delta$ 4.0 is a CH between many neighbors. The broad singlet integrating to 2H at $\delta$ 4.12 appears to be two OH peaks.

The 3-dimensional view shows clearly that the two Hs on a $CH_2$ are not equivalent

Of the two possible structures:

This one has no chiral centers and would not display such complex splitting.

*If the possibility of intramolecular H-bonding is considered, the differences in the environments for each H become even more obvious.*

(b) The formula $C_2H_7NO$ has no elements of unsaturation. The N must be an amine and the O must be an alcohol or ether. Two triplets, each with 2H, are certain to be $-CH_2CH_2-$ . Since there are no carbons left, the N and O, with enough hydrogens to fill their valences, must go on the ends of this chain. The rapidly exchanging OH and $NH_2$ protons appear as a broad, 3H hump at $\delta$ 2.8.

$$HOCH_2CH_2NH_2$$
$$\delta\ 3.7 \qquad \delta\ 2.9$$

13-22

The key is what protons are adjacent to the oxygen. The $CH_3O$ in methyl propionate (IUPAC: methyl propanoate), above, absorbs at $\delta$ 3.9 as a singlet, whereas the $CH_2$ next to the carbonyl absorbs at $\delta$ 2.2 as a quartet. This is in contrast to ethyl acetate, at the left.

**287**

$H_a = \delta\ 2.4$ (singlet, 1H)
$H_b = \delta\ 3.4$ (doublet, 2H)
$H_c = \delta\ 1.8$ (multiplet, 1H)
$H_d = \delta\ 0.9$ (doublet, 6H)

$$\overset{d}{\underset{\underset{d}{CH_3}}{\overset{CH_3}{\underset{|}{H-\underset{a}{O}-\underset{b}{CH_2}-\underset{c}{\overset{|}{C}}-H}}}}$$

13-24

(a) The formula $C_4H_8O_2$ has one element of unsaturation. The 1H singlet at $\delta$ 12.1 indicates a carboxylic acid. The 1H multiplet and the 6H doublet scream isopropyl group.

$$\underset{H_3C}{\overset{H_3C}{>}}CH-\overset{\overset{O}{\|}}{C}-OH$$

(b) The formula $C_9H_{10}O$ has five elements of unsaturation. The 5H pattern between $\delta$ 7.2 and 7.4 indicates monosubstituted benzene. The peak at $\delta$ 9.85 is unmistakably an aldehyde, trying to be a triplet because it is weakly coupled to an adjacent $CH_2$. The two triplets at $\delta$ 2.7-3.0 are adjacent $CH_2$ groups.

(c) The formula $C_5H_8O_2$ has two elements of unsaturation. A 3H singlet at $\delta$ 2.3 is probably a $CH_3$ next to carbonyl. The 2H quartet and the 3H triplet are certain to be ethyl; with the $CH_2$ at $\delta$ 2.7, this also appears to be next to a carbonyl.

$$CH_3CH_2-\overset{\overset{O}{\|}}{C}-\overset{\overset{O}{\|}}{C}-CH_3$$

(d) The formula $C_4H_8O$ has one element of unsaturation, and the signals from $\delta$ 5.0 to 6.0 indicate a vinyl pattern ($CH_2=CH-$). The complex quartet for 1H at $\delta$ 4.3 is a CH bonded to an alcohol, next to $CH_3$. The OH appears as a 1H singlet at $\delta$ 2.5, and the $CH_3$ next to CH is a doublet at $\delta$ 1.3. Put together:

$$CH_2=CH-\overset{\overset{OH}{|}}{CH}-CH_3 \qquad \text{but-3-en-2-ol}$$

(e) The formula $C_7H_{16}O$ is saturated; the oxygen must be an alcohol or an ether, and the broad 1H peak at $\delta$ 1.2 is probably an OH. Let's analyze the spectrum from left to right. The expansion of the 1H multiplet at $\delta$ 1.7 shows seven peaks, a septet—an isopropyl group! Six of the nine H in the pattern at $\delta$ 0.9 must be the doublet from the two methyls from the isopropyl. The 2H quartet at $\delta$ 1.5 must be part of an ethyl pattern, from which the methyl triplet must be the other 3H of the pattern at $\delta$ 0.9. This leaves only the 3H singlet at $\delta$ 1.1 which must be a $CH_3$ with no neighboring Hs.

$$-OH \quad + \quad -CH_2CH_3 \quad + \quad CH_3-\overset{\overset{H}{|}}{\underset{\underset{CH_3}{|}}{C}}- \quad + \quad -CH_3 \ + \ 1\ C$$

These pieces can be assembled in only one way:

$$HO-\overset{\overset{CH_3}{|}}{\underset{\underset{CH_2CH_3}{|}}{C}}-CH(CH_3)_2 \quad \text{2,3-dimethylpentan-3-ol}$$

13-25 Chemical shift values are estimates from Figure 13-41 and from text Appendix 1C, except in (c) and (d), where the values are exact.

13-25 continued

(d) exact values

a
δ 192

b
δ 136

c
δ 135

TMS

200  180  160  140  120  100  80  60  40  20  0
Carbon -13 δ (ppm)

13-26

(a)

δ 37

δ 8

δ 30

δ 209

(b)  30 ÷ 2 = **15**
37 ÷ 2.7 = **14**
8 ÷ 1 = **8**

As a *general* rule,
the 15-20 rule works
fairly well.

$H_3C-C-CH_2-CH_3$
b           a      c

b
3H

a
2H

c
3H

TMS

10   9   8   7   6   5   4   3   2   1   0
δ (ppm)

13-27

(a)

a   a
C=C

C       C
b   O   b

a
δ 130

b
δ 70

TMS

200  180  160  140  120  100  80  60  40  20  0
Carbon -13 δ (ppm)

13-27 continued

(b)

Carbon -13 δ (ppm)

(c)

Carbon -13 δ (ppm)

(d)

Carbon -13 δ (ppm)

13-28 The full carbon spectrum of phenyl propanoate is presented below.

The DEPT-90 will show only the methine carbons, i.e., CH. All other peaks disappear.

The DEPT-135 will show the methyl, $CH_3$, and methine, CH, peaks pointed up, and the methylene, $CH_2$, peaks pointed down.

By analyzing the original spectrum, the DEPT-90, and the DEPT-135, each peak in the carbon NMR can be assigned as to which type of carbon it represents.

13-29

(a) and (b)  Since allyl bromide was the starting material, it is reasonable to expect the allyl group to be present in the impurity: the peak at 115 is a $=CH_2$ , the peak at 138 is $=CH-$ , and the peak at 63 is a deshielded aliphatic $CH_2$ ; assembling the pieces forms an allyl group.  The formula has changed from $C_3H_5Br$ to $C_3H_6O$, so $OH$ has replaced the Br.

$$H_2C=CH-CH_2OH$$

$$\delta\ 115 \quad \delta\ 138 \quad \delta\ 63$$

(c)  Allyl bromide is easily hydrolyzed by water, probably an $S_N1$ process.

$$H_2C=CH-CH_2Br \xrightarrow{\;H_2O\;} H_2C=CH-CH_2OH \;+\; HBr$$

13-30 (a) and (b)

$$a = \delta\ 180$$
$$b = \delta\ 70$$
$$c = \delta\ 28$$
$$d = \delta\ 22$$

Two elements of unsaturation in $C_4H_6O_2$, one of which is a carbonyl, and no evidence of a C=C, prove that a ring must be present.  This product is formed from 4-hydroxybutanoic acid with loss of a molecule of water, as you will see in Chapter 21.

13-31 (a) and (b)

$$a = \delta\ 128$$
$$b = \delta\ 25$$
$$c = \delta\ 23$$

Two elements of unsaturation in $C_6H_{10}$ must be a C=C and a ring.  Only three peaks indicates symmetry.

(c)  Using $PBr_3$ instead of $H_2SO_4/NaBr$ would give a higher yield of bromocyclohexane.

13-32   Compound 2

**Mass spectrum**:  the molecular ion at m/z 136 shows a peak at 138 of about equal height, indicating a bromine atom is present:  136 – 79 = 57.  The fragment at m/z 57 is the base peak; this fragment is most likely a butyl group, $C_4H_9$, so a likely molecular formula is $C_4H_9Br$.
**Infrared spectrum**:  Notable for the absence of functional groups:  no $O-H$, no $N-H$, no $=C-H$, no C=C, no C=O $\Rightarrow$ most likely an alkyl bromide.
**NMR spectrum**:  The 6H doublet at $\delta$ 1.0 suggests two $CH_3$ groups split by an adjacent H—an isopropyl group.  The 2H doublet at $\delta$ 3.2 is a $CH_2$ between a CH and the Br.

Putting the pieces together gives isobutyl bromide.

$$Br-CH_2-CH\begin{smallmatrix}CH_3\\[2pt]CH_3\end{smallmatrix}$$

13-33
The formula $C_9H_{11}Br$ indicates four elements of unsaturation, just enough for a benzene ring.
Here is the most accurate method for determining the number of protons per signal from integration values *when the total number of protons is known*.  Add the integration heights: 44 mm + 130 mm + 67 mm = 241 mm.  Divide by the total number of hydrogens: 241 mm $\div$ 11H = 22 mm/H.  Each 22 mm of integration height = 1H, so the ratio of hydrogens is 2 : 6 : 3.
The 2H singlet at $\delta$ 7.1 means that only two hydrogens remain on the benzene ring, that is, it has four substituents.  The 6H singlet at $\delta$ 2.3 must be two $CH_3$ groups on the benzene ring in identical environments.  The 3H singlet at $\delta$ 2.2 is another $CH_3$ in a slightly different environment from the first two.  Substitution of the three $CH_3$ groups and the Br in the most symmetric way leads to the structures on the next page.

13-33 continued

a = δ 7.1 (singlet, 2H)
b = δ 2.3 (singlet, 6H)
c = δ 2.2 (singlet, 3H)

*A second structure is also possible although it is less likely because the Br would probably deshield the Hs labeled "a" to about 7.3–7.4.*

13-34  The numbers in italics indicate the number of peaks in each signal. Approximate chemical shift values are in bold.

(a)
**1.0  2.2       1.7**
$CH_3 - CH_2 - CCl_2 - CH_3$
*3     4          1*

(b)
**4.0**
*7*  **2-5**
$CH_3 - CH - OH$
**1.2** |         *1*
$CH_3$     (assume OH exchanging
*2*         rapidly—no splitting)

(c)
**1.4**
*10*    **0.9**
$CH_3 - CH - CH_3$
|                *2*
$CH_3$

(d)
*1* **7.2**
**2.3**
$CH_3$
*1*
(All of these benzene H atoms are accidentally equivalent and do not split each other.)

(e)
*3*          O          *4*   *3*
$CH_3 - CH_2 - \overset{\parallel}{C} - O - CH_2 - CH_3$
**1.0**              **4.0  1.2**

(f)
*6* **4.2**
**1.2**
$CH_3$
*2*

13-35  Consult Appendix 1 in the text for chemical shift values. *Your predictions should be in the given range, or within 0.5-1.0 ppm of the given value.*

(a)
—H    all at δ 7.2

(b)
all at δ 1.3

(c)
δ 1.6
$CH_3 - O - CH_2 - CH_2 - CH_2Cl$
δ 3.4      δ 3.8      δ 3.2

(d)
$CH_3 - CH_2 - C \equiv C - H$
δ 1.2    δ 2.2        δ 2.5

(e)
O
$CH_3 - CH_2 - \overset{\parallel}{C} - CH_3$
δ 1.0    δ 2.5    δ 2.0

(f)
$CH_3$  δ 4.3                        δ 2-5
$CH - O - CH_2 - CH_2 - OH$
$CH_3$              δ 3.8    δ 3.8
δ 0.9

(g)
O
C—H
δ 9-10
δ 8.0
H    H
δ7.5  δ8.0

The C=O has its strongest deshielding effect at the adjacent H (ortho) and across the ring (para). The remaining H (meta) is less deshielded.

(h)
δ 6-7            O
$CH_3 - CH = CH - \overset{\parallel}{C} - H$
δ 1.7      δ 5-6      δ 9-10

**294**

13-35 continued

(i)

$$HOOC - CH_2 - CH_2 - \overset{\overset{\displaystyle O}{\|}}{C} - O - \underset{\underset{\displaystyle \delta 1.1}{CH_3}}{\overset{\overset{\displaystyle CH_3}{|}}{CH}} \leftarrow \delta 4.0$$

$\delta\,10\text{-}12 \qquad \delta\,2.3 \qquad \delta\,2.3$

(j)

$\delta\,1.3$ ... $\delta\,4.5$

$\delta\,1.3 \qquad\qquad \delta\,1.7$
allylic

(k)

$\delta\,1.4$

$\delta\,7.2$

$\delta\,2.5$
benzylic

(l)

$\delta\,6\text{-}7$

$\delta\,7.2$

$\delta\,3.0$
allylic and
benzylic

13-36

$$\underset{\underset{\underset{\displaystyle b}{\displaystyle OH}}{|}}{\overset{c}{CH_3} - \overset{a}{CH} - \overset{c}{CH_3}}$$

a $= \delta\,4.0$ (septet, 1H)

b $= \delta\,2.5$ (broad singlet, 1H) (rapidly exchanging)

c $= \delta\,1.2$ (doublet, 6H)

13-37

(a) The chemical shift *in ppm* would not change: $\delta\,4.00$.

(b) Coupling constants do not change with field strength: $J = 7$ Hz, regardless of field strength.

(c) At 60 MHz, $\delta\,4.00 = 4.00$ ppm $= (4.00 \times 10^{-6}) \times (60 \times 10^6 \text{ Hz}) = 240$ Hz

The signal is 240 Hz downfield from TMS in a 60 MHz spectrum.

At 300 MHz, $(4.00 \times 10^{-6}) \times (300 \times 10^6 \text{ Hz}) = 1200$ Hz

The signal is 1200 Hz downfield from TMS in a 300 MHz spectrum. Necessarily, 1200 Hz is exactly 5 times 240 Hz because 300 MHz is exactly 5 times 60 MHz. They are directly proportional.

13-38

$\overset{c}{CH_2} - \overset{b}{CH_2} - O - \overset{\overset{\displaystyle O}{\|}}{C} - \overset{d}{CH_3}$

a

a $= \delta\,7.2\text{-}7.3$ (multiplet, 5H)
b $= \delta\,4.3$ (triplet, 2H)
c $= \delta\,2.9$ (triplet, 2H)
d $= \delta\,2.0$ (singlet, 3H)

(a)

$\underset{b}{CH_3} - O - \underset{a}{CH_2}\underset{c}{CH_3}$

(b)

$\underset{c}{(CH_3)_2}\underset{a}{CH} - \overset{\overset{O}{\|}}{C} - \underset{b}{CH_3}$

(To show the pattern
of peaks, this septet is
larger than it would appear
on a real spectrum.)

(c)

$\underset{a}{ClCH_2}\underset{b}{CH_2}\underset{a}{CH_2}Cl$

13-39 continued

(d)
b
2H

c
1H

a
1H

b     d
H     NH₂

a                    c
H                    H

d
H     NH₂
b

The NH₂ group
shields nearby
protons on benzene,
moving them upfield.

d
4H
very
broad
and
variable
position

TMS

10    9    8    7    6    5    4    3    2    1    0
δ (ppm)

(e)
a
2H

b
2H

a H          H b

O₂N

a H          H b

O—CH c

d
CH₃

CH₃
d

c
1H

(To show the pattern
of peaks, this septet is
larger than it would appear
on a real spectrum.)

d
6H

TMS

10    9    8    7    6    5    4    3    2    1    0
δ (ppm)

(f)
Signal (a) is split into a quartet because of the adjacent
CH₃ with *J* = 7 Hz. Each of those peaks is then split
into a doublet because of the coupling with the trans H,
*J* = 15 Hz. This is called a doublet of quartets, and it is
drawn here as two quartets. In a real spectrum, these
peaks would overlap and would not be a clean doublet
of quartets. See the splitting tree for problem 13-15.

a H          O

d
H₃C

b H

c
CH₃

c
3H

d
3H

a
1H

b

TMS

10    9    8    7    6    5    4    3    2    1    0
δ (ppm)

13-40

(a) The NMR of 1-bromopropane would have three sets of signals, whereas the NMR of 2-bromopropane would have only two sets (a septet and a doublet, the typical isopropyl pattern).

(b) Each spectrum would have a methyl singlet at δ 2. The left structure would show an ethyl pattern (a 2H quartet and a 3H triplet), whereas the right structure would exhibit an isopropyl pattern (a 1H septet and a 6H doublet).

(c) The most obvious difference is the chemical shift of the $CH_3$ singlet. In the compound on the left, the $CH_3$ singlet would appear at δ 2.1, while the compound on the right would show the $CH_3$ singlet at δ 3.8. Refer to the solution of 13-22 for the spectrum of the second compound.

(d) The splitting and integration for the peaks in these two compounds would be identical, so the chemical shift must make the difference. As described in text section 13-5B, the alkyne is not nearly as deshielding as a carbonyl, so the protons in pent-2-yne would be farther upfield than the protons in butan-2-one, by about 0.5 ppm. For example, the methyl on the carbonyl would appear near δ 2.1 while the methyl on the alkyne would appear about δ 1.7.

13-41 The multiplicity in the off-resonance decoupled spectrum is given below each chemical shift: s = singlet; d = doublet; t = triplet; q = quartet. It is often difficult to predict exact chemical shift values; your predictions should be in the right vicinity. There should be no question about the multiplicity and DEPT spectra, however.

The DEPT-90 spectrum for ethyl acetate would have no peaks because there are no CH groups.

(b) actual values

DEPT-90

DEPT-135

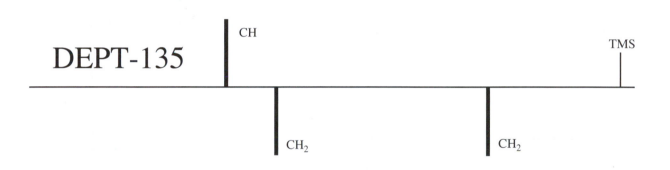

(c) actual values

Carbons b and c are accidentally equivalent.

DEPT-90 and DEPT-135 on next page

13-41 (c) continued

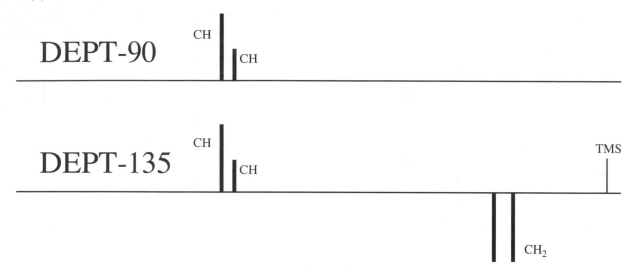

DEPT-90

CH

CH

DEPT-135

CH

CH

TMS

CH₂

(d) estimated from similar structures

$$H_3C-C\equiv C-C-CH_2-CH_3$$

O

||

f    c    b    a    d    e

5    80   90   200  35   10

q    s    s    s    t    q

a
δ 200
C

b
δ 90
C

c
δ 80
C

d
δ 35
CH₂

e
δ 10
CH₃

f
δ 5
CH₃

TMS

200    180    160    140    120    100    80    60    40    20    0

Carbon -13 δ (ppm)

The DEPT-90 spectrum would have no peaks because there are no CH groups.

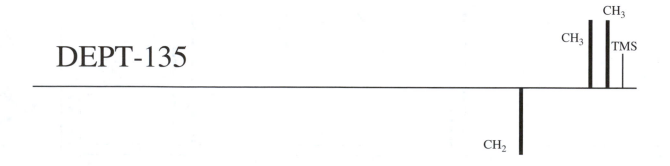

DEPT-135

CH₃

CH₃

TMS

CH₂

Copyright © 2013 Pearson Education, Inc.

**13-42** The multiplicity of the peaks in this off-resonance-decoupled spectrum show two different CHs and a $CH_3$. There is only one way to assemble these pieces with three chlorines. (The multiplet at 0 is TMS.)

$$\delta\ 20 \longrightarrow \underset{q}{CH_3} - \underset{\underset{\delta\ 60}{\underset{d}{\uparrow}}}{CH} - \underset{\underset{\delta\ 75}{\underset{d}{\nwarrow}}}{CH}$$

with Cl and Cl on the carbons.

**13-43** There is no evidence for vinyl hydrogens, so the double bond is gone. Integration gives eight hydrogens, so the formula must be $C_4H_8Br_2$, and the four carbons must be in a straight chain because the starting material was but-2-ene. From the integration, the four carbons must be present as one $CH_3$, one CH, and two $CH_2$ groups. The methyl is split into a doublet, so it must be adjacent to the CH. The two $CH_2$ groups must follow in succession, with two bromine atoms filling the remaining valences. (The spectrum is complex because the asymmetric carbon atom causes the neighboring protons to be diastereotopic.)

$$\underset{\underset{d}{\underbrace{\phantom{xx}}}\ \underset{a}{\phantom{x}}\ \underset{c}{\phantom{x}}\ \underset{b}{\underbrace{\phantom{xxxx}}}}{H-\overset{H}{\underset{H}{C}}-\overset{Br}{\underset{H}{C}}-\overset{H}{\underset{H}{C}}-\overset{Br}{\underset{H}{C}}-H}$$

a = δ 4.3 (sextet, 1H)
b = δ 3.6 (triplet, 2H)
c = δ 2.3 (multiplet, 2H)
d = δ 1.7 (doublet, 3H)

**13-44** There is no evidence for vinyl hydrogens, so the compound must be a small, saturated, oxygen-containing molecule. Starting upfield (toward TMS), the first signal is a 3H triplet; this must be a $CH_3$ next to a $CH_2$. The $CH_2$ is the signal at δ 1.5, but it has six peaks: it must have five neighboring hydrogens, a $CH_3$ on one side and a $CH_2$ on the other side. The third carbon must therefore be a $CH_2$; its signal is a quartet at δ 3.6, split by a $CH_2$ and an OH. To be so far downfield, the final $CH_2$ must be bonded to oxygen. The remaining 1H signal must be from an OH. The compound must be propan-1-ol.

$$\underset{b\quad\ a\quad c\quad d}{HOCH_2CH_2CH_3}$$

a = δ 3.6 (quartet, 2H)
b = δ 3.2 (triplet, 1H)
c = δ 1.5 (6 peaks, 2H)
d = δ 0.9 (triplet, 3H)

**13-45**

$$\underset{Cl}{\overset{CH_3}{CH_3-\underset{|}{\overset{|}{C}}-CH_2CH_3}} \xrightarrow{\text{base}} \text{Isomer A} + \text{Isomer B}$$

Isomer A:
$$\overset{CH_3}{\underset{CH_3}{C}} = \overset{CH_3}{\underset{H}{C}}$$

Isomer B:
$$\overset{H}{\underset{H}{C}} = \overset{CH_3}{\underset{CH_2CH_3}{C}}$$

(a)

Isomer A:
$$\overset{\overset{b}{CH_3}}{\underset{\underset{c}{CH_3}}{C}} = \overset{\overset{d}{CH_3}}{\underset{\underset{a}{H}}{C}}$$

Isomer B:
$$\overset{\overset{d}{H}}{\underset{\underset{d}{H}}{C}} = \overset{\overset{f}{CH_3}}{\underset{\underset{e\ \ g}{CH_2CH_3}}{C}}$$

a = δ 5.2 (quartet, 1H)
b = δ 1.7 (singlet, 3H)
c = δ 1.6 (singlet, 3H)
d = δ 1.5 (doublet, 3H)

d = δ 4.7 (singlet, 2H)
e = δ 2.0 (quartet, 2H)
f = δ 1.7 (singlet, 3H)
g = δ 1.0 (triplet, 3H)

(b) With NaOH as base, the more highly substituted alkene, Isomer A, would be expected to predominate—the Zaitsev Rule. With KO-*t*-Bu as a hindered, bulky base, the less substituted alkene, Isomer B, would predominate (the Hofmann product).

**13-46** "Nuclear waste" is composed of radioactive products from either nuclear reactions—for example, from electrical generating stations powered by nuclear reactors—or residue from medical or scientific studies using radioactive nuclides as therapeutic agents (like iodine for thyroid treatment) or as molecular tracers (carbon-14, tritium H-3, phosphorus-32, nitrogen-15, and many others). The physical technique of *nuclear* magnetic resonance neither uses nor generates any radioactive elements, and does not generate "nuclear waste". (Some people assume that the medical application of NMR, **m**agnetic **r**esonance **i**maging or MRI, purposely dropped the word "nuclear" from the technique to avoid the confusion between *nuclear* and *radioactive*.)

**13-47**

**Mass spectrum**: The molecular ion of m/z 117 suggests the presence of an odd number of nitrogens.

**Infrared spectrum**: No NH or OH appears. Hydrogens bonded to both $sp^2$ and $sp^3$ carbon are indicated around 3000 cm$^{-1}$. The characteristic C≡N peak appears at 2250 cm$^{-1}$ and aromatic C=C is suggested by the peak at 1600 cm$^{-1}$.

**NMR spectrum**: Five aromatic protons are shown in the NMR at δ 7.3. A CH$_2$ singlet appears at δ 3.7. Assemble the pieces:

**13-48** This is a challenging problem, despite the molecule being relatively small.

**Mass spectrum**: The molecular ion at 96 suggests no Cl, Br, or N. The molecule must have seven carbons or fewer.

**Infrared spectrum**: The dominant functional group peak is at 1685 cm$^{-1}$, a carbonyl that is conjugated with C=C (lower wave number than normal, very intense peak). The presence of an oxygen and a molecular ion of 96 lead to a formula of C$_6$H$_8$O, with three elements of unsaturation, a C=O and one or two C=C.

**Carbon NMR spectrum**: The six peaks show, by chemical shift, one carbonyl carbon (196), two alkene carbons (129, 151), and three aliphatic carbons (23, 26, 36). From the DEPT information at the top of the spectrum, the groups are: three CH$_2$ groups, two alkene CH groups, and carbonyl.

Since the structure has one carbonyl and only two alkene carbons, the third element of unsaturation must be a ring.

CH$_2$ + CH$_2$ + CH$_2$ + 1 ring

Since the structure has no methyl group, and no H$_2$C=, all of the carbons must be included in the ring. The only way these pieces can fit together is in cyclohex-2-enone. Notice that the proton NMR was unnecessary to determine the structure, fortunately, since the HNMR was not easily interpreted except for the two alkene hydrogens; the hydrogen on carbon-2 appears as the doublet at 6.0.

cyclohex-2-enone

The mystery mass spec peak at m/z 68 comes from a fragmentation that will be discussed later; it is called a retro-Diels-Alder fragmentation.

**302**

13-49  The key to the carbon NMR lies in the symmetry of these structures.

*ortho*-xylene          *meta*-xylene          *para*-xylene

(a)  In each molecule, the methyl carbons are equivalent, giving one signal in the CNMR.  Considering the ring carbons, the symmetry of the structures shows that *ortho*-xylene would have 3 carbon signals from the ring (total of 4 peaks), *meta*-xylene would have 4 carbon signals from the ring (total of 5 peaks), and *para*-xylene would have only 2 carbon signals from the ring (total of 3 peaks).  These compounds would be instantly identifiable simply by the number of peaks in the carbon NMR.

(b)  The proton NMR would be a completely different problem.  Unless the substituent on the benzene ring is moderately electron-withdrawing or donating, the ring protons absorb at roughly the same position.  A methyl group has essentially no electronic effect on the ring hydrogens, so while the *para* isomer would give a  clean singlet because all its ring protons are equivalent, the *ortho* and *meta* isomers would have only slightly broadened singlets for their proton signals.  (Only a very high field NMR, 500 MHz or higher, would be able to distinguish these isomers in the proton NMR.)

13-50  (a), (b) and (c)  The six isomers are drawn here.  Below each structure is the number of proton signals and the number of carbon signals.  (Note that splitting patterns would give even more clues in the proton NMR.)

| 1xH | 3xH* | 2xH | 2xH | 2xH | 5xH |
| 1xC | 3xC | 3xC | 2xC | 2xC | 4xC |

*Rings always present challenges in stereochemistry.  When viewed in three dimensions, it becomes apparent that the two hydrogens on a $CH_2$ are not equivalent:  on each $CH_2$, one H is *cis* to the methyl and one H is *trans* to the methyl.  These are diastereotopic protons.  A more correct answer to part (b) would be four types of protons; whether all four could be distinguished in the NMR is a harder question to answer.  For the purpose of this problem, whether it is 3 or 4 types of H does not matter because either one, in combination with three types of carbon, will distinguish it from the other 5 structures.

(d)  Two types of H and three types of C can be only one isomer:  2-methylpropene (isobutylene).  (The only isomers that would not be distinguished from each other would be *cis*- and *trans*-but-2-ene.)

13-51

| | peaks in CNMR | peaks in DEPT-90 | peaks in DEPT-135 |
|---|---|---|---|
| | 6 peaks | 1 peak (c) | up: 1 CH (c), 1 CH$_3$ (a)<br>down: 4 CH$_2$ (b,d,e,f) |
| | 5 peaks | 2 peaks (b,c)<br>of equal height | up: 2 CH (b,c), 1 CH$_3$ (a)<br>down: 2 CH$_2$ (d,e) |
| | 5 peaks | 1 peak (c) | up: 1 CH (c), 1 CH$_3$ (a)<br>down: 2 CH$_2$ (d,e) |
| | 5 peaks | 2 peaks (b,c) of<br>unequal height | up: 2 CH (b,c), 2 CH$_3$ (a,e)<br>down: 1 CH$_2$ (d) |

This example clearly shows the utility of DEPT NMR. These four compounds cannot be distinguished easily by regular CNMR. The DEPT-90 in combination with the regular CNMR could distinguish them. However, the easiest method that would give an unambiguous assignment for each structure would be the DEPT-135 where each structure has a unique pattern of peaks.

13-52

(a) MS or IR could not easily distinguish these isomers: same molecular weight and same functional group. They would give dramatically different proton and carbon NMRs however.

2 singlets in HNMR
3 singlets in CNMR

4 signals, all with splitting, in HNMR
4 singlets in CNMR

(b) The only technique that would not readily distinguish these isomers would be MS because they have the same molecular weight and would have similar, though not identical, fragmentation patterns.

IR: C=O about 1730 cm$^{-1}$
HNMR: methyl singlet and ethyl pattern
CNMR: ester C=O about δ 170

IR: C=O about 1710 cm$^{-1}$
HNMR: 3 singlets
CNMR: ketone C=O about δ 200

13-52 continued

(c) The big winner here is MS: they have different molecular weights, plus the Cl has the two isotope peaks that make a Cl atom easily distinguished. The other techniques would have minor differences and would require having a detailed table of frequencies or chemical shifts to determine which is which.

F—⬡—OH

$M^+$ = m/z 112

Cl—⬡—OH

$M^+$ = m/z 128, 130

(d) The only technique that would not readily distinguish these isomers would be IR because they have the same ester functional group. In the CNMR, the DEPT-135 would be most helpful.

MS: main fragment at m/z 43
HNMR: methyl singlet at δ 2.0
CNMR—DEPT 135: O—CH$_2$
near δ 70 would point down

MS: main fragment at m/z 57
HNMR: methyl singlet at δ 3.8
CNMR—DEPT 135: O—CH$_3$
near δ 70 would point up

13-53 Symmetry and chemical shift of the CH$_3$ signals in the CNMR can distinguish these 5 compounds.

four types of peaks in the sp$^3$ region; short peak at δ 45

only structure with a peak at δ 70; four types of peaks in the sp$^3$ region

three types of peaks in the sp$^3$ region

four types of peaks in the sp$^3$ region; tall peak at δ 45

only two types of peaks in the sp$^3$ region

13-54

(a) IR only: all have benzene and C=O; 1 has aldehyde C—H at 2710 and 2810; 2 has OH at 3100-3500; 3 has characteristic COOH from 2500-3500 with shoulder around 2700 cm$^{-1}$

(b) HNMR only: all have benzene para pattern (two doublets): 1 and 2 have peaks at δ 6.8, 3 does not; 1 has sharp singlet from δ 9-10; 3 has broad singlet from δ 10-14; the CH$_3$ in 3 is at δ 2.5, while in 1 and 2 the OCH$_3$ is at δ 3.7.

(c) CNMR only: 1 has the C=O around δ 190; 2 and 3 have C=O around δ 170; 2 has a benzene peak at δ 160 from C—O; all of the benzene signals in 3 are from δ 120-140. Also, the CH$_3$ peaks in 1 and 2 are from δ 50-60 whereas 3 has CH$_3$ signal around δ 20.

# CHAPTER 14—ETHERS, EPOXIDES AND THIOETHERS

**14-1**
The four solvents decrease in polarity in this order: water, ethanol, ethyl ether, and dichloromethane. The three solutes decrease in polarity in this order: sodium acetate (ionic, most polar), 2-naphthol, and naphthalene (no electronegative atoms). The guiding principle to determine solubility is, "Like dissolves like." Compounds of similar polarity will dissolve (in) each other. Thus, sodium acetate will dissolve in water, will dissolve only slightly in ethanol, and will be virtually insoluble in ethyl ether and dichloromethane. 2-Naphthol will be insoluble in water, somewhat soluble in ethanol, and soluble in ether and dichloromethane. Naphthalene will be insoluble in water, partially soluble in ethanol, and soluble in ethyl ether and dichloromethane. (Actual solubilities are difficult to predict, but you should be able to predict *trends*.)

**14-2**    Oxygen shares one of its electron pairs with aluminum; oxygen is the Lewis base, and aluminum is the Lewis acid. An oxygen atom with three bonds and one unshared pair has a positive formal charge. An aluminum atom with four bonds has a negative formal charge.

**14-3** The crown ether has two effects on $KMnO_4$: first, it makes $KMnO_4$ much more soluble in benzene; second, it holds the potassium ion tightly, making the permanganate more available for reaction. Chemists call this a "naked anion" because it is not complexed with solvent molecules.

18-crown-6
a "crown ether"

**14-4** IUPAC name first; then common name (see Appendix 1 in this Solutions Manual for a summary of IUPAC nomenclature). *Current IUPAC recommendations place the position number immediately before the group that it modifies.*

(a) methoxyethene; methyl vinyl ether
(b) 2-ethoxypropane; ethyl isopropyl ether
(c) 1-chloro-2-methoxyethane; 2-chloroethyl methyl ether
(d) 2-ethoxy-2,3-dimethylpentane; no common name
(e) 1,1-dimethoxycyclopentane; no common name
(f) *trans*-2-methoxycyclohexan-1-ol; also possible: (1R,2R)-2-methoxycyclohexan-1-ol; no common name
(g) methoxycyclopropane; cyclopropyl methyl ether
(h) 1-methoxybut-2-yne; no common name
(i) (Z)-2-methoxypent-2-ene; no common name

**14-5**

(a)

The alcohol is ethane-1,2-diol; the common name is ethylene glycol.

14-5 continued

(b)

The mechanism shows that the acid catalyst is regenerated at the end of the reaction.

14-6
(a) dihydropyran
(b) 2-chloro-1,4-dioxane
(c) 3-isopropylpyran
(d) *trans*-2,3-diethyloxirane; *trans*-3,4-epoxyhexane; *trans*-hex-3-ene oxide
(e) 3-bromo-2-ethoxyfuran
(f) 3-bromo-2,2-dimethyloxetane

14-7

m/z 116

CH₃CH₂CH₃CH₂⁺
m/z 57

m/z 73

m/z 101

m/z 43

**14-8** $S_N2$ reactions, including the Williamson ether synthesis, work best when the nucleophile attacks a 1° or methyl carbon. Instead of attempting to form the bond from oxygen to the 2° carbon on the ring, form the bond from oxygen to the 1° carbon of the butyl group.

The OH must first be transformed into a good leaving group: either a tosylate, or one of the halides (not fluoride).

$$\text{(butanol)} \quad \xrightarrow[\text{pyridine}]{\text{TsCl}} \quad \text{(butyl OTs)}$$

$$\text{(cyclohexanol, OH)} \quad \xrightarrow{\text{Na}} \quad \overset{\ominus}{\text{O}} \; \text{Na}^+ \quad \xrightarrow{\text{butyl OTs}} \quad \text{(ether product)}$$

3-butoxy-1,1-dimethylcyclohexane

**14-9** Always put the leaving group on the less substituted carbon to maximize substitution.

(a) $\text{(cyclohexanol, OH)} \xrightarrow{\text{Na}} \xrightarrow{CH_3CH_2CH_2Br} \text{(cyclohexyl propyl ether)}$

(b) $\text{(isopropanol, OH)} \xrightarrow{\text{Na}} \xrightarrow{CH_3I} \text{(OCH}_3\text{ product)}$

(c) $\text{(4-nitrophenol, OH, NO}_2\text{)} \xrightarrow{\text{NaOH}} \xrightarrow{CH_3I} \text{(OCH}_3\text{, NO}_2\text{ product)}$

(d) $CH_3CH_2CH_2OH \xrightarrow{\text{Na}} \xrightarrow{CH_3CH_2Br} CH_3CH_2CH_2OCH_2CH_3$

$CH_3CH_2OH \xrightarrow{\text{Na}} \xrightarrow{CH_3CH_2CH_2Br} CH_3CH_2CH_2OCH_2CH_3$

} Both carbons bonded to oxygen are 1°.

(e) $CH_3-\overset{\displaystyle CH_3}{\underset{\displaystyle CH_3}{C}}-OH \xrightarrow{\text{Na}} \xrightarrow{PhCH_2Br} CH_3-\overset{\displaystyle CH_3}{\underset{\displaystyle CH_3}{C}}-OCH_2Ph$

**14-10**

(a) (1) $\text{(propene)}$ or $\text{(propene)}$ $\xrightarrow[CH_3OH]{Hg(OAc)_2 \quad NaBH_4} \text{(OCH}_3 \text{ product)}$

(2) $\text{(isopropanol, OH)} \xrightarrow{\text{Na}} \xrightarrow{CH_3I} \text{(OCH}_3 \text{ product)}$

**14-10 continued**

(b) (1)   [cyclohexene] $\xrightarrow[\text{CH}_3\text{CH}_2\text{OH}]{\text{Hg(OAc)}_2}$ $\xrightarrow{\text{NaBH}_4}$ [cyclohexyl–OCH$_2$CH$_3$]

(2) [cyclohexanol, OH] $\xrightarrow{\text{Na}}$ $\xrightarrow{\text{CH}_3\text{CH}_2\text{Br}}$ [cyclohexyl–OCH$_2$CH$_3$]

(c) (1) Alkoxymercuration is not practical here; the product does not have Markovnikov orientation.

(2) [isopropyl cyclopentanol, OH] $\xrightarrow{\text{Na}}$ $\xrightarrow{\text{CH}_3\text{I}}$ [isopropyl cyclopentyl–OCH$_3$]

(d) (1) [isopropylidene cyclopentane] $\xrightarrow[\text{CH}_3\text{OH}]{\text{Hg(OAc)}_2}$ $\xrightarrow{\text{NaBH}_4}$ [product with OCH$_3$]   ( [methylenecyclopentane] also possible as a starting material )

(2) [tertiary cyclopentanol, OH] $\xrightarrow{\text{Na}}$ $\xrightarrow{\text{CH}_3\text{I}}$ [product with OCH$_3$]

(e) (1) [isopropylidene cyclopentane] $\xrightarrow[\text{OH (isopropanol)}]{\text{Hg(OAc)}_2}$ $\xrightarrow{\text{NaBH}_4}$ [product with O–isopropyl]   ( [methylenecyclopentane] also possible as a starting material )

(2) Williamson ether synthesis would give a poor yield of product as the halide is on a 2° carbon.

[cyclopentane with C(CH$_3$)$_2$OH] $\xrightarrow{\text{Na}}$ [isopropyl bromide, Br] 2° → [product O–isopropyl]

(f) (1) [isobutylene, =<] $\xrightarrow[\text{phenol –OH}]{\text{Hg(OAc)}_2}$ $\xrightarrow{\text{NaBH}_4}$ [phenyl–O–C(CH$_3$)$_3$]

(2) Williamson ether synthesis is not feasible here. S$_N$2 does not work on either a benzene or a 3° halide.

**14-11** An important principle of synthesis is to avoid mixtures of isomers wherever possible; minimizing separations increases recovery of products. Bimolecular condensation is a random process, assuming similar structures for the two alkyl groups on the ether. Heating a mixture of ethanol and methanol with acid will produce all possible combinations: dimethyl ether, ethyl methyl ether, and diethyl ether. This mixture would be troublesome to separate.

## 14-12

__Ether formation__

$$+ \ H_3O^+$$

__Dehydration__

Remember $\Delta G = \Delta H - T\Delta S$ ? Thermodynamics of a reaction depend on the sign and magnitude of $\Delta G$. As temperature increases, the entropy term grows in importance. In ether formation, the $\Delta S$ is small because two molecules of alcohol give one molecule of ether plus one molecule of water—no net change in the number of molecules. In dehydration, however, one molecule of alcohol generates one molecule of alkene plus one molecule of water—a large increase in entropy. So $T\Delta S$ is more important for dehydration than for ether formation. As temperature increases, the competition will shift toward more dehydration (elimination).

## 14-13

(a) This symmetrical ether at 1° carbons could be produced in good yield by bimolecular condensation.

(b) This unsymmetrical ether could not be produced in high yield by bimolecular condensation. Williamson synthesis would be preferred.

(c) Even though this ether is symmetrical, both carbons are 2°, so bimolecular condensation would give low yields. Unimolecular dehydration to give alkenes would be the dominant pathway. Alkoxymercuration-demercuration is the preferred route.

## 14-14

**14-15**

(a)

(b)   (c)

no $S_N$ at $sp^2$ carbon

(d)

no $S_N$ at $sp^2$ carbon

(e)

no $S_N$ at $sp^2$ carbon

**14-16**

Two proton transfers happen here: one $H^+$ comes off of oxygen, another $H^+$ goes on the other oxygen; several different scenarios of how this happens could be proposed.

Nucleophilic attack on $CH_3$ is faster than on 1° carbon.

BuOH

HOBBr$_2$

$\downarrow$ **2 H$_2$O**

B(OH)$_3$ + **2 HBr**

**14-17** Begin by transforming the alcohols into good leaving groups like halides or tosylates:

**14-18** The sulfur at the center of mustard gas is an excellent nucleophile, and chloride is a decent leaving group. Sulfur can do in *internal* nucleophilic subsitution to make a reactive sulfonium salt and the sulfur equivalent of an epoxide.

(a)

very reactive alkylating agent

inactivated enzyme

**311**

14-18 continued

(b) NaOCl is a powerful oxidizing agent. It oxidizes sulfur to a sulfoxide or more likely a sulfone, either of which is no longer nucleophilic, preventing formation of the cyclic sulfonium salt.

sulfoxide          sulfone

14-19

new bond

starting from

desired target          starting material

Analysis of the desired target structure shows that the new bond can be made from an acetylide ion plus the given starting material, 4-bromobutan-1-ol. However, an acetylide ion is a strong base and will remove H⁺ from the alcohol, making an alkyne that is no longer nucleophilic. The alcohol must be protected before attempting the nucleophilic substitution.

protected—
no reaction

$S_N2$

remove silyl group

14-20

Generally, chemists prefer the peroxyacid method of epoxide formation to the halohydrin method. Reactions (a) and (b) show the peroxyacid method, but the halohydrin method could also be used.

(a)          MCPBA / CH₂Cl₂

(b)          H₂SO₄ / Δ          MCPBA / CH₂Cl₂

(c)          BH₃ • THF          H₂O₂ / HO⁻          2,6-lutidine          2,6-Lutidine is a non-nucleophilic base.

(d)          Hg(OAc)₂ / H₂O          NaBH₄          2,6-lutidine

(e)          NaOH          NaOH can also do unwanted S_N2 and E2 with the chloride.

## 14-21

(a) 1) *tert*-Butyl hydroperoxide is the oxidizing agent. The $(CH_3)_3COOH$ contains the O—O bond just like a peroxyacid. 2) Diethyl tartrate has two asymmetric carbons and is the source of asymmetry; its function is to create a chiral transition state that is of lowest energy, leading to only one enantiomer of product. This process is called *chirality transfer*. 3) The function of the titanium(IV) isopropoxide is to act as the glue that holds all of the reagents together. The titanium holds an oxygen from each reactant—geraniol, *t*-BuOOH, and diethyl tartrate—and tethers them so that they react together, rather than just having them in solution and hoping that they will eventually collide.

(b) All three reactants are required to make Sharpless epoxidation work, but the key to *enantioselective* epoxidation is the chiral molecule, diethyl tartrate. When it complexes (or *chelates*) with titanium, it forms a large structure that is also chiral. As the *t*-BuOOH and geraniol approach the complex, the steric requirements of the complex allow the approach in one preferred orientation. When the reaction between the alkene and *t*-BuOOH occurs, it occurs preferentially from one face of the alkene, leading to one major stereoisomer of the epoxide. Without the chiral diethyl tartrate in the complex, the alkene could approach from one side just as easily as the other, and a racemic mixture would be formed.

(c) Using the enantiomer of diethyl L-tartrate, called diethyl D-tartrate, would give exactly the opposite stereochemical results, *i.e.*, the enantionmer of the first product.

## 14-22

*trans*-but-2-ene

IDENTICAL—MESO

stereochemistry shown in Newman projections:

*trans*

MESO

*See cis-but-2-ene on next page.*

14-22 continued

*cis*-but-2-ene

ENANTIOMERS

stereochemistry shown in Newman projections:

*cis* → → rotate →

CHIRAL—
RACEMIC
MIXTURE

14-23

$H_2C{=}CH_2 \xrightarrow[\text{H}_2\text{O}]{\text{H}_2\text{SO}_4} CH_3CH_2OH$

$H_2C{=}CH_2 \xrightarrow{RCO_3H} H_2C{-}CH_2$ (epoxide)

$\xrightarrow{\text{H}_2\text{SO}_4}$ $CH_3CH_2OCH_2CH_2OH$

Cellosolve®

A peroxyacid is far too expensive to use on the industrial scale. Ethylene oxide is produced in large quantity from ethylene and oxygen using a metal catalyst.

14-24 The cyclization of squalene via the epoxide is an excellent (and extraordinary) example of how Nature uses organic chemistry to its advantage. In one enzymatic step, Nature forms four rings and eight chiral centers! Out of 256 possible stereoisomers, only one is formed!

$\xi$ = new bonds formed

14-25

plus the other enantiomer

**314**

Copyright © 2013 Pearson Education, Inc.

14-26

(a) ... Na⁺

(b) H₂N ... Na⁺

(c) ... Na⁺

(d) ...

(e) N≡C ... K⁺

(f) ... Na⁺

14-27

(a) H¹⁸O ... OH

(b) HO ... ¹⁸OH

(c) Basic nucleophiles react at the less substituted carbon.

(d) In acid, nucleophiles react at the more substituted carbon.

14-28   Newly formed bonds are shown in bold. ▬

(a)

(b)

(c)

14-29

(a) at least three possible methods:

Method 1:   CH₃OH $\xrightarrow{Na}$ CH₃O⁻ Na⁺

... OH $\xrightarrow[\text{py}]{\text{TsCl}}$ ... OTs

Method 2:   PBr₃ (or HBr, Δ)

... Br       CH₃O⁻ Na⁺

PBr₃

CH₃Br ← CH₃OH

Method 3:   ... OH $\xrightarrow{Na}$ ... O⁻ Na⁺

(b) more practical way:

... OH + HOCH₂CH₃ $\xrightarrow[\text{S}_N1]{\text{H}_2\text{SO}_4}$ ... OCH₂CH₃

3°—forms carbocation

Bimolecular condensation works well on 3° carbons.

less practical method by S_N2, requiring more steps:

... O⁻ Na⁺ + ... LG

LG = leaving group like Br or OTs

Recall that ethoxide plus *tert*-butyl bromide gives only E2!  See Solved Problem 14-1 in the text.

14-29 continued
(c) more practical way:

benzylic—forms
carbocation

less practical method by $S_N2$,
requiring more steps:

LG = leaving group
like Br or OTs

Also possible is alkoxymercuration-demercuration:

(d)

See the solution to 14-25.

plus the
enantiomer

(e)

(f) Alkoxymercuration-demercuration is the more practical method.
Williamson ether synthesis on 2° carbons would not give good yields.

14-30

(a)

(b)

(c)

(d)

(e)

(f)

(g)

(h)

(i)

14-31

(a) *sec*-butyl isopropyl ether
(b) *tert*-butyl isobutyl ether
(c) ethyl phenyl ether
(d) chloromethyl propyl ether
(e) methyl *trans*-2-hydroxycyclohexyl ether
(f) cyclopentyl methyl ether
(g) propylene oxide
(h) cyclopropyl vinyl ether
(i) cyclopentene oxide
(j) 2-methyltetrahydrofuran

14-32

(a) 2-methoxypropan-1-ol
(b) ethoxybenzene or phenoxyethane
(c) methoxycyclopentane
(d) 2,2-dimethoxycyclopentan-1-ol
(e) *trans*-1-methoxy-2-methylcyclohexane
(f) *trans*-3-chloro-1,2-epoxycycloheptane
(g) *trans*-1,2-epoxy-1-methoxybutane; or,
    *trans*-2-ethyl-3-methoxyoxirane
(h) 3-bromooxetane
(i) 1,3-dioxane

14-33

(a)

(b)

(c) and (d) no reaction—
ethers are cleaved only
under acid conditions

(e) $\phantom{x}$ no $S_N$ at sp$^2$ carbon

(f)

(g) plus the enantiomer

(h)

(i) Keep at low temperature to
minimize the competing E2
reaction.

(j)

(k)

(l)

(m) plus the enantiomer

(n) plus the enantiomer

14-34

(a) On long-term exposure to air, ethers form peroxides. Peroxides are explosive when concentrated or heated. (For exactly this reason, ethers should *never* be distilled to dryness.)
(b) Peroxide formation can be prevented by excluding oxygen. Ethers can be checked for the presence of peroxides, and peroxides can be destroyed safely by treatment with reducing agents.

14-35

(a) Beginning with (R)-butan-2-ol and producing the (R)-sulfide requires two inversions of configuration.

An alternative approach would be to make the tosylate, displace with chloride or bromide ($S_N2$ with inversion), then do a second inversion with $NaSCH_3$.

14-35 continued

(b) Synthesis of the (S) isomer directly requires only one inversion.

14-36

(a)

molecular ion m/z 102 → m/z 73

$CH_3CH_2CH_2^{\oplus}$

m/z 43

(b)

molecular ion m/z 102

m/z 31 (after H migration)

m/z 59

m/z 71

m/z 87

14-37

+ $H_3O^+$

## 14-38

(a) You could also make hexan-2-ol, then Na followed by CH$_3$I.

(b) You could also make hexan-1-ol, then Na followed by CH$_3$I, similar to the solution to 14-29(a).

(c)

(d)

(e)

(e)

## 14-39  All chiral products in this problem are racemic.

**14-40** The student turned in the wrong product! Three pieces of information are consistent with the desired product: molecular formula $C_4H_{10}O$; O—H stretch in the IR at 3300 cm$^{-1}$ (although it should be strong, not weak); and mass spectrum fragment at m/z 59 (loss of $CH_3$). The NMR of the product should have a 9H singlet at δ 1.0 and a 1H singlet between δ 2 and δ 5. Instead, the NMR shows $CH_3CH_2$ bonded to oxygen. The student isolated diethyl ether, *the typical **solvent** used in Grignard reactions.*

<u>Predicted product</u>

59 CH$_3$

$CH_3$—C—OH

CH$_3$

$C_4H_{10}O$

strong O—H at 3300 cm$^{-1}$

<u>Isolated product</u>

59

$H_3C$—$CH_2$—O—$CH_2$—$CH_3$

$C_4H_{10}O$

weak O—H at 3300 cm$^{-1}$ due to water contamination

**14-41**

(a)

(b)

(c)

(d)

(e)

(f) The *trans* isomer would be the major product from the dehydration because *trans* is more stable than *cis*. Epoxidation with MCPBA is stereospecific and will retain the stereochemistry of the starting double bond. The major product will be *trans*, with some *cis* as a contaminant.

**320**

**14-42**  In the first sequence, no bond is broken to the chiral center, so the configuration of the product is the same as the configuration of the starting material.

$[\alpha]_D = -8.24°$

(Assume the enantiomer shown is levorotatory.)

$[\alpha]_D = -15.6°$

In the second reaction sequence, however, a bond to the chiral carbon is broken once, so the stereochemistry of the process will be a net inversion.

Undoubtedly, some E2 products will also form in this reaction.

$CH_3CH_2\overset{..}{\underset{..}{O}}:\overset{\ominus}{}$  $Na^+$

$S_N2$—INVERSION

RETENTION OF CONFIGURATION

The second sequence involves *retention* followed by *inversion*, thereby producing the *enantiomer* of the 2-ethoxyoctane generated by the first sequence. The optical rotation of the final product will have equal magnitude but opposite sign, $[\alpha]_D = +15.6°$ .

**14-43**

(a)

<u>anhydrous HBr</u>—only Br⁻ nucleophiles present

$BrCH_2CH_2Br$  +  $H_2O$

<u>aqueous HBr</u>—many more $H_2O$ nucleophiles than Br⁻ nucleophiles

+ $H_3O^+$

**14-43 continued**

(b)

**14-44** HA is an acid; A⁻ is the conjugate base.

*Note that all three carbocation intermediates are 3°!*

This process resembles the cyclization of squalene oxide to lanosterol. (See the solution to problem 14-24.) In fact, pharmaceutical synthesis of steroids uses the same type of reaction called a *biomimetic cyclization*.

**14-45**

(a)

(b)

Attack of water gave *inversion* of configuration at the chiral center; *R* became *S*.

14-45 continued

(c)

No bond to the chiral center was broken. Configuration is retained; R stays as R.

(d) The difference in these mechanisms lies in where the nucleophile attacks. Attack at the chiral carbon gives inversion; attack at the achiral carbon retains the configuration at the chiral carbon. These products are enantiomers and must necessarily have optical rotations of opposite sign. The lower enantiomeric excess in the acid catalyzed mechanism probably comes from some opening of the epoxide ring to form a secondary carbocation that can be attacked from either top or bottom by the water nucleophile, producing a mixture of R and S configuration.

14-46    methyl cellosolve®    $CH_3OCH_2CH_2OH$

To begin, what can be said about methyl cellosolve®? Its molecular weight is 76; its IR would show C—O in the 1000–1200 cm$^{-1}$ region and a strong O—H around 3300 cm$^{-1}$; and its NMR would show four sets of signals in the ratio of 3:2:2:1.

The unknown has molecular weight 134; this is double the weight of methyl cellosolve®, minus 18 (water). The IR shows no OH, only ether C—O. The NMR shows no OH, only H—C—O in the ratio of 3:2:2. Apparently, two molecules of methyl cellosolve® have combined in an acid-catalyzed, bimolecular condensation.

(This compound is called *diethylene glycol dimethyl ether*, or a shortened, common name is *diglyme*.)

14-47

The formula $C_8H_8O$ has five elements of unsaturation (enough (4) for a benzene ring). The IR is useful for what it does *not* show. There is neither OH nor C=O, so the oxygen must be an ether functional group.

The HNMR shows a 5H signal at δ 7.2, a monosubstituted benzene. No peaks in the δ 4.5-6.0 range indicate the absence of an alkene, so the remaining element of unsaturation must be a ring. The three protons are non-equivalent, with complex splitting.

+ 2 C + O + 3 H + ring
      ether

These pieces can be assembled in only one manner consistent with the data.

(Note that the CH$_2$ hydrogens are not equivalent (one is *cis* and one is *trans* to the phenyl) and therefore have distinct chemical shifts.)

323

14-48 The key concept is that reagents always go to the less hindered side of a molecule first. In this case, the "underneath" side is less hindered; the "top" side has a $CH_3$ hovering over the double bond and approach from the top will be much more difficult, and therefore slower, than approach from underneath.

Epoxides **B** and **C** are diastereomers; they will have different chemical and physical properties. Nucleophiles can react with **C** much faster than with **B** for precisely the same reason that explained their formation: approach from underneath is less hindered and is faster than approach from the top.

14-49 The product has a new pi bond so this must be an elimination reaction. The strong base *tert*-butoxide points to E2 as the probable mechanism.

The epoxide oxygen is the leaving group in this elimination. Ordinarily, ether oxygens are not leaving groups; in epoxides, however, the severe ring strain makes the ring open fairly readily.

**14-50** As is often true when explaining the properties of molecules, hydrogen bonding is the key.

glycerol
mol. wt. 92 g/mol
b.p. 290 °C
d 1.24 g/mL

mol. wt. 309 g/mol
b.p. 180 °C
d 0.88 g/mL

Glycerol has extremely strong intermolecular hydrogen bonding because of the three OH groups per molecule. Overcoming these intermolecular forces requires a lot of energy: thus, glycerol has a high boiling point despite its fairly low molecular weight, and it flows slowly because hydrogen bonding must be overcome in order for molecules to slide past each other. The density is high because these molecules pack together tightly to maximize hydrogen bonding.

In contrast, the TMS (trimethylsilyl) ether of glycerol not only has no hydrogen bonding, but on each end of the molecule, there is a nonpolar and essentially spherical group—this is like putting on boxing gloves and trying to pick up a dime, or anything! So in spite of the high molecular weight, these molecules tend to stay far apart, explaining the ease of flow, the low density, and the relatively low boiling point.

For a boiling point comparison, look up the structure of "isocetane," which is a highly branched alkane of molecular weight 226 g/mol, with boiling point 240 °C. The van der Waals forces in the TMS ether above must be even lower.

**14-51** Approaching a good synthesis problem begins with comparing the product to the starting material. If new carbons appear in the product, then the synthesis must include a carbon-carbon bond-forming reaction, of which there are very few.

new bond

The new bond is shown in bold. ▬▬

At first glance, this appears to require a simple Grignard reaction, but then we recall that a Grignard reagent cannot coexist with an OH group in the same molecule. Aha! The OH group needs to be protected before the Grignard can proceed.

This could be isolated.

Grignard reagents are stable only in solution; they cannot be isolated.

Fluoride removes the Si group, and water protonates the oxygens.

$Bu_4N^+$ $F^-$

$H_2O$

**325**

15-1   Look for:  1) the number of double bonds to be hydrogenated—the fewer C=C, the smaller the $\Delta H$; 2) conjugation—the more conjugated, the more stable, the lower the $\Delta H$;  3) degree of substitution of the alkenes—the more substituted, the more stable, the lower the $\Delta H$.

(a)

smallest $\Delta H$

biggest $\Delta H$

(b)

smallest $\Delta H$
most stable
of the dienes

triene
biggest $\Delta H$

15-2  Reminder:  H—A is used to symbolize the general form for an acid, that is, a protonated base; $A^-$ is the conjugate base.

NOT conjugated

CONJUGATED

15-3   (You may wish to refer to Problem 2-6.)          orbital picture

(a)

The shaded orbitals are perpendicular to the unshaded orbitals.

The central carbon atom makes two π bonds with two p orbitals.  These p orbitals must necessarily be perpendicular to each other, thereby forcing the groups on the ends of the allene system perpendicular.

(b)

mirror

non-superimposable mirror images  =  enantiomers

15-4 Carbocation stability depends first on conjugation (benzylic, allylic), then on degree of carbon carrying the positive charge.

less significant contributor

more significant contributor

equivalent to the first resonance form

15-5

Two carbons are electron deficient so the nucleophile can attack either.

The most basic species in the reaction mixture will remove this proton. The oxygen of ethanol is more basic than bromide ion.

**15-6**

Two carbons are electron deficient so the nucleophile can attack either.

**15-7**

(a)

Two carbons are electron deficient so the nucleophile can attack either.

(b)

same carbocation as in (a)

Two carbons are electron deficient so the nucleophile can attack either.

**15-7 continued**

(c)

Though the bromonium ion mechanism is typical for isolated alkenes, the greater stability of the resonance-stabilized carbocation will make it the lower energy intermediate for conjugated systems.

Two carbons are electron deficient so the nucleophile can attack either.

(d)

Two carbons are electron deficient so the nucleophile can attack either.

same carbocation as in (d)

(e)

Two carbons are electron deficient so the nucleophile can attack either.

**329**

**15-8**

(a)

$$\underset{\mathbf{A}}{\overset{\overset{\displaystyle Br}{|}\;\overset{\displaystyle Br}{|}}{H_2C-C-C=CH_2}}\;\;+\;\;\underset{\mathbf{B}}{\overset{\overset{\displaystyle Br}{|}\qquad\overset{\displaystyle Br}{|}}{H_2C-C=C-CH_2}}$$

Two carbons are electron deficient so the nucleophile can attack either.

(b)

$$H_2C=C-C=CH_2 \xrightarrow{Br-Br} \left\{ \underset{\mathbf{A^+}}{\overset{Br}{H_2C-\overset{\oplus\,\mathbf{2°}}{C}-C=CH_2}} \longleftrightarrow \underset{\mathbf{B^+}}{\overset{Br}{H_2C-C=C-\overset{\oplus\,\mathbf{1°}}{CH_2}}} \right\} \quad\text{allylic}$$

:Br:⁻

$$\underset{\mathbf{A}}{\overset{\overset{Br}{|}\;\overset{Br}{|}}{H_2C-C-C=CH_2}}\;+\;\underset{\mathbf{B}}{\overset{\overset{Br}{|}\qquad\overset{Br}{|}}{H_2C-C=C-CH_2}}$$

(c) The resonance form **A⁺** , which eventually leads to product **A**, has positive charge on a 2° carbon and is a more significant resonance contributor than structure **B⁺** . With greater positive charge on the 2° carbon than on the 1° carbon, we would expect bromide ion attack on the 2° carbon to have lower activation energy. Therefore, **A** must be the *kinetic* product. At higher temperature, however, the last step becomes reversible, and the stability of the products becomes the dominant factor in determining product ratios. As **B** has a disubstituted alkene whereas **A** is only monosubstituted, it is reasonable that **B** is the major, *thermodynamic* product at 60° C.

(d) At 60° C, ionization of **A** would lead to the same allylic carbocation as shown in (b), which would give the same product ratio as formation of **A** and **B** from butadiene.

$$\underset{\mathbf{A}}{\overset{\overset{Br}{|}\,\overset{Br}{|}}{H_2C-C-C=CH_2}} \xrightarrow{-\,Br^-} \left\{ \overset{Br}{H_2C-\overset{\oplus\,\mathbf{2°}}{C}-C=CH_2} \longleftrightarrow \overset{Br}{H_2C-C=C-\overset{\oplus\,\mathbf{1°}}{CH_2}} \right\} \quad\text{allylic}$$

:Br:⁻

Two carbons are electron deficient so the nucleophile can attack either.

$$\underset{\mathbf{A}}{\overset{\overset{Br}{|}\;\overset{Br}{|}}{H_2C-C-C=CH_2}}\;+\;\underset{\mathbf{B}}{\overset{\overset{Br}{|}\qquad\overset{Br}{|}}{H_2C-C=C-CH_2}}$$

**10%**      **90%**

**15-9**

(a)

15-9 continued

(b)

This would probably be the major product because the double bond is in a more stable position.

recycles in chain mechanism

15-10
(a)

+ ...trans... + ...cis...

(b)    ("Pr" is the abbreviation for propyl, used below.)

NBS + HBr generate a low concentration of Br₂

initiation    Br—Br →(hv) 2 Br•

propagation
The first step is abstraction of allylic hydrogen to generate allylic radical.

radical will be a mixture of *cis* and *trans*

Br• + Pr—C—C=CH₂ (hex-1-ene) → HBr + { Pr—Ċ—C=CH₂ ←→ Pr—C=C—ĊH₂ }

recycles in chain mechanism

Br• + Pr—C—C=CH₂ + Pr—C=C—CH₂

*cis + trans*

## 15-11

(a)

(b) +

These are the major products from abstraction of a 1° allylic H.

(c)

Benzylic radicals are even more stable than allylic.

## 15-12  Both halides generate the same allylic carbanion.

## 15-13

(a)

(b)

(c)  $CH_3CH_2CH_2Br \xrightarrow[\text{ether}]{Mg} CH_3CH_2CH_2MgBr$

This synthesis could also be performed sequentially.

*add one-half equivalent*

dec-5-ene

## 15-14  Chiral products from achiral reactants produce racemic mixtures.  New bonds are in bold. ▬

(a)

(b)

(c)

(d)

(e)

(f)

**332**
Copyright © 2013 Pearson Education, Inc.

15-15

(a)

(b) structure + structure

(c) structure + structure

(d) structure + structure

(e) structure + structure

(f) structure + structure

15-16 Chiral products from achiral reactants produce racemic mixtures.

(a)

(b)

(c) all *cis*

15-17 These structures show the alignment of diene and dienophile in the Diels-Alder transition state, leading to 1,4-orientation in (a) and 1,2-orientation in (b).

(a)

This left structure is a **VERY** minor resonance contributor; however, it explains the orientation for the diene as carbon-1 is more negative and carbon-2, a 3° carbon, is slightly more positive because of methyl group stabilization.

(b)

15-18 For clarity, the sigma bonds formed in the Diels-Alder reaction are shown in bold. Chiral products from achiral reactants produce racemic mixtures.

(a) CH₃O ... COOCH₃

(b) OCH₃ ... H ... H

(c) CHO ... CH₃O

(d) CH₃ ... CN ... CH₃

**15-19** For a photochemically *allowed* process, one molecule must use an excited state in which an electron has been promoted to the first antibonding orbital. All orbital interactions between the excited molecule's HOMO* and the other molecule's LUMO must be bonding for the interaction to be allowed; otherwise, it is a forbidden process.

HOMO*
excited state of
diene HOMO

bonding interaction
(constructive overlap)

$\pi_3$*

antibonding interaction
(destructive overlap)

$\pi$*

LUMO of
dienophile

In the Diels-Alder cycloaddition, the LUMO of the dienophile and the excited state of the HOMO of the diene (labeled HOMO*) produce one bonding interaction and one antibonding interaction. Thus, this is a **photochemically forbidden** process.

**15-20** For a [4 + 4] cycloaddition:

(a)

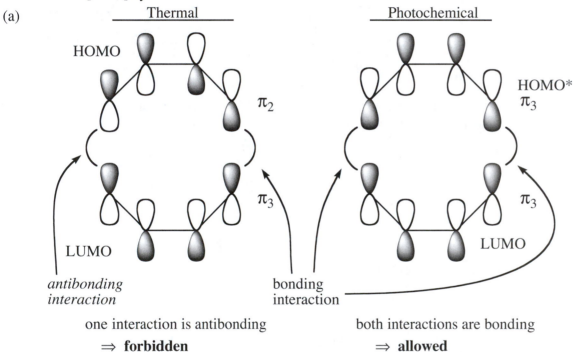

Thermal

HOMO

$\pi_2$

$\pi_3$

LUMO

*antibonding interaction*

bonding interaction

one interaction is antibonding
⇒ **forbidden**

Photochemical

HOMO*
$\pi_3$

$\pi_3$

LUMO

both interactions are bonding
⇒ **allowed**

(b) A [4 + 4] cycloaddition is not thermally allowed, but a [4 + 2] (Diels-Alder) is!

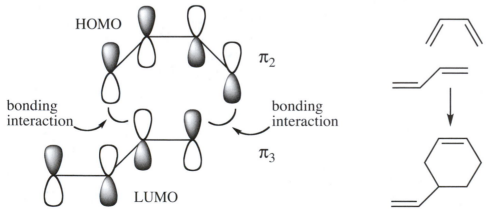

HOMO

$\pi_2$

bonding
interaction

$\pi_3$

bonding
interaction

LUMO

**334**

**15-21**

$$A = \varepsilon c\, l \qquad \varepsilon = \frac{A}{c\, l} \qquad l = 1 \text{ cm} \qquad A = 0.50$$

convert mass to moles:  $1 \text{ mg} \times \dfrac{1 \text{ g}}{1000 \text{ mg}} \times \dfrac{1 \text{ mole}}{160 \text{ g}} = 6 \times 10^{-6} \text{ moles}$

$$c = \frac{6 \times 10^{-6} \text{ moles}}{10 \text{ mL}} \times \frac{1000 \text{ mL}}{1 \text{ L}} = 6 \times 10^{-4} \text{ M}$$

$$\varepsilon = \frac{0.50}{(6 \times 10^{-4})(1)} = 833 \approx \boxed{800}$$

**15-22**

(a) 353 nm:  a conjugated tetraene—must have highest absorption maximum among these compounds;
(b) 313 nm:  closest to the bicyclic conjugated triene in Table 15-2; the diene is in a more substituted ring, so it is not surprising for the maximum to be slightly higher than 304 nm;
(c) 232 nm:  similar to 3-methylenecyclohexene in Table 15-2;
(d) 292 nm:  like cyclohexa-1,3-diene (256 nm) + 1 additional conjugated double bond (30 nm) + 1 alkyl group (5 nm) = predicted value of 291 nm

**15-23**

A  *colorless*         B  *red*

*Extended conjugation* is required for an organic compound to be colored.  The acid form of phenolphthalein **A** has an sp³ carbon disrupting the conjugation, whereas the basic form **B** has a central carbon with sp² hybridization, permitting conjugation over three rings; the conjugated sp² carbons are indicated with a large dot in this picture.

**15-24**

(a) isolated      (b) conjugated      (c) cumulated      (d) conjugated      (e) conjugated

(c) The sp carbon would require linear geometry.

(f) cumulated (1,2) and conjugated (2,4)

$$H_2C = C = C \underset{\underset{1 \quad 2}{}}{\overset{3}{\underset{H}{}}} \overset{H \ 4}{\underset{5}{C}} \overset{5}{\underset{CH_2}{}}$$

**15-25**

(a)          (b)          Cl          (c)          Br          (d)

one equivalent of HCl

(e)          OH

+          Br ⌇⌇⌇ OH

(f)          Br ⌇⌇⌇          +          Br ⌇⌇⌇          +          ⌇⌇⌇ Br

Br          Br   minor— not conjugated          Br

15-25 continued

(g)

substitution          elimination

(h)

COOCH₃

(i)

15-26 Grignard reactions are performed in ether solvent. For clarity, the bonds formed are shown in bold.

(a)

(b)

(c)

15-27

(a)

(b)

(c)

(d)

15-27 continued

(e)

most significant contributor—
negative charge on more electronegative atom

(f)

(g)

Most significant
contributor—all atoms
have full octets.

(h)

Most significant
contributor—all atoms
have full octets.

15-28

(a)    $A = \varepsilon c\, l$        $\varepsilon = \dfrac{A}{c\, l}$        $l = 1$ cm        $A = 0.74$

convert mass to moles:   $0.0010$ g x $\dfrac{1 \text{ mole}}{255 \text{ g}} = 3.9 \times 10^{-6}$ moles

$c = \dfrac{3.9 \times 10^{-6} \text{ moles}}{100 \text{ mL}}$ x $\dfrac{1000 \text{ mL}}{1 \text{ L}} = 3.9 \times 10^{-5}$ M

$\varepsilon = \dfrac{0.74}{(3.9 \times 10^{-5})(1)} \approx \boxed{19{,}000}$

(b)  This large value of $\varepsilon$ could only
come from a conjugated system,
eliminating the first structure.  The
absorption maximum at 235 nm is
most likely a diene rather than a triene.
 The most reasonable structure is:

compare with dienes:

$\lambda_{max} = 232$ nm
Solved Problem 15-3

$\lambda_{max} = 232$ nm
Table 15-2

337

Copyright © 2013 Pearson Education, Inc.

**15-29**

(a) align δ+ and δ−

new sigma bonds
shown in bold

methyl benzoate

(b) The second reaction is called a retro (reverse) Diels-Alder reaction. It is also an electrocyclic rearrangement of electrons, in this case breaking sigma bonds, forming new pi bonds, and creating an *aromatic* ring in the process. The stability of the aromatic ring accounts for the thermodynamic preference for the benzene product.

**15-30** For clarity, the sigma bonds formed in the Diels-Alder reaction are shown in bold. ▬
Chiral products from achiral reactants produce racemic mixtures.

(a)　COOH

(b)　H　COOCH₃

(c)　COOH
*Endo Rule*

(d)　OCH₃　CHO

(e)　CH₃　CN　CN　CH₃　*methyls cis*

(f)　CH₃　CN　CN　CH₃　*methyls trans*

**15-31** For clarity, the sigma bonds formed in the Diels-Alder reaction are shown in bold. ▬
Chiral products from achiral reactants produce racemic mixtures.

(a)　COOCH₃　COOCH₃

(b)　*Endo Rule*

(c)　CH₃　CHO　*Endo Rule*

(d)　H₃CO　*Endo Rule*

(e)　N(CH₃)₂　*Endo Rule*　OEt

(f)　OCH₃　O
*Ring juncture is cis.*

15-32

(a) The absorption at 1630 cm$^{-1}$ suggests a conjugated alkene. The higher temperature allowed for migration of the double bond.

(b)

desired — CH$_2$CH$_2$CH$_3$

actual — CH$_2$CH$_2$CH$_3$

The UV $\lambda_{max}$ at 261 nm verifies that the two double bonds have become conjugated. Isolated double bonds absorb at < 200 nm.

(c)
expected:

[ CH$_2$CH$_2$CH$_3$ ]$^{+\cdot}$ → { CH / HC }  m/z 79  + $\cdot$CH$_2$CH$_2$CH$_3$  mass 43

doubly allylic
m/z 122

actual:

[ CH$_2$—CH$_2$CH$_3$ ]$^{+\cdot}$ → { CH$_2$ / C—CH$_2$ }  m/z 93  + $\cdot$CH$_2$CH$_3$  mass 29

m/z 122

15-33

(a) [butadiene] + [acrolein: CH$_2$=CH—CHO]

(b) [cyclopentadiene] + [Cl—CH=CH—Cl]

(c) CH$_3$—C(=CH$_2$)—CH$_3$ (2,3-dimethylbutadiene)  + [C(COOCH$_3$)≡CH]

(d) [butadiene] + [benzoquinone]

(e) [cyclopentadiene] + [maleic anhydride]

(f) [furan] + [CH(CN)=CH(CN)]

(g) Cl / Cl / Cl / Cl / Cl (hexachlorocyclopentadiene) + [cyclopentene with Cl, Cl]

(h) Cl / Cl / Cl / Cl / Cl (hexachlorocyclopentadiene) + [norbornadiene]

(i) [cyclic sulfone S(=O)=O] + [NC—CH=CH—CN]

15-34

(a)

(b)  The *endo* isomer is usually preferred because of secondary p orbital overlap of C=O with the diene in the transition state.

(c)  The reasoning in (b) applies to stabilization of the transition state of the reaction, not the stability of the product.  Arguments based on transition state stability apply to the rate of reaction, inferring that the *endo* product is the kinetic product.

(d)  At 25 °C, the reaction cannot easily reverse, or at least not very rapidly.  The *endo* product is formed faster and is the major product because its transition state is lower in energy—the reaction is under kinetic control.  At 90 °C, the reverse reaction is not as slow and equilibrium is achieved.  The *exo* product is less crowded and therefore more stable—equilibrium control gives the *exo* as the major product.

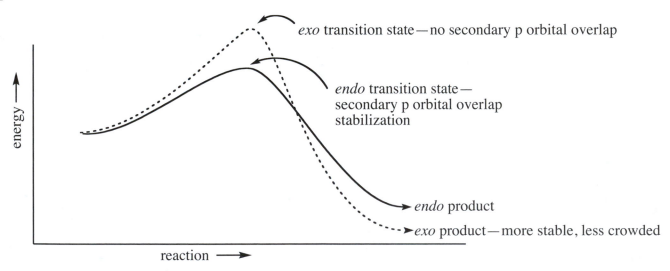

15-35 Nodes are represented by dashed lines.

(a)

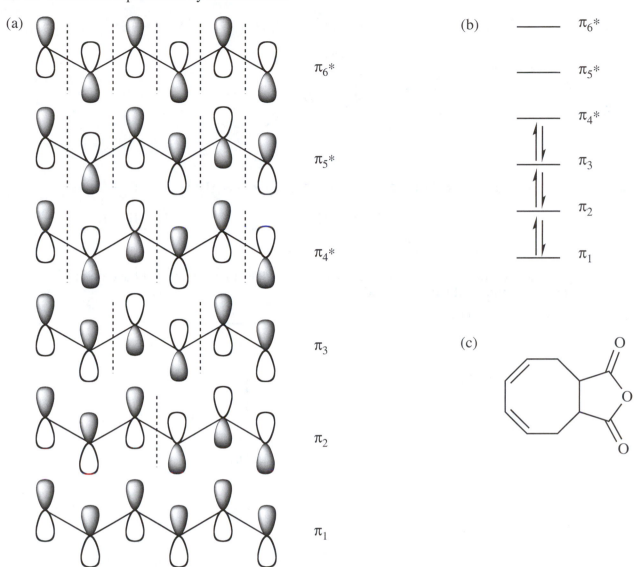

$\pi_6^*$

$\pi_5^*$

$\pi_4^*$

$\pi_3$

$\pi_2$

$\pi_1$

(b)

———— $\pi_6^*$

———— $\pi_5^*$

———— $\pi_4^*$

⇅ $\pi_3$

⇅ $\pi_2$

⇅ $\pi_1$

(c)

(d) Whether the triene is the HOMO and the alkene is the LUMO, or *vice versa*, the answer will be the same.

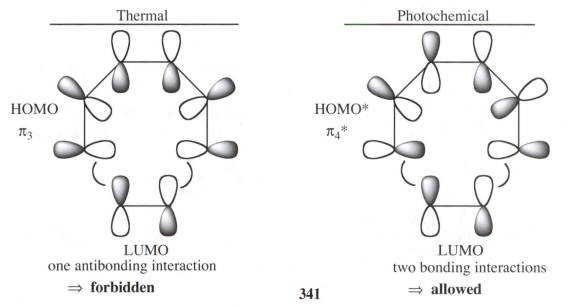

Thermal

HOMO

$\pi_3$

LUMO
one antibonding interaction

⇒ **forbidden**

Photochemical

HOMO*

$\pi_4^*$

LUMO
two bonding interactions

⇒ **allowed**

15-35 continued

(e)

15-36

(a)  $\left\{ \begin{array}{l} H_2C{=}C{-}C{=}C{-}\dot{C}H_2 \longleftrightarrow H_2C{=}C{-}\dot{C}{-}C{=}CH_2 \longleftrightarrow H_2\dot{C}{-}C{=}C{-}C{=}CH_2 \\ \quad\quad | \quad | \quad | \quad\quad\quad\quad\quad\quad | \quad | \quad | \quad\quad\quad\quad\quad\quad | \quad | \quad | \\ \quad\quad H \quad H \quad H \quad\quad\quad\quad\quad\quad H \quad H \quad H \quad\quad\quad\quad\quad\quad H \quad H \quad H \end{array} \right\}$

(b) Five p atomic orbitals will generate five pi molecular orbitals.

(c) The lowest energy molecular orbital has no nodes. Each higher molecular orbital will have one more node, so the fifth molecular orbital will have four nodes.

(d) Nodes are represented by dashed lines.

(e)

$\pi_5^*$

$\pi_4^*$

$\pi_3$

$\pi_2$

$\pi_1$

$\pi_5^*$

$\pi_4^*$

$\pi_3$ (nonbonding)

(f) The HOMO, $\pi_3$, contains an unpaired electron giving this species its radical character. The HOMO is a nonbonding orbital with lobes only on carbons 1, 3, and 5, consistent with the resonance picture.

$\pi_2$

$\pi_1$

15-36 continued

(g) ⎯⎯ $\pi_5^*$

⎯⎯ $\pi_4^*$

⎯⎯ $\pi_3$

⥮ $\pi_2$

⥮ $\pi_1$

Again, it is $\pi_3$ that determines the character of this species. When the single electron in $\pi_3$ of the neutral radical is removed, positive charge appears only in the position(s) which that electron occupied. That is, the positive charge depends on the now *empty* $\pi_3$, with *empty* lobes (positive charge) on carbons 1, 3, and 5, consistent with the resonance description.

$$\left\{ H_2\overset{\oplus}{C}{\sim}{\sim} \longleftrightarrow {\sim}\overset{\oplus}{\underset{H}{C}}{\sim} \longleftrightarrow {\sim}{\sim}\overset{CH_2}{\underset{\oplus}{}} \right\}$$

(h) ⎯⎯ $\pi_5^*$

⎯⎯ $\pi_4^*$

⥮ $\pi_3$

⥮ $\pi_2$

⥮ $\pi_1$

Again, it is $\pi_3$ that determines the character of this species. The negative charge depends on the *filled* $\pi_3$, with lobes (negative charge) on carbons 1, 3, and 5, consistent with the resonance description.

$$\left\{ H_2\overset{\ominus}{\underset{\cdot\cdot}{C}}{\sim}{\sim} \longleftrightarrow {\sim}\overset{\cdot\cdot}{\underset{\ominus H}{C}}{\sim} \longleftrightarrow {\sim}{\sim}\overset{CH_2}{\underset{\ominus\,\cdot\cdot}{}} \right\}$$

15-37
(a)

CH₃ ring with Br → Mg, ether → CH₃ ring with MgBr → acetone (O=C(CH₃)₂) → H₃O⁺ → CH₃ ring, **A**, with H₃C–C–OH attached and CH₃ below

(b) Use text Section 15-13D to predict $\lambda_{max}$ values. Alkyl substituents are circled.

transoid cyclic diene with one
exocyclic double bond = 232 nm
2 extra alkyl groups = 10 nm
TOTAL = 242 nm
desired product—**not
consistent with UV data**

*trisubstituted*

cisoid cyclic diene = 256 nm
1 extra alkyl group = 5 nm
TOTAL = 261 nm

*disubstituted*

*both
trisubstituted
C=C*

cisoid cyclic diene = 256 nm
2 extra alkyl groups = 10 nm
TOTAL = 266 nm—AHA!—
closest to observed value

actual product **B**

## 15-37 continued

(c)

**B** is produced in preference to the other *cisoid* diene above because **B**'s diene system is more highly substituted and therefore more stable.

**B** *both trisubstituted C=C*

---

## 15-38  It is stunningly clever reactions like this that earned E. J. Corey his Nobel Prize.

*Diels-Alder*

Imagine the diene on top of the dienophile in the transition state, leading to the *endo* adduct.

*endo* product

*retro-Diels-Alder*

O=C=O   dienophile

diene

same as

amazing!

15-39 Look for *extended conjugation*, meaning an unbroken string of sp² hybridized atoms, noted here with a large dot on carbon atoms. Other sp² atoms, like O and N, are designated with a *.

(a) Basic Blue 6,
aka Meldola Blue,
a biological stain

Cl⁻

(b) NOT COLORED—
no extended conjugation

(c) kermesic acid,
a red dye from insects

(d) Acid Alizarin
Violet N, a
violet dye

(e) NOT COLORED—
no extended conjugation

(f) p-naphtholbenzein,
a red indicator

(g) food dye:
FD&C Red #3

(h) NOT COLORED—
no extended conjugation

(i) rhodoxanthin, a purple dye
found in plants and bird feathers

(j) food dye:
FD&C Yellow #5

All 4 N and COO⁻ are also sp².

(k) NOT COLORED—
no extended conjugation

**345**

Copyright © 2013 Pearson Education, Inc.

Note: The representation of benzene with a circle to represent the π system is fine for questions of nomenclature, properties, isomers, and reactions. For questions of mechanism or reactivity, however, the representation with three alternating double bonds (the Kekulé picture) is more informative. For clarity and consistency, this Solutions Manual will use the Kekulé form exclusively.

equivalent representations of benzene

Kekulé form used in
the Solutions Manual

16-1

16-2   All values are per mole.

(a)
| | benzene | −208 kJ | −49.8 kcal |
|---|---|---|---|
| − | cyclohexa-1,4-diene | −240 kJ | −57.4 kcal |
| | | $\Delta H = +32$ kJ | +7.6 kcal |

(b)
| | benzene | −208 kJ | −49.8 kcal |
|---|---|---|---|
| − | cyclohexene | −120 kJ | −28.6 kcal |
| | | $\Delta H = -88$ kJ | −21.2 kcal |

(c)
| | cyclohexa-1,3-diene | −232 kJ | −55.4 kcal |
|---|---|---|---|
| − | cyclohexene | −120 kJ | −28.6 kcal |
| | | $\Delta H = -112$ kJ | −26.8 kcal |

16-3

(a)

The biggest failure of resonance theory is the inability to show why benzene is highly stabilized and the other two structures are not.

**16-3 continued**

(b)

**16-4**

**16-5** Figure 16-8 shows that the first three pairs of electrons are in three bonding molecular orbitals of cyclooctatetraene. Electrons 7 and 8, however, are located in two different nonbonding orbitals. As in cyclobutadiene, a planar cyclooctatetraene is predicted to be a diradical, a particularly unstable electron configuration.

**16-6**

Models show that the angles between p orbitals on adjacent π bonds approach 90°.

**16-7** Planarity of a real structure is difficult to predict just by looking at a planar drawing.
(a) nonaromatic: internal hydrogens prevent planarity
(b) nonaromatic: not all atoms in the ring have a p orbital, as one carbon is $sp^3$ hybridized
(c) aromatic if planar: [14]annulene
(d) aromatic: also a [14]annulene in the outer ring: the internal alkene is not part of the aromatic system

**16-8** Azulene satisfies all the criteria for aromaticity, and it has a Huckel number of π electrons: 10. Both heptalene (12 π electrons) and pentalene (8 π electrons) are antiaromatic.

**16-9**

This electronic configuration is antiaromatic. It would be aromatic if it lost 2 electrons to make the double-positive ion.

16-9 continued

(c)
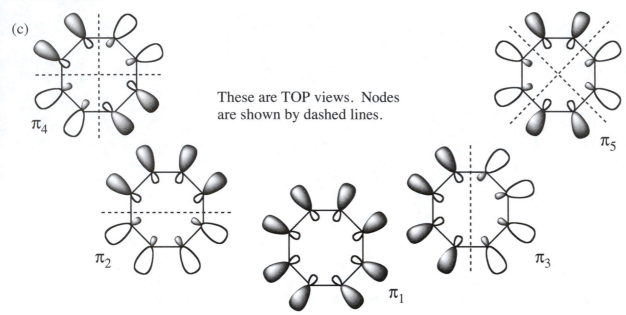
$\pi_4$

$\pi_5$

These are TOP views. Nodes are shown by dashed lines.

$\pi_2$

$\pi_1$

$\pi_3$

16-10 Nodes in these orbital pictures are indicated by dashed lines.

(a)
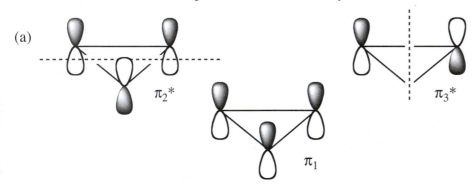
$\pi_2^*$

$\pi_1$

$\pi_3^*$

(b)

——— $\pi_2^*$           ——— $\pi_3^*$
- - - - - - - - - - - - - - - - - - - - - - - - - - - - - - - - - nonbonding
              ——— $\pi_1$

$\pi_1$ is bonding; $\pi_2^*$ and $\pi_3^*$ are antibonding.

(c)
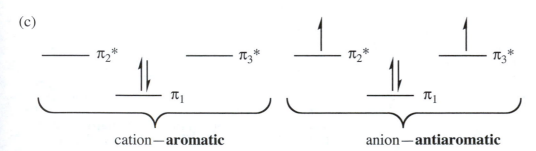

cation—**aromatic**          anion—**antiaromatic**

16-11  Nodes in these orbital pictures are indicated by dashed lines.

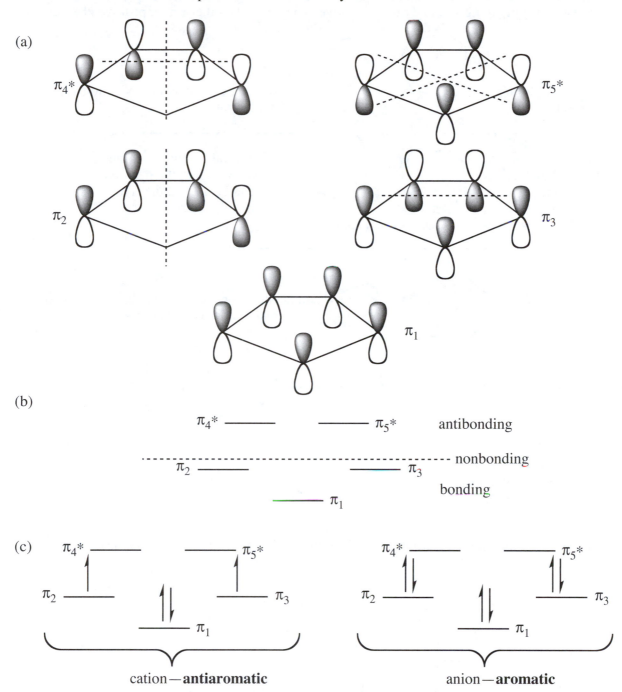

16-12  Assume planarity in these ring systems for the purpose of determining aromaticity.
(a)  antiaromatic:  8 π electrons (4n), not a Huckel number
(b)  aromatic:  10 π electrons (4n + 2), a Huckel number
(c)  aromatic if planar:  18 π electrons (4n + 2), a Huckel number
(d)  antiaromatic:  20 π electrons (4n), not a Huckel number
(e)  nonaromatic:  no cyclic π system
(f)  aromatic if planar:  20 − 2 = 18 π electrons (4n + 2), a Huckel number

**16-13** The reason for the dipole can be seen in a resonance form distributing the electrons to give each ring 6 π electrons. This resonance picture gives one ring a negative charge and the other ring a positive charge.

*aromatic* tropylium ion
6 π electrons

*aromatic* cyclopentadienyl anion
6 π electrons

several resonance forms delocalizing the positive charge around the seven-membered ring

several resonance forms delocalizing the negative charge around the five-membered ring

Composite resonance picture shows that both rings are aromatic with charge separation.

**16-14**

$$\text{(cyclopropene Cl)} + \text{AgBF}_4 \longrightarrow \text{AgCl} + \left\{ \ldots \right\} \text{BF}_4^-$$

*aromatic* cyclopropenium ion

The crystalline material soluble in polar organic solvents is cyclopropenium tetrafluoroborate.

KCl

Cl—⊲| + KBF₄

**16-15** Draw resonance forms showing the carbonyl polarization, leaving a positive charge on the carbonyl carbon.

2 π electrons —
AROMATIC

6 π electrons —
AROMATIC

4 π electrons —
ANTIAROMATIC

The three- and seven-membered rings are aromatic; the five-membered ring is antiaromatic and, not surprisingly, very reactive.

16-16

(a)

cyclopentadienyl anion    pyrrole

Both pyrrole and cyclopentadienyl anion are aromatic systems with 6 pi electrons, two electrons coming from either the carbanion carbon of cyclopentadienyl anion or the nitrogen atom in pyrrole.

(b) Since the two structures are isoelectronic (same number of electrons), the only difference is the identity of the "fifth" atom of the ring, C or N, and the difference between those is one proton: N has one more proton than C, which is why pyrrole does not have a negative charge. It is likely that pyrrole has two extra neutrons as $^{14}N$ is the most abundant isotope of nitrogen and $^{12}C$ is the most abundant isotope of C, but this could vary depending on the particular isotopes present. The extra proton in pyrrole makes the biggest difference.

(c)

most significant contributor

Note that all of the minor resonance contributors with charge separation have (+) charge on the nitrogen atom.

16-17

not basic

The structure of purine shows two types of nitrogens. One type (N-1, N-3, and N-7) has an electronic structure like the nitrogen in pyridine; the pair of electrons is in an sp$^2$ orbital planar with the ring. These electrons are available for bonding, and these three nitrogens are basic. The other type of nitrogen at N-9 has an electronic structure like the nitrogen of pyrrole; its electron pair is in a p orbital, perpendicular to the ring system, and more importantly, an essential part of the aromatic pi system. With this pair of electrons, the pi system is aromatic and has 10 electrons, a Huckel number, so the electron pair is not available for bonding and N-9 is not a basic nitrogen.

16-18

(a) The proton NMR of benzene shows a single peak at δ 7.2; alkene hydrogens absorb at δ 4.5–6. The chemical shifts of 2-pyridone are more similar to benzene's absorptions than they are to alkenes. It would be correct to infer that 2-pyridone is aromatic.

(b)

The lone pair of electrons on the nitrogen in the first resonance form is part of the cyclic pi system. The second resonance form shows three alternating double bonds with 6 electrons in the cyclic pi system, consistent with an aromatic electronic system.

*continued on next page*

16-18(b) continued

These resonance forms show that the hydrogens at positions with greater electron density are shielded, decreasing chemical shift.

δ 6.15     δ 6.57

These resonance forms show that the hydrogens at positions with greater positive charge are deshielded, increasing chemical shift.

H δ 7.26     δ 7.31

(c)

This resonance form of thymine shows a cyclic pi system with 6 pi electrons, consistent with an aromatic system. Four of these electrons came from the two lone pairs of electrons on nitrogens in the first resonance form.

(d)

The ring portion of 5-fluorouracil is identical to the ring structure of thymine in part (c). Both structures have the aromatic ring in common.

16-19

(a)

aromatic: 4 π electrons in double bonds plus one pair from oxygen

(b)

aromatic: 4 π electrons in double bonds plus one pair from sulfur

(c)     sp³

not aromatic: no cyclic π system because of sp³ carbon

(d)

aromatic: Cation on carbon-4 indicates an empty p orbital; two π bonds plus a pair of electrons from oxygen makes a 6 π electron system.

16-19 continued

(e)

aromatic: Resonance form shows "push-pull" of electrons from one O to the other, making a cyclic π system with 6 electrons.

(f)

not aromatic: no cyclic π system because of sp³ carbon

(g)

aromatic: Resonance form shows electron pair from N making a cyclic π system with 6 electrons.

(h)

not aromatic: If 2 π electrons from oxygen are included, then 8 π electrons; if oxygen does not contribute electrons, then oxygen is sp³ and there is no cyclic π system.

16-20

Borazole is a non-carbon equivalent of benzene. Each boron is hybridized in its normal sp². Each nitrogen is also sp² with its pair of electrons in its p orbital. The system has 6 π electrons in 6 p orbitals—aromatic! Although borazole is isoelectronic with benzene, benzene boils at 80 °C whereas borazole boils at 161 °C because of its polar nature.

16-21
(a) resonance braces omitted

anthracene

phenanthrene

(b) anthracene

plus many other resonance forms

Bromide attack at C-10 leaves two aromatic rings (*cis* + *trans*).

*continued on next page*

16-21(b) continued

phenanthrene

cis and trans

Bromide attack at C-9
leaves two aromatic rings.

plus many other
resonance forms

(c) A typical addition of bromine occurs with a bromonium ion intermediate, which can give only anti addition. Addition of bromine to phenanthrene, however, generates a free carbocation because the carbocation is benzylic, stabilized by resonance over two rings. In the second step of the mechanism, bromide nucleophile can attack either side of the carbocation, giving a mixture of *cis* and *trans* products.

(d)

resonance-stabilized

16-22

(a)

The second resonance form shows that the two rings in the dashed box form a 10 electron pi system, isoelectronic with naphthalene. Cipro is aromatic.

(b) For a nitrogen (or any atom) to be basic, it must have an electron pair to share by forming a sigma bond with a proton. The nitrogen atoms labeled 1 and 2 are basic because they have an unshared electron pair, but nitrogen 3 does not have an unshared pair; its electron pair is delocalized in the aromatic pi system.

(c) The three H atoms in bold are on the aromatic ring and will appear in the HNMR between δ 6–8.

16-23

chlorobenzene

o-dichlorobenzene
(1,2-dichlorobenzene)

m-dichlorobenzene
(1,3-dichlorobenzene)

p-dichlorobenzene
(1,4-dichlorobenzene)

1,2,3-trichlorobenzene

1,2,4-trichlorobenzene

1,3,5-trichlorobenzene

1,2,3,4-tetrachloro-
benzene

1,2,3,5-tetrachloro-
benzene

1,2,4,5-tetrachloro-
benzene

pentachlorobenzene
(numbers not needed)

hexachlorobenzene
(numbers not needed)

16-24 (a) fluorobenzene
(d) o-nitrostyrene

(g) 3,4-dinitrophenol

(b) 4-phenylbut-1-yne
(e) p-bromobenzoic acid, or
4-bromobenzoic acid

(h) benzyl ethyl ether, or
benzoxyethane, or
(ethoxymethyl)benzene, or
α-ethoxytoluene

(c) m-methylphenol, or
3-methylphenol
(common name: m-cresol)

(f) isopropoxybenzene, or
isopropyl phenyl ether

16-25

$\lambda_{max}$  220 nm (strong)
258 nm (weak)

In dilute $H_2SO_4$(aq), the
acid is $H_3O^+$.

$-H_2O$

plus resonance forms with positive
charge on the benzene ring

$\lambda_{max}$  250 nm (strong)
290 nm (weak)

Electronic systems with extended
conjugation absorb at longer wavelength.

**355**

Copyright © 2013 Pearson Education, Inc.

16-26

(a) OCH$_3$ / NO$_2$ (benzene ring)

(b) OH / OCH$_3$ / OCH$_3$ (benzene ring)

(c) COOH / NH$_2$ (benzene ring)

(d) NH$_2$ / NO$_2$ (benzene ring)

(e) CH$_3$ / Cl (benzene ring)

(f) HC=CH$_2$ / HC=CH$_2$ (benzene ring)

(g) HC=CH$_2$ / Br (benzene ring)

(h) CHO / CH$_3$O / OCH$_3$ (benzene ring)

(i)  cycloheptatrienyl cation, Cl$^-$

(j) cyclopentadienyl anion, Na$^+$

(k)

(l) CH$_2$OCH$_3$ (benzene ring)

(m) SO$_3$H / CH$_3$ (benzene ring)

(n) CH$_3$ / CH$_3$ (benzene ring)

(o)

16-27 The IUPAC system recommends using position numbers for substituted benzenes. The terms *ortho*, *meta*, and *para* for disubstituted benzenes are commonly used; they are presented in parentheses.

(a) 1,2-dichlorobenzene (*ortho*)
(b) 4-nitroanisole (*para*)
(c) 2,3-dibromobenzoic acid
(d) 2,7-dimethoxynaphthalene
(e) 3-chlorobenzoic acid (*meta*)
(f) 2,4,6-trichlorophenol
(g) 2-*sec*-butylbenzaldehyde (*ortho*)
(h) cyclopropenium tetrafluoroborate

16-28

CH$_3$ (benzene ring) — toluene

CH$_3$ / CH$_3$ (benzene ring) — *o*-xylene

CH$_3$ / CH$_3$ (benzene ring) — *m*-xylene

CH$_3$ / CH$_3$ (benzene ring) — *p*-xylene

CH$_3$ / CH$_3$ / CH$_3$ (benzene ring) — 1,2,3-trimethylbenzene

CH$_3$ / CH$_3$ / CH$_3$ (benzene ring) — 1,2,4-trimethylbenzene

CH$_3$ / CH$_3$ / CH$_3$ (benzene ring) — 1,3,5-trimethylbenzene (common name: mesitylene)

16-29 The key concept in parts (a)–(c) is that an aromatic product is created.

(a)

The protonated carbonyl gives a resonance-stabilized cation. Protonation of the singly bonded oxygen does not generate a resonance-stabilized product.

**B**

*AROMATIC*

The last resonance form shows that the cation produced is aromatic and therefore more stable than the corresponding nonaromatic ion. In this second reaction, the products are more favored than in the first reaction, which is interpreted as the reactant **B** being more basic than **A**.

(b)

**C**

**D**

*AROMATIC*

The product from ionization of **C** is stabilized by resonance. The ionization product of **D** is not only resonance-stabilized but is also aromatic and therefore more stable. A reaction that produces a more stable product will usually happen faster under milder conditions because the transition state leading to that product will be stabilized, leading to a lower activation energy.

16-29 continued

(c)

E

acid catalyst → $H_2O$ +

F

acid catalyst → $H_2O$ +

AROMATIC

Dehydration of **F** produces an aromatic product that is more stable than the product from **E**. A reaction that produces a more stable product will usually happen faster under milder conditions because the transition state leading to that product will be stabilized, leading to a lower activation energy.

(d)

phenol

base

Resonance stabilization of the phenoxide anion shows the negative charge distributed over the one para and two ortho carbons.

umbelliferone

(Umbelliferone is one of about 1000 coumarins isolated from plants, primarily the families of Angiosperms: Fabaceae, Asteraceae, Apiaceae, and Rutaceae.)

Resonance stabilization of the anion of umbelliferone gives not only the same three forms as the phenoxide anion, but in addition, gives an extra resonance form with (−) charge on a carbon, and the most significant resonance contributor, another form with the (−) charge on the other carbonyl oxygen. This anion is much more stable than phenoxide, which we interpret as enhanced acidity of the starting material, umbelliferone. In fact, the pK$_a$ of umbelliferone is 7.7 whereas phenol is about 10.

base

**358**

Copyright © 2013 Pearson Education, Inc.

**16-30** Aromaticity is one of the strongest stabilizing forces in organic molecules. The cyclopentadienyl system is stabilized in the anion form where it has 6 π electrons, a Huckel number. The question then becomes: which of the four structures can lose a proton to become aromatic?

While the first, third, and fourth structures can lose protons from sp³ carbons to give resonance-stabilized anions, only the second structure can make a cyclopentadienide anion. It will lose a proton most easily of these four structures which, by definition, means it is the strongest acid.

**16-31**

**(a)**

**(b)**

**(c)**

different positions
of double bonds

(ignoring enantiomers)

16-31 continued

(d)  The only structure consistent with three isomers of dibromobenzene is the prism structure, called Ladenburg benzene.  It also gives a negative test for alkenes, consistent with the behavior of benzene.  (Kekulé defended his structure by claiming that the "two" structures of *ortho*-dibromobenzene were rapidly interconverted, equilibrating so quickly that they could never be separated.)

(e)  We now know that three- and four-membered rings are the least stable, but this fact was unknown to chemists during the mid-1800s when the benzene controversy was raging.  Ladenburg benzene has two three-membered rings and three four-membered rings (of which only four of the rings are independent), which we would predict to be unstable.  (In fact, the structure has been synthesized.  Called *prismane*, it is NOT aromatic, but rather, is very reactive toward addition reactions.)

16-32

(a)

nonaromatic

antiaromatic—
4 π electrons

aromatic—
2 π electrons

(b)

aromatic—
6 π electrons

aromatic—
6 π electrons

nonaromatic

aromatic—
6 π electrons

nonaromatic

(c)

nonaromatic

aromatic—
6 π electrons

antiaromatic—
4 π electrons

antiaromatic—
4 π electrons

(B is sp², but donates no electrons to the π system.)

(d)

aromatic—
6 π electrons

aromatic—
6 π electrons

nonaromatic

aromatic—
6 π electrons

(e)

nonaromatic

if oxygen is sp² ⇒ antiaromatic (8 π electrons); if oxygen is sp³ ⇒ nonaromatic ⇒ not aromatic in either case

aromatic—
6 π electrons

(f)

aromatic—
6 π electrons

aromatic—
6 π electrons

nonaromatic

16-32 continued

(g)

antiaromatic—
12 π electrons

aromatic—
10 π electrons
Experiment shows the outer
carbons to be close to planar
but not exactly planar.

This is a tough call—it has 10 π electrons so
it could be aromatic, but internal H's might
force it out of planarity.

(h)

(B is sp², but
donates no
electrons to the
π system.)

nonaromatic          aromatic—          antiaromatic—          aromatic—
                     6 π electrons        8 π electrons          6 π electrons

16-33  The clue to azulene is recognition of the five- and seven-membered rings. To attain aromaticity, a
seven-membered carbon ring must have a positive charge; a five-membered carbon ring must have a
negative charge.  Drawing a resonance form of azulene shows this:

other
resonance
forms

composite picture

The composite picture shows that the negative charge is concentrated in the five-membered ring, giving
rise to the dipole.

16-34  Whether a nitrogen is strongly basic or weakly basic depends on the location of its electron pair.  If
the electron pair is needed for an aromatic π system, the nitrogen will not be basic (shown here as "weak
base").  If the electron pair is in either an sp² or sp³ orbital, it is available for bonding, and the nitrogen is a
"strong base".

(a)  HN⧸⎺⧹N:
        weak base      strong base

(b)  strong base

(c)  strong base

(d)  weak base   strong base   strong base

(e)  strong base   weak base

**361**
Copyright © 2013 Pearson Education, Inc.

16-35 Where a resonance form demonstrates aromaticity, the resonance form is shown.

(a) aromatic

(b) aromatic

(c) aromatic

(d) BAD ... BAD

NOT aromatic—although it can be drawn, the resonance form on the left is NOT a significant contributor because the oxygen does not have a full octet; the form on the right shows the correct polarization of the carbonyl, but it's still not aromatic because of only 4 π electrons.

(e) aromatic

(f) aromatic; not basic

(g) aromatic: N is not basic but the O⁻ is a weak base.

(h) not basic / basic — aromatic

(i) basic / basic $sp^3$ — NOT aromatic because of $sp^3$ carbon; both Ns are basic, although the top one is conjugated, lowering its basicity.

(j) basic / basic — aromatic: Bottom N is not basic because it has donated its electron pair to make the ring aromatic.

(k) aromatic; not basic

(l) NOT aromatic: if both N and O are $sp^2$, then the pi system has 8 e⁻; N will be basic.

(m) NOT aromatic: if N is $sp^2$, then the pi system has 8 e⁻.

(n) aromatic; 6 pi electrons in this resonance form

(o) NOT aromatic for the same reason as in part (d)

(2) Nitrogens whose electrons are needed to complete the aromatic π system will not be basic. The only N that are more basic than water are the ones indicated in parts (h), (i), (j), and (l).

16-36

(a)

continued on next page

16-36 (a) continued

(b) <u>initiation</u>

$$Br - Br \xrightarrow{h\nu} 2\ Br\bullet$$

<u>propagation</u>

$$Br\bullet\ +\ H-CH_2 \longrightarrow HBr\ +\ \bullet CH_2$$

resonance-stabilized

$$Br - Br\ +\ \bullet CH_2 \longrightarrow Br\bullet\ +\ Br - CH_2$$

(c) Both reactions are $S_N2$ on primary carbons, but the one at the benzylic carbon occurs faster. In the transition state of $S_N2$, as the nucleophile is approaching the carbon and the leaving group is departing, the electron density resembles that of a p orbital. As such, it can be stabilized through overlap with the $\pi$ system of the benzene ring.

stabilization through overlap

16-37

(a)

3 isomers

same as

## 16-37 continued

(b)

**only 1 isomer**

(c) The original compound had to have been *meta*-dibromobenzene as this is the only dibromo isomer that gives three mononitrated products.

## 16-38

(a) The formula $C_8H_7OCl$ has five elements of unsaturation, probably a benzene ring (4) plus either a double bond or a ring. The IR suggests a conjugated carbonyl at 1690 cm$^{-1}$ and an aromatic ring at 1602 cm$^{-1}$. The NMR shows a total of five aromatic protons, indicating a monosubstituted benzene. A 2H singlet at $\delta$ 4.7 is a deshielded methylene.

(b) The mass spectral evidence of molecular ion peaks of 1 : 1 intensity at 184 and 186 shows the presence of a bromine atom. The m/z 184 minus 79 for bromine gives a mass of 105 for the rest of the molecule, which is about a benzene ring plus two carbons and a few hydrogens. The NMR shows four aromatic hydrogens in a typical *para* pattern (two doublets), indicating a *para*-disubstituted benzene. The 2H quartet and 3H triplet are characteristic of an ethyl group.

## 16-39

(a)

like the ends of a conjugated diene

(b)

Diels-Alder product

new sigma bonds shown in bold

16-40
(a) No, biphenyl is not fused. The rings must share two atoms to be labeled "fused".
(b) There are 12 $\pi$ electrons in biphenyl compared with 10 for naphthalene.
(c) Biphenyl has 6 "double bonds". An isolated alkene releases 120 kJ/mole upon hydrogenation.

    predicted: 6 x 120 kJ/mole (28.6 kcal/mole) $\approx$ 720 kJ/mole (172 kcal/mole)

    observed:                           418 kJ/mole (100 kcal/mole)

           resonance energy:          302 kJ/mole (72 kcal/mole)

(d) On a "per ring" basis, biphenyl is 302 ÷ 2 = 151 kJ/mole, the same as the value for benzene. Naphthalene's resonance energy is 252 kJ/mole (60 kcal/mole); on a "per ring" basis, naphthalene has only 126 kJ/mole of stabilization per ring. This is consistent with the greater reactivity of naphthalene compared with benzene. In fact, the more fused rings, the lower the resonance energy per ring, and the more reactive the compound. (Refer to Problem 16-21.)

16-41 Two protons are removed from $sp^3$ carbons to make $sp^2$ carbons and to generate a $\pi$ system with 10 $\pi$ electrons.

16-42

(a)

(b)

(c)

(d)

(e)

(f)

**16-43** These four bases can be aromatic, partially aromatic, or aromatic in a tautomeric form. In other words, aromaticity plays an important role in the chemistry of all four structures. (Only electron pairs involved in the important resonance are shown.) To answer part (b), nitrogens that are basic are denoted by **B**. Those that are not basic are shown as **NB**. Note that some nitrogens change depending on the tautomeric form.

(a) and (c)

cytosine

aromatic to the extent that this resonance form contributes

tautomer—aromatic

uracil

aromatic to the extent that this resonance form contributes

tautomer—aromatic

guanine

aromatic to the extent that this resonance form contributes

tautomer—fully aromatic

adenine

fully aromatic

**16-44**  (a)  Antiaromatic—only 4 π electrons.

(b)  This molecule is electronically equivalent to cyclobutadiene. Cyclobutadiene is unstable and undergoes a Diels-Alder reaction with another molecule of itself. The *tert*-butyl groups prevent dimerization by blocking approach of any other molecule.

(c)  Yes, the nitrogen should be basic. The pair of electrons on the nitrogen is in an $sp^2$ orbital parallel to the ring and is perpendicular to the π system. It cannot be part of the π system as that would require two p orbitals occupying the same space.

(d)

Analysis of structure **1** shows the three *tert*-butyl groups in unique environments in relation to the nitrogen. We would expect three different signals in the NMR, as is observed at –110° C. Why do signals coalesce as the temperature is increased? Two of the *tert*-butyl groups become equivalent—which two? Most likely, they are **a** and **c** that become equivalent as they are symmetric around the nitrogen. But they are *not* equivalent in structure **1**—what is happening here?

What must happen is an equilibration between structures **1** and **2**, very slow at –110° C, but very fast at room temperature, faster than the NMR can differentiate. So the signal that has coalesced is an average of **a** and **a'** and **c** and **c'**. (This type of low-temperature NMR experiment is also used to differentiate axial and equatorial hydrogens on a cyclohexane.)

The NMR data prove that **1** is not aromatic, and that **1** and **2** are isomers, not resonance forms. If **1** were aromatic, then **a** and **c** would have identical NMR signals at all temperatures.

**16-45**

Mass spectrum:  Molecular ion at 150; base peak at 135, M – 15, is loss of methyl.

Infrared spectrum:  The broad peak at 3500 cm$^{-1}$ is OH; thymol must be an alcohol. The peak at 1620 cm$^{-1}$ suggests an aromatic compound.

NMR spectrum:  The singlet at δ 4.8 is OH; it disappears upon shaking with $D_2O$. The 6H doublet at δ 1.2 and the 1H multiplet at δ 3.2 are an isopropyl group, apparently on the benzene ring. A 3H singlet at δ 2.3 is a methyl group, also on the benzene ring.

 Analysis of the aromatic protons suggests the substitution pattern. The three aromatic hydrogens confirm that there are three substituents. The singlet at δ 6.6 is a proton between two substituents (no neighboring Hs). The doublets at δ 6.75 and δ 7.1 are ortho hydrogens, splitting each other.

$+$  $CH_3$  $+$  $OH$  $+$  $CH(CH_3)_2$

*continued on next page*

16-45 continued

Several isomeric combinations are consistent with the spectra (although the single H giving δ 6.6 suggests that either Y or Z is the OH group—an OH on a benzene ring shields hydrogens ortho to it, moving them upfield).

The structure of thymol is:        These structures fit the data equally well.

The final question is how the molecule fragments in the mass spectrometer:

Resonance stabilization of this benzylic cation includes forms
with positive charge on three ring carbons and on oxygen (shown).

16-46

Mass spectrum:  Molecular ion at 170; two prominent peaks are M − 15 (loss of methyl) and M − 43 (as we shall see, most likely the loss of acetyl, $CH_3CO$).

Infrared spectrum:  The two most significant peaks are at 1680 cm$^{-1}$ (conjugated carbonyl) and 1600 cm$^{-1}$ (aromatic C=C).

NMR spectrum:  A 3H singlet at δ 2.7 is methyl next to a carbonyl, shifted slightly downfield by an aromatic ring.  The other signals are seven aromatic protons.  The 1H at δ 8.7 is a deshielded proton next to a carbonyl.  Since there is only one, the carbonyl can have only one neighboring hydrogen.

Conclusions:

The fragment $C_{10}H_7$ is almost certainly a naphthalene.  The correct isomer (box) is indicated by the NMR.

This isomer is a less good answer as it would have two deshielded protons in the NMR, one of which would be a singlet (it's a doublet in the spectrum).

Although all carbons in hexahelicene are sp$^2$, the molecule is not flat. Because of the curvature of the ring system, one end of the molecule has to sit on top of the other end—the carbons and the hydrogens would bump into each other if they tried to occupy the same plane. In other words, the molecule is the beginning of a spiral. An "upward" spiral is the nonsuperimposable mirror image of a "downward" spiral, so the molecule is chiral and therefore optically active.

The magnitude of the optical rotation is extraordinary: it is one of the largest rotations ever recorded. In general, alkanes have small rotations and aromatic compounds have large rotations, so it is reasonable to expect that it is the interaction of plane-polarized light (electromagnetic radiation) with the electrons in the twisted pi system (which can also be considered as having wave properties) that causes this enormous rotation.

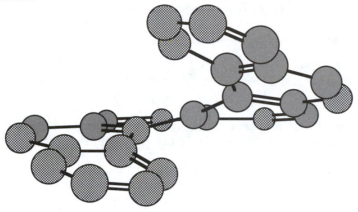

This three-dimensional picture of hexahelicene shows the twist or spiral in the system of six rings.

16-48 The cycloheptatrienyl cation has six pi electrons that just fill the bonding molecular orbitals, making this an aromatic system.

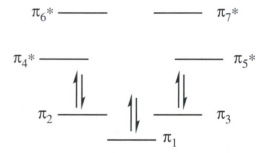

This electronic configuration is aromatic. Electrons fill the bonding molecular orbitals.

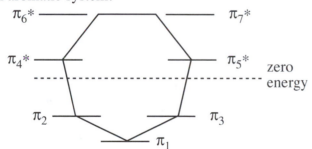

This node is in the same position as the "zero energy" line above.

**16-49** Are you familiar with the concept "tough love", that is, sometimes you have to be stern with someone you love for their own good? The concept is that sometimes to demonstrate one emotion, the behavior has to appear exactly the opposite. Granted, this is a stretch to apply it to atoms, but the point is that sometimes atoms like oxygen and nitrogen can have conflicting effects: they can withdraw electron density by their strong electronegativity (an inductive effect) but at the same time, they can donate electron density through their resonance effect. This phenomenon will prove important in the reactivity of substituted benzenes described in Chapter 17.

Nitrogen is electronegative, so it exerts a deshielding effect on H-2 in pyridine. The effect diminishes with distance as expected with an inductive effect, although it is harder to explain why H-4 is deshielded more than H-3.

In pyridine N-oxide, with an even more electronegative oxygen attached to the N, it would be reasonable to expect that the hydrogens would be deshielded. This is the case with H-3; we can infer that the effect on H-3 is purely an inductive effect.

Even more interesting is the *shielding* effect on H-2 and H-4; these chemical shifts are shifted *upfield*. This must reflect the other side of oxygen's personality, the donation of electron density through a resonance effect. Drawing the resonance forms clearly shows that the electron density at H-2 and H-4 (and presumably H-6) increases through this donation by resonance, in perfect agreement with the NMR results. As we will see in Chapter 17, in most cases, resonance effect trumps inductive effect.

--------------------------------------------------------------------------------

Note: The practice of applying human emotions to inanimate objects is called "anthropomorphizing". For example, we say that an atom is "happy" when it has a full octet of electrons. In casual conversation, this gets a point across, but it is not appropriate in rigorous scientific terms, for example, on exams. Under more formal conditions, we are expected to use the specific terms of science because they are well defined and do not permit sloppy or fuzzy concepts.

As our equipment technician says: "Don't anthropomorphize computers. They hate that."

The representation of benzene with a circle to represent the π system is fine for questions of nomenclature, properties, isomers, and reactions. For questions of mechanism or reactivity, however, the representation with three alternating double bonds (the Kekulé picture) is more informative. For clarity and consistency, this Solutions Manual will use the Kekulé form exclusively.

**17-1**

sigma complex

addition product— NOT AROMATIC

Though the addition of water to the sigma complex can be shown in a reasonable mechanism, the product is not aromatic. Thus, it has lost the 151 kJ/mol (36 kcal/mole) of resonance stabilization energy. The addition reaction is not favorable energetically, and substitution prevails.

**17-2**

**17-3**

2°     2°     3°

Benzene's sigma complex has positive charge on three 2° carbons. The sigma complex above shows positive charge in one resonance form on a 3° carbon, lending greater stabilization to this sigma complex. The more stable the intermediate, the lower the activation energy required to reach it, and the faster the reaction will be.

**17-4** delocalization of the positive charge on the ring

delocalization of the negative charge on the sulfonate group

("Ar" is the general abbreviation for an *aromatic* or *aryl* group, in this case, benzene; "R" is the general abbreviation for an *aliphatic* or *alkyl* group. In cases where the identity of the R group does not matter, it has been used to represent alkyl or aryl groups.)

**17-5**

(a) The key to electrophilic aromatic substitution lies in the stability of the sigma complex. When the electrophile bonds at ortho or para positions of ethylbenzene, the positive charge is shared by the 3° carbon with the ethyl group. Bonding of the electrophile at the meta position lends no particular advantage because the positive charge in the sigma complex is never adjacent to, and therefore never stabilized by, the ethyl group. (Unshared electron pairs on halogen not shown here.)

(b) Electrophilic attack on *p*-xylene gives an intermediate in which only one of the three resonance forms is stabilized by a substituent (see the solution to Problem 17-3). *m*-Xylene, however, is stabilized in two of its three resonance forms. A more stable intermediate gives a faster reaction.

*m*-xylene

17-6  For ortho and para attack, the positive charge in the sigma complex can be shared by resonance with the vinyl group. This cannot happen with meta attack because the positive charge is never adjacent to the vinyl group. (Ortho attack is shown; para attack gives an intermediate with positive charge on the same carbons.)

*"extra" resonance form*

17-7  Attack at only ortho and para positions (not meta) places the positive charge on the carbon with the ethoxy group, where the ethoxy group can stabilize the positive charge by resonance donation of a lone pair of electrons. (Ortho attack is shown; para attack gives a similar intermediate.) (Unshared electron pairs on halogen not shown here.)

*"extra" resonance form*

17-8

ortho

*"extra" resonance form*

meta

para

*"extra" resonance form*

17-9

Substitution: (benzene with OCH(CH$_3$)$_2$) + Br$_2$ $\longrightarrow$ (benzene with OCH(CH$_3$)$_2$ and Br) + **HBr** (g)

Addition: (cyclohexene) + Br$_2$ $\longrightarrow$ (cyclohexane with Br, Br trans)

Substitution generates HBr whereas the addition does not. If the reaction is performed in an organic solvent, bubbles of HBr can be observed, and HBr gas escaping into moist air will generate a cloud. If the reaction is performed in water, adding moist litmus paper to test for acid will differentiate the results of the two compounds.

17-10

(a) Nitration is performed with nitric acid and a sulfuric acid catalyst. In strong acid, amines in general, including aniline, are protonated.

(aniline, NH$_2$) + H$_2$SO$_4$ $\longrightarrow$ (anilinium, $\overset{\oplus}{N}H_3$) + HSO$_4^-$

(b) The NH$_2$ group is a strongly activating ortho,para-director. In acid, however, it exists as the protonated ammonium ion—a strongly **deactivating meta-director**. The strongly acidic nitrating mixture itself forces the reaction to be slower.

(c) The acetyl group removes some of the electron density from the nitrogen, making it much less basic; the nitrogen of this amide is not protonated under the reaction conditions. The N retains enough electron density to share with the benzene ring, so the NHCOCH$_3$ group is still an activating ortho,para-director, though weaker than NH$_2$.

$$\left\{ \text{Ph}-\overset{H}{\underset{}{N}}-\overset{:\ddot{O}:}{\overset{\|}{C}}-CH_3 \longleftrightarrow \text{Ph}-\overset{H}{\underset{}{\overset{\oplus}{N}}}=\overset{:\ddot{O}:^{\ominus}}{\underset{}{C}}-CH_3 \right\}$$

17-11 Nitronium ion attack at the ortho and para positions places positive charge on the carbon adjacent to the bromine, allowing resonance stabilization by an unshared electron pair from the bromine. Meta attack does not give a stabilized intermediate.

ortho

(resonance structures with Br and NO$_2$, including "extra" resonance form)

"extra" resonance form

meta

(resonance structures with Br and NO$_2$)

## 17-11 continued
para

"extra"
resonance form

## 17-12

(a) *o,p-director* CH$_3$ / *m-director* NO$_2$ ... NO$_2$ + O$_2$N ... NO$_2$ CH$_3$

(b) CH$_3$ *o,p-director* ... Cl / NO$_2$ + O$_2$N ... CH$_3$ Cl *o,p-director* + CH$_3$ ... NO$_2$ Cl

trace
Squeezing between two groups is difficult.

(c) *m-director* COOH / Br *o,p-director* ... NO$_2$ + COOH Br ... O$_2$N

(d) *m-director* COOH ... NO$_2$ OCH$_3$ *o,p-director*

(e) *o,p-director* OH ... CH$_3$ NO$_2$ *o,p-director* + OH O$_2$N ... CH$_3$ + OH ... NO$_2$ CH$_3$

trace
Squeezing between two groups is difficult.

(f) *o,p-director* OH O ... *m-director* NO$_2$ + O$_2$N OH O

## 17-13

(a) OCH$_3$ *strong o,p-director* ... NO$_2$ CH$_3$ *weak o,p-director*

(b) Cl ... NO$_2$ NO$_2$ + O$_2$N Cl ... NO$_2$ + Cl *weak o,p-director* ... NO$_2$ NO$_2$ *strong m-director*

trace

(c) OH *strong o,p-director* ... NO$_2$ Cl *weak o,p-director*

(d) OCH$_3$ ... NO$_2$ NO$_2$ + O$_2$N OCH$_3$ ... NO$_2$ + OCH$_3$ *strong o,p-director* ... NO$_2$ NO$_2$ *strong m-director*

trace

**375**

17-13 continued

(e)

NO₂

—NH—C—CH₃  +  O₂N——NH—C—CH₃
*strong o,p-director*

CH₃ *weak o,p-director*                         CH₃

(f)

O₂N

CH₃—C—N̈H——C—NH₂
*strong o,p-director*        *strong m-director*

⎧‾‾‾‾‾‾‾‾⎫  ⎧‾‾‾‾‾‾‾‾⎫
  activating        deactivating

only product

17-14

(a)  Sigma complex of ortho attack—the phenyl substituent stabilizes positive charge by resonance (braces omitted):

Para attack gives similar stabilization.  Meta attack does not permit delocalization of the positive charge on the phenyl substituent.

(b)  The nitronium ion electrophile will prefer to attack the more activated ring, or the less deactivated, ring.

(i)

—NO₂  +

(ii)

trace

(iii)

O₂N———C—CH₃  +  ———C—CH₃

(iv)

——NO₂  +

NO₂                                   NO₂  minor

(v)

O₂N                      O₂N

+

(vi)

NO₂

Major product—minor amounts of nitration would occur on the outer rings; the middle ring has two activating substituents while the outer rings have only one.

17-15

(a)

(b) Multiple alkylation can also occur.

Para isomer is
also formed by
similar mechanism.

(c)

AlCl$_4^-$

methyl
shift

3°  2°

Only a small amount of the ortho isomer might be produced, as steric
interactions will discourage this approach path of the electrophile.

17-16

(a) [structure: cyclohexene + HF → cyclohexyl cation → benzene → phenylcyclohexane]

(b) $CH_3-\underset{\underset{CH_3}{|}}{\overset{\overset{CH_3}{|}}{C}}-OH$ + $BF_3$ ⟶ $CH_3-\underset{\underset{CH_3}{|}}{\overset{\overset{CH_3}{|}}{C}}\oplus$ ⟶ [benzene] ⟶ $CH_3-\underset{\underset{CH_3}{|}}{\overset{\overset{CH_3}{|}}{C}}$—[phenyl]

(c) $CH_2=\underset{\underset{CH_3}{|}}{\overset{\overset{CH_3}{|}}{C}}$ + HF ⟶ $CH_3-\underset{\underset{CH_3}{|}}{\overset{\overset{CH_3}{|}}{C}}\oplus$ ⟶ [tert-butylbenzene] ⟶ [di-tert-butylbenzene: $CH_3-\overset{\overset{CH_3}{|}}{\underset{\underset{CH_3}{|}}{C}}$—[phenyl]—$\overset{\overset{CH_3}{|}}{\underset{\underset{CH_3}{|}}{C}}-CH_3$]

(d) $HO-\underset{\underset{CH_3}{|}}{CHCH_3}$ + $BF_3$ ⟶ $\overset{\oplus}{\underset{\underset{CH_3}{|}}{C}}HCH_3$ ⟶ [toluene] ⟶ [para-isopropyltoluene] + [ortho-isopropyltoluene]

17-17  In (a), (b), and (d), the electrophile has rearranged.

(a) $CH_3-\underset{\underset{CH_3}{|}}{\overset{\overset{CH_3}{|}}{C}}$—[phenyl]
rearranged

(b) [para-sec-butyltoluene] + [ortho-sec-butyltoluene]
rearranged

(c) No reaction: nitrobenzene is too deactivated for the Friedel-Crafts reaction to succeed.

(d) [phenyl]—$\underset{\underset{CH_3}{|}}{\overset{\overset{CH_3}{|}}{C}}-CH(CH_3)_2$
rearranged

17-18

(a) [benzene] + $CH_3CH_2CH_2CH_2Br$ $\xrightarrow{AlCl_3}$ [sec-butylbenzene]
rearranged—not the desired product
plus poly-alkylation products

(b) Gives desired product: activated ring plus 3° carbon in the electrophile gives para as the major product.
(c) Gives desired product plus ortho isomer; use excess bromobenzene to avoid overalkylation.
(d) No reaction; benzamide is too deactivated for a Friedel-Crafts reaction.
(e) Gives desired product: methyl is slightly activating; putting three deactivating nitro groups requires forcing conditions, and great care to avoid detonation!  BOOM!

Copyright © 2013 Pearson Education, Inc.

17-19

(a)

(CH₃)₃CCl / AlCl₃ → (excess)

HNO₃ / H₂SO₄ →

(Separate from *ortho*, although steric hindrance would prevent much *ortho* substitution.)

(b)

CH₃I / AlCl₃ → (excess)

SO₃ / H₂SO₄ →

(Separate from *ortho*.)

The sequence of reactions is important.

(c)

CH₃I / AlCl₃ → (excess)

Cl₂ / AlCl₃ →

(Separate from *ortho*.)

17-20 Friedel-Crafts alkylation is typically followed by a water workup to remove AlCl₃ or other catalyst.

(a)

$$\bigcirc + Cl-\overset{O}{\overset{\|}{C}}-CH_2CH(CH_3)_2 \xrightarrow[\substack{2)\ H_2O \\ workup}]{1)\ AlCl_3} \bigcirc-\overset{O}{\overset{\|}{C}}-CH_2CH(CH_3)_2$$

(b)

$$\bigcirc + Cl-\overset{O}{\overset{\|}{C}}-C(CH_3)_3 \xrightarrow[\substack{2)\ H_2O \\ workup}]{1)\ AlCl_3} \bigcirc-\overset{O}{\overset{\|}{C}}-C(CH_3)_3$$

(c)

$$\bigcirc + Cl-\overset{O}{\overset{\|}{C}}-\bigcirc \xrightarrow[\substack{2)\ H_2O \\ workup}]{1)\ AlCl_3} \bigcirc-\overset{O}{\overset{\|}{C}}-\bigcirc$$

(d)

OCH₃-benzene, CO, HCl / AlCl₃/CuCl → H₃CO—⟨ ⟩—CHO

Making a new bond from benzene to a 1° carbon requires Friedel-Crafts acylation followed by reduction.

(e)

$$\bigcirc + \overset{O}{\underset{Cl}{\overset{\|}{C}}}\text{—} \xrightarrow[\substack{2)\ H_2O \\ workup}]{1)\ AlCl_3} \xrightarrow[\text{Clemmensen}]{Zn(Hg),\ HCl}$$

(f)

$$\bigcirc + Cl-\overset{O}{\overset{\|}{C}}-C(CH_3)_3 \xrightarrow[\substack{2)\ H_2O \\ workup}]{1)\ AlCl_3} \bigcirc-\overset{O}{\overset{\|}{C}}-C(CH_3)_3 \xrightarrow[\text{Clemmensen}]{Zn(Hg),\ HCl} \bigcirc-CH_2C(CH_3)_3$$

**379**

17-20 continued

(g)

![benzene + Cl–C(=O)–CH2CH2CH3 with 1) AlCl3 2) H2O workup gives phenyl propyl ketone, then Zn(Hg), HCl Clemmensen gives butylbenzene]

(h)

![benzene with HNO3/H2SO4 gives nitrobenzene, then Fe/HCl gives aniline, then "Ac2O" (acetic anhydride) gives acetanilide; This reaction appears in text section 19-12—sorry! Another example of this amide reacting by E.A.S. is Problem 17-10(c). Then with acetyl chloride / AlCl3, H2O workup gives para-acetyl acetanilide]

17-21  Another way of asking this question is this:  Why is fluoride ion a good leaving group from **A** but not from **B** (either by $S_N1$ or $S_N2$)?   Nuc = a nucleophile

![structure A: anionic sigma complex with F and Nuc on same carbon, NO2 groups; structure B: F–C with Nuc attacking]

Formation of the anionic sigma complex **A** is the rate-determining (slow) step in nucleophilic aromatic substitution.  The loss of fluoride ion occurs in a subsequent fast step where the nature of the leaving group does not affect the overall reaction rate.  In the $S_N1$ or $S_N2$ mechanisms, however, the carbon-fluorine bond is breaking in the rate-determining step, so the poor leaving group ability of fluoride does indeed affect the rate.

nucleophilic aromatic
substitution

$S_N1$

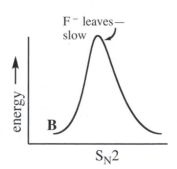

$S_N2$

**17-22**

**17-23**
**(a)**

For simplicity, the $NO_2$ abbreviation for the nitro group has been used where the nitro is not shown distributing the negative charge.

17-23 continued

(b)

(c)

17-23 continued
(d)

$O_2N$ ... Cl ... $NH_2NH_2$ ... $NO_2$

[reaction intermediates with resonance structures]

$:O:$ Cl $NH_2NH_2$ ... $NO_2$

$:O:$ Cl $NH_2NH_2$ ... $NO_2$

Cl $NH_2NH_2$ $O_2N$ ... N ... $:O:$ $:O:$

$NHNH_2$ $O_2N$ ... $NO_2$

H—$NHNH_2$

$O_2N$ $NHNH_2$ ... $NH_2NH_2$ ... $NO_2$

– Cl⁻

Cl $NH_2NH_2$ $O_2N$ ... C ... H ... $NO_2$

Cl $NH_2NH_2$ $O_2N$ ... $:O:$ N $:O:$

17-24 Assume an acidic workup to each of these reactions to produce the phenol, not the phenoxide ion.

(a)

$\xrightarrow[\text{AlCl}_3]{\text{Cl}_2}$ Cl $\xrightarrow[\text{H}_2\text{SO}_4]{\text{HNO}_3}$ Cl ... $NO_2$ + ortho $\xrightarrow[\Delta]{\text{NaOH}}$ OH ... $NO_2$

from addition-elimination mechanism; only this isomer

(b)

$\xrightarrow[\text{AlCl}_3]{\text{Cl}_2}$ Cl $\xrightarrow[350° \text{C}]{\text{NaOH}}$ OH

strongly activated; no catalyst needed

$\xrightarrow{3 \text{ Br}_2}$ OH, Br, Br, Br

(c)

$\xrightarrow[\text{AlCl}_3]{2 \text{ Cl}_2}$ Cl ... Cl + ortho $\xrightarrow[\Delta]{\text{NaOH}}$ Cl ... OH + Cl ... OH

via benzyne mechanism

17-24 continued

(d)

CH₃ → (Cl₂/AlCl₃) → CH₃ with Cl → (NaOH, Δ) → CH₃ with OH (meta) + CH₃ with OH (para)

+ ortho

via benzyne mechanism

(e)

benzene + CH₃CH₂CH₂—C(=O)—Cl / AlCl₃ → O=CCH₂CH₂CH₃ (phenyl ketone) → Zn(Hg)/HCl → CH₂CH₂CH₂CH₃ (butylbenzene)

Cl₂ ↓ AlCl₃

CH₂CH₂CH₂CH₃ with OH (meta) + CH₂CH₂CH₂CH₃ with OH (para) ← NaOH, Δ ← CH₂CH₂CH₂CH₃ with Cl (para) + ortho

via benzyne mechanism

17-25

from chlorobenzene
+ NaOH + heat

17-26 The new carbon-carbon bonds are shown in bold. ▬

(a) vinyl–aryl with OCH₃ and OCH₃ substituents

(b) hexenyl–Ph (cis alkene)

The *cis* stereochemistry of the reactant is retained in the product, a stereospecific reaction.

17-27

(a) (alkene)₂ CuLi

(b) (alkene)₂ CuLi

17-28 The new carbon-carbon bonds are shown in bold. ▬

(a) O (acetyl-aryl with trans alkenyl isobutyl chain)

*trans*

The *trans* isomer is the major product in the Heck reaction.

(b) *trans* (diene with aryl-OCH₃)

Stereochemistry around this double bond is retained.

17-29

(a)

(b)

17-30 The new carbon-carbon bonds are shown in bold. ▬

(a)

(b)

17-31 Any palladium catalyst and base can be proposed for these reactions.

(a) or ester

+ Pd catalyst + a base like NaOH

(b) or ester

+ Pd catalyst + a base like NaOH

17-32

(a) First, the carboxylic acid proton is neutralized.

$$\text{PhCOOH} \xrightarrow{NH_3} NH_4^+ + \text{PhCOO}^-$$

mechanism:

+ Na•

+ Na⁺

plus other resonance forms

H—OEt

plus resonance forms

Na•

plus resonance forms

+ Na⁺

plus resonance forms

$CH_3CH_2O-H$

(b)

The negative charge avoids the carbon bearing the electron-donating methoxy group.

Li•

+ Li⁺

plus resonance forms

H—O-t-Bu

plus resonance forms

Li•

+ Li⁺

plus resonance forms

H—O-t-Bu

**17-33**

(a)

CCl₃ [benzene ring with CCl₃]

first product formed from benzylic substitution

→ more Cl₂, Δ, pressure →

[cyclohexane with Cl, CCl₃, Cl, Cl, Cl, Cl substituents]

Addition of chlorine to the π system of the ring gives a mixture of stereoisomers.

(b) [cyclohexadiene ring with CH(–)–C(=O)–NH₂ group]

(c) [cyclohexane with two CH₃ groups]

CH₃
CH₃

*cis + trans*

(d) [cyclohexadiene ring with CH₃ at top and CH₃ at bottom]

CH₃

CH₃

(e) CH₃O [naphthalene-type bicyclic system] OCH₃

**17-34**

(a) [benzene ring with COOH]

COOH

(b) HOOC —[benzene ring]— COOH

(c) [benzene ring with two COOH groups ortho]

COOH

COOH

**17-35**

Br — Br  $\xrightarrow{h\nu}$  2 Br •   initiation

Br • + [toluene: benzene ring with CH₃ top, H–C–H] $\xrightarrow{\text{propagation step 1}}$ { [•C–H with CH₃ on benzene ring] ↔ [radical resonance structure] ↔ [radical resonance structure] }

[CH₃, C–H on ring radical resonance structures]

Br • + [benzene ring with Br–C–H and CH₃] $\xleftarrow{\text{propagation step 2}}$ { [•C–H with CH₃ on ring] ↔ [radical resonance structure] } + Br — Br

**17-36** A statistical mixture would give 2 : 3 or 40% : 60% α to β. To calculate the relative reactivities, the percents must be corrected for the numbers of each type of hydrogen.

α:  $\dfrac{56\%}{2H}$ = 28 relative reactivity

β:  $\dfrac{44\%}{3H}$ = 14.7 relative reactivity

The reactivity of α to β is  $\dfrac{28}{14.7}$ = 1.9 to 1

**386**

**17-37** Replacement of aliphatic hydrogens with bromine can be done under free radical substitution conditions, but reaction at aromatic carbons is unfavorable because of the very high energy of the aryl radical. Benzylic substitution is usually the only product observed.

(2) produced under extreme conditions

All four benzylic hydrogens are replaced.

**17-38**

**17-39**

(a) Benzylic cations are stabilized by resonance and are much more stable than regular alkyl cations. The product is 1-bromo-1-phenylpropane.

(b)

more stable than

2° benzylic—
*resonance-
stabilized*

2°

1-bromo-1-phenylpropane

17-39 continued

(c) The combination of HBr with a free-radical initiator generates bromine radicals and leads to anti-Markovnikov orientation. (Recall that whatever species adds *first* to an alkene determines orientation.) The product will be 2-bromo-1-phenylpropane.

(d) Assume the free-radical initiator is a peroxide.

Initiation

$$RO \overset{\Delta}{-} OR \longrightarrow 2\ RO\bullet$$

$$RO\bullet\ +\ H-Br \longrightarrow ROH\ +\ Br\bullet$$

Propagation

more stable than

2° benzylic—resonance-stabilized

2°

(Br• recycles in chain mechanism)   Br• +

2-bromo-1-phenylpropane

17-40

(a)

(b)

+ ortho

(c)

+ ortho

**388**

**17-41**

**17-42**

(a) OCH₂CH₃ — 3-methylphenyl ethyl ether

(b)

(c)

+

If excess Br₂ is used, the 2,4,6-tribromo isomer would be produced.

(d)

(e)

(f)

**17-43**

(a)

(b)

**17-44**

(a) C(CH₃)₃

(b) CH₃—CHCH₂CH₃    rearrangement

(c) C(CH₃)₃    rearrangement

(d) Br

(e) C(CH₃)₃

(f) SO₃H

(g) CH₃—C(CH₃)—CH₂CH₃ with phenyl    rearrangement

(h) O (benzophenone)

(i) I

(j) NO₂

(k) CHO

(l)

**17-45**

(a)

major—benzylic        minor—2°

## 17-45 continued

(b)

(c) Gas chromatography will be the optimum technique for separating the six compounds. By itself, GC does not give any structural information, so the more expensive GC instruments have a mass spectrometer as the detector. As the peaks come out of the GC, they go directly into the MS, so a mass spectrum is collected for each GC peak. GC-MS will distinguish the monochloro from the dichloro isomers, but it will not easily identify the specific monochloro or dichloro isomers. (A good GC could also separate *cis-trans* isomers of **1** and **2**.)

(d) The two types of NMR would easily distinguish these four dichloro isomers (ignoring geometric isomers). In HNMR, a Cl—C—H would appear $\approx \delta$ 4 if not benzylic and $\approx \delta$ 5 if benzylic. In CNMR, a C—Cl will appear $\approx \delta$ 65 and a $CCl_2$ will appear $\approx \delta$ 80; benzylic C appears $\approx \delta$ 40, and nonbenzylic $CH_2$ comes $\approx \delta$ 20.

| structure | HNMR | CNMR |
|---|---|---|
| 1 | 2H $\approx \delta$ 5 | 1 tall peak $\approx \delta$ 65, 1 medium peak $\approx \delta$ 20 |
| 2 | 1H $\approx \delta$ 4, 1H $\approx \delta$ 5 | 2 peaks $\approx \delta$ 65, 1 medium peak $\approx \delta$ 40 |
| 3 | 2 signals < $\delta$ 4 | 1 peak $\approx \delta$ 80, 1 peak $\approx \delta$ 40, 1 peak $\approx \delta$ 20 |
| 4 | 1 singlet $\approx \delta$ 2.5-3.0 | 1 peak $\approx \delta$ 80, 1 tall peak $\approx \delta$ 40 |

## 17-46

(a)

(b)

(c)

(d)

17-46 continued

(e)

$\xrightarrow[\text{AlCl}_3]{\text{Cl}_2}$ (Cl) $\xrightarrow[\text{H}_2\text{SO}_4]{\text{HNO}_3}$ (Cl, NO$_2$) $\xrightarrow[\text{AlCl}_3]{\text{Cl}_2}$ (Cl, Cl, NO$_2$)

(f) three methods

from (c) $\xrightarrow[\text{H}_2\text{O}]{\text{Hg(OAc)}_2}$ $\xrightarrow{\text{NaBH}_4}$ (benzene-CH$_2$-CH(OH)-CH$_3$)

OR

$\text{CH}_2\text{MgBr}$ from (c)

$\xrightarrow{\text{H}^+}$

$\text{CH}_3-\overset{\text{O}}{\overset{\|}{\text{CH}}}$

$\xleftarrow{\text{H}_3\text{O}^+}$ (epoxide O)

OR

$\xleftarrow[\text{ether}]{\text{Mg}}$ (Cl-benzene) from (d)

(g)

$\text{CH}_3$ $\xrightarrow[\text{H}_2\text{SO}_4]{\text{HNO}_3}$ $\text{CH}_3$ / $\text{NO}_2$ $\xrightarrow[\Delta]{\text{KMnO}_4}$ $\text{COOH}$ / $\text{NO}_2$ $\xrightarrow[\text{HCl(aq)}]{\text{Fe or Sn or Zn}}$ $\text{COOH}$ / $\text{NH}_2$

(h) two possible methods

$\text{CH}_3$ $\xrightarrow[\text{or NBS}\ h\nu]{\text{Br}_2\ ,\ h\nu}$ $\text{CH}_2\text{Br}$ $\xrightarrow[\text{ether}]{\text{Mg}}$ $\text{CH}_2\text{MgBr}$ + (ketone, O) $\longrightarrow$ $\xrightarrow{\text{H}_2\text{O}}$ (product, OH)  new bond in bold

OR

$\xrightarrow[\text{AlCl}_3]{\text{Br}_2}$ (Br) $\xrightarrow[\text{ether}]{\text{Mg}}$ (MgBr) + (epoxide) $\longrightarrow$ $\xrightarrow{\text{H}_2\text{O}}$ (product, OH)  new bond in bold

(i)

$\text{CH}_3$ $\xrightarrow[\text{AlCl}_3]{\text{Cl}_2}$ $\text{CH}_3$ / Cl  major $\xrightarrow[\text{H}_2\text{SO}_4]{\text{HNO}_3}$ $\text{NO}_2$ / $\text{CH}_3$ / Cl  major $\xrightarrow[\text{HCl(aq)}]{\text{Fe or Sn or Zn}}$ $\text{NH}_2$ / $\text{CH}_3$ / Cl

(j)

$\text{CH}_3$ $\xrightarrow[\text{AlCl}_3]{\text{Br}_2}$ $\text{CH}_3$ / Br $\xrightarrow[\Delta]{\text{KMnO}_4}$ $\text{COOH}$ / Br $\xrightarrow[\text{H}_2\text{SO}_4]{\text{HNO}_3}$ $\text{COOH}$ / $\text{NO}_2$ / Br

## 17-46 continued

(k)

(l)

(m)

OR

17-47  (a)

(b)

(c)

(d) No reaction: Friedel-Crafts acylation does not occur on rings with strong deactivating groups like $NO_2$.

(e)

(f)

or

(g)

(h)

Basic workup isolates free amine.

(i)

(j)

(k)

(l)

**17-48** Major products are shown. Other isomers are possible.

(a)

NO$_2$ *activated* ... *deactivated* ... Br ... OCH$_3$

(b)

*deactivated* ... *activated* ... Br

(c)

COOCH$_3$ ... OCH$_3$ ... *activated* ... Br

**17-49** The new carbon-carbon bonds are shown in bold. ▬

(a)

(b)

*trans*

(c)

*cis—retention of stereochemistry*

(d)

*trans*

(e)

*trans* ... NO$_2$

**17-50**

$$\xrightarrow[\Delta]{AlCl_3}$$ product

CH$_2$CH$_2$COOH

starting material
molecular weight 150

molecular weight 132 =
loss of 18 = loss of H$_2$O

<u>IR spectrum</u>: The dominant peak is the carbonyl at 1710 cm$^{-1}$. No COOH stretch.

<u>NMR spectrum</u>: The splitting is complicated but the integration is helpful. In the region of δ 2.6–3.2, there are two signals, each with an integration value of 2H; these must be the two adjacent methylenes, CH$_2$CH$_2$ . The aromatic region from δ 7.3 to 7.8 has integration of 4H, so the ring must be disubstituted.

<u>Carbon NMR</u>: Of the six aromatic signals, four are C—H and two are C indicating a disubstituted benzene. Also indicated are carbons in a carbonyl and two methylenes.

lose
lose 1 H
retain
molecular weight 132
or
common name: indanone

The product must be the cyclized ketone, formed in an intramolecular Friedel-Crafts acylation.

**17-51**

**A**     **B** (NO$_2$)     **C**     **D** (COOH)

continued on next page

## 17-51 continued

Br / E   OCH₃ / F   G   Br / H (same as E)

## 17-52

a substituted benzyne

### Attack A

product A

### Attack B

product B

## 17-53 The electron-withdrawing carbonyl group stabilizes the adjacent negative charge.

plus resonance forms

plus other resonance forms

plus resonance forms

17-54

(a)
NO₂

(b)
Br

(c)
O
‖
(propionyl group)

(d)
C(CH₃)₃

(e)
(cyclohexyl-substituted naphthalene)

(f)
SO₃H

17-55 Professor Suzuki himself carried out this synthesis of bombykol in 1983.

$$HC \equiv C \underset{\phantom{x}}{\diagdown\diagup\diagdown\diagup\diagdown\diagup} OH \quad + \quad H-B\begin{smallmatrix} OR \\ \\ OR \end{smallmatrix} \longrightarrow$$

↓ H₂O

cis
[cis-1-bromo-1-pentene]
Br
+
RO–B(OR)–CH=CH–...–OH
trans   borate ester

⎰⎱
↓ Pd(PPh₃)₄
a base like NaOH

cis
[product with new bond]
OH
new bond ↗   trans

17-56

benzene
→ 3 Cl₂ / AlCl₃ →

Cl
Cl

Cl
(Why is this
the major
isomer?)

→ NaOH / Δ →

O⁻ Na⁺
Cl

Cl
(Why is this
the major
isomer?)

→ ClCH₂COO⁻ Na⁺ →  H–A →

OCH₂COOH
Cl

Cl

The sodium salt of the carboxylic acid is
typically used in substitutions under basic
conditions to prevent the COOH from
neutralizing, and thereby deactivating, the
phenoxide nucleophile.

17-57

(a)

bromination at C-2

three resonance forms

bromination at C-3

only two resonance forms

full octets

(b)  Attack at C-2 gives an intermediate stabilized by three resonance forms, as opposed to only two resonance forms stabilizing attack at C-3.  Bromination at C-2 will occur more readily than at C-3, and both will be faster than benzene.

17-58  Benzenedicarboxylic acid isomers have the common name of phthalic acid isomers.

(a)

phthalic acid

isophthalic acid

terephthalic acid

$$HNO_3$$
$$H_2SO_4$$

$$HNO_3$$
$$H_2SO_4$$

$$HNO_3$$
$$H_2SO_4$$

continued on next page

17-58  continued
(b)

HNO₃
H₂SO₄

Two isomers possible—
phthalic acid must have
m.p. 210 °C.

HNO₃
H₂SO₄

Three isomers possible—
isophthalic acid must
have m.p. 343 °C.

HNO₃
H₂SO₄

Only one isomer
possible—
terephthalic acid must
have m.p. 427 °C.

17-59

plus
other
resonance
forms

Quick!  Turn the page!

**397**

17-59 continued

You may have heard about bisphenol A (BPA) in the news. There is some concern about the presence of BPA in products that are used for babies, like milk bottles, nipples, and pacifiers. Some states are considering legislation to prohibit BPA in such products.

bisphenol A

plus other resonance forms

17-60

(a) This is an example of kinetic versus thermodynamic control of a reaction. At low temperature, the kinetic product predominates: in this case, almost a 1 : 1 mixture of ortho and para. These two isomers must be formed at approximately equal rates at 0 °C. At 100 °C, however, enough energy is provided for the *desulfonation* to occur rapidly; the large excess of the para isomer indicates the para is more stable, even though it is formed initially at the same rate as the ortho.

(b) The product from the 0 °C reaction will equilibrate as it is warmed, and at 100 °C will produce the same ratio of products as the reaction that was run initially at 100 °C.

(c) The principle of a blocking group is to attach it at the position that you want blocked, perform the other reaction, then remove the blocking group. This is accomplished on benzene very effectively with the sulfonic acid group.

**17-61** Solve the problem by writing the mechanism. (See the solution to problem 17-23(a) for an identical mechanism.)

Hydroxide attack on C-1 puts the negative charge on carbons that do not have the $NO_2$ groups, so these anions are not stabilized.

Hydroxide attack on C-2 puts the negative charge on carbons with nitro groups, thereby increasing the stabilization by delocalizing the negative charge. This intermediate is formed preferentially.

only product formed

## 17-62

reactive sites

from chloro-
benzene +
NaOH at 350 °C

As we saw in Chapter 16, the carbons of the center ring of anthracene are susceptible to electrophilic addition, leaving two isolated benzene rings on the ends. Benzyne is such a reactive dienophile that the reluctant anthracene is forced into a Diels-Alder reaction.

## 17-63

(a)

2,4,5-T

(b)

TCDD

(This is the compound used to poison Ukrainian political leader, Boris Yushchenko, in 2004.)

Two nucleophilic aromatic substitutions form a new six-membered ring. (Though not shown here, this reaction would follow the standard addition-elimination mechanism.)

(c) To minimize formation of TCDD during synthesis: 1) keep the solutions dilute; 2) avoid high temperature; 3) replace chloroacetate with a more reactive molecule like bromoacetate or iodoacetate; 4) add an excess of the haloacetate.

To separate TCDD from 2,4,5-T at the end of the synthesis, take advantage of the acidic properties of 2,4,5-T. The 2,4,5-T will dissolve in an aqueous solution of a weak base like $NaHCO_3$. The TCDD will remain insoluble and can be filtered or extracted into an organic solvent like ether or dichloromethane. The 2,4,5-T can be precipitated from aqueous solution by adding acid.

## 17-64

$C_6H_3OBr_3$

+ **3** HBr

$C_6H_2OBr_4$

+ HBr

**17-65**

Put meta-director
on first.

**17-66** A benzyne must have been generated from the Grignard reagent.

Diels-Alder

**17-67** The intermediate anion forces the loss of hydroxide.

For simplicity, abbreviate

**17-68** Make a boronate ester from one of the halides first, then use Suzuki coupling to link them.

Alternatively, the other boronate ester could be made to couple with 4-bromoisopropylbenzene.

**17-69**

BHA, butylated hydroxyanisole

plus minor isomer with *tert*-butyl *ortho* to OCH₃

hydroquinone

BHT, butylated hydroxytoluene

**17-70**

colorless

conjugated— yellow

Concentrated sulfuric acid "dehydrates" the alcohol, producing a highly conjugated, colored carbocation, and protonates the water to prevent the reverse reaction. Upon adding more water, however, there are too many water molecules for the acid to protonate, and triphenylmethanol is regenerated.

17-71

(a)

plus three resonance
forms with positive
charge on the
benzene ring

plus resonance
forms with
positive charge
on the ring and
on the oxygen
of the phenol

plus resonance forms
on both benzene rings,
the oxygen of the phenol,
and the oxygen in the ring

plus resonance forms with positive charge
on the ring and on the oxygen of the phenol

17-71 continued

(b) Color comes from highly conjugated molecules and ions. See text Section 15-14 and Problem 15-23.

red dianion

numerous
resonance forms

(c)

numerous resonance
forms

# CHAPTER 18—KETONES AND ALDEHYDES

**18-1**

(a) 5-hydroxyhexan-3-one; ethyl β-hydroxypropyl ketone

(b) 3-phenylbutanal; β-phenylbutyraldehyde

(c) *trans*-2-methoxycyclohexanecarbaldehyde (or (1R,2R) if you named this enantiomer); no common name

(d) 6,6-dimethylcyclohexa-2,4-dienone; no common name

**18-2**

(a) $C_9H_{10}O \Rightarrow$ 5 elements of unsaturation

1H doublet (very small coupling constant) at $\delta$ 9.7 $\Rightarrow$ aldehyde hydrogen, next to CH

5H multiple peaks at $\delta$ 7.2-7.4 $\Rightarrow$ monosubstituted benzene

1H multiplet at $\delta$ 3.6 and 3H doublet at $\delta$ 1.4 $\Rightarrow$ $CHCH_3$

The splitting of the hydrogen on carbon-2, next to the aldehyde, is worth examining. In its overall shape, it looks like a quartet due to the splitting from the adjacent $CH_3$. A closer examination of the peaks shows that each peak of the quartet is split into two peaks: this is due to the splitting from the aldehyde hydrogen. The aldehyde hydrogen and the methyl hydrogens are not equivalent, so it is to be expected that the coupling constants will not be equal. If a hydrogen is coupled to different neighboring hydrogens by different coupling constants, they must be considered separately, just as you would by drawing a splitting tree for each type of adjacent hydrogen.

(b) $C_8H_8O \Rightarrow$ 5 elements of unsaturation

cluster of 4 peaks at $\delta$ 128–145 $\Rightarrow$ mono- or para-substituted benzene ring

peak at $\delta$ 197 $\Rightarrow$ carbonyl carbon (the small peak height suggests a ketone rather than an aldehyde)

peak at $\delta$ 26 $\Rightarrow$ methyl next to carbonyl or benzene

This structure is also possible from the chemical shift values, but the DEPT information about the type of carbons present proves the monosubstituted benzene.

**18-3** A compound has to have a hydrogen on a γ carbon (or other atom) in order for the McLafferty rearrangement to occur. Butan-2-one has no γ-hydrogen.

18-4

m/z 128

m/z 43

m/z 85

m/z 113

McLafferty rearrangement

m/z 58

18-5  Cholesterol has one isolated double bond with π to π* transition at < 200 nm.  Therefore, it will show no UV spectrum in the range of 200–400 nm.

Cholest-4-en-3-one has a double bond conjugated with a ketone.  The ketone's n to π* transition will be weak, but there will be a large π to π* transition predicted at 240 nm (base value of 210 nm plus 3 alkyl substituents).  The actual $\lambda_{max}$ for cholest-4-en-3-one is 241 nm with $\log_{10}\varepsilon$ of 4.26, a very strong absorption.

REMINDERS ABOUT SYNTHESIS PROBLEMS:
1.  There may be more than one legitimate approach to a synthesis, especially as the list of reactions gets longer.
2.  Begin your analysis by comparing the target to the starting material.  If the product has more carbons than the reactant, you will need to use one of the small number of reactions that form carbon-carbon bonds.
3.  Where possible, work backwards from the target back to the starting material.
4.  KNOW THE REACTIONS.  There is no better test of whether you know the reactions than attempting synthesis problems.  See Appendix 3 at the back of this Solutions Manual for one example of organizing reactions for effective studying.

18-6  All three target molecules in this problem have more than six carbons, so all answers will include carbon-carbon bond-forming reactions.  So far, the main types of reactions that form carbon-carbon bonds: at $sp^3$ carbon:  the Grignard reaction, $S_N2$ substitution by an acetylide ion or cyanide ion, and the Friedel-Crafts alkylation; at $sp^2$ carbon:  Grignard reaction, Friedel-Crafts acylation, olefin metathesis, and organometallic coupling (organocuprate, Heck, Suzuki).

(a)

18-6 continued

(b)

identical sequence as in part (a) →

OR

→ This method using Friedel-Crafts acylation is more efficient as it is only one step.

(c)

Br → Mg/ether → MgBr → 1) epoxide 2) H₃O⁺ → alcohol → H₂CrO₄ → ketone

OR

Br → Na⁺ ⁻C≡CH → C≡CH → Hg²⁺, H₂O / H₂SO₄ → ketone

18-7  New sigma bonds are shown in bold. ▬

The first equivalent of R—Li reacts with the H⁺ of COOH to produce R—H; the second R—Li adds to the C=O.

(a)

OH → 1) 2 CH₃Li 2) H₃O⁺ → CH₃ + CH₄

(b)

Br → 2 Li → Li + LiBr → HO CH₃ (0.5 equiv.) → H₃O⁺ → CH₃

(c)

OH → 1) 2 CH₃CH₂Li 2) H₃O⁺ →

CH₃Li, CH₃CH₂Li, PhLi, BuLi, and t-BuLi are commercially available.

(d)

OH → 1) 2 (CH₃)₃CLi 2) H₃O⁺ →

18-8  New sigma bonds are shown in bold. ▬

(a) same product as 18-7(c)

(b)

(c) PhCH₂━CN
(simple S_N2)

(d) PhCH₂C cyclopentane (with C=O)

(e) Ph H (with C=O)

**18-9**

(a) 

(b)   $CH_3CH_2C\equiv N$   +   $BrMgCH_2CH_2CH_2CH_3$   $\xrightarrow{H_3O^+}$

(c)

(d)

chain lengthened by 1 C

**18-10**

(a)

(b)

(c)

(d)

(e)

(f)

a cyclic ester
called a lactone

**18-11** Review the reminders on p. 406 of this Manual. There is often more than one correct way to do a synthesis, although a more direct route with fewer steps is usually better.

(a)

(b)

(c)

also reduces ketones:   $\xrightarrow[\text{2) } H_3O^+]{\text{1) } LiAlH_4}$

**18-11 continued**

(d)

**18-12** The triacetoxyborohydride ion is similar to borohydride, $BH_4^-$, where three acetoxy groups have replaced three hydrides.

(a)

$NaBH(OAc)_3$

(b)

**18-13** Trimethylphosphine has α-hydrogens that could be removed by butyllithium, generating undesired ylides.

desired ylide      wrong ylide

**18-14**

(a)

*trans*

*cis*-but-2-ene

The stereochemistry is inverted. The nucleophile triphenylphosphine must attack the epoxide in an *anti* fashion, yet the triphenylphosphine oxide must eliminate with *syn* geometry.

**18-14 continued**

(b)

**18-15**

(a) $CH_2=CHCH_2Br \xrightarrow[\text{2) BuLi}]{\text{1) Ph}_3\text{P}} CH_2=C \overset{\overset{\ominus}{HC}-\overset{\oplus}{PPh_3}}{\underset{H}{|}} \xrightarrow{\text{PhCHO}} PhCH=CH-CH=CH_2$

OR $\quad PhCH_2Br \xrightarrow[\text{2) BuLi}]{\text{1) Ph}_3\text{P}} Ph\overset{\ominus}{CH}-\overset{\oplus}{PPh_3} \xrightarrow{O=CH-CH=CH_2} PhCH=CH-CH=CH_2$

(b) $PhCH=CH-CH_2Br \xrightarrow[\text{2) BuLi}]{\text{1) Ph}_3\text{P}} PhCH=CH-\overset{\ominus}{CH}-\overset{\oplus}{PPh_3} \xrightarrow{CH_2O} PhCH=CH-CH=CH_2$

OR $CH_3I \xrightarrow[\text{2) BuLi}]{\text{1) Ph}_3\text{P}} \overset{\ominus}{CH_2}-\overset{\oplus}{PPh_3} \xrightarrow{PhCH=CH-CH=O} PhCH=CH-CH=CH_2$

**18-16** Many alkenes can be synthesized by two different Wittig reactions (as in the previous problem). The ones shown here form the phosphonium salt from the less hindered alkyl halide.

(a) $PhCH_2Br \xrightarrow[\text{2) BuLi}]{\text{1) Ph}_3\text{P}} Ph\overset{\ominus}{CH}-\overset{\oplus}{PPh_3} + $ $\longrightarrow PhCH=C(CH_3)_2$

(b) $CH_3I \xrightarrow[\text{2) BuLi}]{\text{1) Ph}_3\text{P}} \overset{\ominus}{CH_2}-\overset{\oplus}{PPh_3} + $

(c) $PhCH_2Br \xrightarrow[\text{2) BuLi}]{\text{1) Ph}_3\text{P}} Ph\overset{\ominus}{CH}-\overset{\oplus}{PPh_3} \xrightarrow{PhCH=CH-CH=O} PhCH=CH-CH=CHPh$

(d) $CH_3CH_2Br \xrightarrow[\text{2) BuLi}]{\text{1) Ph}_3\text{P}} CH_3\overset{\ominus}{CH}-\overset{\oplus}{PPh_3} + $

**18-17**

(a)

**410**

Copyright © 2013 Pearson Education, Inc.

18-17 continued

(b)

18-18  Two general principles apply to hydrate formation. Sterically, less hindered C=O form more hydrate, so aldehydes are more likely to form a hydrate than ketones are. Electronically, electron-withdrawing substituents at the alpha carbon intensify the (+) charge on the C=O, making it more likely to form a hydrate.

least amount
of hydrate

greatest amount
of hydrate

18-19

(a)

(b)

(c)

18-20

(a)

(b)

(c)

**411**

18-21 Imine formation has optimum pH 4-5 where protonation of the carbonyl oxygen is the first step. All steps in imine formation are equilibria.

(a)

Imine formation has six steps: four are proton transfers, one is a nucleophilic attack, and one is a leaving group leaving.

**A** proton on (resonance-stabilized cation)
**B** nucleophile attacks
**C** proton off
**D** proton on
**E** leaving group leaves (resonance-stabilized cation)
**F** proton off

(b)

18-21 continued

(c)

plus resonance forms with
(+) charge on benzene ring

Follow the six steps as described on the previous page.

18-22 Whenever a double bond is formed, stereochemistry must be considered. The two compounds are the Z and E isomers. (An electron pair has lower priority than H in the Cahn-Ingold-Prelog system.)

$E$        $Z$

18-23

(a)        (b)        (c)        (d)        (e)        (f)

+ $CH_3NH_2$        + $NH_3$

18-24 This mechanism is the reverse of the one shown in 18-21(c) above.

plus resonance forms with
(+) charge on benzene ring

plus resonance forms with
(+) charge on benzene ring

$- CH_3NH_2$

**413**

Copyright © 2013 Pearson Education, Inc.

**18-25**

Follow the six steps as described in the solution to 18-21(a).

**18-26**

(a)
(b)
(c)
(d)

**18-27**

(a)  PhCHO  +  $H_2NNH-\overset{O}{\overset{\|}{C}}-NH_2$

(b)  +  $H_2NOH$

(c)  +  $H_2N-NHPh$

(d)  +  $NH_2-NH$

(e)

(f)  +  $\overset{NH_2}{\underset{NH_2}{|}}$

18-28

*All steps in acetal formation are equilibria.*

hemiacetal

acetal

plus resonance forms with (+) charge on benzene ring

Acetal formation has seven steps, the first five of which are similar to imine formation: four are proton transfers, two are nucleophilic attacks, and one is a leaving group leaving.

**A** proton on (resonance-stabilized cation)
**B** nucleophile attacks
**C** proton off
**D** proton on
**E** leaving group leaves (resonance-stabilized cation)
**F** second nucleophile attacks (this is the step that is unique to acetal formation)
**G** proton off

18-29

These seven steps are the exact reverse of those in the previous solution.

**415**

## 18-30

(a)

+ 2 CH₃CH₂OH

(b)

$$CH_3-\overset{\overset{\displaystyle O}{\|}}{C}H \quad + \quad 2 \quad \overset{}{\underset{OH}{\diagup\!\!\diagdown}}$$

(c)

+ 2 HO⌒OH

(d)

=O + HO⌒⌒⌒OH

(e)

+

(f)

## 18-31

## 18-32

(a)

Follow the seven steps as described in the solution to problem 18-28.

**416**

Copyright © 2013 Pearson Education, Inc.

18-32 continued
(b)

(c) The mechanisms of formation of an acetal and hydrolysis of an acetal are identical, just in reverse order. This has to be true because this process is an equilibrium: if the forward steps follow a minimum energy path, then the reverse steps have to follow the identical minimum energy path. This is the famous Principle of Microscopic Reversibility, text section 8-4A.

(d)

18-33

(a)

Sodium triacetoxyborohydride is the optimum reagent to reduce aldehydes selectively in the presence of ketones; see the solution to problem 18-12. It does not reduce ketones even with an excess of the reagent.

18-33 continued

(b) *Protect the more reactive functional group, then react at the other group.*

(c) OR WITTIG
1. Ph₃P
2. BuLi
3. PhCHO
4. H₃O⁺

*The last step protonates the oxygen, dehydrates the alcohol, and hydrolyzes the acetal.*

(d)

less-hindered ketone

(e)

(f)

18-34

(a) + Ag⁰

after adding acid

(b)

(c) + Ag⁰

after adding acid

(d)

**18-35** Imines can be formed in either acid or base conditions. Here, the hydrazone formation is shown in acid conditions.

Follow the six steps **A**–**F** of imine formation in the solution to 18-21(a).

reduction of the hydrazone

$N_2$ is nature's best leaving group: it is very stable, and a gas, so it bubbles out of the reaction mixture.

$$+ \quad :N \equiv N:$$

**18-36**

(a)

common name: indane

(b)

(c)

(d)

Aqueous acid first hydrolyzes the acetal; the Zn(Hg)/HCl then reduces both ketones.

18-37

(a) O₂N ... NH ... NO₂

(b) O ‖ N-N-C-NH₂ H

(c) NOH

(d)

(e) OCH₃ | CH₃-CH | OCH₃

(f) OH | H-CH | OCH₃

(g) N-CH₂CH₃ =CCH₂CH₃

(h) O ‖ ... H ... OH → OH ... O

5-hydroxypentanal    hemiacetal

**18-38**  IUPAC names first; then common names.  Recall that IUPAC recommends placing the position numbers as close as possible to the groups they describe.  Parts (i)–(l) have no common names.

(a) heptan-2-one; methyl pentyl ketone
(c) heptanal; no simple common name
(e) butanal; butyraldehyde
(g) 4-bromo-2-methylhexanal; no common name
(i) hexa-2,4-dienal
(k) 3-oxocyclopentanecarbaldehyde

(b) heptan-4-one; dipropyl ketone
(d) benzophenone; diphenyl ketone
(f) propanone; acetone (IUPAC accepts "acetone")
(h) 3-phenylprop-2-enal; cinnamaldehyde
(j) 2-hydroxycyclohexane-1,3-dione
(l) *cis*-2,4-dimethylcyclopentanone

**18-39**

(a) N-NHPh
(b) O ... H
(c) O ... H
(d) O

(e) H ... O ... HO ... + CH₃OH
(f) H ... O
(g) O
(h)

(i) OH ... CH₂OH + H ... O ... H
(j) CH₃ ... N

**18-40**  In order of increasing equilibrium constant for hydrate formation:

$$CH_3-\overset{O}{\overset{\|}{C}}-CH_3 \quad < \quad CH_3-\overset{O}{\overset{\|}{C}}-CH_2Cl \quad < \quad CH_3-\overset{O}{\overset{\|}{C}}-H \quad < \quad ClCH_2-\overset{O}{\overset{\|}{C}}-H \quad < \quad H-\overset{O}{\overset{\|}{C}}-H$$

least amount of hydration                                                greatest amount of hydration

Steric hindrance decreases hydrate formation: electron-withdrawing substituents at the alpha carbon increase hydrate formation.  Without more information, it would be hard to say if the CH₂Cl group is better or worse than H in formation of a hydrate.

18-41

18-42

18-43 $C_6H_{10}O_2$ indicates two elements of unsaturation.

The IR absorption at 1708 cm$^{-1}$ suggests a ketone, or possibly two ketones since there are two oxygens and two elements of unsaturation. The NMR singlets in the ratio of 2 : 3 indicate a highly symmetric molecule. The singlet at $\delta$ 2.15 is probably methyl next to carbonyl, and the singlet at $\delta$ 2.67 integrating to two is likely to be CH$_2$ on the other side of the carbonyl.

Since the molecular formula is double this fragment, the molecule must be twice the fragment.

Two questions arise. Why is the integration 2 : 3 and not 4 : 6? Integration provides a *ratio*, not absolute numbers, of hydrogens. Why don't the two methylenes show splitting? Adjacent, *identical* hydrogens, with identical chemical shifts, do not split each other; the signals for ethane or cyclohexane appear as singlets.

18-44 The formula $C_{10}H_{12}O$ indicates five elements of unsaturation. A solid 2,4-DNP derivative suggests an aldehyde or a ketone, but a negative Tollens test precludes the possibility of an aldehyde; therefore, the unknown must be a ketone.

The NMR shows the typical ethyl pattern at $\delta$ 1.0 (3H, triplet) and $\delta$ 2.5 (2H, quartet), and a monosubstituted benzene at $\delta$ 7.3 (5H, multiplet). The singlet at $\delta$ 3.7 is a CH$_2$, but quite far downfield, apparently deshielded by two groups. Assemble the pieces:

**18-45**

(a)

$$\left[\begin{array}{c} \text{structure} \end{array}\right]^{+\cdot} \quad \longrightarrow \quad \left[\begin{array}{c} \text{structure} \end{array}\right]^{+\cdot} \;+\; \text{structure}$$

m/z 86 → m/z 44 + mass 42

(b)

m/z 114 → m/z 72 + mass 42

(c)

m/z 114 → m/z 58 + mass 56

**18-46**

A molecular ion of m/z 70 means a fairly small molecule. A solid semicarbazone derivative and a negative Tollens test indicate a ketone. The carbonyl (CO) has mass 28, so $70 - 28 = 42$, enough mass for only three more carbons. The molecular formula is probably $C_4H_6O$ (mass 70); with two elements of unsaturation, we can infer the presence of a double bond or a ring in addition to the carbonyl.

The IR shows a strong peak at 1790 cm$^{-1}$, indicative of a ketone in a small ring. No peak in the 1600–1650 cm$^{-1}$ region shows the absence of an alkene. The only possibilities for a small ring ketone containing four carbons are these:

The HNMR can distinguish these. No methyl doublet appears in the NMR spectrum, ruling out **B**. The NMR does show a 4H triplet at $\delta$ 3.1; this signal comes from the two methylenes (C-2 and C-4) adjacent to the carbonyl, split by the two hydrogens on C-3. The signal for the methylene at C-3 appears at $\delta$ 2.0, roughly a quintet because of splitting by four neighboring protons.

The unknown is cyclobutanone, **A**. The symmetry indicated by the carbon NMR rules out structure **B**. The IR absorption of the carbonyl at 1790 cm$^{-1}$ is characteristic of small ring ketones; ring strain strengthens the carbon-oxygen double bond, increasing its frequency of vibration. (See Section 12-9 in the text.)

**18-47**

(a)

(b)

para position is blocked       remove SO₃H

While this Gatterman-Koch formylation reaction (text section 17-11C) would give a decent yield of the ortho product, to avoid reaction at the para position, the SO₃H blocking group could be used. See the solution to problem 17-60(c) for a refresher on how to use the SO₃H as a blocking group.

(c)

(d)

(e)

(f)

(g) two methods:

(h)

**18-48** Recall that "dilute acid" means an aqueous solution, and aqueous acid will remove acetals.

**18-49** Historically, acetals have come from aldehydes and ketals have come from ketones. IUPAC has dropped the *ketal* designation and now recommends just the use of *acetal*.

(a) acetal (old: ketal)

$$CH_3CH_2CH_2-\overset{O}{\overset{\|}{C}}-CH_3 + 2\ CH_3OH$$

(b) hemiacetal (old: hemiketal)

+ CH₃CH₂OH

(c) acetal

+ HO

(d) acetal (old: ketal)

O= + HO OH

(e) acetal

OH / OH + O=CH / H

(f) diether, inert to hydrolysis

(g) imine(s)

NH₂ / NH₂ + O=CH / CH=O

(h) hydrazone (a type of imine)

=O + H₂N–NH₂

18-50

(a)

For a description of steps **A–G**, see the solution to Problem 18-28.

acetal   hemiacetal

plus resonance forms that show the (+) charge on the benzene ring

plus resonance forms that show the (+) charge on the benzene ring

(b)

Follow the six steps as shown in the solution to 18-21(a).

(c)

Ph₃PO +

(d)

(e)

Amines are bases and will be
protonated in acid solution.

18-50 continued

(f) For simplicity, $CH_3CH_2NH_2$ will be abbreviated $EtNH_2$. $EtNH_3^+$ serves as the proton source (acid) here.

18-51

(a)

(b)

(c)

Sodium triacetoxyborohydride selectively reduces the aldehyde; see the solutions to problems 18-12 and 18-33(a).

(d)

(e)

**427**

18-51 continued

(f)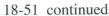

$$\underset{\text{Pt}}{\overset{H_2}{\longrightarrow}}$$

(g)

$$\underset{CH_3OH}{\overset{NaBH_4}{\longrightarrow}}$$

LiAlH₄ in ether
could also be
used—with
caution!

18-52 All of these reactions would be acid-catalyzed.

(a)

+ H₂NOH

(b)

CHO

+

H₂N

(c)

NH₂  +

O

(d)

O

+  HO        OH

(e)

NH₂  +

$$O=C\begin{array}{c}CH_3\\CH_3\end{array}$$

(f)

O

+  **2** CH₃OH

18-53

(a)

NCH₃

(b)  CH₃O    OCH₃

(c)

NOH

(d)

O        O

(e)

N–N–Ph
   |
   H

(f)

Ph    OH

(g)  no reaction

(h)

HO    C≡C–H

(i)

(j)

CH₂

(k)  Na⁺  O⁻    C≡N

(l)

HO    COOH

18-54

(a)

OH
|
C–H
|
Ph

(b)

O
‖
C–O⁻

+ Ag⁰

(c)

H  O
|   ‖
N–N–C–NH₂
‖
C–H

(d)

OCH₂CH₃
|
C–H
|
OCH₂CH₃

(e)

O        O
  \    /
   C
   |
   H

(f)

CH₃

**428**
Copyright © 2013 Pearson Education, Inc.

**18-55**

(a)

CH₃MgI / ether, H₃O⁺ → ... OH → H₂CrO₄ → ...

Swern or DMP or PCC or NaOCl would work too.

(b) ... C≡CH → (HgSO₄, H₂O, H₂SO₄) → ...

(c) ... → 1) O₃, 2) Me₂S → ... + ...

(d) ... OH → H₂CrO₄ → ...

Swern or DMP or PCC or NaOCl would work too.

(e) ... OH (COOH) → 1) excess CH₃Li, 2) H₃O⁺ → ...

(f) ... C≡N → CH₃MgI / ether, H₃O⁺ → ...

**18-56**

(a) ... OH → PCC → ... H

Swern or DMP are OK too.

(b) ... CH₂ → 1) O₃, 2) Me₂S → ... H

(c) ... C≡CH → 1) Sia₂BH, 2) H₂O₂, HO⁻ → ... H

(d) ... Br → NaCN → ... CN → 1) DIBAL-H, 2) H₃O⁺ → ... H

(e) ... Br → Mg, (epoxide), H₃O⁺ → ... OH → PCC → ... H

need to add two carbons

OR HC≡C⁻ Na⁺ → ... C≡CH → 1) Sia₂BH, 2) H₂O₂, HO⁻ →

**429**

## 18-56 continued

(f)

OR

(g)

**18-57** The new bond to carbon comes from the NaBH$_4$ or NaBD$_4$, shown in bold below. The new bond to oxygen comes from the protic solvent.

(a)

(b)

(c)

**18-58** While hydride is a small group, the actual chemical species supplying it, AlH$_4^-$, is fairly large, so it prefers to approach from the less hindered side of the molecule, that is, the side opposite the methyl. This forces the oxygen to go to the same side as the methyl, producing the *cis* isomer as the major product.

**18-59**

(a)

An exocyclic double bond is less stable than an endocyclic double bond.

methylenecyclohexane

18-59 continued

(b) The difficulty in synthesizing methylenecyclohexane from cyclohexanone without using the Wittig reaction rests in the stability of the double bond inside the ring (endocyclic) versus outside the ring (exocyclic).

A dehydration following the E1 mechanism passing through a carbocation intermediate will give the more substituted, endocyclic double bond as the major product. The only chance of making the exocyclic double bond as the major product is to do an E2 elimination using a bulky base to give Hofmann orientation (Chapter 7).

18-60

(a)

(b)

(c)

(d)

**18-61** The key to this problem is understanding that the relative proximity of the two oxygens can dramatically affect their chemistry.

1,2-dioxane

The "second" isomer described: two oxygens connected by a sigma bond are a peroxide. The O—O bond is easily cleaved to give radicals. In the presence of organic compounds, radical reactions can be explosive.

1,3-dioxane

The "third" isomer described: two oxygens bonded to the same $sp^3$ carbon constitute an acetal, which is hydrolyzed in aqueous acid. See the mechanism below.

1,4-dioxane

The "first" isomer described: an excellent solvent (although toxic), these oxygens are far enough apart to act independently. It is a simple ether.

Mechanism of acetal hydrolysis

**18-62** Building a model will help visualize this problem. Ignore stereochemistry at chiral carbons for this problem.

(a)

open-chain form

cyclic form—hemiacetal

(b) Yes, the cyclic form of glucose will give a positive Tollens test. In the basic solution of the Tollens test, the hemiacetal is in equilibrium with the open-chain aldehyde with the cyclic form in much larger concentration. However, it is the open-chain aldehyde that reacts with silver ion, so even though there is only a small amount of open-chain form present at any given time, as more of the open-chain form is oxidized by silver ion, more cyclic form will open to replace the consumed open-chain form. Eventually all of the cyclic form will be dragged kicking and screaming through the open-chain form to be oxidized to the carboxylate. Le Châtelier's Principle strikes again!

cyclic form
larger concentration
at equilibrium

open-chain form
smaller concentration
at equilibrium

$Ag^+$
NOT
reversible

oxidized to
carboxylate

+ $Ag°$
silver
metal

**18-63**

(a)

Any carbon with two oxygens bonded to it with single bonds belongs to the acetal family. If one of the oxygen groups is an OH, then the functional group is a hemiacetal. Thus, the functional group at C-2 is a hemiacetal. (The old name for this group is hemiketal as it came from a ketone.)

18-63 continued
(b) Models will help. Ignore stereochemistry for the mechanism.

hemiacetal (hemiketal)                    same as

18-64

(a)

(b) +

(c)

(d)

(e) +

18-65
(a) ketone: no reaction
(b) aldehyde: positive
(c) enol of an aldehyde—tautomerizes to aldehyde in base: positive
(d) hemiacetal of an aldehyde in equilibrium with the aldehyde in base: positive
(e) acetal—stable in base: no reaction
(f) hemiacetal of an aldehyde in equilibrium with the aldehyde in base: positive

18-66   The structure of **A** can be deduced from its reaction with **J** and **K**. What is common to both products of these reactions is the heptan-2-ol part; the reactions must be Grignard reactions with heptan-2-one, so **A** must be heptan-2-one.

**D**

O
‖
OH

$SOCl_2$

O
‖
Cl

**E**

$CH \equiv CH$

$H_2SO_4$
$HgSO_4$   **F**
$H_2O$

R   R

**G**

1) $O_3$
2) $Me_2S$

MgBr

1) $CH_3CHO$   **B**

2) $H_3O^+$

$(CH_3)_2CuLi$

OH

**C**

$Na_2Cr_2O_7$

$H_2SO_4$

O
‖

**A**

HCN

HO   CN

**H**

$H_3O^+$

HO   COOH

**I**

1) PhMgBr

**J**

2) $H_3O^+$

OH

Ph

MgBr

O

1)

**K**

O

2) $H_3O^+$

OH

O

18-67  The very strong π to π* absorption at 225 nm in the UV spectrum suggests a conjugated ketone or aldehyde.  The IR confirms this:  strong, conjugated carbonyl at 1690 cm$^{-1}$ and small alkene at 1610 cm$^{-1}$. The absence of peaks at 2700–2800 cm$^{-1}$ shows that the unknown is not an aldehyde.

The molecular ion at 96 leads to the molecular formula:

$$\begin{array}{l} 96 \\ \underline{-\ 64} \\ 32 \text{ mass units} \Rightarrow \text{add 2 carbons and 8 hydrogens} \end{array}$$

molecular formula = $C_6H_8O$ = 3 elements of unsaturation

Two elements of unsaturation are accounted for in the enone.  The other one is likely a ring.

The NMR shows two vinyl hydrogens.  The doublet at δ 6.0 says that the two hydrogens are on neighboring carbons (two peaks = one neighboring H).

No methyls are apparent in the NMR, so the 6H group of peaks at δ 2.0–2.4 is most likely 3 $CH_2$ groups. Combining the pieces:

The mass spectral fragmentation can be explained by a "retro" or reverse Diels-Alder fragmentation:

In the HNMR, one of the vinyl hydrogens appears at δ 7.0.  This is typical of an α,β-unsaturated carbonyl because of the resonance form that shows deshielding of the β-hydrogen.

**18-68**

**(a)**

**(b)** Aminoacetal linkages are in the dashed boxes.

deoxyadenosine      deoxycytidine

**(c)**
The first step in the mechanism in part (a) is protonation of the amine's electron pair. The nitrogens of the DNA nucleosides, however, are part of aromatic rings, and the electron pairs are required for the aromaticity of the ring. (See the solution to Problem 16-43 for a description of the aromaticity of these nucleoside bases.) Protonation of the nitrogen will not occur unless the acid is extremely strong; dilute acids will not protonate the N and therefore the nucleoside will be stable.

**18-69**

First, deduce what functional groups are present in **A** and **B**. The IR of **A** shows no alkene and no carbonyl: the strongest peak is at 1210 cm$^{-1}$, possibly a C—O bond. After acid hydrolysis of **A**, the IR of **B** shows a carbonyl at 1715 cm$^{-1}$: a ketone. (If it were an aldehyde, it would have aldehyde C—H around 2700–2800 cm$^{-1}$, absent in the spectrum of **B**.) What functional group has C—O bonds and is hydrolyzed to a ketone? An acetal (ketal)!

**A**
mol. wt. 116 $\Rightarrow$
$C_6H_{12}O_2$

**B**
mol. wt. 72 $\Rightarrow$
$C_4H_8O$

There is only one ketone of formula $C_4H_8O$: butan-2-one.

*continued on next page*

**437**

**A** must have the same alkyl groups as **B**. **A** has one element of unsaturation and is missing only $C_2H_4$ from the partial structure above. The most likely structure is the ethylene ketal. Is this consistent with the NMR?

δ 3.9 (singlet, 4H)

H₂C — CH₂
O     O      **A**

δ 1.3 (singlet, 3H) { H₃C     CH₂ — CH₃ } δ 0.9 (triplet, 3H)

δ 1.6
(quartet, 2H)

What about the peaks in the MS at m/z 87 and 101? The 87 peak is the loss of 29 from the molecular ion at 116.

m/z 116

plus two resonance forms with positive charge on the oxygen atoms

m/z 87

The peak at m/z 101 comes from loss of $CH_3$ from the molecular ion at 116.

plus two resonance forms with positive charge on the oxygen atoms

m/z 101

**18-70** The strong UV absorption at 220 nm indicates a conjugated aldehyde or ketone. The IR shows a strong carbonyl at 1690 cm⁻¹, alkene at 1625 cm⁻¹, and two peaks at 2720 cm⁻¹ and 2810 cm⁻¹ — aldehyde!

O
‖
C=C—C—H

The NMR shows the aldehyde proton at δ 9.5 split into a doublet, so it has one neighboring H. There are only two vinyl protons, so there must be an alkyl group coming off the β carbon:

H  H  O
|   |   ‖
R—C=C—C—H

The only other NMR signal is a 3H doublet: R must be methyl.

H  H  O
|   |   ‖
CH₃—C=C—C—H

"crotonaldehyde"

**19-1** These compounds satisfy the criteria for aromaticity (planar, cyclic π system, and the Huckel number of 4n + 2 π electrons): pyrrole, imidazole, indole, pyridine, 2-methylpyridine, pyrimidine, and purine. The systems with 6 π electrons are: pyrrole, imidazole, pyridine, 2-methylpyridine, and pyrimidine. The systems with 10 π electrons are: indole and purine. The other nitrogen heterocycles shown are not aromatic because they do not have cyclic π systems.

**19-2**

(a)

(b)
$$NH_2$$
$$CH_3-CHCHO$$

(c)

(d)
H
N
CH$_3$

(e)
H$_3$C   CH$_2$CH$_3$
N

(f)
NH$_2$

Cl

**19-3**

(a) pentan-2-amine (old: 2-pentylamine)
(b) *N*-methylbutan-2-amine
(c) 3-aminophenol (or *meta*-)
(d) 3-methylpyrrole
(e) *trans*-cyclopentane-1,2-diamine (or 1*R*,2*R*)
(f) *cis*-3-aminocyclohexanecarbaldehyde (or 1*S*,3*R*)

**19-4**
(a) Resolvable: there are two asymmetric carbons; carbon does not invert.
(b) Not resolvable: the nitrogen is free to invert.
(c) Not resolvable: it is symmetric.
(d) Not resolvable: even though the nitrogen is quaternary, one of the groups is a proton which can exchange rapidly, allowing for inversion.
(e) Resolvable: the nitrogen is quaternary and cannot invert when bonded to carbons.

**19-5** In order of increasing boiling point (increasing intermolecular hydrogen bonding):
(a) triethylamine and propyl ether have the same b.p. < dipropylamine
(b) dimethyl ether < dimethylamine < ethanol
(c) trimethylamine < diethylamine < diisopropylamine (increased number of carbons)

**19-6** Listed in order of increasing basicity. (See Appendix 2 in this Manual for a discussion of acidity and basicity.)

(a) PhNH$_2$ < NH$_3$ < CH$_3$NH$_2$ < NaOH
(b) *p*-nitroaniline < aniline < *p*-methylaniline (*p*-toluidine)
(c) pyrrole < aniline < pyridine < piperidine
(d) 3-nitropyrrole < pyrrole < imidazole

**19-7**

(a) secondary amine: one peak in the 3200–3400 cm$^{-1}$ region, indicating NH

(b) primary amine: two peaks in the 3200–3400 cm$^{-1}$ region, indicating NH$_2$

(c) alcohol: strong, broad peak around 3400 cm$^{-1}$

**19-8** A compound with formula $C_4H_{11}N$ has no elements of unsaturation. The proton NMR shows five types of H, with the $NH_2$ appearing as a broad peak at $\delta$ 1.15, meaning that there are four different groups of hydrogens on the four carbons. The carbon NMR also shows four carbons, so there is no symmetry in this structure; that is, it does not contain a *tert*-butyl group or an isopropyl group.

The multiplet farthest downfield is a CH deshielded by the nitrogen; integration shows it to be one H. There is a 2H multiplet at $\delta$ 1.35, the broad $NH_2$ peak at $\delta$ 1.15, a 3H doublet at $\delta$ 1.05, and a 3H triplet at $\delta$ 0.90. The latter two signals must represent methyl groups next to a CH and a $CH_2$ respectively. So far:

The pieces shown above have one carbon too many, so there must be one carbon that is duplicated: the only possible one is the CH, and the structure reveals itself.

**19-9**

(a)

(b) 

(c)

(d)

**19-10**

(a)

(b)

(c) The fragmentation in (a) occurs more often than the one in (b) because of stability of the radicals produced along with the iminium ions. Ethyl radical is much more stable than methyl radical, so pathway (a) is preferred.

**19-11** Nitration at the 4-position of pyridine is not observed for the same reason that nitration at the 2-position is not observed: the intermediate puts some positive character on an electron-deficient nitrogen, and electronegative nitrogen hates that. (It is important to distinguish this type of positive nitrogen without a complete octet of electrons, from the quaternary nitrogen, also positively charged but with a full octet. It is the number of electrons around atoms that is most important; the charge itself is less important.)

GOOD: (structure)     VERY BAD: (structure)

mechanism

VERY BAD—
N does not have
an octet of electrons.

not
produced
because of
unfavorable
intermediate

**19-12** Any electrophilic attack, including sulfonation, is preferred at the 3-position of pyridine because the intermediate is more stable than the intermediate from attack at either the 2-position or the 4-position. (Resonance forms of the sulfonate group are not shown, but remember that they are important!)

The N of pyridine is basic, and in the strong acid mixture, it will be protonated as shown here. That is part of the reason that pyridine is so sluggish to react: the ring already has a positive charge, so attack of an electrophile is slowed.

Copyright © 2013 Pearson Education, Inc.

**19-13** The middle resonance form demonstrates how the negative charge is distributed onto the more electronegative nitrogen atom.

**19-14**
**(a)**

**(b)**

Amide ion will not attack at C-2 because the anion at C-3 is not as stable as the anion at C-2.

This is a benzyne-type mechanism. (For simplicity above, two steps of benzyne generation are shown as one step: first, a proton is abstracted by amide anion, followed by loss of bromide.) Amide ion is a strong enough base to remove a proton from 3-bromopyridine as it does from a halobenzene. Once a benzyne is generated (two possibilities), the amide ion reacts quickly, forming a mixture of products.

Why does the 3-bromo follow this extreme mechanism while the 2-bromo reacts smoothly by the addition-elimination mechanism? Stability of the intermediate! Negative charge on the electronegative nitrogen makes for a more stable intermediate in the 2-bromo substitution. No such stabilization is possible in the 3-bromo case.

19-15

19-16

(a)   $PhCH_2NH_2$ + excess $CH_3I$   $\xrightarrow{NaHCO_3}$   $PhCH_2\overset{\oplus}{N}(CH_3)_3$   $I^-$

OR   $PhCH_2I$ + $N(CH_3)_3$

(b)   excess $NH_3$ +   Br⌒⌒⌒   ⟶   $H_2N$⌒⌒⌒

(c)   excess $NH_3$ + $PhCH_2Br$   ⟶   $PhCH_2NH_2$

19-17

(a) $CH_3\overset{O}{\overset{\|}{C}}-NHCH_2CH_3$

(b)

(c)

19-18  If the amino group were not protected, it would do a nucleophilic substitution on chlorosulfonic acid.  Later in the sequence, this group could not be removed without cleaving the other sulfonamide group.

19-19

sulfathiazole

continued on next page

## 19-19 continued

sulfapyridine

## 19-20

(a) [structure: hex-1-ene chain]

(b) H₃C–N–CH₃ [structure with chain and terminal alkene] + H₃C–N–CH₃ [ring structure with CH₃]

(c) [piperidine ring with N–CH₃] + H₂C=CH₂

(d) H₃C–N–CH₃ [bicyclic structure with allyl group]

(e) [methylenecyclohexane with N(CH₃)₂ substituent] H₃C–N–CH₃ [cyclohexene with CH₂ linker to N]

(f) [vinyl-substituted ring with N–CH₃]

## 19-21
Orientation of the Cope elimination is similar to Hofmann elimination: the *less* substituted alkene is the major product.

(a) [alkene chain structure]

  + (CH₃)₂NOH

(b) H₂C=CH₂  +  [structure: HO–N with ethyl and chain]    +  [alkene chain] minor
   major

  + (CH₃CH₂)₂NOH

(c) [cyclohexene]  +  (CH₃)₂NOH

(d) H₂C=CH₂  +  [piperidine with N–OH]  +  [ring with N–OH and CH₃] minor

## 19-22
The key to this problem is to understand that Hofmann elimination occurs via an E2 mechanism requiring *anti* coplanar stereochemistry, whereas Cope elimination requires *syn* coplanar stereochemistry.

(a) [structure: H₂C=C with H and CH(CH₃)₂, CH₂, H]

Hofmann orientation loses a hydrogen from the CH₃ as well as the N(CH₃)₃ group to make the less substituted double bond.

19-22 continued

(b) Hofmann elimination

$E$
(Zaitsev product—more highly substituted)

Cope elimination

$Z$
(Zaitsev product—more highly substituted)

19-23

(a)

Aliphatic diazonium ions are very unstable, rapidly decomposing to carbocations.

(b)

(c)

(d)

Aryl diazonium ions are relatively stable if kept cold.

19-24   The diazonium ion can do aromatic substitution like any other electrophile.

most significant resonance contributor

plus two other resonance forms with positive charge on the ring

$:\overset{..}{\underset{..}{Cl}}:^{\ominus}$ (or some other base)

methyl orange

**445**

19-25

(a)

(b)

from (a)

(c)

Aniline must be acylated
before undergoing Friedel-
Crafts reactions.

(d)

from (a)

(e)

(f)

(g)

(h)

**19-26** General guidelines for choice of reagent for reductive amination: use $LiAlH_4$ when the imine or oxime is isolated. Use $NaBH(CH_3COO)_3$, abbreviated $NaBH(OAc)_3$, in solution when the imine or iminium ion is not isolated. Alternatively, catalytic hydrogenation works in most cases.

(a)

$$+ \; CH_3NH_2 \xrightarrow[\text{CH}_3\text{COOH}]{\text{NaBH(OAc)}_3}$$

1° amine

2° amine
$CH_2NHCH_3$

(b)

piperidine

$$+ \; Ph\text{—CHO} \xrightarrow[\text{CH}_3\text{COOH}]{\text{NaBH(OAc)}_3}$$

$CH_2Ph$

(c)

$$+ \; \text{aniline} \xrightarrow{\text{acid catalyst}} \xrightarrow[\text{2) H}_2\text{O}]{\text{1) LiAlH}_4}$$

(d)

$$\xrightarrow[\substack{\text{acid}\\\text{catalyst}}]{H_2NOH} \xrightarrow[\text{2) H}_2\text{O}]{\text{1) LiAlH}_4}$$

(e)

$$PhCH_2\text{—C(O)—}CH_3 \xrightarrow[\substack{\text{acid}\\\text{catalyst}}]{H_2NOH} PhCH_2\text{—C(NOH)—}CH_3 \xrightarrow[\text{2) H}_2\text{O}]{\text{1) LiAlH}_4} PhCH_2\text{—CH(NH}_2)\text{—}CH_3$$

(f)

piperidine

$$NH \; + \; O\text{=}\square \xrightarrow[\text{CH}_3\text{COOH}]{\text{NaBH(OAc)}_3}$$

**19-27**

(a)

$$\xrightarrow[\text{2) H}_2\text{O}]{\text{1) LiAlH}_4}$$

(b)

$$\xrightarrow[\text{2) H}_2\text{O}]{\text{1) LiAlH}_4}$$

**19-28 (a)**

$$\text{(CH}_3\text{(CH}_2)_5\text{CH}_2\text{Br}) + :NH_3 \xrightarrow{S_N2} \text{R–}\overset{+}{N}H_3 \; Br^- \quad + \; :NH_3$$

Good nucleophile—alkylation process can continue.

**(b)** Use a large excess of ammonia to avoid multiple alkylations of each nitrogen.

**19-29**

**(a)**

$$\text{phthalimide anion} \xrightarrow{BrCH_2Ph} \text{N–CH}_2\text{Ph} \xrightarrow[\Delta]{NH_2NH_2} H_2NCH_2Ph$$

**(b)**

$$\text{phthalimide anion} \xrightarrow{Br(CH_2)_5CH_3} \text{N–(CH}_2)_5\text{CH}_3 \xrightarrow[\Delta]{NH_2NH_2} H_2N(CH_2)_5CH_3$$

**(c)**

$$\text{phthalimide anion} \xrightarrow{Br(CH_2)_3COO^-} \text{N–(CH}_2)_3\text{COO}^- \xrightarrow[\text{2) H–A}]{\text{1) NH}_2\text{NH}_2, \Delta} H_2N(CH_2)_3COOH$$

Must use anion to avoid protonating the phthalimide anion.

**19-30** Assume that $LiAlH_4$ or $H_2$/catalyst can be used interchangeably.

**(a)**

$$PhCH_2Br + NaN_3 \longrightarrow PhCH_2N_3 \xrightarrow{\frac{H_2}{Pt}} PhCH_2NH_2$$

**(b)**

$$PhCH_2CH_2Br \xrightarrow{KCN} PhCH_2CH_2C\equiv N \xrightarrow[\text{2) H}_2\text{O}]{\text{1) LiAlH}_4} PhCH_2CH_2CH_2NH_2$$

**(c)**

$$\text{CH}_3(CH_2)_3COOH \xrightarrow[\text{2) H}_2\text{O}]{\text{1) LiAlH}_4} \text{CH}_3(CH_2)_4OH \xrightarrow[\text{pyridine}]{TsCl} \text{CH}_3(CH_2)_4OTs$$

$$\downarrow NaN_3$$

$$\text{CH}_3(CH_2)_4NH_2 \xleftarrow[\text{2) H}_2\text{O}]{\text{1) LiAlH}_4} \text{CH}_3(CH_2)_4N_3$$

**OR**

$$\text{CH}_3(CH_2)_3COOH \xrightarrow{SOCl_2} \text{CH}_3(CH_2)_3COCl \xrightarrow{NH_3} \text{CH}_3(CH_2)_3CONH_2$$

$$\downarrow \text{1) LiAlH}_4 \quad \text{2) H}_2\text{O}$$

$$\text{CH}_3(CH_2)_4NH_2$$

**19-30 continued**

(d)

from (c)

(e)

(f)

(g)

**19-31** To reduce nitroaromatics, the reducing reagents ($H_2$ plus a metal catalyst, or a metal plus HCl) can be used virtually interchangeably. Assume a workup in base to give the free amine final product.

(a)

(b)

+ ortho

(c)

from (a)

(d)

19-32

(a) primary amine; 2,2-dimethylpropan-1-amine, or neopentylamine
(b) secondary amine; N-methylpropan-2-amine, or isopropylmethylamine
(c) tertiary heterocyclic amine and a nitro group; 3-nitropyridine
(d) quaternary heterocyclic ammonium ion; *N,N*-dimethylpiperidinium iodide
(e) tertiary aromatic amine oxide; *N*-ethyl-*N*-methylaniline oxide
(f) tertiary aromatic amine; *N*-ethyl-*N*-methylaniline
(g) tertiary heterocyclic ammonium ion; pyridinium chloride
(h) secondary amine; *N*,4-diethylhexan-3-amine

19-33 Shown in order of increasing basicity. In sets a–c, the aliphatic amine is the strongest base.

(a)
$$\underset{H}{Ph-\overset{H}{N}-Ph} \quad < \quad PhNH_2 \quad < \quad \text{(cyclohexyl)}-NH_2 \quad \text{The aliphatic amine is the strongest base.}$$

(b)
pyrrole $<$ pyridine $<$ piperidine   The aliphatic amine is the strongest base; pyrrole's aromaticity would be lost if protonated.

(c)
pyrrole $<$ imidazole (HN=N) $<$ pyrrolidine (NH)   The aliphatic amine is the strongest base; pyrrole's aromaticity would be lost if protonated.

(d)
$O_2N$-C$_6$H$_4$-$NH_2$ $<$ C$_6$H$_5$-$NH_2$ $<$ $H_3C$-C$_6$H$_4$-$NH_2$   Basicity is a measure of the ability to donate the pair of electrons on the amine. Electron-withdrawing groups like $NO_2$ decrease basicity, while electron-donating groups like $CH_3$ increase basicity.

(e)
benzamide (C=O, $NH_2$) $<$ aniline ($NH_2$) $<$ benzylamine ($CH_2NH_2$)   The aliphatic amine is the strongest base (amides are not basic).

19-34

(a) not resolvable: planar
(b) resolvable: asymmetric carbon
(c) not resolvable: symmetric
(d) resolvable: nitrogen inversion very slow
(e) not resolvable: symmetric
(f) resolvable: asymmetric nitrogen, unable to invert
(g) not resolvable in conditions where the proton on N can exchange
(h) resolvable: asymmetric nitrogen, unable to invert

19-35 The values of $pK_b$ of amines or $pK_a$ of the conjugate acids can be obtained from text Table 19-3. The side of the reaction with the weaker acid and base will be favored at equilibrium. (See Appendix 2 in this Manual for a discussion of acidity and basicity.)

(a)
pyridine + acetic acid (OH, $pK_a$ 4.74) $\rightleftharpoons$ pyridinium (N-H$^+$, $pK_a$ 8.75) + acetate (O$^-$)   *Products are favored.*

19-35 continued

(b) [pyrrole] + [acetic acid] OH  pK$_a$ 4.74  ⇌  [protonated pyrrole] + [acetate] O⁻  *Reactants are favored.*
H ⊕ H  pK$_a$ ≈ −1

(c) [pyridinium] N⊕ H  pK$_a$ 8.75  + [piperidine] N H  →  [pyridine] N  + [piperidinium] H ⊕ H  pK$_a$ 11.12  *Products are favored.*

(d) [anilinium] NH$_3$ ⊕  pK$_a$ 4.60  + [pyrrolidine] N H  →  [aniline] NH$_2$  + [pyrrolidinium] H ⊕ H  pK$_a$ 11.27  *Products are favored.*

19-36 Assume all reductions have aqueous acid workup.

(a) PhCH$_2$CH$_2$CH$_2$NH$_2$  (b) [pentyl] NH$_2$  (c) [piperidine N-oxide] CH$_3$ N⊕ O⊖  (d) [chain] CH$_3$ N OH

(e) [quinolizidine with alkenyl] N CH$_3$  (f) [bicyclic diene] H$_3$C CH$_3$  (g) [N-nitrosopiperidine] N NO  (h) [aniline] NH$_2$  after workup with base

(i) [cyclohexyl acetamide] NHCH$_3$ O  (j) [cyclohexyl ethyl amine] CH$_2$CH$_2$NHCH$_3$  (k) NHCH$_3$ CH$_3$(CH$_2$)$_3$CHCH$_2$CH$_3$

(l) CH$_2$NH$_2$ PhCH$_2$CHCH$_3$  (m) [triethylamine derivative] N  (n) [pyridine] N OCH$_2$CH$_3$  For a reminder on why fluoride is such a good leaving group in nuc. aromatic subst., see the solution to 17-21 or 19-50(b).  (o) [benzene] Br NO$_2$

(p) [alcohol nitrile] HO CN  →(1) LiAlH$_4$ (2) H$_2$O→  HO CH$_2$NH$_2$  (q) [N-cyclopentylaniline] H N  (r) [pentene] major product— Hofmann orientation

19-37

(a) CH$_3$—[benzene]—NH$_2$  →(NaNO$_2$ / HCl)→  CH$_3$—[benzene]—N$_2$⊕ Cl⁻  →(CuCN)→  CH$_3$—[benzene]—CN

## 19-37 continued

(b)

$CH_3$—(m-substituted benzene)—CN  →  (1) $LiAlH_4$  (2) $H_2O$  →  $CH_3$—(m-substituted benzene)—$CH_2NH_2$

from (a)

(c)

$CH_3$—(m-substituted benzene)—$\overset{\oplus}{N_2}$ $Cl^-$  →  KI  →  $CH_3$—(m-substituted benzene)—I

from (a)

(d)

$CH_3$—(m-substituted benzene)—$\overset{\oplus}{N_2}$ $Cl^-$  →  $H_3O^+$, $\Delta$  →  $CH_3$—(m-substituted benzene)—OH

from (a)

(e)

$CH_3$—(m-substituted benzene)—$NH_2$  +  $CH_3\overset{O}{\overset{\|}{C}}-Cl$  →  $CH_3$—(m-substituted benzene)—$NHCOCH_3$  →  $HNO_3$ / $H_2SO_4$  →  $CH_3$—(substituted benzene, $O_2N$)—$NHCOCH_3$ (+ isomer)

after workup with base ←  $H_3O^+$, $\Delta$  ←  $CH_3$—(substituted benzene, $O_2N$)—$NH_2$

(f)

$CH_3$—(m-substituted benzene)—$NH_2$  +  $O{=}$(cyclopentane)  →  $NaBH(OAc)_3$ / $CH_3COOH$  →  $CH_3$—(m-substituted benzene)—$\overset{H}{N}$-cyclopentyl

## 19-38

This fragmentation is favorable because the iminium ion produced is stabilized by resonance. Also, there are three possible cleavages that give the same ion. Both factors combine to make the cleavage facile, at the expense of the molecular ion.

$$\left[ CH_3-\overset{\overset{\displaystyle CH_3}{|}}{\underset{\underset{\displaystyle CH_3}{|}}{C}}-NH_2 \right]^{+\cdot} \longrightarrow \bullet CH_3 + \left\{ \overset{\overset{\displaystyle CH_3}{|}}{\underset{\underset{\displaystyle CH_3}{|}}{\oplus C}}-\overset{\cdot\cdot}{N}H_2 \longleftrightarrow \overset{\overset{\displaystyle CH_3}{|}}{\underset{\underset{\displaystyle CH_3}{|}}{C}}{=}\overset{\oplus}{N}H_2 \right\}$$

58 / m/z 73   mass 15   m/z 58

## 19-39

(a)

(benzene)  →  $HNO_3$ / $H_2SO_4$  →  (benzene)—$NO_2$  →  Fe / HCl  →  (benzene)—$NH_2$  +  $Cl-\overset{O}{\overset{\|}{C}}$-cyclopentyl  →  (benzene)—$NH-\overset{O}{\overset{\|}{C}}$-cyclopentyl

(b)

(benzene)—$NH_2$  +  $Cl-\overset{\overset{O}{\|}}{\underset{\underset{O}{\|}}{S}}$—(benzene)—$CH_3$  →  (benzene)—$\overset{H}{N}-\overset{\overset{O}{\|}}{\underset{\underset{O}{\|}}{S}}$—(benzene)—$CH_3$

**452**

**19-39 continued**

(c)

(d)

(e)

(f)

(g)

**19-40** The problem restricts the starting materials to six carbons or fewer. Always choose starting materials with as many of the necessary functional groups as possible.

(a)

phenacetin

(b)

methamphetamine

(c)

dopamine

19-41

(a)

(b)

19-42

(a)

$SOCl_2$ $NH_3$

1) $LiAlH_4$
2) $H_2O$

$CH_2NH_2$

(b)

$+ \ NH_3$

$\dfrac{NaBH(OAc)_3}{CH_3COOH}$

$CH_2NH_2$

(c)

$Cl-\overset{O}{\overset{\|}{C}}-CH_3$

$N-\overset{O}{\overset{\|}{C}}-CH_3$

1) $LiAlH_4$
2) $H_2O$

$N-CH_2CH_3$

(d)

$HN$ $\dfrac{NaBH(OAc)_3}{CH_3COOH}$

(e)

$SOCl_2$ $NH_3$

1) $LiAlH_4$ 2) $H_2O$

$H_2N$ $NH_2$

19-43

(a)

$Br$ $+$ $\dfrac{NH_2NH_2}{\Delta}$ $NH_2$

(b)

$CH_2CH_3$ $\dfrac{HNO_3}{H_2SO_4}$ $CH_2CH_3$ $\dfrac{Sn}{HCl}$ $CH_2CH_3$

$NO_2$ $NH_2$

(c)

$Br$ $+$ $\dfrac{NH_2NH_2}{\Delta}$ $NH_2$

$Ph$ $Ph$

(d)

$Br$ $\xrightarrow{KCN}$ $CN$ $\begin{array}{l}1) \ LiAlH_4 \\ 2) \ H_2O\end{array}$ $NH_2$

$Ph$ $Ph$ $Ph$

**455**

Copyright © 2013 Pearson Education, Inc.

19-44

(a) When guanidine is protonated, the cation is greatly stabilized by resonance, distributing the positive charge over all atoms (except H):

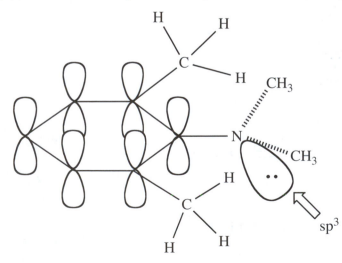

(b) The unprotonated molecule has a resonance form shown below that the protonated molecule cannot have. Therefore, the unprotonated form is stabilized relative to the protonated form. This greater stabilization of the unprotonated form is reflected in weaker basicity.

(c) Anilines are weaker bases than aliphatic amines because the electron pair on the nitrogen is shared with the ring, stabilizing the system. There is a steric requirement, however: the p orbital on the N must be parallel with the p orbitals on the benzene ring in order for the electrons on N to be distributed into the $\pi$ system of the ring.

If the orbital on the nitrogen is forced out of this orientation (by substitution on C-2 and C-6, for example), the electrons are no longer shared with the ring. The nitrogen is hybridized $sp^3$ (no longer any reason to be $sp^2$), and the electron pair is readily available for bonding $\Rightarrow$ increased basicity.

As surprising as it sounds, this aniline is about as basic as a tertiary aliphatic amine, except that the aromatic ring substituent is electron-withdrawing *by induction*, decreasing the basicity slightly. This phenomenon is called *steric inhibition of resonance*. We will see more examples in future chapters. Also, it is the last topic in Appendix 2 in this Manual.

**19-45** In this problem, sodium triacetoxyborohydride, NaBH(OAc)$_3$, will be used for reductive alkylation. Other synthetic sequences may be equally effective. (Acetic acid is abbreviated HOAc.)

(a)

$$\text{CH}_3\text{CH}_2\text{CH}_2\text{CH}_2\text{OH} \xrightarrow[\text{pyridine}]{\text{TsCl}} \text{OTs} \xrightarrow{\text{KCN}} \text{CN} \xrightarrow[\text{2) H}_2\text{O}]{\text{1) LiAlH}_4} \text{NH}_2$$

(b)

$$\text{OH} \xrightarrow{\text{PCC}} \overset{\text{H}}{\underset{\text{O}}{\text{C}}} \xrightarrow[\substack{\text{NaBH(OAc)}_3 \\ \text{HOAc}}]{\text{CH}_3\text{NH}_2} \text{NHCH}_3$$

(c)

$$\text{OH} \xrightarrow[\text{pyridine}]{\text{TsCl}} \text{OTs} \xrightarrow{\text{NH}_3} \text{NH}_2$$

$$\underset{\text{OH}}{\text{}} \xrightarrow{\text{H}_2\text{CrO}_4} \underset{\text{O}}{\text{}} + \text{NH}_2 \xrightarrow[\text{HOAc}]{\text{NaBH(OAc)}_3}$$

$$\xleftarrow[\text{HOAc}]{\text{NaBH(OAc)}_3} \quad \overset{\text{O}}{\underset{}{\text{CH}_3-\text{CH}}} \uparrow \text{PCC} \quad \text{CH}_3\text{CH}_2\text{OH}$$

(d)

The Gatterman-Koch reaction would make benzaldehyde from benzene.

$$\text{CH}_3 \xrightarrow[\text{h}\nu]{\text{NBS}} \text{CH}_2\text{Br} \xrightarrow{\text{NaOH}} \text{CH}_2\text{OH} \xrightarrow{\text{PCC}} \overset{\text{O}}{\underset{\text{H}}{\text{C}}}$$

$$\text{CH}_2\text{NHCH}_2\text{CH}_2\text{CH}_3 \xleftarrow[\text{HOAc}]{\text{NaBH(OAc)}_3} \quad \text{NH}_2 \text{ from (c)}$$

(e)

$$\text{benzene} \xrightarrow[\text{H}_2\text{SO}_4]{\text{HNO}_3} \text{NO}_2 \xrightarrow[\text{HCl}]{\text{Fe}} \text{NH}_2 \xrightarrow[\text{HCl}]{\text{NaNO}_2} \overset{\oplus}{\text{N}_2}\ \text{Cl}^- \xrightarrow{\text{H}_3\text{O}^+} \text{OH}$$

Combine phenol and phenyldiazonium ion.

$$\text{N}=\text{N} \quad \text{OH}$$

(f)

$$\text{benzene} \xrightarrow[\text{AlCl}_3]{} \overset{\text{O}}{\underset{}{\text{}}} \xrightarrow[\text{H}_2\text{SO}_4]{\text{HNO}_3} \overset{\text{O}}{\underset{\text{NO}_2}{\text{}}} \xrightarrow[\substack{\text{HCl} \\ \text{Clemmensen} \\ \text{reduces both}}]{\text{Zn(Hg)}} \underset{\text{NH}_2}{\text{}}$$

after neutral workup

$$\text{OH} \xrightarrow{\text{H}_2\text{CrO}_4} \xrightarrow{\text{SOCl}_2} \overset{\text{O}}{\underset{\text{Cl}}{\text{}}}$$

## 19-45 continued

(g)

## 19-46

(a)

(b)

Hofmann elimination

(c)

## 19-47

coniine

*N,N*-dimethyloct-7-ene-4-amine

19-48
(a)

<u>unknown **X**</u>

—fishy odor $\Rightarrow$ amine

—molecular weight 101 $\Rightarrow$ odd number of nitrogens

$\Rightarrow$ If one nitrogen and no oxygen, the remainder is $C_6H_{15}$.

<u>Mass spectrum:</u>

—fragment at 86 = M − 15 = loss of methyl $\Rightarrow$ the compound is likely to have this structural piece: $CH_3 {-}|{-} C {-} N$
                                                    └ α-cleavage

<u>IR spectrum:</u>

—no OH, no NH $\Rightarrow$ must be a 3° amine

—no C=O or C=C or C≡N or NO$_2$

<u>NMR spectrum:</u>

—only a triplet and quartet, integration about 3 : 2 $\Rightarrow$ ethyl group(s) only

Assemble the evidence: $C_6H_{15}N$, 3° amine, only ethyl in the NMR:

$$CH_3CH_2 - N - CH_2CH_3$$
$$|$$
$$CH_2CH_3$$

(b) React the triethylamine with HCl. The pure salt is solid and odorless.

$$(CH_3CH_2)_3N \ + \ HCl \ \longrightarrow \ (CH_3CH_2)_3\overset{\oplus}{N}H \quad Cl^-$$
                                                          salt

(c) Washing her clothing in dilute acid like vinegar (dilute acetic acid) or dilute HCl would form a water-soluble salt as shown in (b). Normal washing will remove the water-soluble salt.

19-49
(a)

(b) Both answers can be found in the resonance forms of the intermediate, in particular, the resonance form that shows the positive charge on the N. This is the major resonance contributor; what is special about it is that every atom has a full octet, the best of all possible conditions. That does not arise in the benzene intermediate, so it must be easier to form the intermediate from pyrrole than from benzene. Also, acylation at the 3-position puts positive charge at the 2-position and on the N, but never on the other side of the ring, so this substitution has only two resonance forms. The intermediate from acylation at the 3-position is therefore not as stable as the intermediate from acylation at the 2-position.

**19-50**

**(a)**

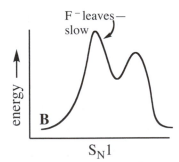

(The mechanism in part (a) shows nucleophilic aromatic substitution of fluoride by $NH_2R$ where $R = CH_3$, proceeding through anionic sigma complexes with resonance structures delocalizing negative charge onto the ortho and para $NO_2$ groups, followed by loss of $F^-$ to give the product $O_2N$-substituted N-methylaniline with $NO_2$.)

R = CH₃  →  R = $CH_3$

**(b)** Why is fluoride ion a good leaving group from **A** but not from **B** (either by $S_N1$ or $S_N2$)?

**A**  **B**

Formation of the anionic sigma complex **A** is the rate-determining (slow) step in nucleophilic aromatic substitution. The loss of fluoride ion occurs in a subsequent fast step where the nature of the leaving group does not affect the overall reaction rate. In the $S_N1$ or $S_N2$ mechanisms, however, the carbon-fluorine bond is breaking in the rate-determining step, so the poor leaving group ability of fluoride does indeed affect the rate.

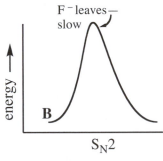

(Three energy diagrams: "nucleophilic aromatic substitution" with "F⁻ leaves — fast" at complex A; "$S_N1$" with "F⁻ leaves — slow" at complex B; "$S_N2$" with "F⁻ leaves — slow" at complex B.)

**(c)** Amines can act as nucleophiles as long as the electron pair on the N is available for bonding. The initial reactant, methylamine, $CH_3NH_2$, is a very reactive nucleophile. However, once the N is bonded to the benzene ring, the electron pair is delocalized onto the ring, especially with such strong electron-withdrawing groups like $NO_2$ in the ortho and para positions. The electrons on N are no longer available for bonding so there is no danger of it acting as a nucleophile in another reaction.

delocalized

(Structure showing $HN-CH_3$ with $O_2N$ and $NO_2$ groups on benzene ring, electron pair delocalized.)

19-51    **Compound A**

**Mass spectrum:**

—molecular ion at 73 = odd mass = odd number of nitrogens;

   if one nitrogen and no oxygen present ⇒ molecular formula $C_4H_{11}N$

—base peak at 44 is M – 29 ⇒ This fragment must be present:

$$\left.\begin{array}{l} -N- \\ \ \ | \\ --C+CH_2CH_3 \\ \ \ | \\ \ \ 44 \underline{\hspace{1cm}} \\ \alpha\text{-cleavage} \end{array}\right\} \text{EITHER} \quad CH_3--\overset{\displaystyle NH_2}{\underset{\displaystyle H}{\overset{|}{\underset{|}{C}}}}-CH_2CH_3 \ \mathbf{1} \quad OR \quad H-\overset{\displaystyle CH_3-NH}{\underset{\displaystyle H}{\overset{|}{\underset{|}{C}}}}-CH_2CH_3 \ \mathbf{2}$$

**IR spectrum:**

—two peaks around 3300 cm⁻¹ indicate a 1° amine so **2** is not correct; no indication of oxygen

**NMR spectrum:**

—two exchangeable protons suggest $NH_2$ present

—1H multiplet at δ 2.8 means a CH—$NH_2$

The structure of **A** must be the same as **1** above:

$$\mathbf{A} \quad CH_3-\overset{\displaystyle NH_2}{\underset{\displaystyle H}{\overset{|}{\underset{|}{C}}}}-CH_2CH_3$$

**Compound B**

an isomer of **A**, so its molecular formula must also be $C_4H_{11}N$

**IR spectrum:**

—only one peak at 3300 cm⁻¹ ⇒ 2° amine

**NMR spectrum:**

—one exchangeable proton ⇒ NH

—two ethyls present

The structure of **B** must be:

$$\mathbf{B} \quad CH_3CH_2\diagdown_{\underset{\displaystyle H}{\overset{\displaystyle |}{N}}}\diagup^{CH_2CH_3}$$

$$\left[ CH_3 \dashv CH_2-\overset{\displaystyle 58}{\underset{\displaystyle H}{\overset{|}{\underset{|}{N}CH_2CH_3}}} \right]^{\stackrel{+}{\cdot}} \longrightarrow CH_2=\overset{\displaystyle \oplus}{\underset{\displaystyle H}{\overset{|}{\underset{|}{N}CH_2CH_3}}} \text{ resonance-stabilized}$$
$$m/z \ 58$$

19-52

(a)  The acid-catalyzed condensation of P2P (a controlled substance) with methylamine hydrochloride gives an imine which can be reduced to **methamphetamine**.  The suspect was probably planning to use zinc in muriatic acid (dilute HCl) for the reduction.

phenyl-2-propanone, **P2P**                                                        methamphetamine
   (phenylacetone)

(b)  The jury acquitted the defendant on the charge of attempted manufacture of methamphetamine.  There were legal problems with possible entrapment, plus the fact that he had never opened the bottle of the starting material.  The defendant was convicted on several possession charges, however, and was awarded four years of institutional time to study organic chemistry.

Mass spectrum:

—molecular ion at 87 = odd mass = odd number of nitrogens present

—if one nitrogen and no oxygens $\Rightarrow$ molecular formula $C_5H_{13}N$

—base peak at m/z 30 $\Rightarrow$ structure must include this fragment

$$R \text{---} CH_2NH_2 \longrightarrow \begin{array}{c} H \quad\quad H \\ \backslash \quad / \\ C = N \oplus \\ / \quad\quad \backslash \\ H \quad\quad H \end{array}$$

$$\underbrace{}_{30} \qquad\qquad m/z\ 30$$

IR spectrum:

—two peaks in the 3300–3400 $cm^{-1}$ region $\Rightarrow$ 1° amine

NMR spectrum:

—singlet at $\delta$ 0.9 for 9H must be a $t$-butyl group

—2H signal at $\delta$ 1.0 exchanges with $D_2O$ $\Rightarrow$ must be protons on N or O

$$\boxed{\begin{array}{c} CH_3 \\ | \\ CH_3 \text{---} C \text{---} CH_2 \text{---} NH_2 \\ | \\ CH_3 \end{array}}$$

$$\delta\ 0.9 \left\{ \begin{array}{c} \delta\ 1.0 \\ \downarrow \\ CH_3 \\ | \\ CH_3 \text{---} C \text{---} CH_2 \text{---} NH_2 \\ | \\ CH_3 \uparrow \\ \delta\ 2.4 \end{array} \right.$$

Note that the base peak in the MS arises from cleavage to give these two, relatively stable fragments:

$$\begin{array}{c} CH_3 \\ | \\ CH_3 \text{---} C\bullet \\ | \\ CH_3 \end{array} \quad + \quad \begin{array}{c} H \quad\quad H \\ \backslash \quad / \\ C = N \oplus \\ / \quad\quad \backslash \\ H \quad\quad H \\ m/z\ 30 \end{array}$$

19-54 (a tough problem)

Molecular formula $C_{11}H_{16}N_2$ has 5 elements of unsaturation, enough for a benzene ring; no oxygens precludes $NO_2$ and amide; if C≡N is present, there are not enough elements of unsaturation left for a benzene ring, so benzene and C≡N are mutually exclusive.

IR spectrum:

—one spike around 3300 $cm^{-1}$ suggests a 2° amine

—no C≡N

—CH and C=C regions suggest an aromatic ring

Proton NMR spectrum:

—5H multiplet at $\delta$ 7.3 indicates a monosubstituted benzene ring (the fact that all the peaks are huddled around 7.3 precludes N being bonded to the ring)

—1H singlet at $\delta$ 2.0 is exchangeable $\Rightarrow$ NH of secondary amine

—2H singlet at $\delta$ 3.5 is $CH_2$; the fact that it is so strongly deshielded and unsplit suggests that it is between a nitrogen and the benzene ring

fragments so far:

$CH_2 \text{-} N$ + NH + 4 C + 8 H + 1 element of unsaturation

continued on next page

Carbon NMR spectrum:

—four signals around $\delta$ 125–138 are the aromatic carbons

—the signal at $\delta$ 65 is the $CH_2$ bonded to the benzene

—the other four carbons come as two signals at $\delta$ 46 and $\delta$ 55; each is a $CH_2$, so there are two sets of two equivalent $CH_2$ groups, each bonded to N to shift it downfield

fragments so far:

There is no evidence for an alkene in any of the spectra, so the remaining element of unsaturation must be a ring. The simplicity of the NMR spectra indicates a fairly symmetric compound.

Assemble the pieces:

19-55 A neutral, aqueous workup is required in each synthesis.

(a)                                                                                        (b)

**19-56** Not only is substitution at C-2 and C-4 the major products, but substitution occurs under surprisingly mild conditions.

Begin by drawing the resonance forms of pyridine N-oxide:

Resonance forms show that the electron density from the oxygen is distributed at C-2 and C-4; these positions would be the likely places for an electrophile to attack.

Resonance forms from electrophilic attack at C-2

N does not have an octet—not a significant resonance contributor.

GOOD! All atoms have octets.

Resonance forms from electrophilic attack at C-3

These two resonance forms place two positive charges on adjacent atoms—not good.

Resonance forms from electrophilic attack at C-4

N does not have an octet—not a significant resonance contributor.

GOOD! All atoms have octets.

continued on next page

19-56 continued

The resonance forms from electrophilic attack at C-3 are bad; only one of the three is a significant contributor, which means that there is not much resonance stabilization. When the electrophile attacks at C-2 or C-4, however, there are two forms that are good plus one great one that has all atoms with full octets. Clearly, attacks at C-2 and C-4 give the most stable intermediates and will be the preferred sites of attack.

19-57

(a)

$A^-$ is the conjugate base of the acid HA.

called a hemiaminal

19-57 continued

(b)

A⁻ is the conjugate
base of the acid HA.

called a hemiaminal

enamine

To this point, everything is the same in the two mechanisms.
But now, there is no H on the N to remove to form the imine.
The only H that can be removed to form a neutral intermediate
is the H on C next to the carbocation.

(c)  A secondary amine has only one H to give, which it loses in the first half of the mechanism to form the
neutral intermediate called a hemiaminal, equivalent to a hemiacetal.  In the second half of the mechanism,
the H on an adjacent carbon is removed to form the neutral product, the enamine.  The type of product
depends entirely on whether the amine begins with one or two hydrogen atoms.

20-1

(a) [structure: CH3CH2CH(CH3)COOH with methyl branch]

(b) [structure: CH3CH2CH(Br)COOH]

(c) [structure: CH3CH(NH2)CH2CH2COOH]

(d) [structure: benzyl-CH2-CH=CH-COOH, with H and H shown]

(e) [structure: cyclohexane with COOH (H wedge) and CH3 (H), "or the enantiomer"]

(f) [structure: (Z or E) H3C, COOH / HOOC, CH3 alkene]

(g) [structure: benzoic acid with COOH and Cl meta]

(h) [structure: benzene with COOH, COOH, CH3]

(i) [structure: HOOC-CH2CH2-CH(NH2)-CH2-COOH]

(j) HOOC [structure: HOOC-CH2CH2CH2-CH(Cl)-COOH]

(k) [structure: CH3CH2CH2-C(=O)-CH2CH2-COOH]

(l) [structure: benzene-CH2-C(=O)-OH]

20-2  IUPAC name first; then common name
(a) 2-iodo-3-methylpentanoic acid; α-iodo-β-methylvaleric acid
(b) (Z)-3,4-dimethylhex-3-enoic acid; no common name
(c) 2,3-dinitrobenzoic acid; no common name
(d) *trans*-cyclohexane-1,2-dicarboxylic acid (or 1R,2R); (*trans*-hexahydrophthalic acid)
(e) 2-chlorobenzene-1,4-dicarboxylic acid; 2-chloroterephthalic acid
(f) 3-methylhexanedioic acid; β-methyladipic acid


20-3  Listed in order of increasing acid strength (weakest acid first).  (See Appendix 2 in this Manual for a review of acidity.)

(a)  $CH_3CH_2COOH$  <  $CH_3-CHCOOH$  <  $CH_3-CCOOH$
                                    |                    (Br/Br)
                                    Br                   Br

The greater the **number** of electron-withdrawing substituents, the greater the stabilization of the carboxylate ion.

(b)  $CH_3CHCH_2CH_2COOH$  <  $CH_3CH_2CHCH_2COOH$  <  $CH_3CH_2CH_2CHCOOH$
         |                          |                         |
         Br                         Br                        Br

The closer the electron-withdrawing group, the greater the stabilization of the carboxylate ion.

(c) $CH_3CH_2COOH$  <  $CH_3-CHCOOH$  <  $CH_3-CHCOOH$  <  $CH_3-CHCOOH$
                           |                 |                 |
                           Cl                C≡N               NO_2

The stronger the electron-withdrawing effect of the substituent, the greater the stabilization of the carboxylate ion.

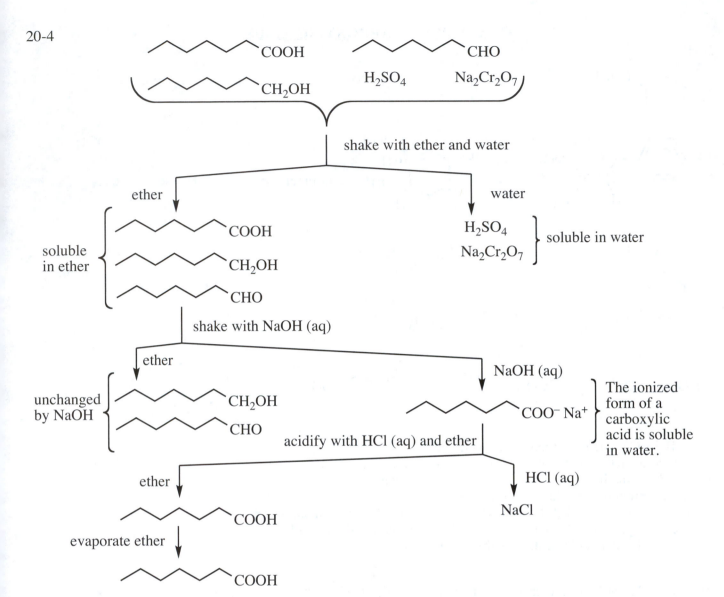

**20-5** The principle used to separate a carboxylic acid (a stronger acid) from a phenol (a weaker acid) is to neutralize with a weak base ($NaHCO_3$), a base strong enough to ionize the stronger acid but not strong enough to ionize the weaker acid.

Copyright © 2013 Pearson Education, Inc.

**20-6** The reaction mixture includes the initial reactant, reagent, desired product, and the overoxidation product—not unusual for an organic reaction mixture.

(a)

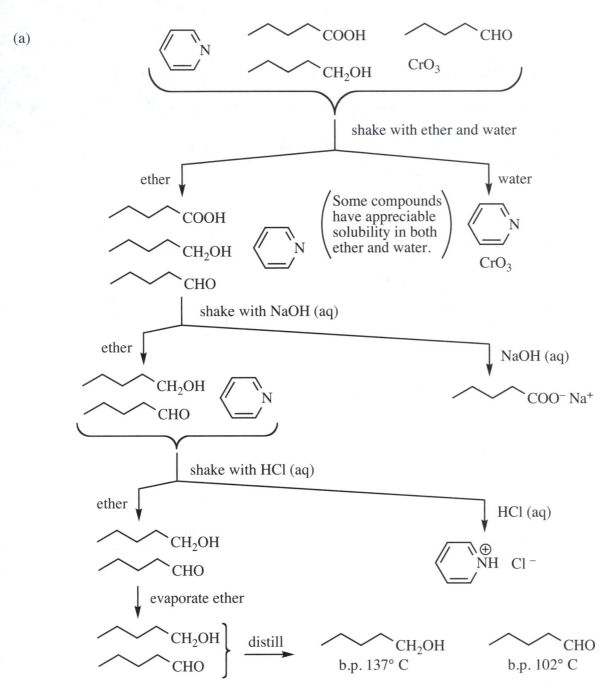

(b) Pentan-1-ol cannot be removed from pentanal by acid-base extraction. These two remaining products can be separated by distillation, the alcohol having the higher boiling point because of hydrogen bonding.


**20-7** The COOH has a characteristic IR absorption: a broad peak from 3400 to 2400 cm$^{-1}$, with a "shoulder" around 2700 cm$^{-1}$. The carbonyl stretch at 1695 cm$^{-1}$ is a little lower than the standard 1710 cm$^{-1}$, suggesting conjugation. The strong alkene absorption at 1650 cm$^{-1}$ also suggests it is conjugated.

(a) The ethyl pattern is obvious: a 3H triplet at δ 1.15 and a 2H quartet at δ 2.4. The only other peak is the COOH at δ 11.9 (a 2.1 δ unit offset added to 9.8).

$$CH_3CH_2-\overset{\displaystyle O}{\overset{\displaystyle \|}{C}}-OH$$

(b)

The multiplet between δ 2 and δ 3 is drawn as a pentet as though it were split equally by the aldehyde proton and the CH₂ group. These coupling constants are probably unequal, in which case the actual splitting pattern will be a complex multiplet.

(c) The chemical shift of the aldehyde proton is between δ 9 and 10, not as far downfield as the carboxylic acid proton. Also, the aldehyde proton is split into a triplet by the CH₂, unlike the COOH proton, which always appears as a singlet. Finally, the CH₂ is split by an extra proton, so it will give a multiplet with complex splitting, instead of the quartet shown in the acid.

20-9

20-10
(a) The linear carboxylic acids typically show a McLafferty peak at m/z 60:

(b)

plus resonance forms as shown in 20-9

continued on next page

20-10 (b) continued

<u>McLafferty rearrangement</u>

mass 42

m/z 116                                    m/z 74

(c)  The peak at m/z 60 shows up only if there are no substituents at the alpha or beta position, i.e., in linear carboxylic acids.  Any substituent group alpha or beta to the COOH will be present in the McLafferty fragment and will increase the mass of the fragment.  The m/z 60 fragment is the lowest possible mass for the McLafferty peak.

20-11

(a)

conc. KMnO$_4$

or 1) O$_3$, 2) H$_2$O

(b)

conc. KMnO$_4$

Δ, H$_3$O$^+$

(c)

Mg
ether

H$_3$O$^+$

H$_2$CrO$_4$

(d)

PBr$_3$

Mg
ether

CO$_2$

H$_3$O$^+$

(e)

conc. KMnO$_4$

Δ, H$_3$O$^+$

(f)

Mg
ether

CO$_2$

H$_3$O$^+$

OR

KCN

H$_3$O$^+$
Δ

20-12 (a)

first intermediate:

$$\left\{ \begin{array}{ccc} \overset{\oplus}{\underset{\|}{:O}}{-}H & \overset{:\ddot{O}{-}H}{\underset{|}{}} & \overset{:\ddot{O}{-}H}{\underset{|}{}} \\ RC{-}\ddot{O}H & RC{-}\ddot{O}H & RC{=}\overset{\oplus}{\ddot{O}}H \\ & \underset{\oplus}{} & \end{array} \right\}$$

second intermediate:

$$\left\{ \begin{array}{ccc} :\ddot{O}{-}H & :\ddot{O}{-}H & \overset{\oplus}{:}\overset{:}{O}{-}H \\ RC\oplus & RC & RC \\ :\ddot{O}R' & \underset{\oplus}{\ddot{O}}R' & :\ddot{O}R' \end{array} \right\}$$

(b) The mechanism of acid-catalyzed nucleophilic acyl substitution may seem daunting, but it is simply a succession of steps that are already very familiar to you.

Typically, these mechanisms have six steps: four proton transfers (two on, two off), a nucleophilic attack, and a leaving group leaving, with a little resonance stabilization thrown in that makes the whole thing work. The six steps are labeled in the mechanism below:

**Step A** proton on (resonance stabilization)
**Step B** nucleophile attacks
**Step C** proton off
**Step D** proton on
**Step E** leaving group leaves (resonance stabilization)
**Step F** proton off

20-12 (c) All steps are reversible, which is the reason the Principle of Microscopic Reversibility applies. Applying the steps as outlined on the previous page (abbreviating $OCH_2CH_3$ as OEt):

**Step A**  proton on (resonance stabilization)
**Step B**  nucleophile attacks
**Step C**  proton off
**Step D**  proton on
**Step E**  leaving group leaves (resonance stabilization)
**Step F**  proton off

H — A  is the acid catalyst

$:A^\ominus$  is the conjugate base, although in hydrolysis reactions, water usually removes $H^+$.

*The resonance forms with positive charge on the benzene ring are usually omitted. However, they still stabilize the (+) charge by delocalization, even though we may not draw them.*

same series
of resonance
forms as above

20-13 For the sake of space in this problem, resonance forms will not be drawn, but remember that they are critical! The generic acid HA is represented here by just H⁺.

**20-14**

(a)

BAD—two adjacent positive charges

(b) Protonation on the OH gives only two resonance forms, one of which is bad because of adjacent positive charges. Protonation on the C=O is good because of three resonance forms distributing the positive charge over three atoms, with no additional charge separation.

(c) The carbonyl oxygen is more "basic" because, by definition, it reacts with a proton more readily. It does so because the intermediate it produces is more stable than the intermediate from protonation of the OH.

**20-15**

(a)

+ $CH_3OH$
Use $CH_3OH$ as solvent.

$\xrightarrow{\text{H—A}}$

+ $H_2O$
Remove water with molecular sieves or by distillation.

(b)

$HC-OH$ + $CH_3OH$
Use $CH_3OH$ as solvent.

$\xrightarrow{\text{H—A}}$

$HC-OCH_3$ + $H_2O$
Remove by distillation.
b.p. 32 °C

(c)

+ $CH_3CH_2OH$
Use $CH_3CH_2OH$ as solvent.

$\xrightarrow{\text{H—A}}$

+ $H_2O$
Remove water with molecular sieves or by distillation.

**20-16** The asterisk ("*") denotes the $^{18}O$ isotope.

(a) and (b)

$$CH_3C(=O)-OH \xrightarrow{H-A} \left\{ CH_3C(=\overset{\oplus}{O}H)-\ddot{O}H \longleftrightarrow CH_3\overset{\oplus}{C}(\ddot{O}-H)-\ddot{O}H \longleftrightarrow CH_3C(\ddot{O}-H)=\overset{\oplus}{O}H \right\}$$

$$H-\overset{\ddot{O}^*}{\phantom{x}}-CH_3$$

$$CH_3C(O-H)(O-H)-\overset{\oplus}{O}^*(H)(CH_3)$$

$$CH_3C(\ddot{O}-H)(OH)(O^*CH_3) \xleftarrow{H-A} CH_3C(O-H)(OH)\overset{\oplus}{O}^*(H)CH_3$$

$$\overset{H-\overset{\oplus}{\ddot{O}}-H}{CH_3C(OH)(O^*CH_3)} \xrightarrow{-H_2O} \left\{ CH_3\overset{\oplus}{C}(\ddot{O}-H)(\ddot{O}^*CH_3) \longleftrightarrow CH_3C(\ddot{O}-H)(\overset{\oplus}{O}^*CH_3) \longleftrightarrow CH_3C(\overset{\oplus}{\ddot{O}}-H)(\ddot{O}^*CH_3) \right\} \xrightarrow{:\ddot{O}H \, CH_3} CH_3C(=O)(O^*CH_3)$$

(c) The $^{18}O$ has two more neutrons, and therefore two more mass units, than $^{16}O$. The instrument ideally suited to analyze compounds of different mass is the mass spectrometer.

**20-17**

(a)

first intermediate:

$$\left\{ H-\overset{\oplus}{C}(\ddot{O}-Et)(\ddot{O}-Et) \longleftrightarrow H-C(\ddot{O}-Et)(=\overset{\oplus}{O}-Et) \longleftrightarrow H-C(=\overset{\oplus}{\ddot{O}}-Et)(\ddot{O}-Et) \right\}$$

second intermediate:

$$\left\{ H-\overset{\oplus}{C}(\ddot{O}-H)(\ddot{O}-Et) \longleftrightarrow H-C(\ddot{O}-H)(=\overset{\oplus}{O}-Et) \longleftrightarrow H-C(=\overset{\oplus}{\ddot{O}}-H)(\ddot{O}-Et) \right\}$$

The more resonance forms that can be drawn to represent an intermediate, the more stable the intermediate. The more stable the intermediate, the more easily it can be formed, that is, under milder conditions. These intermediates are highly stabilized due to delocalization of the positive charge over the carbon and both oxygens. A trace of acid is all that is required to initiate this process.

20-17 continued

(b)

20-18

(a)

(b)

(c)

20-19 $B_2H_6$ and $BH_3 \bullet$ THF can be considered as equivalent reagents.

(a)

$$\xrightarrow{\text{LiAlH}_4} \xrightarrow{\text{H}_3\text{O}^+}$$

or $\xrightarrow{B_2H_6} \xrightarrow{H_3O^+}$

(b)

1) $LiAlH(O\text{-}t\text{-Bu})_3$

2) $H_3O^+$

(c)

$\xrightarrow{B_2H_6} \xrightarrow{H_3O^+}$

$B_2H_6$ (or $BH_3 \bullet$ THF) selectively reduces a carboxylic acid in the presence of a ketone. Alternatively, protecting the ketone as an acetal, reducing the COOH, and removing the protecting group would also be possible but longer.

478

20-20

a hydrate

− H₂O

plus resonance forms

20-21

(a)

$$CH_3CH_2-\overset{O}{\overset{\|}{C}}-OH \;+\; 2\; Li-\text{Ph} \;\xrightarrow{\;H_2O\;}\; CH_3CH_2-\overset{O}{\overset{\|}{C}}-\text{Ph}$$

$$CH_3CH_2-\overset{O}{\overset{\|}{C}}-OH \;\xrightarrow{\;SOCl_2\;}\; CH_3CH_2-\overset{O}{\overset{\|}{C}}-Cl \;+\; \text{benzene} \;\xrightarrow{\;AlCl_3\;}\; CH_3CH_2-\overset{O}{\overset{\|}{C}}-\text{Ph}$$

(b)

$$\text{cyclohexyl}-\overset{O}{\overset{\|}{C}}-OH \;+\; 2\;CH_3Li \;\longrightarrow\; \xrightarrow{\;H_2O\;}\; \text{cyclohexyl}-\overset{O}{\overset{\|}{C}}-CH_3$$

20-22

plus resonance forms
with positive charge
on the benzene ring

plus resonance forms

+ HCl(g)

$$O=C=O \quad gas$$

+ Cl⁻

gas

479

Copyright © 2013 Pearson Education, Inc.

20-23

(a)

Ph—C—Cl    H—Ö—CH₂CH₃  ⟶  Ph—C—Cl  $\xrightarrow{-\text{Cl}^-}$  Ph—C  $\xrightarrow{\text{EtÖH}}$  Ph—C—ÖCH₂CH₃

(b)

CH₃—C—Cl  +  H—N—CH₃  ⟶  CH₃—C—Cl  $\xrightarrow{-\text{Cl}^-}$  CH₃—C  $\xrightarrow{\text{H}_2\ddot{\text{N}}—\text{CH}_3}$  CH₃—C—ṄHCH₃

20-24
(a)

(b)

20-25
(a) pent-2-ynoic acid
(b) 3-amino-2-hydroxybutanoic acid
(c) 3-methylbut-2-enoic acid
(d) *trans*-2-methylcyclopentanecarboxylic acid (or 1*R*,2*R*)
(e) 2,4,6-trinitrobenzoic acid
(f) 5,5-dimethyl-4-oxohexanoic acid

20-26
(a) 3-phenylpropanoic acid; β-phenylpropionic acid
(b) potassium benzoate; no common name
(c) 2-bromo-3-methylbutanoic acid; α-bromo-β-methylbutyric acid
(d) 2-methylbutane-1,4-dioic acid; α-methylsuccinic acid
(e) sodium 3-methylbutanoate; sodium β-methylbutyrate
(f) 3-aminobutanoic acid; β-aminobutyric acid
(g) 2-bromobenzoic acid; *o*-bromobenzoic acid
(h) magnesium ethanedioate; magnesium oxalate
(i) 4-methoxybenzene-1,2-dicarboxylic acid; 4-methoxyphthalic acid

20-27

(a)

$$CH_3-\overset{\overset{\displaystyle O}{\|}}{C}-OH$$

(b)

$$HOOC-\phantom{x}-COOH$$

(c)

$$\left(H-\overset{\overset{\displaystyle O}{\|}}{C}-O^-\right)_2 Mg^{2+}$$

(d)

$$HO-\overset{\overset{\displaystyle O}{\|}}{\phantom{C}}\phantom{x}\overset{\overset{\displaystyle O}{\|}}{\phantom{C}}-OH$$

(e)

$$Cl_2CH-\overset{\overset{\displaystyle O}{\|}}{C}-OH$$

(f)

COOH / OH

(g)

$$\left(\phantom{xxxxxx}-\overset{\overset{\displaystyle \phantom{O}}{}}{C}O^-\right)_2 Zn^{2+}$$

(h)

$$\overset{\overset{\displaystyle O}{\|}}{C}-O^-\ Na^+$$

(i)

$$FCH_2-\overset{\overset{\displaystyle O}{\|}}{C}-O^-\ Na^+$$

20-28

It would be possible to extract the acidic compounds first with NaHCO₃, then with NaOH, before extracting aniline with HCl.

**20-29** Weaker base listed first. (Weaker bases come from *stronger* conjugate acids. Some $pK_a$ values of the conjugate acids are listed in italics.)

(a) $ClCH_2COO^-$ < $CH_3COO^-$ < $PhO^-$
      *2.9*         *4.7*       *10.0*

(b) $CH_3COO^-\ Na^+$ < $HC{\equiv}C^-\ Na^+$ < $NaNH_2$
       *4.7*         *25*        *35*

(c) $PhCOO^-\ Na^+$ < $PhO^-\ Na^+$ < $CH_3CH_2O^-\ Na^+$
   *4.20*       *10.0*       *15.9*

(d) $CH_3COO^-\ Na^+$ <
       *4.7*

< $CH_3CH_2O^-\ Na^+$
   *5.2*       *15.9*

**20-30**

(a) $CH_3COOH\ +\ NH_3\ \longrightarrow\ CH_3COO^-\ {}^+NH_4$

(b)

COOH / COOH + 2 NaOH $\longrightarrow$ $COO^-\ Na^+$ / $COO^-\ Na^+$ + 2 $H_2O$

(c)

$CH_3$—⟨ ⟩—COOH + $CF_3COO^-\ K^+$ $\longrightarrow$ no reaction

(d) $CH_3{-}CHCOOH$ (Br) + $CH_3CH_2COO^-\ Na^+$ $\longrightarrow$ $CH_3{-}CHCOO^-\ Na^+$ (Br) + $CH_3CH_2COOH$

(e)

⟨ ⟩—COOH + $Na^+\ {}^-O$—⟨ ⟩ $\longrightarrow$ ⟨ ⟩—$COO^-\ Na^+$ + HO—⟨ ⟩

**20-31**

$CH_3\overset{O}{\overset{\|}{C}}{-}OCH_2CH_3$ < (b.p. 143 °C structure) < $CH_3CH_2CH_2{-}\overset{O}{\overset{\|}{C}}{-}OH$

lowest b.p. (77 °C)      (b.p. 143 °C)      highest b.p. (162 °C)

The ester cannot form hydrogen bonds and will be the lowest boiling. The alcohol can form hydrogen bonds. The carboxylic acid forms two hydrogen bonds and boils as the dimer, the highest boiling among these three compounds.

**20-32** Listed in order of increasing acidity (weakest acid first):

(a) ethanol < phenol < acetic acid

(b) acetic acid < chloroacetic acid < *p*-toluenesulfonic acid

(c) benzoic acid < *m*-nitrobenzoic acid < *o*-nitrobenzoic acid

(d) butyric acid < β-bromobutyric acid < α-bromobutyric acid

EWG = electron-withdrawing group

Acidity increases with:
1. closer proximity of EWG
2. greater number of EWG
3. increasing strength (electronegativity) of EWG

(e) (Br-cyclopentyl)—COOH < (Cl-cyclopentyl)—COOH < (F-cyclopentyl)—COOH

**20-33** Acetic acid derivatives are often used as a test of electronic effects of a series of substituents: they are fairly easily synthesized (or are commercially available), and $pK_a$ values are easily measured by titration.

Substituents on carbon-2 of acetic acid can express only an inductive effect; no resonance effect is possible because the $CH_2$ is $sp^3$ hybridized and no pi overlap is possible.

Two conclusions can be drawn from the given $pK_a$ values. First, all four substituents are electron-withdrawing because all four substituted acids are stronger than acetic acid. Second, the magnitude of the electron-withdrawing effect increases in the order: OH < Cl < CN < $NO_2$. (It is always a safe assumption that nitro is the strongest electron-withdrawing group of all the common substituents.)

**20-34** See solution to problem 1-55.

Alcohols have pKa ≈ 16–18.

(a) Ascorbic acid is not a carboxylic acid. It is an example of a structure called an ene-diol where one of the OH groups is unusually acidic because of the adjacent carbonyl group. See part (c).

(b) Ascorbic acid, $pK_a$ 4.71, is almost identical to the acidity of acetic acid, $pK_a$ 4.74.

(c) The more acidic H will be the one that, when removed, gives the more stable conjugate base.

start here

THREE resonance forms, two major with (–) on oxygen

major

more acidic

two resonance forms, only one with (–) on oxygen

(d) In the slightly basic pH of physiological fluid, ascorbic acid is present as the conjugate base, ascorbate, whose structure can be represented by any of the three resonance forms on the left of part (c).

**20-35**

(a) [structure: cyclopentane with CH₂OH]

(b) [structure: cyclohexene with CH₂COOH]

(c) [structure: tetralone]

(d) 2 [structure: CH₂CH₂CH₂COOH]

(e) [structure: cyclohexene with COOH]

(f) $CH_3CH_2-\overset{\overset{\displaystyle Ph}{|}}{C}HCH_2OH$

(g) [structure: γ-butyrolactone]

(h) [structure: benzene with two COOH groups ortho]

(i) [structure: cyclohexane with COOH, CHO, CH₂CH₃]

Aqueous acid also removes the acetal.

(j) [structure: aromatic ester with CH₃]

**483**

20-36

(a)

CH₃-CH=CH-CH₂-Br $\xrightarrow[\text{ether}]{\text{Mg}}$ $\xrightarrow{\text{CO}_2}$ $\xrightarrow{\text{H}_3\text{O}^+}$ (E)-CH₃CH=CHCH₂COOH

OR $\xrightarrow{\text{KCN}}$ $\xrightarrow{\text{H}_3\text{O}^+}$

(b)

$\xrightarrow[\Delta, \text{H}_3\text{O}^+]{\text{conc. KMnO}_4}$ **2** CH₃CH₂COOH

(c)

$\xrightarrow[\text{NH}_3 \text{ (aq)}]{\text{Ag}^+}$

OR $\xrightarrow{\text{H}_2\text{CrO}_4}$

(d)

$\xrightarrow{\text{SOCl}_2}$ $\xrightarrow{\text{LiAlH(O-}t\text{-Bu)}_3}$

$\downarrow$ 1) LiAlH₄  2) H₃O⁺

$\xrightarrow{\text{PCC}}$

(e)

COOH $\xrightarrow[\text{H}-\text{A}]{\text{CH}_3\text{OH}}$ COOCH₃

OR $\xrightarrow{\text{CH}_2\text{N}_2}$

OR $\xrightarrow{\text{SOCl}_2}$ COCl $\xrightarrow{\text{CH}_3\text{OH}}$ COOCH₃

(f)

COOH $\xrightarrow{\text{LiAlH}_4}$ $\xrightarrow{\text{H}_3\text{O}^+}$ CH₂OH

OR $\xrightarrow{\text{B}_2\text{H}_6}$ $\xrightarrow{\text{H}_3\text{O}^+}$     B₂H₆ and BH₃ • THF are equivalent reagents.

(g)

CH₂COOH $\xrightarrow{\text{SOCl}_2}$ $\xrightarrow{\text{CH}_3\text{NH}_2}$

(h)

+ $\rightarrow$ $\xrightarrow[\text{Pt}]{\text{H}_2}$

Diels-Alder

20-37  Products are boxed.

(a)  All steps are reversible in an acid-catalyzed ester hydrolysis (abbreviating $OCH_2CH_3$ as OEt).

**Step A**  proton on (resonance stabilization)        **Step D**  proton on
**Step B**  nucleophile attacks                                         **Step E**  leaving group leaves (resonance stabilization)
**Step C**  proton off                                                        **Step F**  proton off

*plus resonance forms
with (+) on benzene ring*

**Step B**

**Step D**

**Step C**

**Step E**     $-$ EtOH

*same series
of resonance
forms as above*

**Step F**

(b)

*essentially
irreversible*

+   HOEt

20-37 continued

A cyclic ester is called a lactone. Lactones form when the OH nucleophile is just a few carbons away from the carbonyl electrophile, especially if a 5- or 6-membered ring can form.

(d)

Esters can be formed from RCOOH only in acid conditions, not in base.

20-38

(a)

(b) Isomers that are *R,S* and *S,S* are diastereomers.

20-39

(a) $PhCH_2CH_2OH$ $\xrightarrow[\text{pyridine}]{\text{TsCl}}$ $\xrightarrow{\text{KCN}}$ $PhCH_2CH_2CN$ $\xrightarrow[\Delta]{H_3O^+}$ $PhCH_2CH_2COOH$

$\downarrow PBr_3$

$PhCH_2CH_2Br$ $\xrightarrow[\text{ether}]{\text{Mg}}$ $\xrightarrow{CO_2}$ $\xrightarrow{H_3O^+}$ $PhCH_2CH_2COOH$

(b)

(c)

(d)

(e)

(f)

20-40

Spectrum A: $C_9H_{10}O_2 \Rightarrow$ 5 elements of unsaturation
$\delta$ 11.8, 1H $\Rightarrow$ COO**H**
$\delta$ 7.3, 5H $\Rightarrow$ monosubstituted benzene ring
$\delta$ 3.8, 1H quartet $\Rightarrow$ C**H**CH$_3$
$\delta$ 1.5, 3H doublet $\Rightarrow$ CHC**H**$_3$

Spectrum B: $C_4H_6O_2 \Rightarrow$ 2 elements of unsaturation
$\delta$ 12.1, 1H $\Rightarrow$ COO**H**
$\delta$ 6.2, 1H singlet $\Rightarrow$ **H**—C=C
$\delta$ 5.7, 1H singlet $\Rightarrow$ **H**—C=C
$\delta$ 1.9, 3H singlet $\Rightarrow$ vinyl CH$_3$ with no H neighbors C**H**$_3$—C=C

Spectrum C: $C_6H_{10}O_2 \Rightarrow$ 2 elements of unsaturation
$\delta$ 12.15, 1H $\Rightarrow$ COO**H**
$\delta$ 7.1, 1H multiplet $\Rightarrow$ **H**−C=C−COOH

$\delta$ 5.8, 1H doublet $\Rightarrow$ C=C**H**−COOH

$\delta$ 2.2-0.9 $\Rightarrow$ CH$_2$CH$_2$CH$_3$

It must be *trans* due to large coupling constant in doublet at $\delta$ 5.8.

20-41

(a)

δ-valerolactone

(b)

Copyright © 2013 Pearson Education, Inc.

**20-42**
**(a)**

$\rightleftharpoons$ $H^+$ + pK$_a$ 11.6

$\rightleftharpoons$ $H^+$ + pK$_a$ 15.7

Hydrogen peroxide is four pK$_a$ units ($10^4$ times) stronger acid than water, so the hydroperoxide anion, HOO$^-$, must be stabilized relative to hydroxide. This is from the *inductive effect* of the electronegative oxygen bonded to the O$^-$; by induction, the negative charge is distributed over both oxygens. The oxygen in hydroxide has to support the full negative charge with no delocalization.

**(b)**

The reason that carboxylic acids are so acidic (over 10 pK$_a$ units more acidic than alcohols) is because of the resonance stabilization of the carboxylate anion with two equivalent resonance forms in which all atoms have octets and the negative charge is on the more electronegative atom—the best of all possible resonance worlds. The peroxyacetate anion, however, cannot delocalize the negative charge onto the carbonyl oxygen; that negative charge is stuck out on the end oxygen like a wet nose on a frigid morning. There is some delocalization of the electron density onto the carbonyl, but with all the charge separation, this second form is a minor resonance contributor. This resonance does explain, however, why peroxyacetic acid is more acidic than hydrogen peroxide. It does not come close to acetic acid, though.

**(c)**

Carboxylic acids boil as the dimer, that is, two molecules are held tightly by hydrogen bonding. The dimer is an 8-membered ring with two hydrogen bonds as shown with dashed lines in the diagram. This works because the carbonyl oxygen has significant negative charge, and the H—O bond is weak because it is a relatively strong acid. The b.p. is 118 °C.

Do peroxyacids boil as the dimer? The author does not know, but there are three reasons to suspect that they do not. First, the b.p. is lower (105 °C) instead of higher, suggesting that they do not boil as a team but rather individually. Second, the dimer shown is a 10-membered ring—still possible but less likely than 8-membered. Third and most important, the electronic nature of the carbonyl group, as implied in the resonance forms in part (b), places less negative charge on the carbonyl oxygen, and the H is less acidic, suggesting that the hydrogen bonding is much less strong.

(a)   Mass spectrum:

—m/z 152 $\Rightarrow$ molecular ion $\Rightarrow$ molecular weight 152

—m/z 107 $\Rightarrow$ M − 45 $\Rightarrow$ loss of COOH

—m/z 77 $\Rightarrow$ monosubstituted benzene ring

IR spectrum:

—3400–2400 cm$^{-1}$, broad $\Rightarrow$ O−H stretch of COOH

—1700 cm$^{-1}$ $\Rightarrow$ C=O

—1240 cm$^{-1}$ $\Rightarrow$ C−O

—1600 cm$^{-1}$ $\Rightarrow$ aromatic C=C

NMR spectrum:

—$\delta$ 6.8–7.3, two signals in the ratio of 2H to 3H $\Rightarrow$ monosubstituted benzene ring

—$\delta$ 4.6, 2H singlet $\Rightarrow$ CH$_2$, deshielded

Carbon NMR spectrum:

—$\delta$ 170, small peak $\Rightarrow$ carbonyl

—$\delta$ 115-157, four peaks $\Rightarrow$ monosubstituted benzene ring; deshielded peak indicates oxygen substitution on the ring

(b)   Fragments indicated in the spectra:

CH$_2$            COOH

m/z 14        m/z 45 and from IR

m/z 77

This appears deceptively simple. The problem is that the mass of these fragments adds to 136, not 152—we are missing 16 mass units $\Rightarrow$ oxygen! Where can the oxygen be? There are only two possibilities:

How can we differentiate? Mass spectrometry!

This structure is consistent with the peak at $\delta$ 157 in the carbon NMR.

phenoxyacetic acid

The m/z 93 peak in the MS confirms the structure is **phenoxyacetic acid**. The CH$_2$ is so far downfield in the NMR because it is between two electron-withdrawing groups, the O and the COOH.

(c)   The COOH proton is missing from the proton NMR. Either it is beyond 10 and the NMR was not scanned (unlikely), or the peak was broadened beyond detection because of hydrogen bonding with DMSO.

20-44 Cyanide substitution is an $S_N2$ reaction and requires a 1° or 2° carbon with a leaving group. The Grignard reaction is less particular about the type of halide, but is sensitive to, and incompatible with, acidic functional groups and other reactive groups.

(a) Both methods will work.

(b) Only Grignard will work. The $S_N2$ reaction does not work on $sp^2$-hybridized carbons, and nucleophilic aromatic substitution does not work well on unactivated benzene rings.

(c) Grignard will fail because of the OH group. The cyanide reaction will work, although an excess of cyanide will need to be added because the first equivalent will deprotonate the phenol.*

(d) Grignard will fail because of the OH group. The cyanide reaction will fail because $S_N2$ does not work on unactivated $sp^2$ carbons. In this case, NEITHER method will work.

(e) Both methods will work, although cyanide substitution on secondary C will be accompanied by elimination.

(f) Grignard will fail because of the OH group. The cyanide reaction will work. Since alcohols are much less acidic than phenols, there is no problem with cyanide deprotonating the alcohol.*

* The $pK_a$ of HCN is 9.1, and the $pK_a$ of phenol is 10.0. Thus cyanide is strong enough to pull off some of the H from the phenol, although the equilibrium would favor cyanide ion and phenol. The $pK_a$ of secondary alcohols is 16–18, so there is no chance that cyanide would deprotonate an alcohol.

*The side with the weaker acid is favored.*

20-45

(a)

stock bottle          students' samples

(b) The spectrum of the students' samples shows the carboxylic acid present. Contact with oxygen from the air oxidized the sensitive aldehyde group to the acid.

(c) Storing the aldehyde in an inert atmosphere like nitrogen or argon prevents oxidation. Freshly prepared unknowns will avoid the problem.

**491**

**20-46** (A more complete discussion of acidity and electronic effects can be found in Appendix 2 in this manual.) A few words about the two types of electronic effects: induction and resonance. Inductive effects are a result of polarized σ bonds, usually because of electronegative atom substituents. Resonance effects work through π systems, requiring overlap of p orbitals to delocalize electrons.

All substituents have an inductive effect compared to hydrogen (the reference). Many groups also have a resonance effect; all that is required to have a resonance effect is that the atom or group have at least one p orbital for overlap.

The most interesting groups have both inductive and resonance effects. In such groups, how can we tell the direction of electron movement, that is, whether a group is electron-donating or electron-withdrawing? And do the resonance and inductive effects reinforce or conflict with each other? We can never "turn off" an inductive effect from a resonance effect; that is, any time a substituent is expressing its resonance effect, it is also expressing its inductive effect. We can minimize a group's inductive effect by moving it farther away; inductive effects decrease with distance. The other side of the coin is more accessible to the experimenter: we can "turn off" a resonance effect in order to isolate an inductive effect. We can do this by interrupting a conjugated π system by inserting an $sp^3$-hybridized atom, or by making resonance overlap impossible for steric reasons (steric inhibition of resonance).

These three problems are examples of separating inductive effects from resonance effects.

**(a) and (b)** In electrophilic aromatic substitution (EAS), the phenyl substituent is an ortho,para-director because it can stabilize the intermediate from electrophilic attack at the ortho and para positions. The phenyl substituent is electron-donating *by resonance*.

BUT:

is a stronger acid than

The greater acidity of phenylacetic acid shows that the phenyl substituent is electron-withdrawing, thereby stabilizing the product carboxylate's negative charge. Does this contradict what was said above? Yes and no. What is different is that, since there is no p-orbital overlap between the phenyl group and the carboxyl group because of the $CH_2$ group in between, the increased acidity must be from a pure *inductive effect*. This structure isolates the inductive effect (which can't be "turned off") from the resonance effect of the phenyl group.

We can conclude three things: (1) phenyl is electron-withdrawing by induction; (2) phenyl is (in this case) electron-donating by resonance; (3) for phenyl, the resonance effect is stronger than the inductive effect (since it is an ortho,para-director).

20-46 continued

(c) The simpler case first—induction only:

$$CH_3O-CH_2-\overset{\overset{\displaystyle O}{\|}}{C}-O-H \quad \text{is a stronger acid than} \quad H-CH_2-\overset{\overset{\displaystyle O}{\|}}{C}-O-H$$

There is no resonance overlap between the methoxy group and the carboxyl group, so this is a pure inductive effect. The methoxy substituent increases the acidity, so methoxy must be electron-withdrawing by induction. This should come as no surprise as oxygen is the second most electronegative element.

The anomaly comes in the decreased acidity of 4-methoxybenzoic acid:

$$CH_3O-\langle\text{ring}\rangle-\overset{\overset{\displaystyle O}{\|}}{C}-O-H \quad \text{is a weaker acid than} \quad H-\langle\text{ring}\rangle-\overset{\overset{\displaystyle O}{\|}}{C}-O-H$$

Through resonance, a pair of electrons from the methoxy oxygen can be donated through the benzene ring to the carboxyl group—a stabilizing effect. However, this electron donation *destabilizes* the carboxylate anion as there is already a negative charge on the carboxyl group; the resonance donation intensifies the negative charge. Since the product of the equilibrium would be destabilized relative to the starting material, the proton donation would be less favorable, which we define as a weaker acid.

Methoxy is another example of a group that is electron-withdrawing by induction but electron-donating by resonance.

(d) This problem gives three pieces of data to interpret.

(1) $$CH_3-CH_2-\overset{\overset{\displaystyle O}{\|}}{C}-O-H \quad \text{is a weaker acid than} \quad H-CH_2-\overset{\overset{\displaystyle O}{\|}}{C}-O-H$$

Interpretation: the methyl group is electron-donating by induction.

(2)

Interpretation: the methyl group is electron-donating by induction. This interpretation is consistent with (1), as expected, since methyl cannot have any resonance effect.

(3)

Interpretation: this is the anomaly. Contradictory to the data in (1) and (2), by putting on two methyl groups, the substituent seems to have become electron-withdrawing instead of electron-donating. How?

Quick! Turn the page!

20-46 continued

Steric inhibition of resonance! In benzoic acid, the phenyl ring and the carboxyl group are all in the same plane, and benzene is able to donate electrons by resonance overlap through parallel p orbitals. This stabilizes the starting acid (and destabilizes the carboxylate anion) and makes the acid weaker than it would be without resonance.

Putting substituents at the 2- and 6-positions prevents the carboxyl or carboxylate from coplanarity with the ring. Resonance is interrupted, and now the carboxyl group sees a phenyl substituent that cannot stabilize the acid through resonance; the stabilization of the acid is lost. At the same time, the *electron-withdrawing inductive effect* of the benzene ring stabilizes the carboxylate anion. These two effects work together to make this acid unusually strong. (Apparently, the slight electron-donating inductive effect of the methyls is overpowered by the stronger electron-withdrawing inductive effect of the benzene ring.)

COOH group is perpendicular to the plane of the benzene ring— no resonance interaction.

This three-dimensional view down the C-C bond between the COOH and the benzene ring shows that COOH is twisted out of the benzene plane.

20-47
(a)

(b) Compound 1 has eight carbons, and Compound 2 has six carbons. Two carbons have been lost: the two carbons of the acetal have been cleaved. (This is the best way to figure out reactions and mechanisms: find out which atoms of the reactant have become which atoms of the product, then determine what bonds have been broken and formed.)
(c) Acetals are stable to base, so the acetal must have been cleaved when acid was added.

20-47 continued

(d) The carbons have been numbered in compounds 1 and 2 on the previous page to help you visualize which atoms in the reactant become which atoms in the product. The overall process requires cleavage of the acetal to expose two alcohols. The 3° alcohol at carbon-3 can be found in the product, so it is the primary alcohol at carbon-5 that reacts with the carboxylic acid to form the lactone.

# CHAPTER 21—CARBOXYLIC ACID DERIVATIVES

**21-1** IUPAC name first; then common name

(a)  isobutyl benzoate (both IUPAC and common)
(b)  phenyl methanoate; phenyl formate
(c)  methyl 2-phenylpropanoate; methyl α-phenylpropionate
(d)  3-methyl-*N*-phenylbutanamide; β-methylbutyranilide
(e)  *N*-benzylethanamide; *N*-benzylacetamide
(f)  3-hydroxybutanenitrile; β-hydroxybutyronitrile
(g)  3-methylbutanoyl bromide; isovaleryl bromide
(h)  dichloroethanoyl chloride; dichloroacetyl chloride
(i)  methanoic 2-methylpropanoic anhydride; formic isobutyric anhydride
(j)  cyclopentyl cyclobutanecarboxylate (both IUPAC and common)
(k)  5-hydroxyhexanoic acid lactone; δ-caprolactone
(l)  *N*-cyclopentylbenzamide (both IUPAC and common)
(m)  propanedioic anhydride; malonic anhydride
(n)  1-hydroxycyclopentanecarbonitrile; cyclopentanone cyanohydrin
(o)  *cis*-4-cyanocyclohexanecarboxylic acid; no common name
(p)  3-bromobenzoyl chloride; *m*-bromobenzoyl chloride
(q)  5-(*N*-methylamino)heptanoic acid lactam; no common name
(r)  *N*-ethanoylpiperidine; *N*-acetylpiperidine

**21-2** An aldehyde has a C—H absorption (usually 2 peaks) at 2700–2800 cm$^{-1}$. A carboxylic acid has a strong, broad absorption between 2400 and 3400 cm$^{-1}$. The spectrum of methyl benzoate has no peaks in this region.

**21-3** Sometimes the C=O stretch alone is not definitive in assigning a functional group. Other peaks in the IR spectrum can give evidence that supports or excludes a functional group.

An aldehyde has a C—H absorption (usually 2 peaks) at 2700–2800 cm$^{-1}$. A carboxylic acid has a strong, broad absorption between 2400–3400 cm$^{-1}$ with a characteristic "shoulder" around 2600 cm$^{-1}$. An ester has a strong C—O peak around 1000–1200 cm$^{-1}$; this peak is often as strong as the C=O. A ketone doesn't have any of these extra peaks, so we deduce a ketone *by exclusion*, that is, by eliminating all other possible functional groups.

**21-4**

(a)  acid chloride:  single C=O peak at 1800 cm$^{-1}$; no other carbonyl comes so high;

(b)  primary amide:  broad C=O at 1650 cm$^{-1}$ and two N—H peaks between 3200–3400 cm$^{-1}$

(c)  anhydride:  two C=O absorptions at 1750 and 1820 cm$^{-1}$

**21-5  (a)** The formula $C_3H_5NO$ has two elements of unsaturation. The IR spectrum shows two peaks between 3200 and 3400 cm$^{-1}$, an $NH_2$ group. The strong peak at 1670 cm$^{-1}$ is a C=O, and the peak at 1610 cm$^{-1}$ is a C=C. This accounts for all of the atoms.
The HNMR corroborates the assignment. The 1H multiplet at δ 5.8 is the vinyl H next to the carbonyl. The 2H multiplet at δ 6.3 is the vinyl hydrogen pair on carbon-3. The 2H singlet at δ 4.8 is the amide hydrogens. The CNMR confirms the structure: two vinyl carbons and a carbonyl.

$$H_2C=CH-\overset{\overset{\displaystyle O}{\|}}{C}-NH_2$$

**(b)** The formula $C_5H_8O_2$ has two elements of unsaturation. The IR spectrum shows no OH (small peak at 3500 is an overtone of the C=O at 1730), so this compound is neither an alcohol nor a carboxylic acid. The strong peak at 1730 cm$^{-1}$ is likely an ester carbonyl. The C—O appears 1050–1250 cm$^{-1}$. The IR shows no C=C absorption, so the other element of unsaturation is likely a ring. The carbon NMR spectrum shows the carbonyl carbon at δ 171, the C—O carbon at δ 69, and three more carbons in the aliphatic region, but no carbons in the vinyl region δ 100-150, so there can be no C=C. The proton NMR shows multiplets of 2H at δ 4.3 and 2.5, most likely $CH_2$ groups next to oxygen and carbonyl respectively.

The only structure with an ester, four $CH_2$ groups, and a ring, is δ-valerolactone:

21-6 Chloride ion is shown as the base in these substitutions on acid chlorides, but any other species with an unshared electron pair could also remove a strongly acidic proton.

(a)

(b)

plus three resonance forms with positive charge delocalized on the benzene ring

*The carbonyl oxygen is more nucleophilic than the single-bonded oxygen because the product is resonance stabilized.*

plus all the resonance forms as above

(c)

*Nucleophilic attack by this oxygen does not generate a resonance-stabilized intermediate and is much less likely than that shown in part (b).*

21-6 continued

(d)

(e)

The leaving group is ethoxide ion, $CH_3CH_2O^-$, a very strong base. Ethoxide would never be a leaving group in an $S_N2$ reaction as it is too strong a base.

21-7  **Figure 21-9 is critical!** Reactions that go from a more reactive functional group to a less reactive functional group ("downhill reactions") will occur readily.
(a)  Amide to acid chloride will NOT occur—it is an "uphill" transformation.
(b)  Acid chloride to amide will occur rapidly.
(c)  Amide to ester will NOT occur—another "uphill" transformation.
(d)  Acid chloride to anhydride will occur rapidly.
(e)  Anhydride to amide will occur rapidly.

21-8

(a)  $CH_3CH_2-\overset{\displaystyle O}{\overset{\|}{C}}-Cl$ + $HOCH_2CH_3$ $\longrightarrow$ $CH_3CH_2-\overset{\displaystyle O}{\overset{\|}{C}}-OCH_2CH_3$ + HCl

(b)

(c)

(d)

## 21-8 continued

(e)

$$CH_3-\overset{O}{\underset{}{C}}-Cl \;+\; HO-\overset{CH_3}{\underset{CH_3}{C}}-CH_3 \longrightarrow \overset{O}{\underset{}{C}}CH_3 \;\; \overset{CH_3}{\underset{CH_3}{O-C-CH_3}} \;+\; HCl$$

This reaction would have to be kept cold to avoid elimination. Esters of *tert*-butyl alcohol are hard to make because 3° alcohols eliminate so easily.

(f)

$$Cl-\overset{O}{\underset{}{C}}-CH_2CH_2-\overset{O}{\underset{}{C}}-Cl \;+\; 2 \;\; \diagup\!\!\diagdown\!OH \longrightarrow \diagup\!\!O-\overset{O}{\underset{}{C}}-CH_2CH_2-\overset{O}{\underset{}{C}}-O\diagdown \;+\; 2 \; HCl$$

## 21-9

(a)

$$H_3C-\overset{O}{\underset{}{C}}-Cl \;+\; HN(CH_3)_2 \longrightarrow H_3C-\overset{O}{\underset{}{C}}-N(CH_3)_2 \;+\; HCl$$

(b)

$$H_3C-\overset{O}{\underset{}{C}}-Cl \;+\; H_2N-\!\!\bigcirc \longrightarrow H_3C-\overset{O}{\underset{}{C}}-NH-\!\!\bigcirc \;+\; HCl$$

(c)

$$\bigcirc\!\!-\overset{O}{\underset{}{C}}-Cl \;+\; NH_3 \longrightarrow \bigcirc\!\!-\overset{O}{\underset{}{C}}-NH_2 \;+\; HCl$$

(d)

$$\bigcirc\!\!-\overset{O}{\underset{}{C}}-Cl \;+\; HN\!\!\bigcirc \longrightarrow \bigcirc\!\!-\overset{O}{\underset{}{C}}-N\!\!\bigcirc \;+\; HCl$$

## 21-10

(a) (i)

$$H_3C-\overset{O}{\underset{}{C}}-O-\overset{O}{\underset{}{C}}-CH_3 \;+\; HOCH_2-\!\!\bigcirc \longrightarrow H_3C-\overset{O}{\underset{}{C}}-OCH_2-\!\!\bigcirc \;+\; HO-\overset{O}{\underset{}{C}}-CH_3$$

(ii)

$$H_3C-\overset{O}{\underset{}{C}}-O-\overset{O}{\underset{}{C}}-CH_3 \;+\; HN\!\!\diagdown \longrightarrow H_3C-\overset{O}{\underset{}{C}}-N\!\!\diagdown \;+\; HO-\overset{O}{\underset{}{C}}-CH_3$$

(b) (i)

$$H_3C-\overset{:\!O\!:}{\underset{}{C}}-O-\overset{O}{\underset{}{C}}CH_3 \;+\; H\overset{..}{O}CH_2Ph \longrightarrow H_3C-\overset{:\overset{\ominus}{O}:}{\underset{\underset{}{H-\overset{\oplus}{O}-CH_2Ph}}{C}}-\overset{..}{O}-\overset{O}{\underset{}{C}}CH_3$$

$$\downarrow$$

$$H_3C-\overset{:O:}{\underset{}{C}}-\overset{..}{O}-CH_2Ph \;+\; H-\overset{..}{O}-\overset{O}{\underset{}{C}}CH_3 \longleftarrow H_3C-\overset{:O:}{\underset{\underset{H-\overset{\oplus}{O}-CH_2Ph}{}}{C}} \;+\; \overset{\ominus}{\underset{}{:}}\overset{..}{O}-\overset{O}{\underset{}{C}}CH_3 \quad \text{plus resonance form}$$

continued on next page

**499**

Copyright © 2013 Pearson Education, Inc.

21-10 (b) continued

(ii)

21-11

21-12

OCH₂CH₃   + ⁻:O—CCH₃ ⇌   OCH₂CH₃   + HO—CCH₃

HOCH₂CH₃ + ⁻:O—CCH₃   ←   CH₃C—O—H  ⁻:OCH₂CH₃ +   ←

21-13

Ph—C—OEt  H—A → { Ph—C—OEt ↔ Ph—C—OEt ↔ Ph—C=OEt }   plus resonance forms with positive charge on benzene ring

CH₃OH

OH H
Ph—C—O⁺—Et ← A—H   Ph—C—OEt ←   Ph—C—OEt
O—CH₃        O—CH₃   CH₃OH   H—O⁺—CH₃

– EtOH

{ Ph—C⁺  ↔  Ph—C  ↔  Ph—C }  CH₃OH →  Ph—C—OCH₃
  O—CH₃      O⁺—CH₃     O—CH₃

plus resonance forms with positive charge on benzene ring

21-15 (a)

Any of the oxygens present in this reaction mixture can serve as base to remove a proton. The base does not have to be $HSO_4^-$.

two rapid proton transfers

Any of the oxygens present in this reaction mixture can serve as base to remove a proton. The base does not have to be $HSO_4^-$.

aspirin

(b) As a true catalyst that speeds the rate of reaction without being consumed, the drop of $H_2SO_4$ is regenerated in the last step; it is used in the mechanism but not consumed. Its function in the mechanism is to make the carbonyl carbon of the anhydride more positive and therefore more susceptible to nucleophilic attack by the OH of the phenol, which is a weak nucleophile. Without the $H_2SO_4$ catalyst, this reaction would still occur but it would be much slower, probably by several factors of 10.

21-16    The asterisk (*) will denote $^{18}O$.

(a)

$$H_3C-\overset{\overset{\displaystyle :O:}{\|}}{C}-O^*R \xrightarrow{\quad \overset{\ominus}{\underset{:OH}{}} \quad} H_3C-\overset{\overset{\displaystyle :\overset{..}{\overset{..}{O}}:}{|}}{\underset{\underset{:O-H}{|}}{C}}-\overset{..}{\overset{..}{O}}^*R \longrightarrow H_3C-\overset{\overset{\displaystyle :O:}{\|}}{C}-\overset{..}{\overset{..}{O}}: \; + \; \overset{\ominus}{:}\overset{..}{\overset{..}{O}}^*R \longrightarrow H_3C-\overset{\overset{\displaystyle O}{\|}}{C}-\overset{\ominus}{O}$$

$$O^*R = \quad ^{18}O-\overset{\overset{\displaystyle CH_2CH_3}{|}}{\underset{\underset{CH_3}{R}}{C}}\!\!\text{""}H$$

+

$$H^{18}O-\overset{\overset{\displaystyle CH_2CH_3}{|}}{\underset{\underset{CH_3}{R}}{C}}\!\!\text{""}H$$

The alcohol product contains the $^{18}O$ label, with none in the carboxylate.  The bond between $^{18}O$ and the tetrahedral carbon with ($R$) configuration did not break, so the configuration is retained.

(b)   The products are identical regardless of mechanism.

$$H_3C-\overset{\overset{\displaystyle :O:}{\|}}{C}-O^*R \xrightarrow{\;H-A\;} \left\{ CH_3\overset{\overset{\displaystyle \overset{\oplus}{:O}-H}{\|}}{C}-\overset{..}{\overset{..}{O}}^*R \longleftrightarrow CH_3\overset{\overset{\displaystyle :\overset{..}{O}-H}{|\oplus}}{C}-\overset{..}{\overset{..}{O}}^*R \longleftrightarrow CH_3\overset{\overset{\displaystyle :\overset{..}{O}-H}{|}}{C}=\underset{\oplus}{\overset{..}{\overset{..}{O}}^*R} \right\}$$

$$H_2\overset{..}{O}: \downarrow$$

$$CH_3\overset{\overset{\displaystyle OH\ H}{|\quad |\oplus}}{\underset{\underset{O-H}{|}}{C}}-\overset{..}{\overset{..}{O}}^*R \xleftarrow{\;H-A\;} CH_3\overset{\overset{\displaystyle OH}{|}}{\underset{\underset{O-H}{|}}{C}}-\overset{..}{\overset{..}{O}}^*R \xleftarrow[\;H_2\overset{..}{O}:\quad H-\overset{\oplus}{\overset{|}{O}}-H\;]{} CH_3\overset{\overset{\displaystyle OH}{|}}{C}-O^*R$$

$$\Big\downarrow -\; HO^*R$$

$$\left\{ H_3C-\overset{\overset{\displaystyle :\overset{..}{O}-H}{|}}{C}\oplus \;\; \underset{:\overset{..}{O}-H}{} \longleftrightarrow H_3C-\overset{\overset{\displaystyle :\overset{..}{O}-H}{|}}{\underset{\underset{\oplus\overset{..}{O}-H}{\|}}{C}} \longleftrightarrow H_3C-\overset{\overset{\displaystyle \oplus\overset{..}{O}-H}{\|}}{\underset{\underset{:\overset{..}{O}-H}{|}}{C}} \right\} \xrightarrow{\;H_2\overset{..}{O}:\;} H_3C-\overset{\overset{\displaystyle :O:}{\|}}{C}-OH$$

acetic acid

(c)  The $^{18}O$ has 2 more neutrons in its nucleus than $^{16}O$.  Mass spectra of these products would show the molecular ion of acetic acid at its standard value of m/z 60, whereas the molecular ion of butan-2-ol would appear at m/z 76 instead of m/z 74, proving that the heavy isotope of oxygen went with the alcohol.  This demonstrates that the bond between oxygen and the carbonyl carbon is broken, not the bond between the oxygen and the alkyl carbon.

   To show if the alcohol was chiral or racemic, measuring its optical activity in a polarimeter and comparing to known values would prove its configuration.  (The heavy oxygen isotope has a negligible effect on optical rotation.)

21-17

(a) A catalyst is defined as a chemical species that speeds a reaction but is not consumed in the reaction. In the acidic hydrolysis, acid is used in the first and fourth steps of the mechanism but is regenerated in the third and last steps. Acid is not consumed; the final concentration of acid is the same as the initial concentration. In the basic hydrolysis, however, the hydroxide that initially attacks the carbonyl is never regenerated. An alkoxide leaves from the carbonyl, but it quickly neutralizes the carboxylic acid. For every molecule of ester, one molecule of hydroxide is consumed; the base *promotes* the reaction but does not *catalyze* the reaction.

(b) Basic hydrolysis is not reversible. Once an ester molecule is hydrolyzed in base, the carboxylate cannot form an ester. Acid catalysis, however, is an equilibrium: the mixture will always contain some ester, and the yield will never be as high as in basic hydrolysis. Second, long chain fatty acids are not soluble in water until they are ionized; they are soluble only as their sodium salts (soap). Basic hydrolysis is preferred for higher yield and greater solubility of the product.

21-18

21-19

21-20

(a)

21-20 continued

(b)

21-21  In the basic hydrolysis (21-20(a)), the step that drives the reaction to completion is the final step, the deprotonation of the carboxylic acid by the amide anion.  In the acidic hydrolysis (21-20(b)), protonation of the amine by acid is exothermic and it prevents the reverse reaction by tying up the pair of electrons on the nitrogen so that the amine is no longer nucleophilic.

21-22

*amide is halfway point*

21-23

Note: species with positive charge on carbon adjacent to benzene also have resonance forms (not shown) with the positive charge distributed over the ring.

21-24

(a) Reduction occurs when a new C—H bond is formed. In ester reduction, a new C—H bond is formed in the first step and in the third step. This can also be seen in this mechanism where the steps are similarly labeled.

(b)

Step 1—reduction

Step 2—group leaves

$-Cl^{\ominus}$

Step 3—reduction

*aldehyde intermediate*

Step 4—workup; alkoxide protonation

21-25

(a)

(b) NHCH$_2$CH$_3$

(c)

(d)

(e)

(f)

CH$_2$NH$_2$

H$_3$C    N    CH$_2$CH$_2$CH$_3$

NH$_2$

21-26

Note: species with positive charge on carbon adjacent to benzene also have resonance forms (not shown) with the positive charge distributed over the ring.

21-27

Grignard reagents plus acid chlorides or esters produce tertiary alcohols where two of the R groups are the same, having both come from the Grignard reagent.

**21-28** The new carbon-carbon bonds are shown in bold. ▬

(a)

These alcohols can also be synthesized from ketones:

OR

(b)

formate ester

Adding Grignard to a formate ester is a very practical way to make 2° alcohols where both R groups are the same.

(c)

CH₃C≡N +  ⟶MgBr ⟶ H₃O⁺ ⟶

$CH_3C{\equiv}N$

OR

$N{\equiv}C$

+ CH₃MgBr ⟶ H₃O⁺ ⟶

**21-29**

good—all atoms have octets

$CH_3O$— ⬡ —$CCH_2CH_3$  + HCl + AlCl₃

**509**

21-30

(a)

benzene + CH$_3$C(=O)–Cl $\xrightarrow{\text{AlCl}_3}$ phenyl–C(=O)–CH$_3$

(b)

benzene + phenyl–C(=O)–Cl $\xrightarrow{\text{AlCl}_3}$ phenyl–C(=O)–phenyl

(c)

benzene + CH$_3$CH$_2$CH$_2$C(=O)–Cl $\xrightarrow{\text{AlCl}_3}$ phenyl–C(=O)–CH$_2$CH$_2$CH$_3$ $\xrightarrow[\text{HCl}]{\text{Zn(Hg)}}$ phenyl–CH$_2$CH$_2$CH$_2$CH$_3$

21-31

(a) (i)

HC(=O)–O–C(=O)CH$_3$ + H$_2$N–C$_6$H$_5$ $\longrightarrow$ HC(=O)–NH–C$_6$H$_5$ + HO–C(=O)CH$_3$

(ii)

HC(=O)–O–C(=O)CH$_3$ + HOCH$_2$–C$_6$H$_5$ $\longrightarrow$ HC(=O)–OCH$_2$–C$_6$H$_5$ + HO–C(=O)CH$_3$

(b) (i)

H–C(=O)–O–CCH$_3$ + H$_2$N Ph $\longrightarrow$

H–C(–Ö$^{\ominus}$)(–O–CCH$_3$)(H–NHPh$^{\oplus}$) $\longrightarrow$

H–C(=Ö) + H–NHPh$^{\oplus}$ + $^{\ominus}$:Ö–CCH$_3$ $\longrightarrow$

H–C(=O)–NHPh + H–Ö–CCH$_3$

(ii)

H–C(=O)–O–CCH$_3$ + HÖCH$_2$Ph $\longrightarrow$

H–C(–Ö$^{\ominus}$)(–O–CCH$_3$)(H–Ö$^{\oplus}$–CH$_2$Ph) $\longrightarrow$

H–C(=Ö) + H–Ö$^{\oplus}$–CH$_2$Ph + $^{\ominus}$:Ö–CCH$_3$ $\longrightarrow$

H–C(=O)–Ö–CH$_2$Ph + H–Ö–CCH$_3$

21-32

(a)

HC–Cl does not exist; it is unstable relative to its decomposition products, CO and HCl. Acetic formic anhydride is the most practical way to formylate the alcohol.

*Do not use formyl chloride on exams—it will be marked incorrect!*

(b)

Acetic anhydride is more convenient, easier to handle, less toxic and less expensive than acetyl chloride.

(c)

The acid chloride would tend to react at both carbonyls instead of just one; only the anhydride will give this product. Recall that a carboxylic acid and $NH_3$ can make an amide only at very high temperature, so once the COOH is formed, the reaction will stop there.

(d)

The acid chloride would tend to react at both carbonyls instead of just one; only the anhydride will give this product. As in part (c), once the COOH is formed, it will not go onto ester with methoxide under basic conditions.

21-33

21-34

(a)

Generally, acetic anhydride is the optimum reagent for the preparation of acetate esters. Acetyl chloride would also react with the carboxylic acid to form a mixed anhydride.

(b)

Fischer esterification works well to prepare simple carboxylic esters. The diazomethane method would also react with the phenol, making the phenyl ether.

(c)

The more powerful reducing agent LiAlH$_4$ is required to reduce esters to primary alcohols.

(d)

Amides can be made directly from esters plus an amine. Aniline is a poor nucleophile, however, so heating this reaction will be required to complete the reaction in a reasonable time.

21-35  Syntheses may have more than one correct approach.

(a)

$$Ph-\overset{\overset{\displaystyle O}{\|}}{C}-OCH_3 + 2\ PhMgBr \xrightarrow{\text{ether}} \xrightarrow{H_3O^+} Ph-\underset{\underset{\displaystyle Ph}{|}}{\overset{\overset{\displaystyle OH}{|}}{C}}-Ph$$

(b)

$$H-\overset{\overset{\displaystyle O}{\|}}{C}-OCH_2CH_3 + 2\ PhCH_2MgBr \xrightarrow{\text{ether}} \xrightarrow{H_3O^+} H-\underset{\underset{\displaystyle CH_2Ph}{|}}{\overset{\overset{\displaystyle OH}{|}}{C}}-CH_2Ph$$

(c)

$$Ph-\overset{\overset{\displaystyle O}{\|}}{C}-OCH_3 + H_2NCH_2CH_3 \xrightarrow{\Delta} Ph-\overset{\overset{\displaystyle O}{\|}}{C}-NHCH_2CH_3$$

(d)

$$H-\overset{\overset{\displaystyle O}{\|}}{C}-OCH_2CH_3 + 2\ PhMgBr \xrightarrow{\text{ether}} \xrightarrow{H_3O^+} H-\underset{\underset{\displaystyle Ph}{|}}{\overset{\overset{\displaystyle OH}{|}}{C}}-Ph$$

**513**

21-35 continued

(e)

$$Ph-\overset{\overset{\displaystyle O}{\|}}{C}-OCH_3 \ + \ LiAlH_4 \ \xrightarrow{\text{ether}} \ \xrightarrow{H_3O^+} \ PhCH_2OH$$

(f)

$$Ph-\overset{\overset{\displaystyle O}{\|}}{C}-OCH_3 \ \xrightarrow{H_3O^+} \ Ph-\overset{\overset{\displaystyle O}{\|}}{C}-OH \ + \ CH_3OH$$

(g)  PhCH$_2$OH  $\xrightarrow[\text{2) KCN}]{\text{1) TsCl, pyridine}}$  PhCH$_2$C≡N  $\xrightarrow[\Delta]{H_3O^+}$  PhCH$_2$$-\overset{\overset{\displaystyle O}{\|}}{C}-$OH  $\xrightarrow[\underset{\displaystyle CH_3}{\overset{\displaystyle CH_3}{HO-}}]{H_2SO_4}$  PhCH$_2$$-\overset{\overset{\displaystyle O}{\|}}{C}-$O

from (e)

(h)

$$PhCH_2-\overset{\overset{\displaystyle O}{\|}}{C}-OCH(CH_3)_2 \ + \ \textbf{2} \ CH_3CH_2MgBr \ \xrightarrow{\text{ether}} \ \xrightarrow{H_3O^+} \ PhCH_2-\overset{\overset{\displaystyle OH}{|}}{\underset{\displaystyle CH_2CH_3}{C}}-CH_2CH_3$$

from (g)

(i) How to make an 8-carbon diol
from an ester that has no more
than 8 carbons?  Make the ester a
lactone!

+ LiAlH$_4$  $\xrightarrow{\text{ether}}$  $\xrightarrow{H_3O^+}$

21-36

(a)

$\xrightarrow{\text{LiAlH}_4}$ $\xrightarrow{\text{H}_2\text{O}}$

(b)

+ H$_2$N  $\xrightarrow{\Delta}$  + HO   Heating the reaction will form the amide faster.

(c)

+ Cl  $\longrightarrow$  + HCl   Acetic anhydride would work as well as acetyl chloride; a base such as pyridine or sodium bicarbonate or triethylamine is usually added to consume the HCl.

(d)

$\xrightarrow[-H_2O]{300\ ^\circ C}$   $\xrightarrow{\text{LiAlH}_4}$ $\xrightarrow{\text{H}_2\text{O}}$

gamma-amino-
butyric acid

This high-temperature synthesis of amides is used industrially but not in normal lab
reactions.  Alternatively, an ester could be produced from the COOH, followed by
milder cyclization to the amide.

**21-37** There may be other correct approaches to these problems.

(a)

(b)

See the comment in the solution to 21-36(c) about using acetic anhydride instead of acetyl chloride.

(c)

**21-38**

(a)

$$PhCH_2-\overset{O}{\overset{\|}{C}}-OH \xrightarrow{SOCl_2} PhCH_2-\overset{O}{\overset{\|}{C}}-Cl \xrightarrow{NH_3} PhCH_2-\overset{O}{\overset{\|}{C}}-NH_2 \xrightarrow{POCl_3} PhCH_2-C\equiv N$$

(b)

$$PhCH_2-\overset{O}{\overset{\|}{C}}-OH \xrightarrow[\text{2) } H_2O]{\text{1) LiAlH}_4} PhCH_2CH_2OH \xrightarrow[\text{pyridine}]{TsCl} PhCH_2CH_2OTs \xrightarrow{NaCN} PhCH_2CH_2CN$$

(c)

**21-39**

(a)

(b)

(c)

PBr$_3$
or make the tosylate

$\xrightarrow{\text{PBr}_3}$ → Br

NaCN

$\xleftarrow[\text{2) H}_3\text{O}^+]{\text{1) CH}_3\text{MgI}}$

CN

21-40

$CH_3-N=C=\overset{..}{\underset{..}{O}}$ → $\left\{ CH_3-\overset{..}{N}=\overset{\ominus}{C}-\overset{..}{\underset{..}{O}}\colon \quad \longleftrightarrow \quad CH_3-\overset{\ominus}{\underset{..}{N}}-C=\overset{..}{O} \right\}$

Ar$-\overset{..}{O}H$

two rapid
proton transfers

Ar =

Ar$-\overset{\oplus}{\underset{..}{O}}H$ | Ar$-\overset{..}{O}H$

$$\underset{\substack{H_3C \\ \phantom{x}}}{\overset{O}{\underset{H}{N}}}\overset{O}{\underset{}{C}}O$$

Sevin®
(carbaryl)

21-41

(a)

$\xrightarrow{\text{H}_3\text{O}^+}$

OH
OH

$+ \ CO_2$

(i) carbonate ester
(iii) not aromatic

(b)

$\xrightarrow{\text{H}_3\text{O}^+}$

HS— (HO, O)

(i) thiolactone
(iii) not aromatic

(c)

$\xrightarrow{\text{H}_3\text{O}^+}$

SH   SH   $+ \ CO_2$

(i) thiocarbonate ester
(iii) not aromatic

(d)

$\left\{ \underset{HN \quad NH}{\overset{O}{\bigcirc}} \quad \longleftrightarrow \quad \underset{HN \quad \overset{\oplus}{N}H}{\overset{\overset{\ominus}{O}}{\bigcirc}} \right\}$ $\xrightarrow{\text{H}_3\text{O}^+}$ $CO_2$ + $H_2N \quad NH_2$ → $H_2N \quad NH$

enediamine          imine

(i) a substituted urea
(iii) AROMATIC—more easily seen in the resonance form shown

(The enediamine product would not be stable in aqueous acid. It would probably tautomerize to an imine, hydrolyze to ammonia and 2-aminoethanal, then polymerize.)

21-41 continued

(e)  At first glance, this AROMATIC compound does not appear to be an acid derivative.  Like any enol, however, its tautomer must be considered.

lactam                           enamine                                       imine                        + NH₃

(f)

(i)  a carbamate or urethane
(iii) AROMATIC—more easily seen in the resonance form

21-42
(a)                              (b)                              (c)

(d)              (e)              (f)              (g)

(h)  OH                    (i)                    (j)                    (k)
     α
     γ    β   C≡N
     "Valero" has 5 carbons.                          Oxygen is on the beta carbon
                                                      of the 4-carbon chain.

(l)              (m)              (n)              (o)
                 α
                 β
                 γ   δ
                 "Capro" has 6 carbons.

"Methanoic" is IUPAC, but most
chemists use "formic".

21-43

(a)  3-methylpentanoyl chloride
(c)  acetanilide; N-phenylethanamide
(e)  phenyl acetate; phenyl ethanoate
(g)  benzonitrile
(i)  dimethyl isophthalate, or dimethyl benzene-1,3-
dicarboxylate
(k)  4-hydroxypentanoic acid lactone; γ-valerolactone

(b)  benzoic formic anhydride
(d)  N-methylbenzamide
(f)  methyl benzoate
(h)  4-phenylbutane nitrile; γ-phenylbutyronitrile
(j)  N,N-diethyl-3-methylbenzamide
(l)  3-aminopentanoic acid lactam; β-valerolactam

## 21-44

(a) $$\underset{\text{PhC-OCH}_2\text{CH}_3}{\overset{\text{O}}{\|}}$$

(b) $$\underset{\text{PhC-O-CCH}_3}{\overset{\text{O}\quad\text{O}}{\|\quad\|}}$$

(c) $$\underset{\underset{\text{H}}{|}}{\overset{\text{O}}{\overset{\|}{\text{PhC-N}}}}\!\!-\text{Ph}$$

(d) 4-methoxyphenyl phenyl ketone ($\text{H}_3\text{CO}$-benzene-$\overset{\text{O}}{\overset{\|}{\text{C}}}$-Ph)

(e) $$\underset{\underset{\text{Ph}}{|}}{\overset{\text{OH}}{\overset{|}{\text{Ph}-\text{C}-\text{Ph}}}}$$

(f) $$\underset{\text{PhC-H}}{\overset{\text{O}}{\|}}$$

## 21-45

(a) Ph-O-$\overset{\text{O}}{\overset{\|}{\text{C}}}$-CH$_3$

(b) Ph-O-$\overset{\text{O}}{\overset{\|}{\text{C}}}$-H

(c) phthalimide-type structure with NH–Ph and OH, C=O groups
**Anhydrides react only once.**

(d) $\text{H}_3\text{CO}$-phenyl-C(=O)-CH$_2$CH$_2$-$\overset{\text{O}}{\overset{\|}{\text{C}}}$-OH

(e) Ph-CH(OH)-CH$_2$-N(H)-$\overset{\text{O}}{\overset{\|}{\text{C}}}$-CH$_3$
**Amines are more nucleophilic than alcohols and amides are more stable than esters.**

(f) Ph-CH(O-C(CH$_3$)=O)-CH$_2$-N(H)-$\overset{\text{O}}{\overset{\|}{\text{C}}}$-CH$_3$

## 21-46

When a carboxylic acid is treated with a basic reagent, the base removes the acidic proton rather than attacking at the carbonyl (proton transfers are much faster than formation or cleavage of other types of bonds). Once the carboxylate anion is formed, the carbonyl is no longer susceptible to nucleophilic attack: nucleophiles do not attack sites of negative charge. By contrast, in acidic conditions, the protonated carbonyl has a positive charge and is activated to nucleophilic attack.

<u>basic conditions</u>

$$\underset{}{\overset{\text{O}}{\overset{\|}{\text{R-C-OH}}}} + {}^{\ominus}\text{OR'} \longrightarrow \underset{}{\overset{\text{O}}{\overset{\|}{\text{R-C-O}}}}{}^{\ominus} + \text{HOR'}$$

anion — not susceptible to nucleophilic attack

<u>acidic conditions</u>

$$\underset{}{\overset{\text{O}}{\overset{\|}{\text{R-C-OH}}}} + \text{H}^+ \longrightarrow \underset{\oplus}{\overset{\text{OH}}{\overset{|}{\text{R-C-OH}}}}$$

rapidly attacked by R'OH nucleophile

## 21-47

Products after adding dilute acid in the workup:

(a) $$\underset{\text{HC-OH}}{\overset{\text{O}}{\|}} + \text{HO-Ph}$$

(b) propanoic acid ($\overset{\text{O}}{\overset{\|}{\text{C}}}$-OH) + HOCH$_2$CH$_3$

(c) benzene ring with CH$_2$CH$_2$COOH and OH substituents

(d) ethylene glycol (OH, OH) + oxalic acid (HO-C(=O)-C(=O)-OH)

## 21-48

(a) benzene-NH$_2$ + $\underset{\text{H}}{\overset{\text{O}}{\overset{\|}{\text{C}}}}$-O-$\overset{\text{O}}{\overset{\|}{\text{C}}}$-CH$_3$ $\longrightarrow$ benzene-NH-$\overset{\text{O}}{\overset{\|}{\text{C}}}$-H + HO-$\overset{\text{O}}{\overset{\|}{\text{C}}}$-CH$_3$

See the solution to 21-32(a), and text Sec. 21-11, for use of acetic formic anhydride.

21-48 continued

(b)

(c)

(d)

(e)

(f)

(g)

Any ester where ethylene glycol displaces methanol (by transesterification) will be reduced with LiAlH$_4$.

Aqueous acid workup removes ketal protecting group.

(h)

**OR**

**519**

21-49

(a)

If propan-2-ol is the solvent, it will also serve as the base instead of chloride ion.

(b)

$HOCH_3$

(c)

Note: species with positive charge on carbon adjacent to benzene also have resonance forms (not shown) with the positive charge distributed over the ring.

(d)

21-49 continued

(e)

(f)

(g)

No bond to the chiral center is broken, so the configuration is retained.

R configuration

still R

21-50

(a) Ph—C(=O)—O—cyclohexane

(b) cyclohexane—C(=O)—NHCH₃

(c) Ph—C(=O)—N(pyrrolidine)

(d)

(e) PhCH₂OH

(f) piperidine N—H

(g)

(h)

(i)

(j)

(k) PhCH₂CHCH₃CH₂NH₂ ... CH₃

(l) transesterification

**21-51**

(a)

(b)

(c)

**21-52**

(a)

glycerol  +  **3** $CH_3(CH_2)_{12}COOH$  $\Longrightarrow$  trimyristin

(b)

$\xrightarrow[\text{2) H}_2\text{O}]{\text{1) LiAlH}_4}$

glycerol  +  **3** $CH_3(CH_2)_{12}CH_2OH$

tetradecan-1-ol

**21-53**

(a)

$\xrightarrow[\text{AlCl}_3/\text{CuCl} \atop \Delta]{\text{CO, HCl}}$  Gatterman-Koch  + para  $\xrightarrow{\text{H}_2\text{CrO}_4}$  $\xrightarrow{\text{Ac}_2\text{O}}$

(b)

$\xrightarrow{\text{HNO}_3 \atop \text{H}_2\text{SO}_4}$  + ortho  $\xrightarrow[\text{HCl}]{\text{Fe}}$  $\xrightarrow{\text{1 equivalent} \atop \text{Ac}_2\text{O}}$

$NH_2$ is more nucleophilic than OH, and an amide is more stable than an ester.

**21-54**

(a)

$\xrightarrow{\text{Ac}_2\text{O}}$

(b)

$\xrightarrow{\text{SOCl}_2}$  $\xrightarrow{\text{NH}_3}$  $\xrightarrow{\text{POCl}_3}$

21-54 continued

(c)

(d)

(e)

(f)

two equivalents

(g)

(h)

21-55 Diethyl carbonate has *two* leaving groups on the carbonyl. It can undergo *two* nucleophilic acyl substitutions, followed by one nucleophilic addition.

(a)

$$\text{EtO}-\overset{\overset{\displaystyle O}{\|}}{\text{C}}-\text{OEt} \xrightarrow[\text{subst. 1}]{\text{PhMgBr}} \left[\text{Ph}-\overset{\overset{\displaystyle O}{\|}}{\text{C}}-\text{OEt}\right] \xrightarrow[\text{subst. 2}]{\text{PhMgBr}} \left[\text{Ph}-\overset{\overset{\displaystyle O}{\|}}{\text{C}}-\text{Ph}\right] \xrightarrow[\text{addition}]{\text{PhMgBr}} \xrightarrow{\text{H}_3\text{O}^+} \text{Ph}-\underset{\underset{\displaystyle \text{Ph}}{|}}{\overset{\overset{\displaystyle \text{OH}}{|}}{\text{C}}}-\text{Ph}$$

(b) $\text{CH}_3\text{CH}_2\text{Br} \xrightarrow[\text{ether}]{\text{Mg}} \text{CH}_3\text{CH}_2\text{MgBr}$

$$3\,\text{CH}_3\text{CH}_2\text{MgBr} + \text{EtO}-\overset{\overset{\displaystyle O}{\|}}{\text{C}}-\text{OEt} \longrightarrow \xrightarrow{\text{H}_3\text{O}^+} \text{CH}_3\text{CH}_2-\underset{\underset{\displaystyle \text{CH}_2\text{CH}_3}{|}}{\overset{\overset{\displaystyle \text{OH}}{|}}{\text{C}}}-\text{CH}_2\text{CH}_3$$

21-55 continued

(c) Lexan® is a polycarbonate that makes bisphenol A (BPA) into its carbonate ester. Phosgene will react faster than diethyl carbonate because acid chlorides react faster than esters, but the chemistry is the same. An acid catalyst would speed this transesterification.

HO—⟨C₆H₄⟩—C(CH₃)₂—⟨C₆H₄⟩—OH  +  EtO—C(=O)—OEt  →[HA]  —[O—⟨C₆H₄⟩—C(CH₃)₂—⟨C₆H₄⟩—O—C(=O)]—ₙ

BPA                                                          Lexan® polycarbonate

Sevin® insecticide can be made from diethyl carbonate by sequential addition of methylamine and 1-naphthol. The sequence is important: make the amide first because the N is a poor leaving group.

EtO—C(=O)—OEt  +  H₂NCH₃  →[Δ]  H₃C—N(H)—C(=O)—OEt  +  (1-naphthol, OH)  →[H⁺]  H₃C—N(H)—C(=O)—O—(naphthyl)

Sevin®

21-56  Triethylamine is nucleophilic, but it has no H on nitrogen to lose, so it forms a salt instead of a stable amide.

CH₃C(=O)—Cl  +  :NEt₃  ⟶  CH₃C(—Cl)(O⁻)(⁺NEt₃)  ⟶[–Cl⁻]  CH₃C(=O)—N⁺(Et)(Et)Et

When ethanol is added, it attacks the carbonyl of the salt, with triethylamine as the leaving group.

CH₃C(=O)—⁺NEt₃  +  HÖ—CH₂CH₃  ⟶  CH₃C(O⁻)(⁺NEt₃)(⁺O(H)—CH₂CH₃)  ⟶  NEt₃  +  CH₂C(O⁻)... H—⁺O—CH₂CH₃

CH₃—C(=O)—O—CH₂CH₃  +  Et₃N⁺H  Cl⁻

21-57

(a)

H₂C=CH—(CH₂)₄—OH  →[1) O₃][2) Me₂S]  O=CH—(CH₂)₄—OH

→[Ag⁺, NH₃ (aq)]  ⁻O—C(=O)—(CH₂)₄—OH

→[HA, Δ]  (seven-membered lactone, oxepanone)

**21-57 continued**

(b) CH₃O—⟨benzene⟩ →(Br₂)→ CH₃O—⟨benzene⟩—Br →(Mg, ether)→ →(CO₂)→ →(H₃O⁺)→ CH₃O—⟨benzene⟩—COOH

→(SOCl₂)↓

CH₃O—⟨benzene⟩—CONH₂ ←(NH₃)— CH₃O—⟨benzene⟩—COCl

(c) ⟨benzene⟩—CH₂Br →(KCN)→ ⟨benzene⟩—CH₂CN →(1) LiAlH₄ / 2) H₂O)→ ⟨benzene⟩—CH₂CH₂NH₂

(d) 

⟨ring with COOH top, HO, OH, OH⟩ →(1) NaOH excess / 2) CH₃I excess)→ ⟨ring with COOH top, CH₃O, OCH₃, OCH₃⟩ →(1) LiAlH₄ / 2) H₂O)→ ⟨ring with CH₂OH top, CH₃O, OCH₃, OCH₃⟩

Any methyl ester formed will be reduced with LAH.

↓ TsCl, pyridine

⟨ring with CH₂CH₂NH₂ top, CH₃O, OCH₃, OCH₃⟩ ←(1) LiAlH₄ / 2) H₂O)— ⟨ring with CH₂CN top, CH₃O, OCH₃, OCH₃⟩ ←(KCN)— ⟨ring with CH₂OTs top, CH₃O, OCH₃, OCH₃⟩

**21-58** The rate of a reaction depends on its activation energy, that is, the difference in energy between starting material and the transition state. The transition state in saponification is similar in structure, and therefore in energy, to the tetrahedral intermediate:

⟨benzene⟩—C(O⁻)(OH)—OCH₃

tetrahedral intermediate

The tetrahedral carbon has no resonance overlap with the benzene ring, so any resonance effect of a substituent on the ring will have very little influence on the energy of the transition state.

What will have a big influence on the activation energy is whether a substituent stabilizes or destabilizes the starting material. Anything that stabilizes the starting material will therefore increase the activation energy, slowing the reaction; anything that destabilizes the starting material will decrease the activation energy, speeding the reaction.

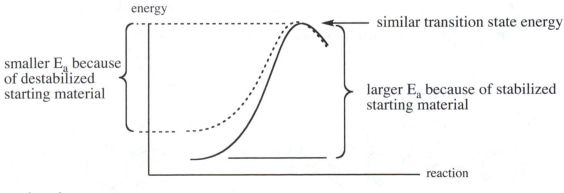

energy

smaller $E_a$ because of destabilized starting material

similar transition state energy

larger $E_a$ because of stabilized starting material

reaction

continued on next page

**525**

21-58 continued

(a) One of the resonance forms of methyl p-nitrobenzoate has a positive charge on the benzene carbon adjacent to the positive carbonyl carbon. This resonance form destabilizes the starting material, decreasing the activation energy, speeding the reaction.

poor resonance contributor, destabilizing the starting material; no effect in the transition state

(b) One of the resonance forms of methyl p-methoxybenzoate has all atoms with full octets, and negative charge on the most electronegative atom. This resonance form stabilizes the starting material, increasing the activation energy, slowing the reaction.

good resonance contributor, stabilizing the starting material; no effect in the transition state

21-59

21-60

(a)

$$CH_3—N=C=O \;+\; H_2O \longrightarrow H_3C—\overset{H}{\underset{}{N}}—\overset{O}{\overset{||}{C}}—OH \longrightarrow CH_3NH_2 \,(g) \;+\; CO_2 \,(g)$$

methyl isocyanate          a carbamic acid — unstable

Both of these reactions are exothermic. In a closed vessel like an industrial reactor, the production of gaseous products causes a large pressure increase, risking an explosion.

(b)

two rapid proton transfers

Decomposition could be proposed as either acid- or base-catalyzed.

(c) Alternatively, the sequence of reactions could be swapped.

phosgene

21-61

(a) (i) The repeating functional group is an ester, so the polymer is a polyester (named Kodel®).
(ii) hydrolysis products:

HO–C(=O)–[benzene ring]–C(=O)–OH  +  HOCH₂–[cyclohexane ring]–CH₂OH

(iii) The monomers could be the same as the hydrolysis products, or else some reactive derivative of the dicarboxylic acid, like an acid chloride or an ester derivative.

(b)(i) The repeating functional group is an amide, so the polymer is a polyamide (named Nylon 6).
(ii) hydrolysis product:

H₂N–[chain]–C(=O)–OH

(iii) The monomer could be the same as the hydrolysis product, but in the polymer industry, the actual monomer used is the lactam shown at the right.

(c) (i) The repeating functional group is a carbonate, so the polymer is a polycarbonate (named Lexan®).
(ii) hydrolysis products:

HO–[benzene ring]–C(CH₃)(CH₃)–[benzene ring]–OH  +  CO₂

(iii) The phenol monomer would be the same as the hydrolysis product; phosgene or a carbonate ester would be the other monomer.

(d) (i) The repeating functional group is an amide, so the polymer is a polyamide.
(ii) hydrolysis product:

H₂N–[benzene ring]–COOH    *p*-aminobenzoic acid, PABA, used in sunscreens

(iii) The monomer could be the same as the hydrolysis product; a reactive derivative of the acid such as an ester could also be used.


21-62  A singlet at $\delta$ 2.15 is H on carbon next to carbonyl, the only type of proton in the compound. The IR spectrum shows no OH, and shows two carbonyl absorptions at high frequency, characteristic of an anhydride. Mass of the molecular ion at 102 proves that the anhydride must be acetic anhydride, a reagent commonly used in aspirin synthesis.

H₃C–C(=O)–O–C(=O)–CH₃

Acetic anhydride can be disposed of by hydrolyzing (carefully! exothermic!) and neutralizing in aqueous base.

21-63

Even-mass fragments come from two-bond cleavages. In carbonyl compounds, the McLafferty rearrangement is the most common type of two-bond cleavage. Butyl butanoate, molecular weight 144 g/mole, can do McLafferty rearrangement on each side of the carbonyl, PLUS it can do them in sequence!

m/z 144          m/z 88          m/z 88          m/z 60

same

21-64

IR spectrum:

—sharp spike at 2250 cm$^{-1}$ $\Rightarrow$ C$\equiv$N

—1750 cm$^{-1}$ $\Rightarrow$ C=O
—1200 cm$^{-1}$ $\Rightarrow$ C—O } maybe an ester

NMR spectrum:

—triplet and quartet $\Rightarrow$ CH$_3$CH$_2$

—this quartet at $\delta$ 4.3 $\Rightarrow$ CH$_3$CH$_2$O

—2H singlet at $\delta$ 3.5 $\Rightarrow$ isolated CH$_2$

$$CH_3CH_2O \ + \ \overset{O}{\overset{||}{C}} \ + \ CH_2 \ + \ C\equiv N$$

sum of the masses is 113, consistent with the MS

The fragments can be combined in only two possible ways:

$$CH_3CH_2O-\overset{O}{\overset{||}{C}}-CH_2C\equiv N \qquad CH_3CH_2OCH_2-\overset{O}{\overset{||}{C}}-C\equiv N$$

A                                      B

The NMR proves the structure to be **A**. If the structure were **B**, the CH$_2$ between oxygen and the carbonyl would come farther downfield than the CH$_2$ of the ethyl (deshielded by oxygen and carbonyl instead of by oxygen alone). As this is not the case, the structure cannot be **B**.

The peak in the mass spectrum at m/z 68 is due to α-cleavage of the ester:

m/z 113          m/z 68

+ CH$_3$CH$_2$O • mass 45

21-65

IR spectrum: A strong carbonyl peak at 1720 cm$^{-1}$, in conjunction with the C—O peak at 1200 cm$^{-1}$, suggests the presence of an ester. An alkene peak appears at 1660 cm$^{-1}$. $C=C$ $\quad \overset{O}{\overset{||}{C}}-O-C$

HNMR spectrum: The typical ethyl pattern stands out: 3H triplet at $\delta$ 1.25 and 2H quartet at $\delta$ 4.2. The chemical shift of the CH$_2$ suggests it is bonded to an oxygen. The other groups are: a 3H doublet at $\delta$ 1.8, likely to be a CH$_3$ next to one H; a 1H doublet at $\delta$ 5.8, a vinyl hydrogen with one neighboring H; and a 1H multiplet at $\delta$ 6.9, another vinyl H with many neighbors, the far downfield chemical shift suggests that it is beta to the carbonyl. The large coupling constant in the doublet at $\delta$ 5.8 shows that the two vinyl hydrogens are *trans*.

**528**

21-65 continued
There is only one possible way to assemble these pieces:

common name: ethyl crotonate

δ 1.8, d H$_3$C H δ 5.8, d
C=C
δ 6.9, m H C—OCH$_2$CH$_3$ δ 1.25, t
‖
O δ 4.2, q

CNMR spectrum: The six unique carbons are unmistakable: the C=O of the ester at δ 166; the two vinyl carbons at δ 144 (beta to C=O) and at δ 123 (alpha to C=O); the CH$_2$—O of the ester at δ 60; and the two methyls at δ 18 and at δ 14.

Mass spectrum: This structure has mass 114, consistent with the molecular ion.
Major fragmentations:

CH$_3$CH$_2$O• + mass 45

plus two other resonance forms

m/z 69

CH$_3$• +

m/z 99

plus one other resonance form

21-66  If you solved this problem, put a gold star on your forehead.

The formula $C_6H_8O_3$ indicates 3 elements of unsaturation.

IR spectrum:  The absence of strong OH peaks shows that the compound is neither an alcohol nor a carboxylic acid.  There are two carbonyl absorptions:  the one about 1770 cm$^{-1}$ is likely a strained cyclic ester (reinforced with the C—O peak around 1150 cm$^{-1}$), while the one at 1720 cm$^{-1}$ is probably a ketone.  (An anhydride also has two peaks, but they are of higher frequency than the ones in this spectrum.)

HNMR spectrum:  The NMR shows four types of protons.  The 2H multiplet at $\delta$ 4.3 is a $CH_2$ group next to an oxygen on one side.  The 1H multiplet at $\delta$ 3.7 is also strongly deshielded (probably by two carbonyls), a CH next to a $CH_2$.  The 3H singlet at $\delta$ 2.45 is a $CH_3$ on one of the carbonyls.  The remaining two hydrogens are highly coupled, a $CH_2$ where the two hydrogens are not equivalent.  There are no vinyl hydrogens (and no alkene carbons in the carbon NMR), so the remaining element of unsaturation must be a ring.

Assemble the pieces:

On each carbon, the "up" hydrogen is *cis* to the acetyl group, while the "down" hydrogen is *trans*.  Thus, the two hydrogens on each of these carbons are not equivalent, leading to complex splitting.

Carbon NMR:

21-67
The formula $C_5H_9NO$ has 2 elements of unsaturation.

IR spectrum:  The strongest peak at 1670 cm$^{-1}$ comes low in the carbonyl region; in the absence of conjugation (no alkene peak observed), a carbonyl this low is almost certainly an amide.  There is one broad peak in the NH/OH region, hinting at the likelihood of a secondary amide.

HNMR spectrum:  The broad peak at $\delta$ 7.55 is exchangeable with $D_2O$; this is an amide proton.  A broad, 2H peak at $\delta$ 3.3 is a $CH_2$ next to nitrogen.  A broad, 2H peak at $\delta$ 2.4 is a $CH_2$ next to carbonyl.  The 4H peak at $\delta$ 1.8 is probably two more $CH_2$ groups.  There appears to be coupling among these protons but it is not resolved enough to be useful for interpretation.  This is often the case when the compound is cyclic, with restricted rotation around carbon-carbon bonds, giving *non-equivalent* (axial and equatorial) hydrogens on the same carbon.

CNMR spectrum:  The peak at $\delta$ 175 is the C=O of the amide.  All of the peaks between $\delta$ 25 and $\delta$ 50 are aliphatic sp$^3$ carbons, no sp$^2$ carbons, so the remaining element of unsaturation cannot be a C=C; it must be a ring.  The carbon peak farthest downfield is the carbon adjacent to N.

The most consistent structure:

$\delta$-valerolactam

22-1

(a)

1                                    2

(b) Enol **1** will predominate at equilibrium as its double bond is conjugated with the benzene ring, making it more stable than **2**.

(c)

basic conditions forming enol **1**

acidic conditions forming enol **1**

basic conditions forming enol **2**

22-1 (c) continued

acidic conditions forming enol **2**

22-2

(a)

This planar enol intermediate has lost all chirality. Protonation can occur with equal probability at either face of the pi bond leading to racemic product.

1 : 1

(plus one other resonance form for each)

racemic mixture

(b)

mixture of diastereomers

**532**

**22-3** If the ketone is not symmetric, more than one enolate may be formed.

(a)

(b)

(c)

(d) Two enolates are possible.

(e) Two enolates are possible.

(f) Three enolates are possible.

continued on next page

## 22-4

(a)  Sodium ethoxide is not a strong enough base to deprotonate cyclohexanone more than a tiny amount:

A difference of 4 $pK_a$ units means an equilibrium constant of $\approx 10^{-4}$, meaning 0.01% of the cyclohexanone is deprotonated.

Since $\approx$ 99.99% of the NaOEt is still present when benzyl bromide is added, these two react by Williamson ether synthesis.  Unreacted cyclohexanone will remain in the reaction solution.

(b)  Two changes are needed to make the desired reaction feasible:  (1)  use a strong, non-nucleophilic base such as LDA that will completely deprotonate cyclohexanone, and (2)  use an aprotic solvent such as THF or diethyl ether so that cyclohexanone cannot find a proton from solvent.  These conditions will work fine.

## 22-5  New carbon-carbon bonds are shown in bold.  ▬

(a)

(b)

(c)

22-6

Steps **A–F** are similar to imine formation, described in the solution to Problem 18-21.

22-7

R = CH₂Ph

22-8

(a) 

$$Ph-\overset{N-CH_3}{\underset{|}{C}}-CH_3$$

(b) 

$$H_3C-\overset{CH_3}{\underset{N}{|}}\\Ph-\overset{|}{C}=CH_2$$

(c) 

(d) 

22-9  Any 2° aliphatic amines can be used to make enamines; pyrrolidine is a typical one.

(a)  pyrrolidine  $\xrightarrow[-H_2O]{HA}$  $\xrightarrow[2) H_3O^+]{1)}$ 

(b)  pyrrolidine  $\xrightarrow[-H_2O]{HA}$  $\xrightarrow[2) H_3O^+]{1) BrCH_2Ph}$  Ph

(c)  Ph  +  pyrrolidine  $\xrightarrow[-H_2O]{HA}$  $\xrightarrow[2) H_3O^+]{1) PhC-Cl}$  $PhC-CH_2-CPh$

22-10

+ the enantiomer

Phenol is like an enol but is an even weaker nucleophile than an enol, yet it readily reacts with bromine.  Only the ortho attack is shown here; the para would be abundant too.

22-11

$Ph-C(=O)-CH_2-H$ ... $\overset{\ominus}{:OH}$ ... $\left\{ Ph-\overset{:O:}{C}=\overset{\ominus}{CH_2} \longleftrightarrow Ph-\overset{:\overset{\ominus}{O}:}{C}=CH_2 \right\}$ ... $Cl-Cl$ ... $Ph-C(=O)-\overset{Cl}{\underset{H}{C}}-H$

$\overset{\ominus}{:OH}$

$Ph-C(=O)-\overset{Cl}{\underset{H}{C}}-Cl$ $\longleftarrow$ $\left\{ Ph-\overset{:O:}{C}-\overset{Cl}{\overset{\ominus}{CH}} \longleftrightarrow Ph-\overset{:\overset{\ominus}{O}:}{C}=\overset{Cl}{CH} \right\}$ ... $Cl-Cl$

$\overset{\ominus}{:OH}$

$\left\{ Ph-\overset{:O:}{C}-\overset{Cl}{\overset{\ominus}{C}}-Cl \longleftrightarrow Ph-\overset{:\overset{\ominus}{O}:}{C}=\overset{Cl}{C}-Cl \right\}$ $\longrightarrow$ $Ph-C(=O)-\overset{Cl}{\underset{Cl}{C}}-Cl$ ... $Cl-Cl$

22-12   The cyclohexyl group is abbreviated "*c*-Hx".   [cyclohexyl ring] = *c*-Hx

$c\text{-Hx}-C(=O)-CH_2-H$ ... $\overset{\ominus}{:OH}$ ... $\left\{ c\text{-Hx}-\overset{:O:}{C}-\overset{\ominus}{CH_2} \longleftrightarrow c\text{-Hx}-\overset{:\overset{\ominus}{O}:}{C}=CH_2 \right\}$ ... $Br-Br$ ... $c\text{-Hx}-C(=O)-\overset{Br}{\underset{H}{C}}-H$

$\overset{\ominus}{:OH}$

$c\text{-Hx}-C(=O)-\overset{Br}{\underset{H}{C}}-Br$ $\longleftarrow$ $\left\{ c\text{-Hx}-\overset{:O:}{C}-\overset{Br}{\overset{\ominus}{CH}} \longleftrightarrow c\text{-Hx}-\overset{:\overset{\ominus}{O}:}{C}=\overset{Br}{CH} \right\}$ ... $Br-Br$

$\overset{\ominus}{:OH}$

$\left\{ c\text{-Hx}-\overset{:O:}{C}-\overset{Br}{\overset{\ominus}{C}}-Br \longleftrightarrow c\text{-Hx}-\overset{:\overset{\ominus}{O}:}{C}=\overset{Br}{C}-Br \right\}$ $\longrightarrow$ $c\text{-Hx}-\overset{:O:}{C}-\overset{Br}{\underset{Br}{C}}-Br$ $\underset{\overset{\ominus}{:OH}}{\rightleftharpoons}$ $c\text{-Hx}-\overset{:\overset{\ominus}{O}:}{C}-CBr_3$ ... $Br-Br$

$\overset{|}{O-H}$

$c\text{-Hx}-\overset{O}{C}-\overset{\ominus}{O}:$ + HCBr₃  $\longleftarrow$  $c\text{-Hx}-\overset{:O:}{C}-O-H$ + $:CBr_3^{\ominus}$

bromoform

**537**

22-13

(a)  COO⁻ Na⁺
+ CHCl₃

(b) COO⁻ Na⁺
+ CHI₃
(precipitate)

(c) Ph—C—C—CH₃ with =O and two Br

22-14 Methyl ketones, and those alcohols which can be oxidized to methyl ketones, will give a positive iodoform test. All of the compounds in this problem except pentan-3-one (part (d)) will give a positive iodoform test.

22-15

HBr +

22-16

$$\underset{\text{CH}_3\text{COOH}}{\overset{\text{Cl}_2}{\longrightarrow}}$$  Cl  $$\underset{\Delta}{\overset{\text{KOH}}{\longrightarrow}}$$

from Solved Problem 22-2    E2 elimination

This is a common and effective way to make carbon-3 reactive where it was not in the starting material.

22-17 Monobromination at the alpha carbon is typical of the HVZ reaction.

(a) Br, Br, O
↓ H₂O
Br, COOH

(b) Br, Br, O
↓ H₂O
Br, COOH

(c) O, Br, OH, Br, O
↓ H₂O
HO, O, OH, Br, O

(d) An acyl bromide forms but no alpha-hydrogen is present, so no H-V-Z reaction takes place.

HO, O, Br, OH, Br, O
possible product using an excess of Br₂

22-18

In general, the equilibrium in aldol condensations of ketones favors reactants rather than products. There is significant steric hindrance at both carbons with new bonds, so it is reasonable to conclude that this reaction of cyclohexanone would also favor reactants at equilibrium.

22-19 New carbon-carbon bonds are shown in bold. ━━

(a)

(b)

(c)

22-20 All the steps in the aldol condensation are reversible. Adding base to diacetone alcohol promoted the reverse aldol reaction. The equilibrium greatly favors acetone.

22-21

Carbons of the electrophile are shown in **bold** just to keep track of which carbons come from which molecule.

22-22

(a) <u>acidic conditions</u>

(b) <u>basic conditions</u>

22-23

22-24  The carbon-carbon double bond is the new bond in each case.

(a)

$E + Z$

(b)

$E + Z$

(c)

22-25

(a)

Step 1:  carbon skeletons

Step 2:  nucleophile generation

(2,2-Dimethylpropanal has no α-hydrogen.)

continued on next page

22-25 (a) continued

Step 3: nucleophilic attack

Step 4: dehydration to final product

Step 5: Combine Steps 2, 3, and 4 to complete the mechanism.

(b)

Step 1: carbon skeletons

Step 2: nucleophile generation

(Benzaldehyde has no α-hydrogen.)

continued on next page

Step 3: nucleophilic attack

Step 4: dehydration to final product

Step 5: Combine Steps 2, 3, and 4 to complete the mechanism.

22-26 This solution presents the sequence of reactions leading to the product, following the format of the Problem-Solving feature. This is not a complete mechanism.

Step 2: generation of the nucleophile

Step 3: nucleophilic attack

continued on next page

22-26 continued

Step 4: dehydration

Remember that this is not an E2 reaction; it goes through an enolate intermediate.

The same sequence of steps subsequently occurs on the other side.

final product

22-27

There are three problems with the reaction as shown:

1. Hydrogen on a 3° carbon (structure **A**) is less acidic than hydrogen on a 2° carbon. The 3° hydrogen will be removed at a slower rate than the 2° hydrogen.

2. Nucleophilic attack by the 3° carbon will be more hindered, and therefore slower, than attack by the 2° carbon. Structure **B** is quite hindered.

3. Once a normal aldol product is formed, dehydration gives a conjugated system that has great stability. The aldol product **C** cannot dehydrate because no α-hydrogen remains. Some **C** will form, but eventually the reverse-aldol process will return **C** to starting materials which, in turn, will react at the other α-carbon to produce the conjugated system. (This reason is the Kiss of Death for **C**.)

**22-28**
(a)
(b)

**22-29**

**22-30**

The formation of a 7-membered ring is unfavorable for entropy reasons: the farther apart the nucleophile and the electrophile, the harder time they will have finding each other. If the molecule has a possibility of forming a 5- or a 7-membered ring, it will almost always prefer to form the 5-membered ring.

**22-31**

22-32

(a)

$CH_3CH_2CH_2\overset{\displaystyle O}{\overset{\|}{C}}-H$ + $\overset{\displaystyle O}{\overset{\|}{CH_2-CH}}$

Not feasible: requires condensation of two different aldehydes, each with α-hydrogens.

$CH_2CH_2CH_3$

(b)

$Ph\overset{\displaystyle O}{\overset{\|}{C}}-CH_2CH_3$ + $CH_3CH_2\overset{\displaystyle O}{\overset{\|}{C}}-Ph$   Feasible: this is a self-condensation.

(c)

$Ph\overset{\displaystyle O}{\overset{\|}{C}}-H$ + $CH_3-\overset{\displaystyle O}{\overset{\|}{C}}-CH_3$

Feasible: only one reactant has α-hydrogen; cannot use excess benzaldehyde as acetone has two reactive sites.

(d)

Feasible; however, the cyclization from the carbon α to the aldehyde to the ketone carbonyl is also possible.

(e)

Feasible: symmetric reagent will give the same product in either direction of cyclization.

22-33  (a) and (b)

starting diketone

546
Copyright © 2013 Pearson Education, Inc.

22-34

(a) The side reaction with sodium methoxide is transesterification. The starting material, and therefore the product, would be a mixture of methyl and ethyl esters.

$$H_3C-\overset{\overset{\displaystyle O}{\|}}{C}-OCH_2CH_3 \; + \; NaOCH_3 \; \rightleftharpoons \; H_3C-\overset{\overset{\displaystyle O}{\|}}{C}-OCH_3 \; + \; NaOCH_2CH_3$$

(b) Sodium hydroxide would irreversibly saponify the ester, completely stopping the Claisen condensation as the carbonyl no longer has a leaving group attached to it.

$$H_3C-\overset{\overset{\displaystyle O}{\|}}{C}-OCH_2CH_3 \; + \; NaOH \; \longrightarrow \; H_3C-\overset{\overset{\displaystyle O}{\|}}{C}-O^{\ominus} \; Na^+ \; + \; HOCH_2CH_3$$

22-35

There are two reasons why this reaction gives a poor yield. The nucleophilic carbon in the enolate is 3° and attack is hindered. More important, the final product has no hydrogen on the α-carbon, so the deprotonation by base that is the driving force in other Claisen condensations cannot occur here. What is produced is an *equilibrium mixture* of product and starting materials; the conversion to product is low.

22-36 Products after mild acid workup are shown. New carbon-carbon bonds are shown in bold. ▬

(a)

(b) Ph

(c)

(d)

22-37

PhCH$_2$C$-$C$-$OMe   ⊖:OMe   →   { PhCH$_2$C$-$C$-$OMe  ⟷  PhCH$_2$C$=$C$-$OMe }

PhCH$_2$CH$_2$$-$C$-$OMe

PhCH$_2$CH$_2$$-$C$-$C$-$C$-$OMe   ⟵ $-$MeO$^-$   PhCH$_2$CH$_2$$-$C$-$C$-$C$-$OMe
            CH$_2$Ph                                    MeO   CH$_2$Ph

SAME

Ph        H                    ⊖:OMe        { Ph                                  }
        OMe                      →                          OMe
        Ph                                                  Ph

{ Ph          OMe   ⟷   Ph          OMe }
       Ph                      Ph

acid workup

Ph          OMe          SAME          PhCH$_2$CH$_2$$-$C$-$C$-$C$-$OMe
       Ph                                             CH$_2$Ph

22-38 Condensed structural formulas and line formulas are shown.

(a) CH$_3$CH$_2$CH$_2$$-$C$-$OCH$_2$CH$_3$

(b) PhCH$_2$C$-$OCH$_3$

(c) CH$_3$CHCH$_2$$-$C$-$OCH$_2$CH$_3$
          H$_3$C

This one would be difficult because the alpha-carbon is hindered.

22-39

This is the final product, after removal of the α-hydrogen by ethoxide, followed by reprotonation during the workup.

$- EtO^-$

- - - - - - - - - - - - - - - - - - - - - - - - - - - - - - - - - - - - - - - - - - - - - - - -

This is the final product, after removal of the α-hydrogen by methoxide, followed by reprotonation during the workup.

$- MeO^-$

22-40

(a) not possible by Dieckmann—not a β-keto ester

(b)

$+ NaOCH_3$ (mixture of products results)

(c)

$+ NaOCH_3$

(d)

$+ NaOCH_2CH_3$

The protecting group is necessary to prevent aldol condensation. Aqueous acid workup removes the protecting group.

22-41

H—C—C—OEt + ⁻:OEt  →  { H—C⁻—C=O:  ↔  H—C=C—O:⁻ —OEt }

Ph—C(=O:)—OEt

Ph—C(=O)—C(H)—C(=O)—OEt  ⁻:O⁻  H  O
with EtO  H  ; − EtO⁻

:O:⁻  H  O
Ph—C—C—C—OEt
         H

SAME ⟹

O   H   O
Ph—C—C—C—OEt   + ⁻:OEt
        H

→ { :O:  :O:          :O:  :O:⁻        :O:⁻  :O:
    Ph—C—C⁻—OEt  ↔  Ph—C=C—OEt  ↔  Ph—C=C—OEt }
          H              H              H

H—A

O   O
Ph—C—C(H)(H)—C—OEt   (SAME)   Ph—C(=O)—C(H)—C(=O)—OEt
        H                              H

22-42  New carbon-carbon bonds are shown in bold. ▬

(a)
O   O
Ph—C—C(Ph)—C—OCH₃
         Ph

(b)
O   O
CH₃—C—C(Ph)—C—OCH₃   +   Ph—CH₂—C(=O)—CH₂—C(=O)—OCH₃

plus two self-condensation products—a poor
choice because both esters have α-hydrogens

(c)
O   O
EtO—C—C(=O)—CH₂—C—OEt

(d)
O   O
EtO—C—C(CH₃)—C—OEt

**22-43**

(a) $Ph-\overset{O}{\overset{\|}{C}}-OEt \ + \ CH_3CH_2-\overset{O}{\overset{\|}{C}}-OEt$

(b) $PhCH_2-\overset{O}{\overset{\|}{C}}-OMe \ + \ MeO-\overset{O}{\overset{\|}{C}}-\overset{O}{\overset{\|}{C}}-OMe$

(c) $EtO-\overset{O}{\overset{\|}{C}}-CH_2Ph \ + \ EtO-\overset{O}{\overset{\|}{C}}-OEt$

(d) $(CH_3)_3C-\overset{O}{\overset{\|}{C}}-OMe \ + \ CH_3CH_2CH_2CH_2-\overset{O}{\overset{\|}{C}}-OMe$

**22-44** These Claisen products would likely be accompanied by aldol by-products.

(a)

(b)

(c)

Unsymmetrical ketones can form two different enolates.

**22-45**

(a) two ways:

(b) $CH_3CH_2-\overset{O}{\overset{\|}{C}}-CH_2CH_3 \ + \ CH_3CH_2O-\overset{O}{\overset{\|}{C}}-\overset{O}{\overset{\|}{C}}-OCH_2CH_3$

(c)

(d) two ways:

**22-46**

(a)

(b)

(c)

$$\left\{ \begin{array}{ccc} \text{N}\!\!\equiv\!\!\text{C}-\overset{\ominus}{\underset{H}{\text{C}}}-\overset{:\text{O}:}{\overset{\|}{\text{C}}}-\text{OEt} & \longleftrightarrow & \overset{\ominus}{:}\!\text{N}\!=\!\text{C}\!=\!\underset{H}{\text{C}}-\overset{:\text{O}:}{\overset{\|}{\text{C}}}-\text{OEt} & \longleftrightarrow & \text{N}\!\equiv\!\text{C}-\underset{H}{\text{C}}\!=\!\overset{:\overset{\ominus}{\text{O}}:}{\text{C}}-\text{OEt} \end{array} \right\}$$

(d)

$$\left\{ \begin{array}{ccc} \overset{\ominus}{:}\!\overset{:\text{O}:}{\underset{:\text{O}:}{\overset{\oplus}{\text{N}}}}-\overset{\ominus}{\underset{H}{\text{C}}}-\overset{:\text{O}:}{\overset{\|}{\text{C}}}-\text{CH}_3 & \longleftrightarrow & \overset{\ominus}{:}\!\overset{:\overset{\ominus}{\text{O}}:}{\underset{:\text{O}:}{\overset{\oplus}{\text{N}}}}=\underset{H}{\text{C}}-\overset{:\text{O}:}{\overset{\|}{\text{C}}}-\text{CH}_3 & \longleftrightarrow & \overset{\ominus}{:}\!\overset{:\text{O}:}{\underset{:\text{O}:}{\overset{\oplus}{\text{N}}}}-\underset{H}{\text{C}}=\overset{:\overset{\ominus}{\text{O}}:}{\text{C}}-\text{CH}_3 \end{array} \right\}$$

(Other resonance forms of the nitro group are not shown.)

22-47 In the products, the wavy lines cross the bonds that must be made by alkylation, before hydrolysis and decarboxylation produce the substituted acetic acid.

(a)

(b)

(c)

(d)

22-48
(a) Only two substituent groups plus a hydrogen atom can appear on the alpha carbon after decarboxylation at the end of the malonic ester synthesis. The product shown has three alkyl groups, so it cannot be made by malonic ester synthesis.

desired product

22-48 continued

(b)

pK$_a$ 24

LDA

pK$_a$ ≈40

With such a large difference in pK$_a$ values, products are favored >> 99%.

(c)

LDA

plus resonance form
as shown in part (b)

S$_N$2

$H_3O^+$
Δ

22-49

(a) PhCH$_2$CH$_2$—C—CH$_3$
+ CO$_2$ + EtOH

(b)          + CO$_2$ + EtOH

(c)          + CO$_2$ + EtOH

22-50 In the products, the wavy lines indicate the bonds that must be made by alkylation, before hydrolysis and decarboxylation produce the substituted acetone.

(a)

1) NaOEt
2) PhCH$_2$Br

$H_3O^+$
Δ

+ CO$_2$ + EtOH

CH$_2$Ph        CH$_2$Ph

(b)

1) 2 NaOEt
2) Br(CH$_2$)$_4$Br

$H_3O^+$
Δ

+ CO$_2$ + EtOH

(c)

1) NaOEt
2) PhCH$_2$Br

CH$_2$Ph

1) NaOEt
2) Br

CH$_2$Ph

Δ | $H_3O^+$

CO$_2$ + EtOH +

Ph

**22-51**

(a) There are two problems with attempting to make this compound by acetoacetic ester synthesis. The acetone "core" of the product is shown in the box. This product would require alkylation at BOTH carbons of the acetone "core" of acetoacetic ester; in reality, only one carbon undergoes alkylation in the acetoacetic ester synthesis. Second, it is not possible to do an $S_N2$ type reaction on an unsubstituted benzene ring, so neither benzene could be attached by acetoacetic ester synthesis.

(b)

plus resonance forms showing
delocalization of $e^-$ into ring and C=O

(c)

**22-52**

forward direction

**22-53** First, you might wonder why this sequence does not make the desired product:

The poor yield in this conjugate addition is due primarily to the numerous competing reactions: the ketone enolate can self-condense (aldol), can condense with the ketone of MVK (aldol), or can deprotonate the methyl of MVK to generate a new nucleophile. The complex mixture of products makes this route practically useless. (continued on next page)

22-53 continued
What permits enamines (or other stabilized enolates) to work are: a) the certainty of which atom is the nucleophile, and b) the lack of self-condensation. Enamines can also do conjugate addition:

HA

MVK

high yield

22-54 The enolate of acetoacetic ester can be used in a Michael addition to an α,β-unsaturated ketone like MVK.

EtO

NaOEt

$H_3O^+$

EtO

$H_3O^+$
Δ

$CO_2$ + EtOH +

heptane-2,6-dione

δ   γ   β   α

22-55

$\{$ :N≡C−CH=CH_2 ⟷ $\overset{\ominus}{:}$N=C−$\overset{\oplus}{C}$H=CH_2 ⟷ $\overset{\ominus}{:}$N=C=CH−$\overset{\oplus}{C}$H_2 $\}$
acrylonitrile

$\overset{\ominus}{}$Nuc:

N≡C−CH_2−CH_2
     |
     Nuc

Nuc−H

$\{$ :N≡C−$\overset{\ominus}{C}$H−CH_2 ⟷ $\overset{\ominus}{:}$N=C=CH−CH_2 $\}$
                |                              |
                Nuc                           Nuc

nitroethylene

(Some resonance forms of the nitro group are not shown.)

$\overset{\ominus}{}$Nuc:

Nuc−H

22-56

(a)

PhCH=CH–C(=O)–OEt

EtO–C(=O)...C(=O)–OEt (with O below each)

(b)  $CH_2=CH-C\equiv N$

EtO–C(=O)...C(=O)–CH₃

followed by hydrolysis
and decarboxylation

(c)  two ways

$CH_2=CH-C\equiv N$

(cyclopentanone with COOEt)

followed by hydrolysis
and decarboxylation

OR

(piperidine enamine of cyclopentene)

$CH_2=CH-C\equiv N$

followed by hydrolysis

(d)

$H_3C$–N–$CH_3$

$CH_2=CH-C(=O)-Ph$
$CH_3$

followed by hydrolysis

(e)

$CH_3-C(=O)-CH_2-C(=O)-OEt$

1) NaOEt   2) $CH_3I$

$CH_3-C(=O)-CH(CH_3)-C(=O)-OEt$   +   (methyl vinyl ketone)

NaOEt

$\Delta$ | $H_3O^+$

$CH_3-C(=O)-CH(CH_3)-CH_2-CH_2-C(=O)-CH_3$   + $CO_2$ + EtOH

(Could also be synthesized by the
Stork enamine reactions.)

(f)

(cyclopentenone)

$\left( H_2C=C\overset{}{\underset{H}{|}} \right)_2 CuLi$

22-57

Step 1: carbon skeleton

Step 2: nucleophile generation

These hydrogen atoms are more acidic than the other alpha hydrogens because the anion can be stabilized by the benzene ring.

plus resonance forms with negative charge on the benzene ring

Step 3: nucleophilic attack (Michael addition)

HO—H

continued on next page

22-57 continued

Step 4: conversion to final product

(nucleophile formation)

plus one other (enolate) resonance form

(nucleophilic attack)

(base-catalyzed dehydration)

plus one other (enolate) resonance form

Step 5: The complete mechanism is the combination of Steps 2, 3, and 4. Notice that this mechanism is simply described by:
1) Enolate formation, followed by Michael addition;
2) Aldol condensation, followed by dehydration.

22-58

Step 1: carbon skeleton

comes from

Step 2: nucleophile generation

continued on next page

22-58 continued

Step 3: nucleophilic attack

$$Ph-\overset{\overset{:O:}{\|}}{C}-H \quad + \quad \overset{\ominus}{:}CH_2-\overset{\overset{O}{\|}}{C}-OCOCH_3 \longrightarrow Ph-\overset{\overset{\ominus}{:O:}}{\underset{}{CH}}-CH_2-\overset{\overset{O}{\|}}{C}-OCOCH_3 \longrightarrow$$

$$Ph-\overset{\overset{OH}{|}}{CH}-CH_2-\overset{\overset{O}{\|}}{C}-OCOCH_3$$

H—OAc

Step 4: conversion to final product

$$Ph-\overset{\overset{OH}{|}}{CH}-\underset{\overset{|}{H}}{CH}-\overset{\overset{O}{\|}}{C}-OCOCH_3 \longrightarrow Ph-\overset{\overset{OH}{|}}{CH}-\overset{\ominus}{CH}-\overset{\overset{:O:}{\|}}{C}-OCOCH_3$$

$$CH_3COO\overset{\ominus}{:}$$

plus one other (enolate) resonance form

$$Ph-CH=CH-\overset{\overset{O}{\|}}{C}-OCOCH_3$$

hydrolysis

$$PhCH=CH-\overset{\overset{:O:}{\|}}{C}-O-\overset{\overset{O}{\|}}{C}-CH_3 \quad \overset{H—A}{\longrightarrow} \quad \left\{ PhCH=CH-\overset{\overset{\oplus}{:O}-H}{\|}{C}-O-\overset{\overset{O}{\|}}{C}-CH_3 \right.$$

$$PhCH=CH-\overset{\overset{OH}{|}}{C}-O-\overset{\overset{:O:}{\|}}{C}-CH_3 \quad \longleftarrow \quad \left\{ PhCH=CH-\overset{\overset{\overset{..}{O}-H}{|}}{\underset{\oplus}{C}}-O-\overset{\overset{O}{\|}}{C}-CH_3 \right.$$

H—A

$$H_2O:$$

plus other resonance forms with:
a) positive on the other oxygen;
b) positive on the benzylic carbon;
c) positive on the benzene ring

two fast proton transfers

$$H_2O:$$

$$PhCH=CH-\overset{\overset{OH}{|}}{\underset{\overset{|}{OH}}{C}}-O-\overset{\overset{\oplus}{:O}-H}{\|}{C}-CH_3 \longrightarrow O=\overset{\overset{O-H}{|}}{\underset{\overset{|}{CH_3}}{C}} \quad + \quad \left\{ PhCH=CH-\overset{\oplus}{\underset{\overset{|}{:OH}}{C}}\overset{H-\overset{..}{O}:}{} \right.$$

plus two other resonance forms

plus other resonance forms with:
a) positive on the other oxygen;
b) positive on the benzylic carbon;
c) positive on the benzene ring

$$PhCH=CH-\overset{\overset{O}{\|}}{C}-OH \quad \longleftarrow \quad \left\{ PhCH=CH-\overset{\overset{H-\overset{..}{O}:}{\oplus}}{\underset{\overset{|}{:OH}}{C}} \right.$$

$$H_2O:$$

Step 5: The complete mechanism is the combination of Steps 2, 3, and 4.

**559**

22-59 The Robinson annulation consists of a Michael addition followed by aldol cyclization with dehydration. In the retrosynthetic direction, disconnect the alkene formed in the aldol/dehydration, then disconnect the Michael addition to discover the reactants.

(a) Aldol and dehydration form the α,β double bond:

Michael addition forms a bond to the β' carbon:

(b) Aldol and dehydration form the α,β double bond:

Michael addition forms a bond to the β' carbon:

**22-60** The most acidic hydrogens are shown in boldface. (See Appendix 2 in this manual for a review of acidity.)

(a)

(b)

(c)

(d)

same enolate as in (b)

(e)

(f)

Because of the stereochemistry of the double bond, the H atoms on the left side of the ring are not identical to those on the right side. However, the two enolates will be equivalent except for the double bond geometry.

22-60 continued

(g)

(h)

same enolate as in (g)

22-61  In order of increasing acidity.  The most acidic protons are shown in boldface.  (The approximate pK$_a$ values are shown for comparison.)  See Appendix 2 in this manual for a review of acidity.

(g)
pK$_a$ 25
least acidic

<

(b)
pK$_a$ 20

<

(f)
pK$_a$ 17-18

<

(a)
pK$_a$ 13

fully deprotonated
by ethoxide ion

<

(e)
pK$_a$ 10

<

(c)
pK$_a$ 9

<

(d)
pK$_a$ 5
most acidic

22-62  The wavy line lies across the bond formed in the aldol condensation.

(a)  $-H_2O$

(b)  $-H_2O$

(c)  $-2 H_2O$

(d)  $-H_2O$

22-62 continued

(e) $- H_2O$

(f) $- H_2O$

22-63 The bold bond is formed in the Claisen condensation.

(a)

(b)

(c)

(d)

(e)

22-64

(a) mechanism of aldol condensation in problem 22-62(a)

mechanism continued on next page

22-64 (a) continued

(b) mechanism of aldol condensation in problem 22-62(b)

mechanism continued on next page

22-64(b) continued

from previous page

(c) mechanism of Claisen condensation in problem 22-63(a)

(This product will be deprotonated by methoxide
but regenerated upon acidic workup.)

(d) mechanism of Claisen condensation in problem 22-63(b)

(This product will be deprotonated by methoxide
but regenerated upon acidic workup.)

**565**

22-65

The enol form is stable because of the conjugation and because of intramolecular hydrogen-bonding in a six-membered ring.

keto          enol

In dicarbonyl compounds in general, the weaker the electron-donating ability of the group G, the more it will exist in the enol form: aldehydes (G = H) are almost completely enolized, then ketones (G = R group), esters (G = OR), and finally amides (G = NR$_2$) which have virtually no enol content.

keto          enol

22-66  All of these Robinson annulations are catalyzed by NaOH.

(a)   CH$_3$          +

Ph          O                    O

(b)          +

CH$_3$

(c)          +

22-67  All products shown are after acidic workup.

(a) aldol self-condensation

CHO          +          CHO          HO⁻

(b) Claisen self-condensation

COOEt          OEt          COOEt          EtO⁻

(c) aldol cyclization

O          HO⁻          O

## 22-67 continued

### (d) mixed Claisen

OR

### (e) mixed aldol

### (f) enamine acylation or mixed Claisen

OR

In practice, the mixed Claisen reactions starting from a ketone enolate plus an ester will give a considerable amount of aldol self-condensation.

## 22-68

(a)

(b) + CHI$_3$

after acid workup

(c)

(d)

(e)

(f)

(g)

(h)

initial product

+ CO$_2$ + CH$_3$OH

(i)

initial product

+ CO$_2$
+ CH$_3$CH$_2$OH

22-69

(a) reagents: $Br_2$, HOAc

(b) reagents: $Br_2$, $PBr_3$, followed by $H_2O$

(c) reagents: excess $I_2$ (or $Br_2$ or $Cl_2$), NaOH

(d)

$$Ph-\overset{\overset{O}{\|}}{C}-H \;+\; \overset{\oplus\;\;\ominus}{Ph_3P-CHCH_3} \longrightarrow PhCH=CHCH_3 \;+\; Ph_3P=O$$

(e)

An enamine avoids self-condensation. Any 2° amine could be used to form the enamine.

(f)

22-70 In the products, the wavy lines indicate the bonds that must be made by alkylation, before hydrolysis and decarboxylation produce the substituted acetic acid.

(a)

(b)

(c)

22-71 In the products, the wavy lines indicate the bonds that must be made by alkylation, before hydrolysis and decarboxylation produce the substituted acetone.

(a)

22-71 continued

(b)

+ CO₂ + EtOH

(c) The acetoacetic ester synthesis makes substituted acetone, so where is the acetone in this product?

substituted acetone

The single bond to this substituted acetone can be made by the acetoacetic ester synthesis. How can we make the α,β double bond? Aldol condensation!

make by conjugate → ← make by aldol
addition        cyclization/dehydration

22-72 These compounds are made by aldol condensations followed by other reactions. The key is to find the skeleton made by the aldol.

(a) Where is the possible α,β-unsaturated carbonyl in this skeleton?

forward synthesis

(b) The aldol skeleton is not immediately apparent in this formidable product. What can we see from it? Most obvious is the β-dicarbonyl (β-ketoester), which we know to be a good nucleophile, capable of substitution or Michael addition. In this case, Michael addition is most likely as the site of attack is β to another carbonyl.

continued on
next page

AHA! The aldol product reveals itself. (See the solution to 22-71 (c).)

## 22-72 continued
### forward synthesis

(c)  The key in this product is the α-nitroketone, the equivalent of a β-dicarbonyl system, capable of doing Michael addition to the β-carbon of the other α,β-unsaturated system.

### forward synthesis

Z + E—each will give
the same final product

## 22-73

(a)

E + Z

22-73 continued

(b)

(c) Robinson annulations are explained most easily by remembering that the first step is a Michael addition, followed by aldol cyclization with dehydration.

**571**

22-73 (c) continued

OMe

$CH_3$

O

$\xleftarrow{\;-HO^{\ominus}\;}$
(dehydration)

from previous page

OMe

HO

$CH_3 - \overset{\ominus}{C}$

O

plus one other
resonance form

(d)

The negative charge in this structure
is stabilized by resonance with the
carbonyl.

two rapid
proton transfers

$H_3O^+$

$H_2\ddot{O}$

two rapid
proton transfers

two rapid
proton transfers

$H_3O^+$

$H_2\ddot{O}$

+

enol
tautomer

plus one other
resonance form

plus one other
resonance form

$H_3O^+$

$H_2\ddot{O}$

two rapid
proton transfers

$\downarrow -H_2O$

mechanism continued on next page

**572**

from previous page

22-74 In this problem, structures of intermediates with positive charge on double-bonded N will have a less-significant resonance contributor with the positive charge on C without an octet of electrons.

(a)

(b)

(c)

(d)

aldol, then dehydration

**573**

22-75

(a)

$$CO_2 + CH_3OH +$$

hydrolysis,
decarboxylation

(b)

(c)

(d)

Doing this reaction first
blocks this side of the ketone.

Anion formed in Claisen is used
without isolation in the next step.

22-76

(a)

plus one other
resonance form

plus one other
resonance form

from previous page

plus one other
resonance form

(b)

plus one other
resonance form

plus one other
resonance form

plus other
resonance forms

22-76 continued

(c)

+ ⟋＝＼CN

(d)

retro-aldol to
this product

crossed Claisen product

Reaction of two molecules of Grignard reagent at an ester is expected. The unexpected product arises from a conjugate addition.

To regenerate the aromaticity of the furan ring, some species will accept an electron pair in the C—H bond. The most likely electron acceptor (the definition of an oxidizing agent) is the magnesium ion.

Once this conjugate addition product is formed, two Grignards will add to the ester in the normal fashion. See the solution to Problem 10-18(a), or text section 10-9D, to review the mechanism of Grignard addition to an ester.

22-78

plus two other
resonance forms

$\Delta \mid H_3O^+$

 mechanism continued on next page

from previous page

$$CO_2 + 2 \text{ EtOH } + \quad \begin{array}{c} Ph \quad H \\ C \\ \| \\ C \\ H \quad COOH \end{array} \quad \longleftarrow \quad \left[ \begin{array}{c} Ph \quad H \\ C \\ \| \\ C \\ HOOC \quad COOH \end{array} \right]$$

22-79

To identify the retro-aldol, we must first locate the HO that is beta to the C=O.

dihydroxyacetone phosphate

glyceraldehyde-3-phosphate

(plus one other resonance form)

22-80  This is an aldol condensation. **P** stands for a protein chain in this problem.

plus one other resonance form

This is another molecule of protonated aldehyde.

enol serving as a nucleophile

plus one other resonance form

two fast proton transfers

$- H_2O$

## 22-81

(a) aldol, Michael, aldol cyclization, decarboxylation

(b) Michael followed by Claisen condensation; hydrolysis and decarboxylation

Students:  use this page for notes or to solve problems.

Remember about Fischer projections, first introduced in Chapter 5, section 5-10: vertical bonds are equivalent to dashed bonds, going behind the plane of the paper, and horizontal bonds are equivalent to wedge bonds, coming toward the viewer.

Fischer projection:

Fischer projections are always drawn with the highest priority functional group at the top.

---

23-1

glucose          mirror image          fructose          mirror image

All four of these compounds are chiral and optically active.

23-2

(a)

two asymmetric carbons ⇒ four stereoisomers (two pairs of enantiomers) if none are meso

**23-2 continued**

(b)

$$\begin{array}{c} CH_2OH \\ | \\ C=O \\ | \\ H\!-\!\!-\!OH \\ | \\ CH_2OH \end{array} \qquad \begin{array}{c} CH_2OH \\ | \\ O=C \\ | \\ HO\!-\!\!-\!H \\ | \\ CH_2OH \end{array}$$

one chiral center $\Rightarrow$ two stereoisomers
(enantiomers)

(c) An aldohexose has four chiral carbons and sixteen stereoisomers. A ketohexose has three chiral carbons and eight stereoisomers.

**23-3**

(a)

$$\begin{array}{c} CH_2OH \\ | \\ C=O \\ | \\ CH_2OH \end{array}$$

(b)

$$\begin{array}{c} CHO \\ | \\ H\!-\!\!-\!OH \\ | \\ CH_2OH \end{array} \qquad \begin{array}{c} CHO \\ | \\ HO\!-\!\!-\!H \\ | \\ CH_2OH \end{array}$$

**23-4**

$$\begin{array}{c} CHO \\ | \\ HO\!-\!\!-\!H \\ | \\ H\!-\!\!-\!OH \\ | \\ HO\!-\!\!-\!H \\ | \\ HO\!-\!\!-\!H \\ | \\ CH_2OH \end{array}$$
L-(−)-glucose

$$\begin{array}{c} CHO \\ | \\ H\!-\!\!-\!OH \\ | \\ HO\!-\!\!-\!H \\ | \\ HO\!-\!\!-\!H \\ | \\ CH_2OH \end{array}$$
L-(+)-arabinose

$$\begin{array}{c} CHO \\ | \\ HO\!-\!\!-\!H \\ | \\ HO\!-\!\!-\!H \\ | \\ CH_2OH \end{array}$$
L-(+)-erythrose

$$\begin{array}{c} CHO \\ | \\ HO\!-\!\!-\!H \\ | \\ CH_2OH \end{array}$$
L-(−)-glyceraldehyde

**23-5**

$$\begin{array}{c} 2 \\ CHO \\ | \\ 4\ H\!\blacktriangleright\!C\!\blacktriangleleft\!OH\ 1 \\ | \\ CH_2OH \\ 3 \end{array}$$

$$\begin{array}{c} 2 \\ CHO \\ | \\ 1\ HO\!\blacktriangleright\!C\!\blacktriangleleft\!H\ 4 \\ | \\ CH_2OH \\ 3 \end{array}$$

D = *R*            L = *S*

**23-6**

$$\begin{array}{c} CH_2OH \\ | \\ HO\!-\!\!-\!H \\ | \\ H\!-\!\!-\!OH \\ | \\ CH_2CH_2CH_3 \end{array} \qquad \begin{array}{c} CH_2OH \\ | \\ H\!-\!\!-\!OH \\ | \\ HO\!-\!\!-\!H \\ | \\ CH_2CH_2CH_3 \end{array}$$

Carbon-1 is shown at the top of a Fischer projection.

**23-7** (a)

Fischer projections:

$$\begin{array}{c} Ph \\ | \\ H\!-\!\!-\!OH \\ | \\ H\!-\!\!-\!NHCH_3 \\ | \\ CH_3 \end{array}$$
**1**

$$\begin{array}{c} Ph \\ | \\ HO\!-\!\!-\!H \\ | \\ H_3CHN\!-\!\!-\!H \\ | \\ CH_3 \end{array}$$
**2**

$$\begin{array}{c} Ph \\ | \\ H\!-\!\!-\!OH \\ | \\ H_3CHN\!-\!\!-\!H \\ | \\ CH_3 \end{array}$$
**3**

$$\begin{array}{c} Ph \\ | \\ HO\!-\!\!-\!H \\ | \\ H\!-\!\!-\!NHCH_3 \\ | \\ CH_3 \end{array}$$
**4**

23-7 (a) continued: stereochemical view

|  | Ph |  | Ph |  | Ph |  | Ph |
|--|-----|--|-----|--|-----|--|-----|
| H | OH | HO | H | H | OH | HO | H |
| H | NHCH₃ | H₃CHN | H | H₃CHN | H | H | NHCH₃ |
| **1** | CH₃ | **2** | CH₃ | **3** | CH₃ | **4** | CH₃ |

(b) Ephedrine is the *erythro* diastereomer, represented by structures **1** and **2**. Structures **3** and **4** represent pseudoephedrine, the *threo* diastereomer.

(c) The Fischer-Rosanoff convention assigns D and L to the configuration of the asymmetric carbon at the bottom of the Fischer projection. Structure **1** is D-ephedrine; structure **2** is L-ephedrine; structure **3** is L-pseudoephedrine; and structure **4** is D-pseudoephedrine.

(d) It is not possible to determine which is the (+) or (–) isomer of any compound just by looking at the structure. The only compound that has a direct correlation between the direction of optical rotation and its D or L designation is glyceraldehyde, about which the D and L system was designed. For all other compounds, optical rotation can only be determined by measurement in a polarimeter.

23-8

(a)

CHO
H — OH
H —3— OH
H — OH
H — OH
CH₂OH

D-allose

(b)

CHO
HO —2— H
HO — H
HO — H
H — OH
CH₂OH

D-talose

(c)

CHO
HO — H
H —3— OH
HO — H
H — OH
CH₂OH

D-idose

(d)

CHO
H — OH
HO — H
H —4— OH
CH₂OH

D-xylose

⟹

CHO
H — OH
HO — H
HO — H
CH₂OH

L-arabinose

← inversion at bottom chiral center ⟹ L-series sugar

23-9   D-mannose

1
CHO
HO —2— H
HO —3— H
H —4— OH
H —5— OH
6 CH₂OH

rotate

⟹

6
CH₂OH
5
H / H
4 OH   HO   H
HO     OH
3     2
H     H
O

**583**

**23-10** D-allose

OH at C-3 is axial.

**23-11** D-talopyranose

OH groups at C-2 and C-4 are axial.

**23-12**

(a)

D-arabinofuranose

(b)

D-ribofuranose

**23-13**

**23-14**

(a) α-D-mannopyranose

axial = α

(b) β-D-galactopyranose

equatorial = β

(c) β-D-allopyranose

equatorial = β

(d) α-D-arabinofuranose

trans to CH₂OH = α

(e) β-D-ribofuranose

cis to CH₂OH = β

Students: This concept map intends to clarify the difference between an acetal and a ketal as mentioned in text section 23-6. The terms *hemiketal* and *ketal* are no longer used by IUPAC for organic nomenclature, having been grouped with *hemiacetal* and *acetal*. However, biochemists still use these terms (as do many organic chemists!) to differentiate between monosaccharides containing the ketone or aldehyde functional group, *i.e.*, ketoses and aldoses.

To summarize: an aldehyde reacts with one molecule of alcohol to form a hemiacetal and with two molecules of alcohol to form an acetal; a ketone reacts with one molecule of alcohol to form a hemiketal and with two molecules of alcohol to form a ketal.

**23-15**  a = fraction of galactose as the α anomer; b = fraction of galactose as the β anomer

a (+ 150.7°) + b (+ 52.8°) = + 80.2°

a + b = 1;  b = 1 − a

a (+ 150.7°) + (1 − a) (+ 52.8°) = + 80.2° $\xrightarrow{\text{solve for "a"}}$ a = 0.28 ;  b = 0.72

The equilibrium mixture contains 28% of the α anomer and 72% of the β anomer.

**23-16**

erythrose

The planar enolate can reprotonate from either side, producing a mixture of erythrose and threose.

erythrose   +   threose

**23-17**

fructose

**586**

23-18

fructose

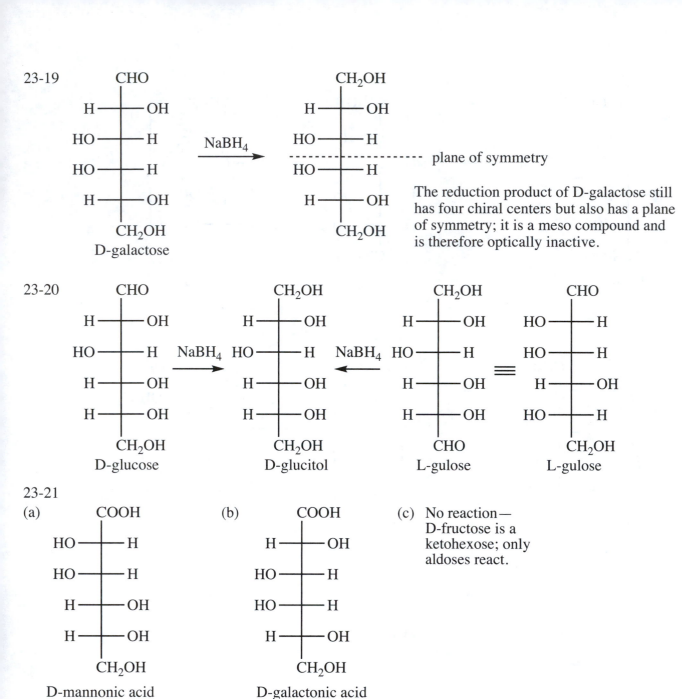

23-19

CHO
H—OH
HO—H
HO—H
H—OH
CH₂OH

D-galactose

NaBH₄ →

CH₂OH
H—OH
HO—H
------------------- plane of symmetry
HO—H
H—OH
CH₂OH

The reduction product of D-galactose still has four chiral centers but also has a plane of symmetry; it is a meso compound and is therefore optically inactive.

23-20

CHO
H—OH
HO—H
H—OH
H—OH
CH₂OH

D-glucose

NaBH₄ →

CH₂OH
H—OH
HO—H
H—OH
H—OH
CH₂OH

D-glucitol

← NaBH₄

CH₂OH
H—OH
HO—H
H—OH
H—OH
CHO

L-gulose

≡

CHO
HO—H
HO—H
H—OH
HO—H
CH₂OH

L-gulose

23-21

(a)

COOH
HO—H
HO—H
H—OH
H—OH
CH₂OH

D-mannonic acid

(b)

COOH
H—OH
HO—H
HO—H
H—OH
CH₂OH

D-galactonic acid

(c) No reaction—
D-fructose is a ketohexose; only aldoses react.

23-22

(a)

COOH
HO—H
HO—H
H—OH
H—OH
COOH

mannaric acid

(b)

COOH
H—OH
HO—H
HO—H
H—OH
COOH

galactaric acid (meso)

## 23-23

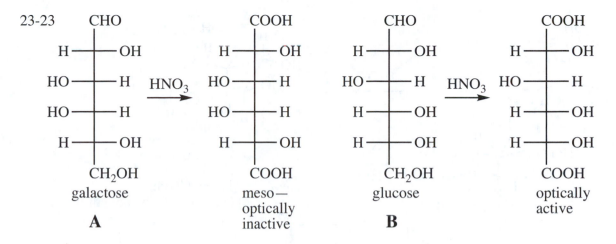

## 23-24

(a) not reducing: an acetal ending in "oside"
(b) reducing: a hemiacetal ending in "ose"
(c) reducing: a hemiacetal ending in "ose"
(d) not reducing: an acetal ending in "oside"
(e) Reducing: one of the rings has a hemiacetal.
(f) Not reducing: all anomeric carbons are in acetal form.

Like many common names, the name "sucrose" is misleading: the "ose" ending suggests a hemiacetal structure, but sucrose is a glycoside, so technically its name should end in "oside".

## 23-25  This problem asks for structures from 23-24 parts (a), (c), and (d).

(a) axial = α

(c) axial = α

(d) cis to CH₂OH = β

## 23-26

**23-27**

HCN is released from amygdalin. HCN is a potent cytotoxic (cell-killing) agent, particularly toxic to nerve cells.

**23-28**

α and β-D-fructofuranose

The aglycone in each product is circled.

ethyl β-D-fructofuranoside

+

ethyl α-D-fructofuranoside

**23-29**

**23-30** Excess CH₃I is needed to form methyl ethers at all OH groups.

(a) 

$CH_3OH_2C$ — $O$ — $CH_2OCH_3$

H  CH₃O

H    OCH₃

OCH₃  H

(b)

$CH_3O$   $OCH_3$

H      H — O

$CH_3O$   $OCH_3$

H   O   H

CH₃

**23-31** Excess acetic anyhydride is needed to acetylate all OH groups.

(a)

H    OAc

H

AcO      H — O

AcO

H  AcO    H

H    OAc

(b)

$AcOH_2C$ — $O$ — $OAc$

H    H

H      H

AcO    OAc

**23-32**

(a) The part of each structure in the box is common to all three monosaccharides.

23-32 continued

(b)

D-Talose must be the C-2 epimer of D-galactose.

23-33   Reagents for the Ruff degradation are:  1. $Br_2$, $H_2O$;  2. $H_2O_2$, $Fe_2(SO_4)_3$ .

23-34   Reagents for the Ruff degradation are:  1. $Br_2$, $H_2O$;  2. $H_2O_2$, $Fe_2(SO_4)_3$ .

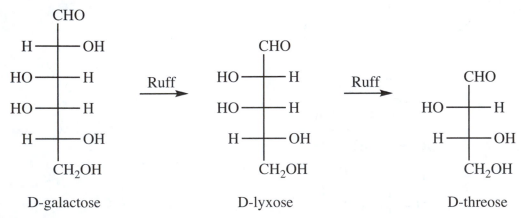

**23-35** Reagents for the Ruff degradation are: 1. $Br_2$, $H_2O$; 2. $H_2O_2$, $Fe_2(SO_4)_3$ .

D-allose          D-altrose

**23-36**

CHO
HO —— H
H —— OH
H —— OH
CH₂OH
D-arabinose

$\xrightarrow[\;H_2O\;]{Br_2}$ $\xrightarrow[\;Fe_2(SO_4)_3\;]{H_2O_2}$

CHO
H —— OH
H —— OH
CH₂OH
D-erythrose

1) HCN/KCN
2) $H_2$/Pd/BaSO₄
3) $H_3O^+$

CHO
HO —— H
H —— OH
H —— OH
CH₂OH
D-arabinose

+

CHO
H —— OH
H —— OH
H —— OH
CH₂OH
D-ribose

**23-37**

CHO
HO —— H
H —— OH
H —— OH
CH₂OH
D-arabinose

$\xrightarrow[(1)]{H_2NOH \cdot HCl}$

H
\
C=NOH
HO —— H
H —— OH
H —— OH
CH₂OH
oxime

$\xrightarrow[(2)]{Ac_2O}$

C≡N
HO —— H
H —— OH
H —— OH
CH₂OH
cyanohydrin

$\xrightarrow[\;H_2O\;(3)]{HO^-}$

$CN^-$ +

CHO
H —— OH
H —— OH
CH₂OH
D-erythrose

23-38 Solve this problem by working backward from (+)-glyceraldehyde.

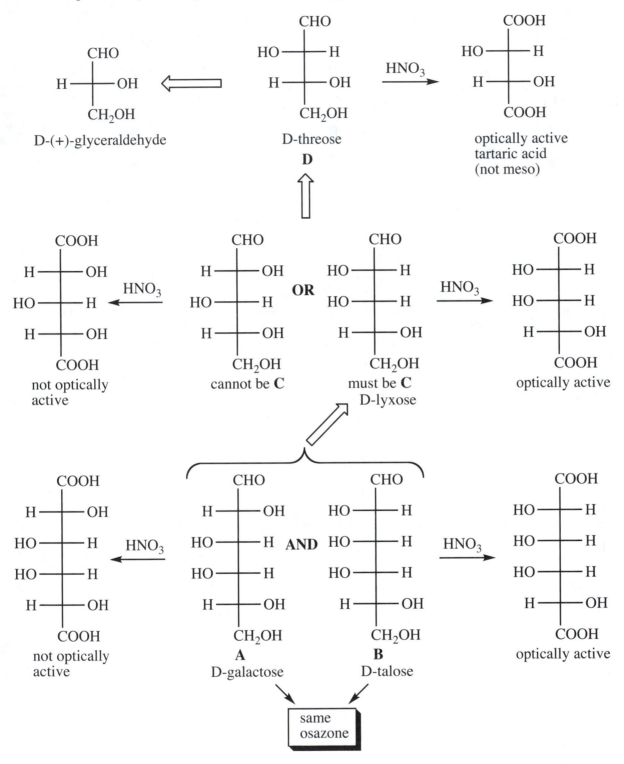

**23-39** This problem requires logic, inference, and working backwards, just as it did for Emil Fischer.

(a) Glucose and mannose degrade by one carbon to give arabinose. Parts (a) and (b) determine the structure of arabinose.

This is the only optically active 5-carbon aldaric acid.

This *could be* arabinose.

This *could be* arabinose.

From the nitric acid oxidation, we do not know which end is the alcohol and which end is the aldehyde.

(This structure is rotated 180° in the second part of part (b) to put the aldehyde at the top, as is customary.)

(b) Once we get to the 4-carbon sugars, then we can tell which **5**-carbon sugar must be arabinose.

This is optically active; it has two asymmetric carbons and no plane of symmetry.

This possible arabinose degrades to this 4-carbon sugar, oxidation of which gives an **optically active** aldaric acid. Conclusion: this starting material is not arabinose.

This is optically inactive; it has two asymmetric carbons and a plane of symmetry. This is *meso*-tartaric acid.

This possible arabinose degrades to this 4-carbon sugar, oxidation of which gives an **optically inactive** aldaric acid. Conclusion: this starting material is arabinose, and the 4-carbon sugar is erythrose.

CONCLUSIONS from (a) and (b):

arabinose

erythrose

23-39 continued

(c) Arabinose came from Ruff degradation of glucose or mannose.

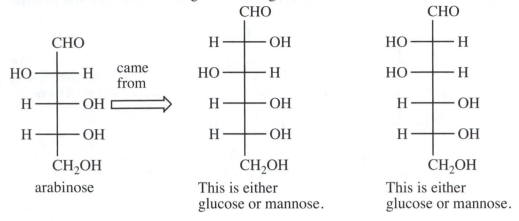

arabinose        This is either        This is either
                 glucose or mannose.   glucose or mannose.

(d) Chemically interchange the aldehyde with the alcohol at C-6. Mannose produces mannose.

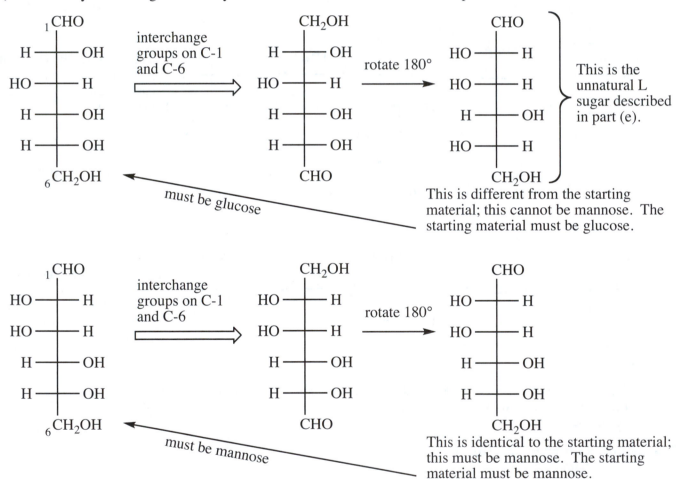

(e) The unnatural L sugar is shown at the end of the first sequence in part (d). This completes Fischer's proof of glucose.

- - - - - - - - - - - - - - - - - - - - - - - - - - - - - - - - - - - - - - - - - - - - - - - - - - - - - - - - - - - - - - -

It is hard for us in the 21st century to appreciate the challenges that Fischer and his students faced while doing this work: they had no spectroscopy, no chromatography—and since the incandescent light bulb had been invented only a decade earlier, they might not have had any electric lights! They might have had to interpret their NMRs by candlelight!

**23-40**

(a)

D-fructose

(b) acidic hydrolysis

$CH_3OH$ + $CH_3O$ —

$^1CH_2OCH_3$

$^2C{=}O$

$CH_3O-\underset{3}{|}-H$

$H-\underset{4}{|}-OCH_3$

$H-\underset{5}{|}-OH$

$^6CH_2OCH_3$

(c) Determining that the open chain form of fructose is a ketone at C-2 with a free OH at C-5 shows that fructose exists as a furanose hemiacetal.

**23-41**

(a)

methyl β-D-fructofuranoside

D-glyceraldehyde

(b)

methyl β-D-fructopyranoside

formic acid

Production of one equivalent of formic acid, and two fragments containing two and three carbons respectively, proves that the glycoside was in a six-membered ring.

**23-41 continued**

(c) In periodic acid oxidation of an aldohexose glycoside, glyceraldehyde is generated from carbons 4, 5, and 6. If configuration at the middle carbon is D, that means that carbon-5 of the aldohexose must have had the D configuration. On the other hand, if the isolated glyceraldehyde had the L configuration, then the original aldohexose must have been an L sugar.

**23-42**

The glycoside linkage does not change—only the stereochemistry at the hemiacetal anomeric carbon.

α anomer of maltose

β anomer of maltose

**23-43**

β anomer of maltose          open chain form of maltose

$Ag^+$, $NH_3$ (aq)

+ $Ag^0$
(mirror)

**23-44** Lactose is a hemiacetal; in water, the hemiacetal is in equilibrium with the open-chain form and can react as an aldehyde. Therefore, lactose can mutarotate and is a reducing sugar.

α anomer of lactose          β anomer of lactose

**23-45** Gentiobiose is a hemiacetal; in water, the hemiacetal is in equilibrium with the open-chain form and can react as an aldehyde. Gentiobiose can mutarotate and is a reducing sugar.

**23-46** Trehalose must be two glucose molecules connected by an α-1,1'-glycoside. Wedge and dash bonds are shown at the anomeric carbons for clarity.

α-D-glucopyranosyl-α-D-glucopyranoside

glucose rotated 180°

**23-47**

raffinose

galactose

glucose

fructose

α-1,6'

α-1,2'

β-2,1'

The lower glycoside linkage is α-1,2' from the glucose point of view, but β-2,1' from the fructose point of view.

invertase

melibiose = 6-O-(α-D-galactopyranosyl)-D-glucopyranose

α-1,6'

+

fructose (D-fructofuranose)

**23-48** cellulose acetate

23-49 cytosine:

uracil:

guanine:

23-50 Aminoglycosides, including nucleosides, are similar to acetals: stable to base, cleaved by acid.

(a)

$R_2NH_2$

3° aliphatic amine — strong base

$H^+$

$H_2O:$

$H_2O:$

hemiacetal form of ribose

(b)

cytidine

site of protonation; weak base

adenosine

See discussion on next page.

**23-50(b) continued**

Nucleosides are less rapidly hydrolyzed in aqueous acid because the site of protonation (the N in adenosine, and in cytidine, the oxygen shown with the negative charge in the second resonance form) is much less basic than the aliphatic amine in an aminoglycoside as shown in part (a). Nucleosides require stronger acid, or longer time and higher temperature, to be hydrolyzed.

This is important in living systems, as it would cause genetic damage or even death of an organism if its DNA or RNA were too easily decomposed. Organisms go to great lengths and expend considerable energy to maintain the structural integrity of their DNA.

**23-51**

guanine / cytosine / adenine / thymine

The polar resonance forms show how the hydrogen bonds are particularly strong. Each oxygen has significant negative charge, and in each pair, one H—N is polarized more strongly because the N has positive charge.

**23-52**

(a) (b) (c)

**23-53**

(a) (b) (c) (d)

**23-54**

(a) D-(−)-ribose    (b) D-(+)-altrose    (c) L-(+)-erythrose    (d) L-(−)-galactose    (e) L-(+)-idose

23-55

(a) D-aldohexose  (D configuration, aldehyde, 6 carbons)
(b) D-aldopentose (D configuration, aldehyde, 5 carbons)
(c) L-ketohexose  (L configuration, ketone, 6 carbons)
(d) L-aldohexose (L configuration, aldehyde, 6 carbons)
(e) D-ketopentose  (D configuration, ketone, 5 carbons)
(f) L-aldotetrose  (L configuration, aldehyde, 4 carbons)
(g) 2-acetamido-D-aldohexose (D configuration, aldehyde, 6 carbons in chain, with acetamido group at C-2)

23-56

(a)

(b) The products are diastereomers with different physical properties. They could be separated by crystallization, distillation, or chromatography.
(c) Both products are optically active. Each has two chiral centers and no plane of symmetry.

23-57

(a)

(b)

(S,S)-(−)-tartaric acid        (R,R)-(+)-tartaric acid        (R,S)-meso-tartaric acid

23-58

(a)
```
        COOH
   H  ——— OH
  HO  ——— H
  HO  ——— H
   H  ——— OH
       CH₂OH
```
$$
\begin{array}{c}
\text{COOH} \\
\text{H}-\!\!\!-\text{OH} \\
\text{HO}-\!\!\!-\text{H} \\
\text{HO}-\!\!\!-\text{H} \\
\text{H}-\!\!\!-\text{OH} \\
\text{CH}_2\text{OH}
\end{array}
$$

(b)
$$
\begin{array}{c}
\text{CHO} \\
\text{HO}-\!\!\!-\text{H} \\
\text{HO}-\!\!\!-\text{H} \\
\text{HO}-\!\!\!-\text{H} \\
\text{H}-\!\!\!-\text{OH} \\
\text{CH}_2\text{OH}
\end{array}
\quad + \quad
\begin{array}{c}
\text{CH}_2\text{OH} \\
\text{C}=\text{O} \\
\text{HO}-\!\!\!-\text{H} \\
\text{HO}-\!\!\!-\text{H} \\
\text{H}-\!\!\!-\text{OH} \\
\text{CH}_2\text{OH}
\end{array}
\quad + \ \text{others}
$$

(c)

(d)
$$
\begin{array}{c}
\text{COO}^{\ominus} \\
\text{H}-\!\!\!-\text{OH} \\
\text{HO}-\!\!\!-\text{H} \\
\text{HO}-\!\!\!-\text{H} \\
\text{H}-\!\!\!-\text{OH} \\
\text{CH}_2\text{OH}
\end{array}
$$

(e)
$$
\begin{array}{c}
\text{CH}_2\text{OH} \\
\text{H}-\!\!\!-\text{OH} \\
\text{HO}-\!\!\!-\text{H} \\
\text{HO}-\!\!\!-\text{H} \\
\text{H}-\!\!\!-\text{OH} \\
\text{CH}_2\text{OH}
\end{array}
$$

(f)

(g)

(h)
$$
\begin{array}{c}
\text{CH}_2\text{OH} \\
\text{H}-\!\!\!-\text{OH} \\
\text{HO}-\!\!\!-\text{H} \\
\text{HO}-\!\!\!-\text{H} \\
\text{H}-\!\!\!-\text{OH} \\
\text{CH}_2\text{OH}
\end{array}
$$

(i)
$$
\begin{array}{c}
\text{CHO} \\
\text{HO}-\!\!\!-\text{H} \\
\text{HO}-\!\!\!-\text{H} \\
\text{H}-\!\!\!-\text{OH} \\
\text{CH}_2\text{OH}
\end{array}
$$

(j)
$$
\begin{array}{c}
\text{CHO} \\
\text{H}-\!\!\!-\text{OH} \\
\text{H}-\!\!\!-\text{OH} \\
\text{HO}-\!\!\!-\text{H} \\
\text{HO}-\!\!\!-\text{H} \\
\text{H}-\!\!\!-\text{OH} \\
\text{CH}_2\text{OH}
\end{array}
\quad + \quad
\begin{array}{c}
\text{CHO} \\
\text{HO}-\!\!\!-\text{H} \\
\text{H}-\!\!\!-\text{OH} \\
\text{HO}-\!\!\!-\text{H} \\
\text{HO}-\!\!\!-\text{H} \\
\text{H}-\!\!\!-\text{OH} \\
\text{CH}_2\text{OH}
\end{array}
$$

(k)
$$
\begin{array}{c}
\text{CHO} \\
\text{H}-\!\!\!-\text{OH} \\
\text{HO}-\!\!\!-\text{H} \\
\text{HO}-\!\!\!-\text{H} \\
\text{H}-\!\!\!-\text{OH} \\
\text{CH}_2\text{OH}
\end{array}
\quad \xrightarrow{\text{excess HIO}_4} \quad
\mathbf{5}\ \ \underset{\substack{\text{from C-1} \\ \text{through C-5}}}{\text{H}-\overset{\displaystyle\text{O}}{\overset{\|}{\text{C}}}-\text{OH}}
\quad + \quad
\mathbf{1}\ \ \underset{\text{from C-6}}{\text{H}-\overset{\displaystyle\text{O}}{\overset{\|}{\text{C}}}-\text{H}}
$$

23-59

(a)

(b)

(c)

(d)

23-60

(a)

(b)

(c)

23-61

(a) methyl β-D-fructofuranoside
(b) 3,6-di-O-methyl-β-D-mannopyranose
(c) 4-O-(α-D-fructofuranosyl)-β-D-galactopyranose
(d) β-D-N-acetylgalactopyranosamine, or 2-acetamido-2-deoxy-β-D-galactopyranose

23-62   These are reducing sugars and would undergo mutarotation:

   —in problem 23-59:  (b) and (c);
   —in problem 23-60:  (a) and (c);
   —in problem 23-61:  (b), (c), and (d).

23-63

(a)

CHO
HO —— H
H —— OH
H —— OH
H —— OH
CH₂OH

D-altrose

(b)

β-D-altropyranose

23-64

galactose {

6-O-(α-D-galactopyranosyl)-D-fructofuranose

α-1,6

fructose

23-65

(a) These two D-aldopentoses will give optically active aldaric acids.

CHO
HO —— H
H —— OH
H —— OH
CH₂OH

D-arabinose

CHO
HO —— H
HO —— H
H —— OH
CH₂OH

D-lyxose

(b) Only D-threose (of the aldotetroses) will give an optically active aldaric acid.

D-threose

CHO
HO —— H
H —— OH
CH₂OH

23-65 continued

(c) **X** is D-galactose.

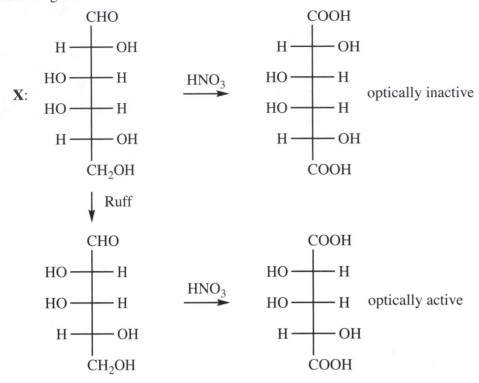

The other aldohexose that gives an optically inactive aldaric acid is D-allose, with all OH groups on the right side of the Fischer projection. Ruff degradation followed by nitric acid gives an optically *inactive* aldaric acid, however, so **X** cannot be D-allose.

(d) The optically active, five-carbon aldaric acid comes from the optically active pentose, not from the optically inactive, six-carbon aldaric acid. The principle is not violated.

(e)

CHO
HO — H
HO — H
H — OH
CH₂OH

→ Ruff →

CHO
HO — H
H — OH
CH₂OH
D-threose

→ HNO₃ →

COOH
HO — H
H — OH
COOH
(S,S)-tartaric acid
optically active

23-66 Tagatose is a monosaccharide, a ketohexose, that is found in the pyranose form.

(a)

CH₂OH
C=O
HO — H
HO — H
H — OH
CH₂OH

(b)

**606**

23-67

(a) The less hindered 1° OH will react first.

(b)

Because Si is larger than C, the Si—O bond length is around 1.6 Angstrom units, whereas a typical C—O bond is about 1.4 Angstrom units.

(c)

23-68

(a)

(c)

The 4- and 6-OH groups of glucose reacted with benzaldehyde to form the acetal.

(c) and (d) The chiral center is marked with (*). This stereoisomer is a diastereomer since only one chiral center is inverted. This diastereomer puts the phenyl group in an axial position, definitely less stable than the structure shown in part (a). Only the product shown in (a) will be isolated from this reaction.

**23-68** continued

(e)

HO, O, base

Only the 2'- and 3'-OH groups are close enough to form this cyclic acetal (ketal) from acetone, called an acetonide.

$H_3C$    $CH_3$

**23-69**

(a)

(b)

(c)

23-70 Bonds from C to H are omitted for simplicity.

23-71

(a) No, there is no relation between the amount of G and A.

(b) Yes, this must be true mathematically.

(c) Chargaff's rule must apply only to double-stranded DNA. For each G in one strand, there is a complementary C in the opposing strand, but there is no correlation between G and C *in the same strand*.

23-72 Bonds from C to H are omitted for simplicity.

(a)

2'-deoxythymidine
(abbreviated dT)

3'-azido-2',3'-dideoxythymidine
(AZT)

No phosphate can attach to the azide group, so
synthesis of the DNA chain is terminated.

(b)

AZT 5'-triphosphate

**23-73** Recall from text section 19-16 that nitrous acid is unstable, generating nitrosonium ion.

(a)

$$H-O-N=O \; + \; H^+ \longrightarrow H_2O \; + \; \overset{\oplus}{N}=O$$

cytosine (C)

two rapid proton transfers

No mechanism for substitution of diazonium ions was presented in Chapter 19. What is shown here is one possibility but not the only one.

− N₂ gas

If this species exists, its lifetime is very short as water will attack quickly.

enol tautomer of uracil

uracil (U)

(b)  In base pairing, cytosine pairs with guanine.  If cytosine is converted to uracil, however, each replication will not carry the complement of cytosine (guanine) but instead will carry the complement of uracil (adenine).  This is the definition of a mutation, where the wrong base is inserted in a nucleic acid chain.

(c)  In RNA, the transformation of cytosine (C) to uracil (U) is not detected as a problem because U is a base normally found in RNA so it goes unrepaired.  In DNA, however, thymine (with an extra methyl group) is used instead of uracil.  If cytosine is diazotized to uracil, the DNA repair enzymes detect it as a mutation and correct it.

23-74

(a)

(b) Trityl groups are specific for 1° alcohols for steric reasons: the trityl group is so big that even a 2° alcohol is too crowded to react at the central carbon. It is possible for a trityl to go on a 2° alcohol, but the reaction is exceedingly slow and in the presence of a 1° alcohol, the reaction is done at the 1° alcohol long before the 2° alcohol gets started.

(c) Reactions happen faster when the product or intermediate is stabilized. Each $OCH_3$ group stabilizes the carbocation by resonance as shown here; two $OCH_3$ groups stabilize more than just one, increasing the rate of removal. The color comes from the extended conjugation through all three rings and out onto the $OCH_3$ groups. Compare the DMT structure with that of phenolphthalein, the most common acid base indicator, that turns pink in its ring open form shown here. Phenolphthalein is simply another trityl group with different substituents.

One of the major resonance contributors shows the delocalization of the positive charge on the oxygen.

One of the resonance contributors shows the delocalization of the positive charge onto the central carbon.

plus many more resonance forms

phenolphthalein

24-1

(a) 

(b) 

(c) 

(d) 

24-2

(a) The configurations around the asymmetric carbons of (*R*)-cysteine and (*S*)-alanine are the same. The *designation* of configuration changes because sulfur changes the priorities of the side chain and the COOH.

(*S*)-alanine with group priorities shown

(*R*)-Cysteine with group priorities shown; note that the COOH and the CH₂SH priorities are reversed compared with (*S*)-alanine.

(b) Fischer projections show that both (*S*)-alanine and (*R*)-cysteine are L-amino acids.

(*S*)-alanine
L-alanine

(*R*)-cysteine
L-cysteine

24-3 In their evolution, plants have needed to be more resourceful than animals in developing biochemical mechanisms for survival. Thus, plants make more of their own required compounds than animals do. The amino acid phenylalanine is produced by plants but required in the diet of mammals. To interfere with a plant's production of phenylalanine is fatal to the plant, but since humans do not produce phenylalanine, glyphosate is virtually nontoxic to us.

24-4 Here is a simple way of determining if a group will be protonated: *at solution pH below the group's pKₐ value, the group will be protonated; at pH higher than the group's pKₐ value, it will not be protonated.*

(a) $H_2N-\underset{\underset{CH(CH_3)CH_2CH_3}{|}}{\overset{\overset{H}{|}}{C}}-COO^{\ominus}$  pH 11 ⟹ Both groups are deprotonated.

(b) pH 2 ⟹ Both groups are protonated.

(c) $\overset{\oplus}{H_3N}-\underset{\underset{CH_2CH_2CH_2NH-C-NH_2}{|}}{\overset{\overset{H}{|}}{C}}-COO^{\ominus}$  pH 7 ⟹ Only the amines remain protonated.

(d) $\overset{\oplus}{H_3N}-\underset{\underset{CH_2CH_2COO^{\ominus}}{|}}{\overset{\overset{H}{|}}{C}}-COO^{\ominus}$  pH 7 ⟹ Only the amine remains protonated.

**613**

**24-4 continued**

(e)

alanine

lysine

aspartic acid

(i) pH 6

(ii) pH 11

(iii) pH 2

**24-5**

Protonation of the guanidino group gives a resonance-stabilized cation with all octets filled and the positive charge delocalized over three nitrogen atoms. Arginine's strongly basic isoelectric point reflects the unusual basicity of the guanidino group due to this resonance stabilization in the protonated form. (See Problems 1-39 and 19-49(a).)

**24-6**

tryptophan

histidine

not basic
(like pyrrole)

basic
(like pyridine)

not basic
(like pyrrole)

The basicity of any nitrogen depends on its electron pair's availability for bonding with a proton. In tryptophan, the nitrogen's electron pair is part of the aromatic $\pi$ system; without this electron pair in the $\pi$ system, the molecule would not be aromatic. Using this electron pair for bonding to a proton would therefore destroy the aromaticity—not a favorable process.

In the imidazole ring of histidine, the electron pair of one nitrogen is also part of the aromatic $\pi$ system and is unavailable for bonding; this nitrogen is not basic. The electron pair on the other nitrogen, however, is in an $sp^2$ orbital available for bonding, and is about as basic as pyridine.

**24-7** At pH 9.7, alanine (isoelectric point (pI) 6.0) has a charge of –1 and will migrate to the anode. Lysine (pI 9.7) is at its isoelectric point and will not move. Aspartic acid (pI 2.8) has a charge of –2 and will also migrate to the anode, faster than alanine.

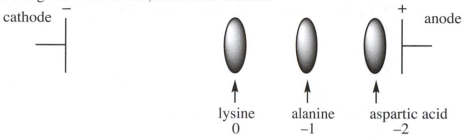

**24-8** At pH 6.0, tryptophan (pI 5.9) has a charge of zero and will not migrate. Cysteine (pI 5.0) has a partial negative charge and will move toward the anode. Histidine (pI 7.6) has a partial positive charge and will move toward the cathode.

**24-9**

(a) $O{=}C{-}COOH$ with $CH_3$ $\xrightarrow[\text{H}_2,\,\text{Pd}]{\text{NH}_3}$ $H_2N{-}CH{-}COOH$ with $CH_3$

(b) $O{=}C{-}COOH$ with $CH_2CH(CH_3)_2$ $\xrightarrow[\text{H}_2,\,\text{Pd}]{\text{NH}_3}$ $H_2N{-}CH{-}COOH$ with $CH_2CH(CH_3)_2$

(c) $O{=}C{-}COOH$ with $CH_2OH$ $\xrightarrow[\text{H}_2,\,\text{Pd}]{\text{NH}_3}$ $H_2N{-}CH{-}COOH$ with $CH_2OH$

(d) $O{=}C{-}COOH$ with $CH_2CH_2CONH_2$ $\xrightarrow[\text{H}_2,\,\text{Pd}]{\text{NH}_3}$ $H_2N{-}CH{-}COOH$ with $CH_2CH_2CONH_2$

**24-10** All of these reactions use: first arrow: (1) $Br_2/PBr_3$, followed by $H_2O$ workup; second arrow: (2) excess $NH_3$, followed by neutralizing workup.

(a) 

$$\underset{\underset{H}{|}}{H_2C}-COOH \xrightarrow{(1)} \underset{\underset{H}{|}}{Br-CH}-COOH \xrightarrow{(2)} \underset{\underset{H}{|}}{H_2N-CH}-COOH$$

(b)

$$\underset{\underset{CH_2CH(CH_3)_2}{|}}{H_2C}-COOH \xrightarrow{(1)} \underset{\underset{CH_2CH(CH_3)_2}{|}}{Br-CH}-COOH \xrightarrow{(2)} \underset{\underset{CH_2CH(CH_3)_2}{|}}{H_2N-CH}-COOH$$

(c)

$$\underset{\underset{CH_2CH_2COOH}{|}}{H_2C}-COOH \xrightarrow{(1)} \underset{\underset{CH_2CH_2COOH}{|}}{Br-CH}-COOH \xrightarrow{(2)} \underset{\underset{CH_2CH_2COOH}{|}}{H_2N-CH}-COOH$$

In part (c), care must be taken to avoid reaction α to the other COOH. In practice, this would be accomplished by using less than one-half mole of bromine per mole of the diacid.

**24-11**

(a)

(b)

(c)

salt, not
acid — why?

24-11 continued

(d)

$$\underset{\substack{|\\ \text{COOEt}}}{\overset{\substack{\text{COOEt}\\ |}}{\text{N}-\text{CH}}} \xrightarrow[\text{2) BrCH}_2\text{CH(CH}_3)_2]{\text{1) NaOEt}} \xrightarrow[\Delta]{\text{H}_3\text{O}^+} \overset{\oplus}{\text{H}_3}\text{N}-\underset{\substack{|\\ \text{CH}_2\text{CH(CH}_3)_2}}{\text{CH}}-\text{COOH}$$

24-12

$$\text{AcNH}-\underset{\substack{|\\ \text{COOEt}}}{\overset{\substack{\text{COOEt}\\ |}}{\text{CH}}} \xrightarrow[\text{2) BrCH}_2\text{Ph}]{\text{1) NaOEt}} \text{AcNH}-\underset{\substack{|\\ \text{COOEt}}}{\overset{\substack{\text{COOEt}\\ |}}{\text{C}}}-\text{CH}_2\text{Ph} \xrightarrow[\Delta]{\text{H}_3\text{O}^+} \overset{\oplus}{\text{H}_3}\text{N}-\underset{\substack{|\\ \text{CH}_2\text{Ph}}}{\text{CH}}-\text{COOH}$$

acetamidomalonic
ester

24-13

(a)

$$\text{PhCH}_2-\overset{\substack{\text{O}\\ ||}}{\text{CH}} \xrightarrow[\text{H}_2\text{O}]{\text{NH}_3,\ \text{HCN}} \text{PhCH}_2-\underset{\substack{|\\ \text{C}\equiv\text{N}}}{\overset{\substack{\text{NH}_2\\ |}}{\text{CH}}} \xrightarrow[\Delta]{\text{H}_3\text{O}^+} \overset{\oplus}{\text{H}_3}\text{N}-\underset{\substack{|\\ \text{CH}_2\text{Ph}}}{\text{CH}}-\text{COOH}$$

(b) While the solvent for the Strecker synthesis is water, the proton acceptor is ammonia and the proton donor is ammonium ion.

This protonated
imine is rapidly
attacked by
cyanide nucleophile.

(abbreviate as

$$\text{R}-\text{C}\equiv\text{N}$$

on the next page)

continued on next page

**617**

Copyright © 2013 Pearson Education, Inc.

24-13 (b) continued

mechanism of acid hydrolysis of the nitrile

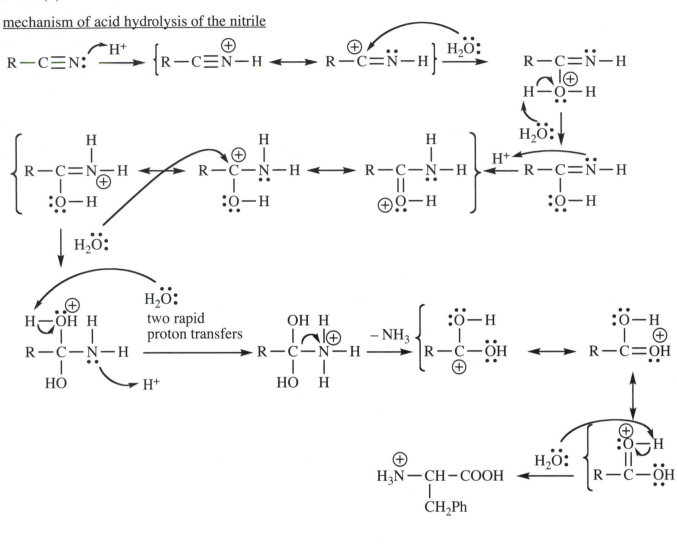

24-14

(a)

$$\underset{\underset{CH_2CH(CH_3)_2}{|}}{\overset{\overset{O}{\|}}{C}} - H \xrightarrow[\text{H}_2\text{O}]{\text{NH}_3, \text{HCN}} \underset{\underset{CH_2CH(CH_3)_2}{|}}{H_2N-CH-CN} \xrightarrow[\Delta]{\text{H}_3\text{O}^+} \underset{\underset{CH_2CH(CH_3)_2}{|}}{\overset{\oplus}{H_3N}-CH-COOH}$$

(b)

$$\underset{\underset{CH(CH_3)_2}{|}}{\overset{\overset{O}{\|}}{C}} - H \xrightarrow[\text{H}_2\text{O}]{\text{NH}_3, \text{HCN}} \underset{\underset{CH(CH_3)_2}{|}}{H_2N-CH-CN} \xrightarrow[\Delta]{\text{H}_3\text{O}^+} \underset{\underset{CH(CH_3)_2}{|}}{\overset{\oplus}{H_3N}-CH-COOH}$$

(c)

$$\underset{\underset{CH_2COOH}{|}}{\overset{\overset{O}{\|}}{C}} - H \xrightarrow[\text{H}_2\text{O}]{\text{NH}_3, \text{HCN}} \underset{\underset{CH_2COO^{\ominus}}{|}}{H_2N-CH-CN} \xrightarrow[\Delta]{\text{H}_3\text{O}^+} \underset{\underset{CH_2COOH}{|}}{\overset{\oplus}{H_3N}-CH-COOH}$$

before acidification

**618**

**24-15** In acid solution, the free amino acid will be protonated, with a positive charge, and probably soluble in water as are other organic ions. The acylated amino acid, however, is not basic since the nitrogen is present as an amide. In acid solution, the acylated amino acid is neutral and not soluble in water. Water extraction or ion-exchange chromatography (Figure 24-12) would be practical techniques to separate these compounds.

**24-16**

abbreviate $\overset{\oplus}{H_3N}-\underset{\underset{CH_2Ph}{|}}{CH}-\overset{\overset{O}{\|}}{C}OEt$ as $R-\overset{\overset{O}{\|}}{C}OEt$

(protonated form in acid solution)

plus two other resonance forms

**24-17**

$\overset{\oplus}{H_3N}-\underset{\underset{CH_2CH_2CONH_2}{|}}{CH}-\overset{\ominus}{COO}$ $\xrightarrow[H^+]{PhCH_2OH}$ $\overset{\oplus}{H_3N}-\underset{\underset{CH_2CH_2CONH_2}{|}}{CH}-COOCH_2Ph$ $\xrightarrow{H_2,\ Pd}$ $\overset{\oplus}{H_3N}-\underset{\underset{CH_2CH_2CONH_2}{|}}{CH}-COOH$

$+\ CH_3Ph$

**24-18**

$\overset{\oplus}{H_3N}-\underset{\underset{CH_2CH_2SCH_3}{|}}{CH}-\overset{\ominus}{COO}$ $\xrightarrow[\text{base like Et}_3N]{PhCH_2O-\overset{\overset{O}{\|}}{C}-Cl}$ $PhCH_2O\overset{\overset{O}{\|}}{C}-\underset{\overset{|}{H}}{N}-\underset{\underset{CH_2CH_2SCH_3}{|}}{CH}-COOH$

$\downarrow H_2,\ Pd$

$PhCH_3\ +\ CO_2\ +\ \overset{\oplus}{H_3N}-\underset{\underset{CH_2CH_2SCH_3}{|}}{CH}-COOH$

These are the most significant resonance contributors in which the electronegative oxygens carry the negative charge. There are also two other forms in which the negative charge is on the carbons bonded to the nitrogen, plus the usual resonance forms involving the alternate Kekulé structures of the benzene rings.

24-20

(a)

Thr — Phe — Met

(b)

seryl — arginyl — glycyl — phenylalanine

(c)

**I** (isoleucine) — **M** (methionine) **Q** (glutamine) **D** (aspartic acid) **K** (lysine)

(d)

**E** (glutamic acid) — **L** (leucine) — **V** (valine) — **I** (isoleucine) — **S** (serine)

*Try spelling your name in peptides!*

**620**

24-21

(a)

(b)

(c)

(d)

24-22

Step 3

H₂N—Ile→Gln→peptide  $\xrightarrow[\text{2) H}_3\text{O}^+]{\text{1) PhNCS}}$  [structure] + H₂N—Gln→peptide

Step 4

H₂N—Gln→peptide  $\xrightarrow[\text{2) H}_3\text{O}^+]{\text{1) PhNCS}}$  [structure] + H₂N—peptide

24-23  Abbreviate the N-terminus of the peptide chain as NH₂R .

(a)  This is a nucleophilic aromatic substitution by the addition-elimination mechanism.  The presence of two nitro groups makes this reaction feasible under mild conditions.

(b)  The main drawback of the Sanger method is that only one amino acid is analyzed per sample of protein.  The Edman degradation can usually analyze more than 20 amino acids per sample of protein.

24-24

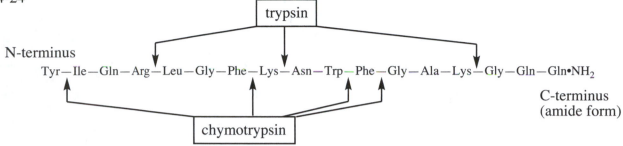

24-25

Phe — Gln — Asn

Pro — Arg — Gly•NH₂

Cys — Tyr — Phe

Asn — Cys — Pro — Arg

Tyr — Phe — Gln — Asn

Cys — Tyr — Phe — Gln — Asn — Cys — Pro — Arg — Gly•NH₂

24-26  abbreviations used in this problem:

mechanism of formation of Z-Ala

continued on next page

24-26 continued

mechanism of ethyl chloroformate activation

The oxygen of C=O
is more nucleophilic
than OH.

plus two other
resonance forms

mechanism of the coupling with valine

24-27

Z       Ala       Val       Phe

$$\underset{\text{CH}_3}{\text{PhCH}_2\overset{\text{O}}{\overset{\|}{\text{C}}}-\text{NH}-\text{CH}-\overset{\text{O}}{\overset{\|}{\text{C}}}-\text{NH}-\underset{\text{CH(CH}_3)_2}{\text{CH}}-\overset{\text{O}}{\overset{\|}{\text{C}}}-\text{NH}-\underset{\text{CH}_2\text{Ph}}{\text{CH}}-\overset{\text{O}}{\overset{\|}{\text{C}}}-\text{OH}}$$

↓   $\text{Cl}-\overset{\text{O}}{\overset{\|}{\text{C}}}-\text{OEt}$

$$\text{Z}-\text{NH}-\underset{\text{CH}_3}{\text{CH}}-\overset{\text{O}}{\overset{\|}{\text{C}}}-\text{NH}-\underset{\text{CH(CH}_3)_2}{\text{CH}}-\overset{\text{O}}{\overset{\|}{\text{C}}}-\text{NH}-\underset{\text{CH}_2\text{Ph}}{\text{CH}}-\overset{\text{O}}{\overset{\|}{\text{C}}}-\text{O}-\overset{\text{O}}{\overset{\|}{\text{C}}}-\text{OEt}$$

↓   $\text{H}_2\text{NCH}_2\text{COOH}$     glycine

$$\text{Z}-\text{NH}-\underset{\text{CH}_3}{\text{CH}}-\overset{\text{O}}{\overset{\|}{\text{C}}}-\text{NH}-\underset{\text{CH(CH}_3)_2}{\text{CH}}-\overset{\text{O}}{\overset{\|}{\text{C}}}-\text{NH}-\underset{\text{CH}_2\text{Ph}}{\text{CH}}-\overset{\text{O}}{\overset{\|}{\text{C}}}-\text{NH}-\underset{\text{H}}{\text{CH}}-\overset{\text{O}}{\overset{\|}{\text{C}}}-\text{OH}$$

↓   $\text{Cl}-\overset{\text{O}}{\overset{\|}{\text{C}}}-\text{OEt}$

$$\text{Z}-\text{NH}-\underset{\text{CH}_3}{\text{CH}}-\overset{\text{O}}{\overset{\|}{\text{C}}}-\text{NH}-\underset{\text{CH(CH}_3)_2}{\text{CH}}-\overset{\text{O}}{\overset{\|}{\text{C}}}-\text{NH}-\underset{\text{CH}_2\text{Ph}}{\text{CH}}-\overset{\text{O}}{\overset{\|}{\text{C}}}-\text{NH}-\underset{\text{H}}{\text{CH}}-\overset{\text{O}}{\overset{\|}{\text{C}}}-\text{O}-\overset{\text{O}}{\overset{\|}{\text{C}}}-\text{OEt}$$

↓   $\underset{\text{CH}_2\text{CH(CH}_3)_2}{\text{H}_2\text{N}-\text{CH}-\text{COOH}}$     leucine

$$\text{Z}-\text{NH}-\underset{\text{CH}_3}{\text{CH}}-\overset{\text{O}}{\overset{\|}{\text{C}}}-\text{NH}-\underset{\text{CH(CH}_3)_2}{\text{CH}}-\overset{\text{O}}{\overset{\|}{\text{C}}}-\text{NH}-\underset{\text{CH}_2\text{Ph}}{\text{CH}}-\overset{\text{O}}{\overset{\|}{\text{C}}}-\text{NH}-\underset{\text{H}}{\text{CH}}-\overset{\text{O}}{\overset{\|}{\text{C}}}-\text{NH}-\underset{\text{CH}_2\text{CH(CH}_3)_2}{\text{CH}}-\text{COOH}$$

↓ $\text{H}_2$, Pd

$$\text{H}_2\text{N}-\underset{\text{CH}_3}{\text{CH}}-\overset{\text{O}}{\overset{\|}{\text{C}}}-\text{NH}-\underset{\text{CH(CH}_3)_2}{\text{CH}}-\overset{\text{O}}{\overset{\|}{\text{C}}}-\text{NH}-\underset{\text{CH}_2\text{Ph}}{\text{CH}}-\overset{\text{O}}{\overset{\|}{\text{C}}}-\text{NH}-\underset{\text{H}}{\text{CH}}-\overset{\text{O}}{\overset{\|}{\text{C}}}-\text{NH}-\underset{\text{CH}_2\text{CH(CH}_3)_2}{\text{CH}}-\text{COOH}$$

24-28

$$PhCH_2O\overset{O}{\underset{\|}{C}}-Cl \quad + \quad \underset{\underset{CH_3-CHCH_2CH_3}{|}}{H_2N-CH-COOH} \quad \longrightarrow \quad \underset{\underset{CH_3-CHCH_2CH_3}{|}}{Z-NH-CH-\overset{O}{\overset{\|}{C}}-OH}$$

isoleucine

$$\downarrow \quad Cl-\overset{O}{\overset{\|}{C}}-OEt$$

$$\underset{\underset{CH_3-CHCH_2CH_3}{|}}{Z-NH-CH-\overset{O}{\overset{\|}{C}}-O-\overset{O}{\overset{\|}{C}}-OEt}$$

$$\downarrow \quad H_2NCH_2COOH \qquad glycine$$

$$\underset{\underset{CH_3-CHCH_2CH_3}{|}}{Z-NH-CH-\overset{O}{\overset{\|}{C}}-NH-}\underset{\underset{H}{|}}{CH-\overset{O}{\overset{\|}{C}}-OH}$$

$$\downarrow \quad Cl-\overset{O}{\overset{\|}{C}}-OEt$$

$$\underset{\underset{CH_3-CHCH_2CH_3}{|}}{Z-NH-CH-\overset{O}{\overset{\|}{C}}-NH-}\underset{\underset{H}{|}}{CH-\overset{O}{\overset{\|}{C}}-O-\overset{O}{\overset{\|}{C}}-OEt}$$

$$\downarrow \quad \underset{\underset{CH_2CONH_2}{|}}{H_2N-CH-COOH} \qquad asparagine$$

$$\underset{\underset{CH_3-CHCH_2CH_3}{|}}{Z-NH-CH-\overset{O}{\overset{\|}{C}}-NH-}\underset{\underset{H}{|}}{CH-\overset{O}{\overset{\|}{C}}-NH-}\underset{\underset{CH_2CONH_2}{|}}{CH-COOH}$$

$$\downarrow \quad H_2, Pd$$

$$\underset{\underset{CH_3-CHCH_2CH_3}{|}}{H_2N-CH-\overset{O}{\overset{\|}{C}}-NH-}\underset{\underset{H}{|}}{CH-\overset{O}{\overset{\|}{C}}-NH-}\underset{\underset{CH_2CONH_2}{|}}{CH-COOH}$$

24-29   We use "*c*-Hx" to stand for "cyclohexyl".

DiCyclohexylCarbodiimide
DCC

$$c\text{-Hx}-N=C=N-c\text{-Hx}$$

mechanism

This is an "activated" carboxyl group. The OH has been converted into an excellent leaving group.

plus other resonance forms

resonance-stabilized

DiCyclohexylUrea
DCU

24-30

Boc | Ala | Val | Phe

$$Me_3COC-NH-CH-C-NH-CH-C-NH-CH-C-O-\underline{P}$$
(with C=O groups; side chains: CH$_3$, CH(CH$_3$)$_2$, CH$_2$Ph)

↓ CF$_3$COOH

$$\overset{\oplus}{H_3N}-CH-C-NH-CH-C-NH-CH-C-O-\underline{P}$$
(side chains: CH$_3$, CH(CH$_3$)$_2$, CH$_2$Ph)

↓ DCC, $Me_3COC-NHCH_2COOH$    Boc—glycine

$$Me_3COC-NH-CH-C-NH-CH-C-NH-CH-C-NH-CH-C-O-\underline{P}$$
(side chains: H, CH$_3$, CH(CH$_3$)$_2$, CH$_2$Ph)

↓ CF$_3$COOH

$$\overset{\oplus}{H_3N}-CH-C-NH-CH-C-NH-CH-C-NH-CH-C-O-\underline{P}$$
(side chains: H, CH$_3$, CH(CH$_3$)$_2$, CH$_2$Ph)

↓ DCC, $Me_3COC-NH-CH-COOH$    Boc—leucine
(side chain: CH$_2$CHMe$_2$)

$$Boc\ NH-CH-C-NH-CH-C-NH-CH-C-NH-CH-C-NH-CH-C-O-\underline{P}$$
(side chains: CH$_2$CHMe$_2$, H, CH$_3$, CH(CH$_3$)$_2$, CH$_2$Ph)

↓ HF

$$\overset{\oplus}{H_3N}-CH-C-NH-CH-C-NH-CH-C-NH-CH-C-NH-CH-C-OH$$
(side chains: CH$_2$CHMe$_2$, H, CH$_3$, CH(CH$_3$)$_2$, CH$_2$Ph)

24-31

Boc—asparagine

Boc—glycine    $Me_3COC$—$NHCH_2COOH$    DCC

Boc—isoleucine    $Me_3COC$—NH-CH-COOH    DCC

$CF_3COOH$

HF

**24-32**

(a) phenylalanine:

at pH 1
fully protonated

at pI = pH 5.5
charges balanced

at pH 11
fully deprotonated

(b) histidine:

pyridine-type ← pyrrole-type

at pH 1
fully protonated

at pH 4
COOH deprotonated

at pI = pH 7.6
charges balanced

at pH 11
fully deprotonated

The imidazole ring is aromatic, with a pyrrole-type N that is not basic and a pyridine-type N that is weakly basic. Aromatic amines like pyridine are weaker bases than aliphatic amines by several pK units.

(c) glutamic acid:

at pH 1
fully protonated

at pI = pH 3.2
charges balanced

at pH 7

at pH 11
fully deprotonated

The $NH_3^+$ group is a strong electron-withdrawing group and strengthens the nearby COOH by 2–3 pK units as compared to the side chain COOH, which is not affected by induction in this molecule.

**24-33**

$$H_2N-CH-\overset{\overset{O}{\|}}{C}-NH-CH-\overset{\overset{O}{\|}}{C}-NH-CH-\overset{\overset{O}{\|}}{C}-NH_2$$

CH_2OH    (CH_2)_2    CH_2CH_2SCH_3

CONH_2

**24-34**

(a)

+ $CO_2$ +

(b)
CH_3
|
$H_2N-CHCOOH$

(c)
$$CH_3\overset{\overset{O}{\|}}{C}-NH-CH-COOH$$
(CH_2)_4NHCOCH_3

(d)
The enzyme does not recognize D amino acids.

COOH

L-proline

+

HOOC

N-acetyl-D-proline    CH_3

24-34 continued

(e) H₂N—CH-CN
     |
     H₃C—CHCH₂CH₃

(f) $\overset{\oplus}{H_3N}$—CH-COOH
     |
     H₃C—CHCH₂CH₃
     *isoleucine*

(g) Br—CH-$\overset{\overset{O}{\|}}{C}$—Br
     |
     CH₂CH(CH₃)₂
     With a water workup, the acid
     bromide would become COOH.

(h) H₂N—CH-$\overset{\ominus}{COO}$ $\overset{\oplus}{NH_4}$   OR
     |
     CH₂CH(CH₃)₂
     *leucine*

H₂N—CH-$\overset{\overset{O}{\|}}{C}$—NH₂   This is another possible answer,
     |                         depending on whether the acid bromide
     CH₂CH(CH₃)₂              is hydrolyzed before adding ammonia.

24-35

(a) $\overset{\overset{O}{\|}}{C}$—COOH   $\xrightarrow[H_2, Pd]{NH_3}$   $\overset{\oplus}{H_3N}$—CH-COOH   valine
     |                                        |                    after acid workup
     CH(CH₃)₂                                 CH(CH₃)₂

(b) H₂C—COOH   $\xrightarrow[PBr_3]{Br_2 \quad H_2O}$   Br—CH-COOH   $\xrightarrow{\overset{excess}{NH_3}}$   $\overset{\oplus}{H_3N}$—CH-COOH
     |                                        |                              |                        after acid workup
     H₃C—CHCH₂CH₃                             H₃C—CHCH₂CH₃                   H₃C—CHCH₂CH₃
                                                                            *isoleucine*

(c) $\overset{\overset{O}{\|}}{C}$—H   $\xrightarrow[H_2O]{NH_3, HCN}$   H₂N—CH-CN   $\xrightarrow[\Delta]{H_3O^+}$   $\overset{\oplus}{H_3N}$—CH-COOH
     |                                        |                                        |
     CH₂CH(CH₃)₂                              CH₂CH(CH₃)₂                              CH₂CH(CH₃)₂
                                                                                      *leucine*

(d)
     [phthalimide]—N-CH(COOEt)₂   $\xrightarrow[2) BrCH_2Ph]{1) NaOEt}$   [phthalimide]—N-C(CH₂Ph)(COOEt)₂   $\xrightarrow[\Delta]{H_3O^+}$   $\overset{\oplus}{H_3N}$—$\overset{\overset{H}{|}}{C}$—COOH
                                                                                                                                          |
                                                                                                                                          CH₂Ph
                                                                                                                                          *phenylalanine*

24-36

(a) H₂N—CH-COOH   +   HOCH(CH₃)₂   $\xrightarrow{H^+}$   $\overset{\oplus}{H_3N}$—CH-COOCH(CH₃)₂
     |                                                    |
     CH₃                                                  CH₃

(b) H₂N—CH-COOH   +   PhC$\overset{\overset{O}{\|}}{}$—Cl   $\xrightarrow{(pyridine)}$   PhC$\overset{\overset{O}{\|}}{}$—NH-CH-COOH
     |                                                                                          |
     CH₃                                                                                        CH₃

(c) H₂N—CH-COOH   +   PhCH₂O$\overset{\overset{O}{\|}}{C}$—Cl   $\xrightarrow{Et_3N}$   PhCH₂O$\overset{\overset{O}{\|}}{C}$—NH—CH-COOH
     |                                                                                          |
     CH₃                                                                                        CH₃

24-36 continued

(d) $H_2N-CH-COOH$ + $Me_3COC-O-COCMe_3$ → $Me_3COC-NH-CH-COOH$
  $\quad\quad\quad\quad\quad |$ $\quad\quad\quad\quad\quad\quad || \quad\quad\quad ||$ $\quad\quad\quad\quad\quad\quad\quad\quad || \quad\quad\quad\quad\quad\quad |$
  $\quad\quad\quad\quad\quad CH_3$ $\quad\quad\quad\quad\quad\quad O \quad\quad\quad\quad O$ $\quad\quad\quad\quad\quad\quad\quad\quad\quad O \quad\quad\quad\quad\quad\quad\quad CH_3$

24-37

D-alanine

24-38

racemic histidine

24-39

racemic tryptophan

24-40

(a)  neutral

(b)  neutral

(c)  basic: two basic side chains and one acidic side chain

**631**

24-40 continued

(d)

$$H_2N-CH\text{-}\overset{\overset{\displaystyle O}{\|}}{C}-NH\text{-}CH\text{-}\overset{\overset{\displaystyle O}{\|}}{C}-NH\text{-}CH\text{-}\overset{\overset{\displaystyle O}{\|}}{C}-OH$$

with side chains: $CH_2$ / $CH_2COOH$ ; $CH_2SH$ ; $CH_2CH_2CONH_2$

acidic: carboxylic acid side chain, and weakly acidic SH

24-41

(a), (b), (c)

isoleucine                    glutamine

$$CH_3CH_2-CH-CH-NH\overset{*}{-}C-CH-CH_2CH_2-C-NH_2$$

with $CH_3$ on the CH; C-terminus → $CONH_2$; $NH\overset{*}{-}CO-CH_2NH_2$ ← N-terminus

glycine

Peptide bonds are denoted with asterisks (*).

(d) glycylglutaminylisoleucinamide; Gly—Gln—Ile • $NH_2$

24-42

Aspartame:

$$H_2N-CH\text{-}\overset{\overset{\displaystyle O}{\|}}{C}-NH-CH\text{-}\overset{\overset{\displaystyle O}{\|}}{C}-OCH_3$$

with side chains $CH_2COOH$ and $CH_2Ph$ } methyl ester

aspartic acid (from Edman degradation)   phenylalanine

no free COOH ⇒ no reaction with carboxypeptidase

Aspartame is aspartylphenylalanine methyl ester.

24-43

$$H_2N-CH\text{-}\overset{\overset{\displaystyle O}{\|}}{C}-NH-CH\text{-}\overset{\overset{\displaystyle O}{\|}}{C}-NH-CH\text{-}\overset{\overset{\displaystyle O}{\|}}{C}-NH-CH\text{-}\overset{\overset{\displaystyle O}{\|}}{C}-NH-CH\text{-}\overset{\overset{\displaystyle O}{\|}}{C}-OH$$

with side chains $CH_2Ph$ ; $CH_3$ ; $H$ ; $CH_2$/$CH_2SCH_3$ ; $CH_3$

phenylalanine   alanine   glycine   methionine   alanine

from Edman degradation              from carboxy-peptidase

**632**

Copyright © 2013 Pearson Education, Inc.

24-44
(a)

protect the N-terminus of the first amino acid

$$\text{PhCH}_2\text{O}\overset{\text{O}}{\overset{\|}{\text{C}}}-\text{Cl} \quad + \quad \text{H}_2\text{N}-\underset{\underset{\text{CH}_2\text{CHMe}_2}{|}}{\text{CH}}-\text{COOH} \quad \longrightarrow \quad \text{Z}-\text{NH}-\underset{\underset{\text{CH}_2\text{CHMe}_2}{|}}{\text{CH}}-\overset{\text{O}}{\overset{\|}{\text{C}}}-\text{OH}$$

leucine

activate the C-terminus

$$\text{Cl}-\overset{\text{O}}{\overset{\|}{\text{C}}}-\text{OEt}$$

$$\text{Z}-\text{NH}-\underset{\underset{\text{CH}_2\text{CHMe}_2}{|}}{\text{CH}}-\overset{\text{O}}{\overset{\|}{\text{C}}}-\text{O}-\overset{\text{O}}{\overset{\|}{\text{C}}}-\text{OEt}$$

add the next amino acid

$$\text{H}_2\text{N}-\underset{\underset{\text{CH}_3}{|}}{\text{CH}}-\text{COOH} \qquad \text{alanine}$$

$$\text{Z}-\text{NH}-\underset{\underset{\text{CH}_2\text{CHMe}_2}{|}}{\text{CH}}-\overset{\text{O}}{\overset{\|}{\text{C}}}-\text{NH}-\underset{\underset{\text{CH}_3}{|}}{\text{CH}}-\overset{\text{O}}{\overset{\|}{\text{C}}}-\text{OH}$$

activate the C-terminus

$$\text{Cl}-\overset{\text{O}}{\overset{\|}{\text{C}}}-\text{OEt}$$

$$\text{Z}-\text{NH}-\underset{\underset{\text{CH}_2\text{CHMe}_2}{|}}{\text{CH}}-\overset{\text{O}}{\overset{\|}{\text{C}}}-\text{NH}-\underset{\underset{\text{CH}_3}{|}}{\text{CH}}-\overset{\text{O}}{\overset{\|}{\text{C}}}-\text{O}-\overset{\text{O}}{\overset{\|}{\text{C}}}-\text{OEt}$$

add the next amino acid

$$\text{H}_2\text{N}-\underset{\underset{\text{CH}_2\text{Ph}}{|}}{\text{CH}}-\text{COOH} \qquad \text{phenylalanine}$$

$$\text{Z}-\text{NH}-\underset{\underset{\text{CH}_2\text{CHMe}_2}{|}}{\text{CH}}-\overset{\text{O}}{\overset{\|}{\text{C}}}-\text{NH}-\underset{\underset{\text{CH}_3}{|}}{\text{CH}}-\overset{\text{O}}{\overset{\|}{\text{C}}}-\text{NH}-\underset{\underset{\text{CH}_2\text{Ph}}{|}}{\text{CH}}-\text{COOH}$$

deprotect the N-terminus  $\downarrow$ H$_2$, Pd

$$\text{H}_2\text{N}-\underset{\underset{\text{CH}_2\text{CHMe}_2}{|}}{\text{CH}}-\overset{\text{O}}{\overset{\|}{\text{C}}}-\text{NH}-\underset{\underset{\text{CH}_3}{|}}{\text{CH}}-\overset{\text{O}}{\overset{\|}{\text{C}}}-\text{NH}-\underset{\underset{\text{CH}_2\text{Ph}}{|}}{\text{CH}}-\text{COOH}$$

24-44 continued
(b)

Boc—phenylalanine

Attach C-terminus of
N-protected amino acid
to polymer support.

CF$_3$COOH    deprotect N-terminus

Boc—alanine

DCC    add next amino acid and couple

CF$_3$COOH    deprotect N-terminus

Boc—leucine

DCC    add next amino acid and couple

HF    deprotect and remove from polymer

## 24-45

protect N-terminus

$$PhCH_2O\overset{O}{\overset{\|}{C}}-Cl \quad + \quad H_2N-\underset{\underset{CH_3}{|}}{CH}-COOH \longrightarrow Z-NH-\underset{\underset{CH_3}{|}}{CH}-\overset{O}{\overset{\|}{C}}-OH$$

alanine

activate the C-terminus

$$Cl-\overset{O}{\overset{\|}{C}}-OEt$$

$$Z-NH-\underset{\underset{CH_3}{|}}{CH}-\overset{O}{\overset{\|}{C}}-O-\overset{O}{\overset{\|}{C}}-OEt$$

$$H_2N-\underset{\underset{CH(CH_3)_2}{|}}{CH}-COOH \qquad \text{valine}$$

add the next amino acid

$$Z-NH-\underset{\underset{CH_3}{|}}{CH}-\overset{O}{\overset{\|}{C}}-NH-\underset{\underset{CH(CH_3)_2}{|}}{CH}-\overset{O}{\overset{\|}{C}}-OH$$

$$\downarrow H_2, Pd \qquad \text{deprotect the N-terminus}$$

$$H_2N-\underset{\underset{CH_3}{|}}{CH}-\overset{O}{\overset{\|}{C}}-NH-\underset{\underset{CH(CH_3)_2}{|}}{CH}-\overset{O}{\overset{\|}{C}}-OH$$

React the N-terminus of the dipeptide at the left with the N-protected, C-activated tripeptide below.

$$Z-NH-\underset{\underset{H_3C-CHCH_2CH_3}{|}}{CH}-\overset{O}{\overset{\|}{C}}-NH-\underset{\underset{CH_2CHMe_2}{|}}{CH}-\overset{O}{\overset{\|}{C}}-NH-\underset{\underset{CH_2Ph}{|}}{CH}-\overset{O}{\overset{\|}{C}}-O-\overset{O}{\overset{\|}{C}}OEt$$

$$Z-NH-\underset{\underset{H_3C-CHCH_2CH_3}{|}}{CH}-\overset{O}{\overset{\|}{C}}-NH-\underset{\underset{CH_2CHMe_2}{|}}{CH}-\overset{O}{\overset{\|}{C}}-NH-\underset{\underset{CH_2Ph}{|}}{CH}-\overset{O}{\overset{\|}{C}}-NH-\underset{\underset{CH_3}{|}}{CH}-\overset{O}{\overset{\|}{C}}-NH-\underset{\underset{CH(CH_3)_2}{|}}{CH}-COOH$$

$$\downarrow H_2, Pd \qquad \text{deprotect the N-terminus}$$

$$H_2N-\underset{\underset{H_3C-CHCH_2CH_3}{|}}{CH}-\overset{O}{\overset{\|}{C}}-NH-\underset{\underset{CH_2CHMe_2}{|}}{CH}-\overset{O}{\overset{\|}{C}}-NH-\underset{\underset{CH_2Ph}{|}}{CH}-\overset{O}{\overset{\|}{C}}-NH-\underset{\underset{CH_3}{|}}{CH}-\overset{O}{\overset{\|}{C}}-NH-\underset{\underset{CH(CH_3)_2}{|}}{CH}-COOH$$

Ile     Leu     Phe     Ala     Val

24-46

(a) There are two possible sources of ammonia in the hydrolysate. The C-terminus could have been present as the amide instead of the carboxyl, or the glutamic acid could have been present as its amide, glutamine.

(b) The C-terminus is present as the amide. The N-terminus is present as the lactam (cyclic amide) combining the amino group with the carboxyl group of the glutamic acid side chain.

(c) The fact that hydrolysis does not release ammonia implies that the C-terminus is not an amide. Yet, carboxypeptidase treatment gives no reaction, showing that the C-terminus is not a free carboxyl group. Also, treatment with phenyl isothiocyanate gives no reaction, suggesting no free amine at the N-terminus. The most plausible explanation is that the N-terminus has reacted with the C-terminus to produce a cyclic amide, a lactam. (These large rings, called macrocycles, are often found in nature as hormones or antibiotics.)

24-47

(a) Lipoic acid is a mild oxidizing agent. In the process of oxidizing another reactant, lipoic acid is reduced.

oxidized form          reduced form

(b)

(c)

24-48

(a)  histidine:

See the solution to Problem 24-6.

24-48 continued

(b)

In the protonated imidazole, the two Ns are similar in structure, and both NH groups are acidic.

(c)

We usually think of protonation-deprotonation reactions occurring in solution where protons can move with solvent molecules. In an enzyme active site, there is no "solvent", so there must be another mechanism for movement of protons. Often, conformational changes in the protein will move atoms closer or farther. Histidine serves the function of moving a proton toward or away from a particular site by using its different nitrogens in concert as a proton acceptor and a proton donor.

24-49 The high isoelectric point suggests a strongly basic side chain as in lysine. The $N—CH_2$ bond in the side chain of arginine is likely to have remained intact during the metabolism. Can you propose a likely mechanism for this reaction? Think of this as a nucleophilic acyl substitution on $C=NH$ instead of $C=O$.

24-50

(a) glutathione:

**Notice that the first peptide bond uses the COOH at carbon-5 of glutamic acid.**
Edman degradation will not cleave the peptide bond with this unusual bonding.

glutamic acid (from 2,4-DNFB, Sanger reagent)

cysteine

glycine (from carboxypeptidase)

(b) reaction:  **2** glutathione  +  $H_2O_2$  $\longrightarrow$  glutathione disulfide  +  **2** $H_2O$

structure of glutathione disulfide:

$\Longleftarrow$ **new S–S bond**

24-51  Peptide A analysis gave $NH_3$ in addition to the amino acids, so the Glu in the analysis must have been Gln in the original peptide.

end groups:  N-terminus  Ala ——————————— Ile  C-terminus

chymotrypsin fragments

A  Glu—Gly—Tyr  (middle)

B  Ala—Lys—Phe
N-terminus

C  Arg—(Ser? Leu?)—Ile
C-terminus

Ala—Lys—Phe—Glu—Gly—Tyr—Arg—(Ser? Leu?)—Ile

trypsin fragments

D  Ala—Lys

E  Phe—Glu—Gly—Tyr—Arg

F  Ser—Leu—Ile

Ala—Lys—Phe—Gln—Gly—Tyr—Arg—Ser—Leu—Ile

24-52

(a) One important factor in ester reactivity is the ability of the alkoxy group to leave, and the main factor in determining leaving group ability is the stabilization of the anion. An NHS ester is more reactive than an alkyl ester because the anion $R_2NO^-$ has an electron-withdrawing group on the $O^-$, thereby distributing the negative charge over two atoms instead of just one. A simple alkyl ester has the full negative charge on the oxygen with nowhere to go, $RO^-$.

This resonance form of the leaving group shows the (+) charge on N that stabilizes the (–) on oxygen.

(b)

reactive anhydride

abbreviate O—Succ

tetrahedral intermediate

$CF_3COO^-$ + R—C(=O)—O—N (NHS ester)

tetrahedral intermediate

NHS trifluoroacetate is an amazing reagent. Not only does it activate the carboxylic acid through a mixed anhydride to form an ester under mild conditions, but the NHS leaving group is also the nucleophile that forms an ester that is both stable enough to work with, yet easily reactive when it needs to be. Perfect!

(c)

$R$—C(=O)—N(H)—R + HO—Succ

**639**

Copyright © 2013 Pearson Education, Inc.

24-53

(a)

(b) The product has the opposite configuration from the starting material because of the stereochemistry of the reactions. Diazotization does not break a bond to the chiral center so the L configuration is retained in Reaction 1. It is well documented that azide substitution is an $S_N2$ process that proceeds with inversion of configuration; this is why this process works. The third reaction does not break or form a bond to the chiral center, so the D configuration is retained.

25-1

trimyristin

25-2

all *cis*

excess
H$_2$
$\xrightarrow{\text{Ni}}$

trilinolein, m.p. < –4 °C
(liquid at room temperature)

tristearin, m.p. 72 °C
(solid at room temperature)

You could predict that the melting point of trilinolein is lower than triolein
(–4 °C) because more double bonds lower the melting point. Sources differ
on the m.p. of trilinolein, ranging from –17 °C to –43 °C.

25-3

triolein, m.p. –4 °C
(liquid at room temperature)

excess
CH$_3$OH
$\xrightarrow{\text{NaOH}}$

The combination of
NaOH and excess
CH$_3$OH produces
NaOCH$_3$, which
transesterifies the fatty
acids.

3  methyl oleate

+ glycerol (glycerine)

25-4

(a)

$$2 \ CH_3(CH_2)_{16}-\overset{\overset{\displaystyle O}{\|}}{C}-O^- \ Na^+ \ + \ Ca^{2+} \longrightarrow \left[ CH_3(CH_2)_{16}-\overset{\overset{\displaystyle O}{\|}}{C}-O \right]_2 Ca \ + \ 2 \ Na^+$$

(b)

$$2 \ CH_3(CH_2)_{16}-\overset{\overset{\displaystyle O}{\|}}{C}-O^- \ Na^+ \ + \ Mg^{2+} \longrightarrow \left[ CH_3(CH_2)_{16}-\overset{\overset{\displaystyle O}{\|}}{C}-O \right]_2 Mg \ + \ 2 \ Na^+$$

(c)

$$3 \ CH_3(CH_2)_{16}-\overset{\overset{\displaystyle O}{\|}}{C}-O^- \ Na^+ \ + \ Fe^{3+} \longrightarrow \left[ CH_3(CH_2)_{16}-\overset{\overset{\displaystyle O}{\|}}{C}-O \right]_3 Fe \ + \ 3 \ Na^+$$

25-5

(a) Both sodium carbonate (its old name is "washing soda") and sodium phosphate will increase the pH above 6, so that the carboxyl group of the soap molecule will remain ionized, thus preventing precipitation.

(b) In the presence of calcium, magnesium, and ferric ions, the carboxylate group of soap will form precipitates called "hard-water scum", or as scientists label it, "bathtub ring". Both carbonate and phosphate ions will form complexes or precipitates with these cations, thereby preventing precipitation of the soap from solution.

25-6 In each structure, the hydrophilic portion is circled. The uncircled part is hydrophobic.

benzalkonium chloride

Nonoxynol-9

Gardol®

## 25-7

pentamer

electrophile

## 25-8

A phosphatidic acid has four different groups on the center carbon of the 3-carbon glycerol portion as shown in the dashed boxes, making that central carbon an asymmetric carbon atom.

**25-9**  Estradiol is a phenol and can be ionized with aqueous NaOH. Testosterone does not have any hydrogens acidic enough to react with NaOH. Treatment of a solution of estradiol and testosterone in organic solvent with aqueous base will extract the phenoxide form of estradiol into the aqueous layer, leaving testosterone in the organic layer. Acidification of the aqueous base will precipitate estradiol, which can be filtered. Evaporation of the organic solvent will leave testosterone.

**25-10**  Models may help. Abbreviations: "ax" = axial; "eq" = equatorial. Note that substituents at *cis*-fused ring junctures are axial to one ring and equatorial to another.

(a)

(b)

25-10 continued

(c)

(d) ax to **B**
eq to **A**

ax
CH₃

ax to **A**
eq to **B**

ax
HO

25-11

geranial

menthol

camphor

abietic acid

camphor shown
in top view

25-12 β-carotene

25-13

α-farnesene
sesquiterpene

limonene
monoterpene

(Limonene can also be circled
in the other direction around
the ring—see menthol in the
solution to problem 25-11.)

OR

α-pinene
monoterpene

OR

zingiberene
sesquiterpene

25-14

(a)  A saturated fat is a solid triglyceride in which all three
chains are saturated fatty acids with an even number of
carbons, most commonly between 12 and 20 carbons.

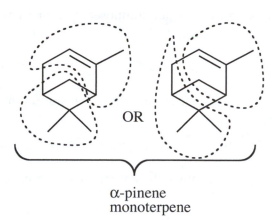

$$CH_2-O-\overset{\overset{\displaystyle O}{||}}{C}-(CH_2)_{12}CH_3$$

$$CH-O-\overset{\overset{\displaystyle O}{||}}{C}-(CH_2)_{12}CH_3$$

$$CH_2-O-\overset{\overset{\displaystyle O}{||}}{C}-(CH_2)_{12}CH_3$$

(b)  A polyunsaturated oil is a liquid
triglyceride in which all three fatty acid
chains contain an even number of
carbons, most commonly between 12
and 20 carbons, with two or more
double bonds distributed among the
three chains.

**645**

25-14 continued

(c) A wax is an ester composed of a long-chain alcohol and a long-chain fatty acid, for example:

(d) A soap is the sodium or potassium salt of a fatty acid.

Na⁺ or K⁺

(e) A detergent can be anionic (the original category), cationic, or nonionic.

Cl⁻

benzalkonium chloride—
a cationic detergent

$-SO_3^-$   Na⁺

SDS, sodium dodecyl sulfate, or
SLS, sodium lauryl sulfate,
an anionic detergent

Nonoxynol-9—
a nonionic detergent

$C_9H_{19}$

(f) A phospholipid is a class of lipid, usually a triglyceride, that contains a phosphate group, for example:

$$CH_2-O-\overset{\overset{\displaystyle O}{\|}}{C}-(CH_2)_{16}CH_3$$

$$CH-O-\overset{\overset{\displaystyle O}{\|}}{C}-(CH_2)_{16}CH_3$$

$$CH_2-O-\overset{\overset{\displaystyle O}{\|}}{\underset{\underset{\displaystyle O^{\ominus}}{|}}{P}}-O^{\ominus}$$

a phosphatidic acid
in ionized form

(g) A prostaglandin is a metabolite of arachidonic acid, a 20-carbon polyunsaturated fatty acid, containing a 5-membered ring. Many prostaglandins are potent mammalian hormones.

HO

COOH

HO

OH

$PGF_{2\alpha}$

$C_{20}H_{34}O_5$

(h) A steroid is a tetracyclic compound originally discovered in mammalian systems, but now found in almost all plants and animals. All contain this four-ring structural unit.

substructure of all steroids

(i) A sesquiterpene is a 15-carbon compound composed of three isoprene units.

α-farnesene
a sesquiterpene

25-15 (a) triglyceride   (b) synthetic detergent   (c) wax   (d) sesquiterpene
(e) prostaglandin   (f) steroid

25-16

(a)   **3**  $CH_3(CH_2)_7CH = CH(CH_2)_7COO^{\ominus} Na^+$  +  $HOCH(CH_2OH)_2$
          soap                                          glycerol

(b)

$$CH_2-O-\overset{\overset{\displaystyle O}{||}}{C}-(CH_2)_{16}CH_3$$
$$CH-O-\overset{\overset{\displaystyle O}{||}}{C}-(CH_2)_{16}CH_3 \quad \text{tristearin}$$
$$CH_2-O-\overset{\overset{\displaystyle O}{||}}{C}-(CH_2)_{16}CH_3$$

(c)

(mixture of diastereomers)

(d)   **3**  $CH_3(CH_2)_7CHO$  +

$$CH_2-O-\overset{\overset{\displaystyle O}{||}}{C}-(CH_2)_7CHO$$
$$CH-O-\overset{\overset{\displaystyle O}{||}}{C}-(CH_2)_7CHO$$
$$CH_2-O-\overset{\overset{\displaystyle O}{||}}{C}-(CH_2)_7CHO$$

(e)   **3**  $CH_3(CH_2)_7COOH$  +

$$CH_2-O-\overset{\overset{\displaystyle O}{||}}{C}-(CH_2)_7COOH$$
$$CH-O-\overset{\overset{\displaystyle O}{||}}{C}-(CH_2)_7COOH$$
$$CH_2-O-\overset{\overset{\displaystyle O}{||}}{C}-(CH_2)_7COOH$$

(f)

(mixture of diastereomers)

25-17

(a)   $CH_3(CH_2)_7CH = CH(CH_2)_7COOH \xrightarrow[\text{Ni}]{H_2} \xrightarrow[\text{2) } H_3O^+]{\text{1) LiAlH}_4} CH_3(CH_2)_{16}CH_2OH$

(b)   $CH_3(CH_2)_7CH = CH(CH_2)_7COOH \xrightarrow[\text{Ni}]{H_2} CH_3(CH_2)_{16}COOH$

**647**

25-17 continued

(c) $CH_3(CH_2)_{16}COOH$ + $HOCH_2(CH_2)_{16}CH_3$ $\xrightarrow[\Delta]{H^+}$ $CH_3(CH_2)_{16}COOCH_2(CH_2)_{16}CH_3$
from (b)         from (a)

(d) $CH_3(CH_2)_7CH=CH(CH_2)_7COOH$ $\xrightarrow[\text{2) Me}_2\text{S}]{\text{1) O}_3}$ $CH_3(CH_2)_7CH=O$ + $O=CH(CH_2)_7COOH$
                                                                          nonanal

(e) $CH_3(CH_2)_7CH=CH(CH_2)_7COOH$ $\xrightarrow[\text{H}_2\text{O, }\Delta]{\text{KMnO}_4}$ $CH_3(CH_2)_7COOH$ + $HOOC(CH_2)_7COOH$
                                                                                      nonanedioic acid

(f) $CH_3(CH_2)_7CH=CH(CH_2)_7COOH$ $\xrightarrow[\text{PBr}_3]{\substack{\text{excess} \\ \text{Br}_2}}$ $\xrightarrow{H_2O}$

(structure showing stereochemistry with Br, H, H, Br and CH$_3$(CH$_2$)$_7$ and (CH$_2$)$_6$CHCOOH with Br)

25-18

(a)

NaOH removes proton from N.

(b)

25-19

$CH_3(CH_2)_{16}CH_2OSO_3^-$ $Na^+$ $\xleftarrow{\text{NaOH}}$ $CH_3(CH_2)_{16}CH_2OSO_3H$

**648**
Copyright © 2013 Pearson Education, Inc.

25-20  Reagents in parts (a), (b), and (d) would react with alkenes. If both samples contained alkenes, these reagents could not distinguish the samples. Saponification (part (c)), however, is a reaction of an ester, so only the vegetable oil would react, not the hydrocarbon oil mixture. With saponification of a vegetable oil, as NaOH is consumed, the pH would gradually drop; with a petroleum oil, the pH would not change.

25-21  (a) Add an aqueous solution of calcium ion or magnesium ion. Sodium stearate will produce a precipitate, while the sulfonate will not precipitate.
(b) Beeswax, an ester, can be saponified with NaOH. Paraffin wax is a solid mixture of alkanes and will not react.
(c) Myristic acid will dissolve (or be emulsified) in dilute aqueous base. Trimyristin will remain unaffected.
(d) Triolein (an unsaturated oil) will decolorize bromine in $CCl_4$, but trimyristin (a saturated fat) will not.

25-22
(a)

$$CH_2-O-\overset{\overset{\displaystyle O}{\|}}{C}-(CH_2)_{12}CH_3$$

asymmetric * $CH-O-\overset{\overset{\displaystyle O}{\|}}{C}-(CH_2)_7CH=CH(CH_2)_7CH_3$   optically active
carbon

$$CH_2-O-\overset{\overset{\displaystyle O}{\|}}{C}-(CH_2)_7CH=CH(CH_2)_7CH_3$$

(b)

$$CH_2-O-\overset{\overset{\displaystyle O}{\|}}{C}-(CH_2)_7CH=CH(CH_2)_7CH_3$$

$CH-O-\overset{\overset{\displaystyle O}{\|}}{C}-(CH_2)_{12}CH_3$    not optically active; symmetric

$$CH_2-O-\overset{\overset{\displaystyle O}{\|}}{C}-(CH_2)_7CH=CH(CH_2)_7CH_3$$

25-23

$$CH_2-O-\overset{\overset{\displaystyle O}{\|}}{C}-(CH_2)_{16}CH_3$$

asymmetric * $CH-O-\overset{\overset{\displaystyle O}{\|}}{C}-(CH_2)_7CH=CH(CH_2)_7CH_3$   optically active
carbon

$$CH_2-O-\overset{\overset{\displaystyle O}{\|}}{C}-(CH_2)_7CH=CH(CH_2)_7CH_3$$

(a)

$$CH_2-O-\overset{\overset{\displaystyle O}{\|}}{C}-(CH_2)_{16}CH_3$$

$\xrightarrow[\text{Ni}]{H_2}$   $CH-O-\overset{\overset{\displaystyle O}{\|}}{C}-(CH_2)_{16}CH_3$   not optically active

$$CH_2-O-\overset{\overset{\displaystyle O}{\|}}{C}-(CH_2)_{16}CH_3$$

25-23 continued

(b)

$\xrightarrow[\text{CCl}_4]{\text{Br}_2}$

[structure: optically active (mixture of diastereomers)]

$CH_3(CH_2)_7$ — with Br, H, H, Br on a chain — $(CH_2)_7C$ —O—$\overset{*}{C}H$

$CH_2-O-\overset{\displaystyle O}{\overset{\|}{C}}(CH_2)_{16}CH_3$

$CH_2-O-\overset{\displaystyle O}{\overset{\|}{C}}-(CH_2)_7$ — with Br, H, H, Br — $(CH_2)_7CH_3$

optically active
(mixture of diastereomers)

(c) Products are not optically active.

$HOCH(CH_2OH)_2$ + $Na^+ \ ^-OOC(CH_2)_{16}CH_3$ + **2** $Na^+ \ ^-OOC(CH_2)_7CH=CH(CH_2)_7CH_3$

glycerol

(d)

$\xrightarrow[\text{2) Me}_2\text{S}]{\text{1) O}_3}$

$CH_2-O-\overset{\displaystyle O}{\overset{\|}{C}}-(CH_2)_{16}CH_3$

$\overset{*}{C}H-O-\overset{\displaystyle O}{\overset{\|}{C}}-(CH_2)_7CHO$

$CH_2-O-\overset{\displaystyle O}{\overset{\|}{C}}-(CH_2)_7CHO$

optically active

+ **2** $O=CH(CH_2)_7CH_3$
not optically active

25-24 The products in (b), (c) and (f) are mixtures of stereoisomers.

(a) [cyclohexane ring with Br and CH₃ at top, t-butyl group with Br at bottom]

(b) [cyclohexane ring with CH₃ and Br at top, isobutyl group with Br at bottom]

(c) [cyclohexane ring with Br and CH₃ at top, Br, with C(CH₃)(Br)CH₂Br group at bottom]

(d) [ring-opened structure with O=, H—, and CHO groups, acetyl group]  + $CH_2O$

(e) [structure with HOOC, O=, acetyl group]  + $CO_2$

(f) [cyclohexane ring with H, CH₃ at top, HO and H, HO–CH₂–CH(CH₃) group]

650
Copyright © 2013 Pearson Education, Inc.

25-26

(a)

eq to **A**
ax to **B**

25-27

(a) sesquiterpene          OR

(b) monoterpene          OR

(c) sesquiterpene          OR

Like a soap or detergent, this cholic acid–glycine combination has a very polar "head" and a slightly polar "tail". The "tail" can dissolve nonpolar molecules and the polar "head" can carry the complex into polar media.

**652**

(d) sesquiterpene

**25-28** The formula $C_{18}H_{34}O_2$ has two elements of unsaturation; one is the carbonyl, so the other must be an alkene or a ring. Catalytic hydrogenation gives stearic acid, so the carbon cannot include a ring; it must contain an alkene. The products from $KMnO_4$ oxidation determine the location of the alkene:

$$HOOC(CH_2)_4COOH \; + \; HOOC(CH_2)_{10}CH_3 \; \Longrightarrow \; HOOC(CH_2)_4CH=CH(CH_2)_{10}CH_3$$

If the alkene were *trans*, the coupling constant for the vinyl protons would be about 15 Hz; a 10 Hz coupling constant indicates a *cis* alkene. The 7 Hz coupling is from the vinyl Hs to the neighboring $CH_2$ groups.

petroselenic acid

H      H

J = 10 Hz

**25-29**

COOH   linolenic acid

Ni ↓ **1** $H_2$

COOH   linoleic acid—known

COOH   NEW

COOH   NEW

Ni ↓ **1** $H_2$

COOH   NEW

COOH   NEW

COOH   oleic acid—known

In addition to the new isomers with different positions of the double bonds, there has been growing concern over the *trans* fatty acids created by isomerizing the naturally occurring *cis* double bonds. More manufacturers are now listing the percent of "*trans* fat" on their food labels.

25-30

(a) Of the two, only nepetalactone is a terpene. The other has only 8 carbons and terpenes must be in multiples of 5 carbons.

(b) Of the two, only the second is aromatic, as can be readily seen in one of the resonance forms.

like benzene    like furan

(c) Each compound is cleaved with NaOH (aq) to give an enolate that tautomerizes to the more stable keto form.

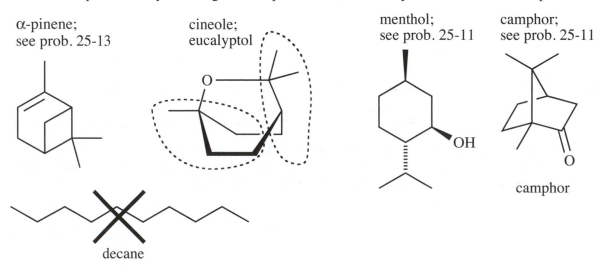

25-31

(a) Four of these components in Vicks Vapo-Rub® are terpenes. The only one that is not is decane; it is not composed of isoprene units despite having the correct number of carbons for a monoterpene.

Three of the four terpenes have had their isoprene units indicated in previous problems; in each of those three, there are two possible ways to assign the isoprene units, so those pictures will not be duplicated here.

α-pinene;
see prob. 25-13

cineole;
eucalyptol

menthol;
see prob. 25-11

camphor;
see prob. 25-11

camphor

decane

(b) Vicks Vapo-Rub® must be optically active, as it contains four optically active terpenes.

25-32

(a) High temperature and the diradical $O_2$ molecule strongly suggest a radical mechanism.

(b) Radical stability follows the order: benzylic > allylic > 3° > 2° > 1°. A radical at C-11 would be *doubly* allylic, making it a prime site for radical reaction.

(c)

*initiation*  ↓ Δ

⟷ *plus two allylic resonance forms*

*propagation step 1*

*propagation step 2*

*recycles to begin propagation step 1*

foul-smelling aldehydes, ketones, and carboxylic acids = "rancidity"

(d) Antioxidants are molecules that stop the free radical chain mechanism. In each of the two cases here, abstraction of the phenolic H makes an oxygen radical that is highly stabilized by resonance, so stable that it does not continue the free radical chain process. It only takes a small amount of antioxidant to prevent the chain mechanism. Interestingly, in breakfast cereals, the antioxidant is usually put in the plastic bag that the cereal is packaged in, rather than in the food itself. BHA and BHT can also be used directly in food as there is no evidence that they are harmful to humans.

BHA radical

BHT radical

Students: use this page for notes or to solve problems.

> Note: In this chapter, the "wavy bond" symbol means the continuation of a polymer chain.
> ∿∿∿∿∿∿∿

**26-1**

1° radical, and *not* resonance-stabilized—
**This orientation is not observed.**

Orientation of addition typically generates the more stable intermediate; the energy difference between a 1° radical (shown above) and a benzylic radical is huge. Moreover, this energy difference accumulates with each repetition of the propagation step. The phenyl substituents must necessarily be on alternating carbons because the orientation of attack is always the same—not a random process.

**26-2**

26-3  The benzylic hydrogen will be abstracted in preference to a 2° hydrogen because the benzylic radical is both 3° and resonance-stabilized, and the 2° radical is neither.

26-4  Addition occurs with the orientation giving the more stable intermediate. In the case of isobutylene, the growing chain will bond at the less substituted carbon to generate the more highly substituted carbocation.

26-5

(a)  Chlorine can stabilize a carbocation intermediate by resonance.

26-5 continued

(b) The —OCOCH₃ group can stabilize the carbocation intermediate by resonance.

(c) Terrible for cationic polymerization: both substituents are electron-withdrawing and would *destabilize* the carbocation intermediate.

H   COOCH₃
|   |
~~~ C — C ⊕       *destabilized* carbocation
|   |
H   C≡N

26-6   benzylic

H   H   H   H
|   |   |   |
~~~ C — C — C — C ~~~       +
|   |   |   |
Ph  H   Ph  H

middle of a polystyrene chain

H   H   H   H
⊕|   |   |   |
C — C — C — C ~~~
|   |   |   |
Ph  H   Ph  H

growing polystyrene chain

hydride transfer →

H   H   H   H
|   |   |   |
H — C — C — C — C ~~~
|   |   |   |
Ph  H   Ph  H

terminated chain

+

H   H   Ph  H
|   |   |   |
~~~ C — C — C — C ~~~
|   |   |   |
Ph  H       H

branch

Ph      Ph
|       |
~~~ C ----- C ~~~
...
Ph   Ph   Ph

H₂C   H
\   /
C
|⊕
Ph

H   H
\   /
C=C
/   \
H   Ph

H   H   Ph  H
|   |   |   |
~~~ C — C — C — C ~~~
|   |  ⊕|   |
Ph  H       H

new benzylic cation

Polystyrene is particularly susceptible to branching because the 3° benzylic cation produced by a hydride transfer is so stable. In poly(isobutylene), there is no hydrogen on the carbon with the stabilizing substituents; any hydride transfer would generate a 2° carbocation at the expense of a 3° carbocation at the end of a growing chain—this is an increase in energy and therefore unfavorable.

2°

H   Me  H   Me
|   |   |   |
~~~ C — C — C — C ~~~       +
|   |   |   |
H   Me  H   Me

middle of a poly(isobutylene) chain

Me  H   Me  H
|   |   |   |
⊕C — C — C — C ~~~       **no hydride transfer**
3° |   |   |   |
Me  H   Me  H

growing poly(isobutylene) chain

**659**

26-7

26-8

*etc.*

This polymerization goes so quickly because the anionic intermediate is highly resonance stabilized by the carbonyl and the cyano groups. A stable intermediate suggests a low activation energy, which translates to a fast reaction.

26-9

(a)

H H H H      H H H H
| | | |      | | | |
~C—C—C—C~  +  :C—C—C—C~
| | | |       ⊖| | |
CN H CN H     CN H CN H

middle of a poly(acrylonitrile) chain     growing poly(acrylonitrile) chain

H H CN H      H H H H
| | ⊖| |      | | | |
~C—C—C—C~  +  H—C—C—C—C~
| | •• |      | | | |
CN H H        CN H CN H

terminated chain

H       H
 \     /
  C = C
 /     \
H       CN

H H CN H
| | | |
~C—C—C—C~
| | | |
CN H | H
   H₂C    H
     \••/
      C⊖
      |
      CN

branch

                    CN        CN
                    |         |
~C—C—C—C~  ⟹  ~~~       ~~~
| | | |              |         |
CN H | H           CN   CN   CN   CN

(b) The chain-branching hydride transfer (from a cationic mechanism) or proton transfer (from an anionic mechanism) ends a less-highly-substituted end of a chain and generates an intermediate on a more-highly-substituted middle of a chain (a 3° carbon in these mechanisms). This stabilizes a carbocation, but greater substitution *destabilizes* a carbanion. Branching can and does happen in anionic mechanisms, but it is less likely than in cationic mechanisms.

26-10   isotactic poly(acrylonitrile)

NC H NC H NC H NC H NC H

syndiotactic polystyrene

Ph H H Ph Ph H H Ph Ph H

26-11

(a)

all *trans*

(b) The *trans* double bonds in gutta-percha allow for more ordered packing of the chains, that is, a higher degree of crystallinity. (Recall how *cis* double bonds in fats and oils lower the melting points because the *cis* orientation disrupts the ordering of the packing of the chains.) The more crystalline a polymer is, the less elastic it is.

26-12  Whether the alkene is *cis* or *trans* is not specified.

(a)

isobutylene          isoprene

(b)

styrene          butadiene

26-13  The repeating unit in each polymer is boxed.
(a)  Nomex®

(b)  Kevlar®

26-14  Kodel® polyester  (only one repeating unit shown)

662
Copyright © 2013 Pearson Education, Inc.

Glycerol is a trifunctional molecule, so not only does it grow in two directions to make a chain, it grows in three directions. All of its chains are cross-linked, forming a three-dimensional lattice with very little motion possible. The more cross-linked the polymer is, the more rigid it is.

26-16 For simplicity in this problem, bisphenol A will be abbreviated as a substituted phenol.

(a)

bisphenol A

mechanism

## 26-16 continued

(b)

bisphenol A     represented as

mechanism

When ethoxide leaves from diethyl carbonate, it immediately deprotonates another phenol, generating ethanol as the small molecule produced in this condensation.

26-17 Bisphenol A is made by condensing two molecules of phenol with one molecule of acetone, with loss of a molecule of water. This is an electrophilic aromatic substitution (more specifically, a Friedel-Crafts alkylation), and would require an acid catalyst to generate the carbocation. While a Lewis acid could be used, the mechanism below shows a protic acid.

There goes the water as a leaving group.

plus four resonance forms

**26-18**

Ph—N=C=O  +  H—O—Et  →  { Ph—N=C—O—Et  ↔  Ph—N—C—O—Et }

two rapid proton transfers

:O:
‖
Ph—N—C—O—Et
|
H

**26-19** Glycerol is a trifunctional alcohol. It uses two of its OH groups in a growing chain. The third OH group cross-links with another chain. The more cross-linked a polymer, the more rigid it is.

**26-20**

urethane linkage

bisphenol A          toluene diisocyanate

**26-21**

(a)

H   Me  H   Me  H   Me
|   |   |   |   |   |
C — C — C — C — C — C
|   |   |   |   |   |
H   Me  H   Me  H   Me

(b) Polyisobutylene is an addition polymer. No small molecule is lost, so this cannot be a condensation polymer.

(c) Either cationic polymerization or free-radical polymerization would be appropriate. The carbocation or free-radical intermediate would be 3° and therefore relatively stable. Anionic polymerization would be inappropriate as there is no electron-withdrawing group to stabilize the anion.

**26-22**

(a) Polychloroprene (Neoprene®) is an addition polymer.
(b) Polychloroprene comes from the diene, chloroprene, just as natural rubber comes from isoprene:

chloroprene

**26-23**

(a) It is a polyurethane.
(b) As with all polyurethanes, it is a condensation polymer.

26-23 continued

(c) $\sim\!\!\sim$ $CH_2CH_2CH_2-\overset{\displaystyle H}{\underset{\displaystyle H}{N}}-\overset{\displaystyle O}{\overset{\|}{C}}-O\!\sim\!\!\sim$ $\xrightarrow{H_2O}$ $HOCH_2CH_2CH_2NH_2$ + $CO_2$

26-24

(a) It is a polyester.

(b) As with all polyesters, it is a condensation polymer.

(c)

$HOCH_2CH_2CH_2CH_2OH$ + $CH_3O-\overset{O}{\overset{\|}{C}}-\!\!\!\bigcirc\!\!\!-\overset{O}{\overset{\|}{C}}-OCH_3$

$H^+ \downarrow \Delta$

$\sim\!\!\sim CH_2CH_2CH_2CH_2-O-\overset{O}{\overset{\|}{C}}-\!\!\!\bigcirc\!\!\!-\overset{O}{\overset{\|}{C}}-O\!\sim\!\!\sim$
$+ CH_3OH$

Using the dicarboxylic acid instead of the ester would produce water as the small neutral molecule lost in this condensation.

26-25

(a) Urylon® is a polyurea.

(b) A polyurea is a condensation polymer.

(c)

$\sim\!\!\sim (CH_2)_9-\overset{\displaystyle H}{\underset{\displaystyle H}{N}}-\overset{O}{\overset{\|}{C}}-\overset{\displaystyle H}{\underset{\displaystyle H}{N}}\!\sim\!\!\sim$ $\xrightarrow{H_2O}$ $H_2N(CH_2)_9NH_2$ + $CO_2$

26-26

(a) Polyethylene glycol, abbreviated PEG, is a polyether.

(b) PEG is usually made from ethylene oxide (first reaction shown). In theory, PEG could also be made by intermolecular dehydration of ethylene glycol (second reaction shown), but the yields are low and the chains are short.

$n$ (ethylene oxide) + $HO^-$ $\longrightarrow$ $HO\frown\frown O\frown\frown O\frown\frown O\sim\!\!\sim$

ethylene
oxide

$HO\frown OH$ (ethylene glycol) $\xrightarrow[-H_2O]{\overset{H^+}{\Delta}}$ $HO\frown\frown O\frown\frown O\frown\frown O\sim\!\!\sim$

ethylene
glycol

(c) Basic catalysts are most likely as they open the epoxide to generate a new nucleophile. Acid catalysts are possible but they risk dehydration and ether cleavage.

**667**

26-26 continued

(d) Mechanism of ethylene oxide polymerization (showing hydroxide as the base):

26-27
The key to determining the starting monomer for a ring-opening metathesis polymer (ROMP) is to "reconnect" the two carbons of the repeating unit. This process is similar to determining the starting material in ozonolysis problem, where the two new C=O were reconnected as an alkene.

(a) 4 C in repeating unit ⟹

(a) cyclohexane plus 2 C in repeating unit ⟹

redraw

bicyclo[2.2.2]octene

26-28

(a)

~~~C—O—C—O—C—O—C—O~~~   Delrin® (polyformaldehyde)

(b) All of these intermediates are resonance-stabilized.

etc.

trimer

(c) Delrin® is an addition polymer; instead of adding across the double bond of an alkene, addition occurs across the double bond of a carbonyl group.

(a) *cis*

*trans*

(b) Each structure has a fully conjugated chain. It is reasonable to expect electrons to be able to be transferred through the π system, just as resonance effects can work over long distances through conjugated systems.

(c) It is not surprising that the conductivity is directional. Electrons must flow along the π system of the chain, so if the chains were aligned, conductivity would be greater in the direction parallel to the polymer chains. (It is possible, though less likely, that electrons could pass from the π system of one chain to the π system of another, that is, perpendicular to the direction of the chain; we would expect reduced conductivity in that direction.)

26-30

(a) A Nylon is a polyamide. Amides can be hydrolyzed in aqueous acid, cleaving the polymer chain in the process.

$$\overset{\quad\;\;\overset{O}{\|}\quad\;\;\overset{O}{\|}\quad}{\sim\!\!\sim\!\!\sim\underset{\underset{H}{|}}{N}-C\sim\!\!\sim\!\!\sim C-\underset{\underset{H}{|}}{N}\sim\!\!\sim\!\!\sim} \xrightarrow{H_3O^+} \sim\!\!\sim\!\!\sim\overset{\oplus}{N}H_3 \;+\; HO-\overset{O}{\overset{\|}{C}}\sim\!\!\sim\!\!\sim\overset{O}{\overset{\|}{C}}-OH \;+\; \overset{\oplus}{H_3N}\sim\!\!\sim\!\!\sim$$

(b) A polyester can be saponified in aqueous base, cleaving the polymer chain in the process.

$$\sim\!\!\sim\!\!\sim O-\overset{O}{\overset{\|}{C}}\sim\!\!\sim\!\!\sim\overset{O}{\overset{\|}{C}}-O\sim\!\!\sim\!\!\sim \xrightarrow{NaOH} \sim\!\!\sim\!\!\sim OH \;+\; \overset{\ominus}{O}-\overset{O}{\overset{\|}{C}}\sim\!\!\sim\!\!\sim\overset{O}{\overset{\|}{C}}-\overset{\ominus}{O} \;+\; HO\sim\!\!\sim\!\!\sim$$

26-31

(a)

poly(vinyl acetate) $\xrightarrow[\text{H}^+ \text{ or HO}^-]{\text{H}_2\text{O}}$ poly(vinyl alcohol)

(b) A polyester is a condensation polymer in which monomer units are linked through ester groups as part of the polymer chain. Poly(vinyl acetate) is really a substituted polyethylene, an **addition** polymer, with only carbons in the chain; the ester groups are in the side chains, not in the polymer backbone.

(c) Hydrolysis of the esters in poly(vinyl acetate) does not affect the chain because the ester groups do not occur in the chain as they do in Dacron®.

(d) Vinyl alcohol cannot be polymerized because it is unstable, tautomerizing to acetaldehyde.

$$H_2C=\overset{\overset{OH}{|}}{CH} \;\rightleftharpoons\; CH_3-\overset{\overset{O}{\|}}{CH}$$

## 26-32

cellulose acetate

(b) Cellulose has three OH groups per glucose monomer, which form hydrogen bonds with other polar groups. Transforming these OH groups into acetates makes the polymer much less polar and therefore more soluble in organic solvents.

(c) The acetone dissolved the cellulose acetate in the fibers. As the acetone evaporated, the cellulose acetate remained but no longer had the fibrous, woven structure of cloth. It recrystallized as white fluff.

(d) Any article of clothing made from synthetic fibers is susceptible to the ravages of organic solvents. The structure of the shoe may disintegrate, and the solvent may penetrate more quickly.

## 26-33

Bakelite® is highly cross-linked through the ortho and para positions of phenol; each phenol can form a chain at two ring positions, then form a branch at the third position.

mechanism

plus resonance forms

plus four resonance forms

Further coupling at ortho positions leads to cross-linked Bakelite®.

670

Copyright © 2013 Pearson Education, Inc.

26-34

This leads to
cross-linking.

– HO⁻

from above

imine

plus another resonance form

**26-35**

| glycolic acid | lactic acid | glycolic acid | lactic acid | glycolic acid | lactic acid |

**26-36**

cellulose = cotton

polypropylene

As we have seen repeatedly through this presentation of organic chemistry, physical and chemical behavior depend on *structure*. The structure of cotton, i.e., cellulose, has multiple oxygen atoms that form hydrogen bonds with water. When cotton gets wet, it holds onto the water tightly, as you have seen if you have put cotton clothes in a clothes dryer—it takes a long time to dry. Polypropylene is a hydrocarbon with no hydrogen-bonding groups; the fiber feels dry because it cannot hold the water the way cotton can. Athletic garments are increasingly using polypropylene because they allow evaporation and cooling during periods of exertion; cotton is just the opposite.

**26-37**

(a) This addition polymer is called polyvinylidene chloride, trade name Saran®. It could be made by any of the three mechanism types: radical, cationic, or anionic.

monomer

(b) This polyester is a condensation polymer of two monomers, a diol and a derivative of phthalic acid, either the anhydride, an ester, the acid chloride, or the acid itself. Heating the monomers will make the polymer; no catalyst is required if done at high temperature.

monomer

one of these is the other monomer

X = OH or Cl or OR

(c) When the substituent is on every fourth carbon, and one double bond in the chain in every 4-carbon unit, the polymer must come from addition across a diene, probably under cationic conditions because the methoxy group stabilizes cationic intermediates by resonance.

monomer

26-37 continued

(d) This polyamide (Nylon) is a condensation polymer made from two monomers, a diamine and a derivative of succinic acid, either the anhydride, an ester, the acid chloride, or the acid itself. Heating the monomers will make the polymer; no catalyst is required if done at high temperature.

$H_2N$ ⁀ $NH_2$

monomer

One of these is the other monomer.

X = OH or Cl or OR

26-38

(a)

hydroxyethyl methacrylate

polymerize using radical or anionic conditions

polymer

(b) This polymer has a few properties that make it useful as the material in soft, extended-wear contact lenses. First, carboxylic acids usually are crystalline solids with high melting points, but esters and alcohols are low melting, often liquids, so the polymer with this ester is softer than the carboxylic acid or even the methyl ester. (The methyl ester, polymethyl methacrylate or Plexiglas, was the first material used in the original hard contact lenses.) Second, the ability of the free OH to form hydrogen bonds with water makes the contact lens more fluid and less irritating to the cornea. Third, a hidden advantage but very important for ocular health: the fluidity of the contact lens also permits oxygen to go through the lens. Because the cornea does not have a large blood flow, it needs to absorb oxygen from the air to maintain its health, and this enhanced gas permeability permits the contact lens to be worn for days at a time without compromising the health of the cornea. Thanks, polymers!

Note to the student: BON VOYAGE!
I hope you have enjoyed your travels
through organic chemistry.
            Jan William Simek

Students: Use this last page to muse on what a great experience you have had in organic chemistry, unlocking the molecular secrets of the natural world.

# Appendix 1—Summary of IUPAC Nomenclature of Organic Compounds

## Introduction

The purpose of the IUPAC system of nomenclature is to establish an international standard of naming compounds to facilitate communication. The goal of the system is to give each structure a unique and unambiguous name, and to correlate each name with a unique and unambiguous structure.

## I. Fundamental Principle

IUPAC nomenclature is based on naming a molecule's longest chain of carbons connected by single bonds, whether in a continuous chain or in a ring. All deviations, either multiple bonds or atoms other than carbon and hydrogen, are indicated by prefixes or suffixes according to a specific set of priorities.

## II. Alkanes and Cycloalkanes (also called "aliphatic" compounds)

Alkanes are the family of saturated hydrocarbons, that is, molecules containing carbon and hydrogen connected by single bonds only. These molecules can be in continuous chains (called linear or acyclic), or in rings (called cyclic or alicyclic). The names of alkanes and cycloalkanes are the root names of organic compounds. Beginning with the five-carbon alkane, the number of carbons in the chain is indicated by the Greek or Latin prefix. Rings are designated by the prefix "cyclo". (In the geometrical symbols for rings, each apex represents a carbon with the number of hydrogens required to fill its valence.)

| | | | | | | |
|---|---|---|---|---|---|---|
| $C_1$ | $CH_4$ | methane | | $C_{12}$ | $CH_3[CH_2]_{10}CH_3$ | dodecane |
| $C_2$ | $CH_3CH_3$ | ethane | | $C_{13}$ | $CH_3[CH_2]_{11}CH_3$ | tridecane |
| $C_3$ | $CH_3CH_2CH_3$ | propane | | $C_{14}$ | $CH_3[CH_2]_{12}CH_3$ | tetradecane |
| $C_4$ | $CH_3[CH_2]_2CH_3$ | butane | | $C_{20}$ | $CH_3[CH_2]_{18}CH_3$ | icosane |
| $C_5$ | $CH_3[CH_2]_3CH_3$ | pentane | | $C_{21}$ | $CH_3[CH_2]_{19}CH_3$ | henicosane |
| $C_6$ | $CH_3[CH_2]_4CH_3$ | hexane | | $C_{22}$ | $CH_3[CH_2]_{20}CH_3$ | docosane |
| $C_7$ | $CH_3[CH_2]_5CH_3$ | heptane | | $C_{23}$ | $CH_3[CH_2]_{21}CH_3$ | tricosane |
| $C_8$ | $CH_3[CH_2]_6CH_3$ | octane | | $C_{30}$ | $CH_3[CH_2]_{28}CH_3$ | triacontane |
| $C_9$ | $CH_3[CH_2]_7CH_3$ | nonane | | $C_{31}$ | $CH_3[CH_2]_{29}CH_3$ | hentriacontane |
| $C_{10}$ | $CH_3[CH_2]_8CH_3$ | decane | | $C_{40}$ | $CH_3[CH_2]_{38}CH_3$ | tetracontane |
| $C_{11}$ | $CH_3[CH_2]_9CH_3$ | undecane | | $C_{50}$ | $CH_3[CH_2]_{48}CH_3$ | pentacontane |

cyclopropane

cyclobutane

cyclopentane

cyclohexane

cycloheptane

cyclooctane

The IUPAC system of nomenclature is undergoing many changes, most notably in the placement of position numbers. The new system places the position number close to the functional group designation; however, you should be able to use and recognize names in either the old or the new style. Ask your instructor which system to use.

## III. Nomenclature of Molecules Containing Substituents and Functional Groups

### A. Priorities of Substituents and Functional Groups

LISTED HERE FROM HIGHEST TO LOWEST PRIORITY, except that the substituents within Group C have equivalent priority.

#### Group A—Functional Groups Named By Prefix Or Suffix

| Functional Group | Structure | Prefix | Suffix |
|---|---|---|---|
| Carboxylic Acid | $R-\overset{\overset{O}{\|\|}}{C}-OH$ | carboxy- | -oic acid (-carboxylic acid) |
| Aldehyde | $R-\overset{\overset{O}{\|\|}}{C}-H$ | oxo- (formyl) | -al (carbaldehyde) |
| Ketone | $R-\overset{\overset{O}{\|\|}}{C}-R$ | oxo- | -one |
| Alcohol | $R-O-H$ | hydroxy- | -ol |
| Amine | $R-N\big\langle$ | amino- | -amine |

#### Group B—Functional Groups Named By Suffix Only

| Functional Group | Structure | Prefix | Suffix |
|---|---|---|---|
| Alkene | $\overset{\diagdown}{\underset{\diagup}{}}C=C\overset{\diagup}{\underset{\diagdown}{}}$ | -------- | -ene |
| Alkyne | $-C\equiv C-$ | -------- | -yne |

#### Group C—Substituent Groups Named By Prefix Only

| Substituent | Structure | Prefix | Suffix |
|---|---|---|---|
| Alkyl (see next page) | $R-$ | alkyl- | -------- |
| Alkoxy | $R-O-$ | alkoxy- | -------- |

(Alkoxy groups take the name of the alkyl group, like methyl or ethyl, drop the "yl", and add "oxy"; $CH_3O$ is methoxy; $CH_3CH_2O$ is ethoxy.)

| Halogen | | | |
|---|---|---|---|
| | $F-$ | fluoro- | -------- |
| | $Cl-$ | chloro- | -------- |
| | $Br-$ | bromo- | -------- |
| | $I-$ | iodo- | -------- |

Miscellaneous substituents and their prefixes

| $-NO_2$ | $-CH=CH_2$ | $-CH_2CH=CH_2$ | phenyl |
|---|---|---|---|
| nitro | vinyl | allyl | |

<u>Common alkyl groups</u>—replace "ane" ending of alkane name with "yl". Alternate names for complex substituents are given in brackets.

—$CH_3$
methyl

—$CH_2CH_3$
ethyl

—$CH_2CH_2CH_3$
propyl (*n*-propyl)

—$CH_2CH_2CH_2CH_3$
butyl (*n*-butyl)

$$—CH\begin{matrix} CH_3 \\ \\ CH_3 \end{matrix}$$
isopropyl
[1-methylethyl]

$$—CH_2—CH\begin{matrix} CH_3 \\ \\ CH_3 \end{matrix}$$
isobutyl
[2-methylpropyl]

$$—CH\begin{matrix} CH_3 \\ \\ CH_2CH_3 \end{matrix}$$
*sec*-butyl
[1-methylpropyl]

$$—\underset{CH_3}{\overset{CH_3}{C}}—CH_3$$
*t*-butyl or
*tert*-butyl
[1,1-dimethylethyl]

## B. Naming Substituted Alkanes and Cycloalkanes—Group C Substituents Only

Organic compounds containing substituents from Group C are named following this sequence of steps, as indicated on the examples below:

•Step 1. Find the longest continuous carbon chain. Determine the root name for this parent chain. In cyclic compounds, the ring is usually considered the parent chain, unless it is attached to a longer chain of carbons; indicate a ring with the prefix "cyclo" before the root name. (When there are two longest chains of equal length, use the chain with the greater number of substituents.)

•Step 2. Number the chain in the direction such that the position number of the first substituent is the smaller number. If the first substituents have the same number, then number so that the second substituent has the smaller number, *etc*.

•Step 3. Determine the name and position number of each substituent. (A substituent on a nitrogen is designated with an "*N*" instead of a number; see Section **III**.D.1. below.)

•Step 4. Indicate the number of identical groups by the prefixes di, tri, tetra, *etc*.

•Step 5. Place the position numbers and names of the substituent groups, in alphabetical order, before the root name. In alphabetizing, ignore prefixes like *sec*-, *tert*-, di, tri, *etc*., but include iso and cyclo. Always include a position number for each substituent, regardless of redundancies. In case of ties, where numbering could begin with either of two carbons, begin with the carbon closer to the one with more substituents, or else the carbon with the substituent whose name is earlier in the alphabet.

3-bromo-2-chloro-5-ethyl-4,4-dimethyloctane

3-fluoro-4-isopropyl-2-methylheptane

1-*sec*-butyl-3-nitrocyclohexane
(Numbering is determined by the alphabetical order of substituents, "b" before "n".)

1-bromo-3-fluorocyclobutane

3-bromo-1,1-difluorocyclobutane

## C. Naming Molecules Containing Functional Groups from Group B—Suffix Only

1. Alkenes—Follow the same steps as for alkanes, except:
a. Number the chain of carbons *that includes the C=C* so that the C=C has the lower position number, since it has a higher priority than any substituents;
b. Change "ane" to "ene" and assign a position number to the first carbon of the C=C; place the position number just before the name of functional group(s);
c. Designate geometrical isomers with a *cis,trans* or *E,Z* prefix.

4,4-difluoro-3-methylbut-1-ene

1,1-difluoro-2-methyl-buta-1,3-diene

5-methylcyclopenta-1,3-diene

Special case: When the chain cannot include an alkene, a substituent name is used. See Section **V.A.2.a.**

3-vinylcyclohex-1-ene

*Numbering must be on EITHER a ring OR a chain, but not both.*

2. Alkynes—Follow the same steps as for alkanes, except:
a. Number the chain of carbons *that includes the C≡C* so that the alkyne has the lower position number;
b. Change "ane" to "yne" and assign a position number to the first carbon of the C≡C; place the position number just before the name of functional group(s).
Note: The Group B functional groups (alkene and alkyne) are considered to have equal priority: in a molecule with both an ene and an yne, whichever is closer to the end of the chain determines the direction of numbering. In the case where each would have the same position number, the alkene takes the lower number. In the name, "ene" comes before "yne" because of alphabetization.

4,4-difluoro-3-methylbut-1-yne

pent-3-en-1-yne
("yne" closer to end of chain)

pent-1-en-4-yne
(The "ene" and "yne" have equal priority unless they have the same position number, when "ene" takes the lower number.)

(Notes: 1. An "e" is dropped if the letter following it is a vowel: "pent-3-en-1-yne" , not "pent-3-ene-1-yne".
2. An "a" is added if inclusion of di, tri, *etc.*, would put two consonants together: "buta-1,3-diene", not "but-1,3-diene".)

## D. Naming Molecules Containing Functional Groups from Group A—Prefix or Suffix

In naming molecules containing one or more of the functional groups in Group A, the group of highest priority is indicated by suffix; the others are indicated by prefix, with priority equivalent to any other substituents. The table in Section **III.A.** defines the priorities; they are discussed on the following pages in order of increasing priority.

Now that the functional groups and substituents from Groups A, B, and C have been described, a modified set of steps for naming organic compounds can be applied to all simple structures:

•Step 1. Find the highest priority functional group. Determine and name the longest continuous carbon chain that includes this group.

•Step 2. Number the chain so that the highest priority functional group is assigned the lower number. (The number "1" is often omitted when there is no confusion about where the group must be. Aldehydes and carboxylic acids must be at the first carbon of a chain, so a "1" is rarely used with those functional groups.)

•Step 3. If the carbon chain includes multiple bonds (Group B), replace "ane" with "ene" for an alkene or "yne" for an alkyne. Designate the position of the multiple bond with the number of the first carbon of the multiple bond.

•Step 4. If the molecule includes Group A functional groups, replace the last "e" with the suffix of the highest priority functional group, and include its position number just before the name of the highest priority functional group.

•Step 5. Indicate all Group C substituents, and Group A functional groups of lower priority, with a prefix. Place the prefixes, with appropriate position numbers, in alphabetical order before the root name.

1. Amines: prefix: amino-; suffix: -amine—substituents on nitrogen denoted by "*N*"

$CH_3CH_2CH_2$—$NH_2$

propan-1-amine

3-methoxycyclohexan-1-amine
("1" is optional in this case.)

*N,N*-diethylbut-3-en-2-amine

2. Alcohols: prefix: hydroxy-; suffix: -ol

$CH_3CH_2$—OH

ethanol

but-3-en-2-ol

2-aminocyclobutan-1-ol
("1" is optional in this case.)

3. Ketones: prefix: oxo-; suffix: -one (pronounced "own")

3-hydroxybutan-2-one

cyclohex-3-en-1-one
("1" is optional in this case.)

4-(*N,N*-dimethylamino)pent-4-en-2-one

4. Aldehydes: prefix: oxo-, or formyl- (O=CH-); suffix: -al (abbreviation: —CHO)
An aldehyde can only be on carbon 1, so the "1" is generally omitted from the name.

methanal;
formaldehyde

ethanal;
acetaldehyde

4-hydroxybut-2-enal

4-oxopentanal

**679**

Appendix 1, Summary of IUPAC Nomenclature, continued

Special case:  When the chain cannot include the carbon of the aldehyde, the suffix "carbaldehyde" is used:

cyclohexanecarbaldehyde

5. Carboxylic Acids: prefix: carboxy-; suffix: -oic acid (abbreviation: —COOH)
A carboxylic acid can only be on carbon 1, so the "1" is generally omitted from the name.
(Note: Chemists traditionally use, and IUPAC accepts, the names "formic acid" and "acetic acid" in place of "methanoic acid" and "ethanoic acid".)

methanoic acid;
formic acid

ethanoic acid;
acetic acid

2-amino-3-phenylpropanoic acid

2,2-dimethyl-3,4-dioxobutanoic acid

Special case:  When the chain numbering cannot include the carbon of the carboxylic acid, the suffix "carboxylic acid" is used:

2-formyl-4-oxocyclohexanecarboxylic acid
("Formyl" is used to indicate an aldehyde as a substituent when its carbon cannot be in the chain numbering.)

E.  Naming Carboxylic Acid Derivatives
The six common groups derived from carboxylic acids are, in decreasing priority after carboxylic acids:  salts, anhydrides, esters, acyl halides, amides, and nitriles.

1.  Salts of Carboxylic Acids
Salts are named with cation first, followed by the anion name of the carboxylic acid, where "**ic acid**" is replaced by "**ate**" :

| | | |
|---|---|---|
| acet**ic acid** | becomes | acet**ate** |
| butano**ic acid** | becomes | butano**ate** |
| cyclohexanecarboxyl**ic acid** | becomes | cyclohexanecarboxyl**ate** |

$CH_3$—CHCOO⁻ Li⁺

lithium 2-aminopropanoate

$ClCH_2COO^-$ Na⁺

sodium chloroacetate

ammonium 2-methoxy-cyclobutanecarboxylate

2. Anhydrides:  "oic acid" is replaced by "oic anhydride"

R—C—OH
alkanoic acid

R—C—O—C—R
alkanoic anhydride

benzoic anhydride

**680**

### 3. Esters

Esters are named as "organic salts", that is, the alkyl name comes first, followed by the name of the carboxylate anion. (common abbreviation: —COOR)

carboxylate    alkyl

"alkanoate"    "alkyl"

"alkyl alkanoate"

ethyl acetate

isopropyl 2,2-dimethylpropanoate

vinyl prop-2-enoate

methyl 3-hydroxycyclo-pentanecarboxylate

cyclohexyl 2-phenylacetate

### 4. Acyl Halides: "oic acid" is replaced by "oyl halide"

R—C—OH ⟹ R—C—Cl
alkanoic acid      alkanoyl chloride

butanoyl chloride

benzoyl chloride

### 5. Amides: "oic acid" is replaced by "amide"

R—C—OH ⟹ R—C—NH₂
alkanoic acid      alkanamide

butanamide

benzamide

Amides are notable for their role in biochemistry, i.e., the special amide bond between two amino acids is called a peptide bond.

### 6. Nitriles: "oic acid" is replaced by "enitrile"

R—C—OH ⟹ R—C≡N
alkanoic acid      alkanenitrile

butanenitrile

benzonitrile

(common spelling differs from IUPAC)

## IV. Nomenclature of Aromatic Compounds

"Aromatic" compounds are those derived from benzene and similar ring systems. As with aliphatic nomenclature described above, the process is: determining the root name of the parent ring; determining priority, name, and position number of substituents; and assembling the name in alphabetical order. *Functional group priorities are the same in aliphatic and aromatic nomenclature.* See p. 676 for the list of priorities.

### A. Common Parent Ring Systems

or

benzene

naphthalene

anthracene

681

Copyright © 2013 Pearson Education, Inc.

Appendix 1, Summary of IUPAC Nomenclature, continued

B. Monosubstituted Benzenes

1. Most substituents keep their designation, followed by the word "benzene":

chlorobenzene         nitrobenzene         ethylbenzene

2. Some common substituents change the root name of the ring. IUPAC accepts these as root names, listed here in decreasing priority (same as Group A, p. 676):

benzoic acid    benzene-sulfonic acid    benzaldehyde    phenol    aniline    anisole    toluene

C. Disubstituted Benzenes

1. Designation of substitution—only three possibilities:

common:    *ortho-*   (*o-*)      *meta-*   (*m-*)      *para-*   (*p-*)
IUPAC:    1,2-           1,3-           1,4-

2. Naming disubstituted benzenes—Priorities from Group A, p. 676, determine root name and substituents.

*p*-dibromobenzene    *m*-aminobenzoic acid    *o*-methoxybenzaldehyde    *m*-methylphenol
1,4-dibromobenzene    3-aminobenzoic acid    2-methoxybenzaldehyde    3-methylphenol

D. Polysubstituted Benzenes—must use numbers to indicate substituent position

3,4-dichloro-*N*-methylaniline     2,4,6-trinitrotoluene (TNT)     ethyl 4-amino-3-hydroxybenzoate

**682**

Appendix 1, Summary of IUPAC Nomenclature, continued

## E. Aromatic Ketones

A special group of aromatic compounds are ketones where the carbonyl is attached to at least one benzene ring. Such compounds are named as "phenones"; the prefix depends on the size and nature of the group on the other side of the carbonyl. These are the common examples:

acetophenone

propiophenone

butyrophenone

benzophenone

## V. Nomenclature of Bicyclic Compounds

"Bicyclic" compounds are those that contain two rings. There are four possible arrangements of two rings that depend on how many atoms are shared by the two rings. The first arrangement in which the rings do not share any atoms does not use any special nomenclature, but the other types require a method to designate how the rings are put together. Once the ring system is named, then functional groups and substituents follow the standard rules described above.

Type 1. Two rings with no common atoms
These follow the standard rules of choosing one parent ring system and describing the other ring as a substituent.

Ketone is the highest priority functional group, phenyl is substituent.

⟹ 3-phenylcyclohexan-1-one  ("1" could be omitted here.)

Benzene is the parent ring system as it is larger than cyclopentane and it has three substituents.

⟹ 1-cyclopentyl-2,3-dinitrobenzene

Type 2. Two rings with one common atom—spiro ring system
The ring system in spiro compounds is indicated by the word "spiro" (instead of "cyclo"), followed by brackets indicating how many atoms are contained in each path around the rings, ending with the alkane name describing how many carbons are in the ring systems including the spiro carbon. (If any atoms are not carbons, see section VI.) Numbering follows the smaller path first, passing through the spiro carbon and around the second ring.

spiro[3.4]octane

spiro[4.5]decane

**683**

Appendix 1, Summary of IUPAC Nomenclature, continued

Substituents and functional groups are indicated in the usual ways. Spiro ring systems are always numbered smaller before larger, and numbered in such a way as to give the highest priority functional group the lower position number.

spiro[3.4]oct-5-ene

7,7-dimethylspiro[4.5]decan-2-one

## Type 3. Two rings with two common atoms—fused ring system

Two rings that share two common atoms are called fused rings. This ring system and the next type called bridged rings share the same designation of ring system. Each of the two common atoms is called a bridgehead atom, and there are three paths between the two bridgehead atoms. In contrast with naming the spiro rings, the *longer* path is counted first, then the shorter, then the shortest. In fused rings, the shortest path is always a zero, meaning zero atoms between the two bridgehead atoms. Numbering starts at a bridgehead, continues around the largest ring, through the other bridgehead and around the shorter ring. (In these structures, bridgeheads are marked with a dark circle for clarity.)

bicyclo[2.1.0]pentane
(path of 2 atoms and a path of 1 atom)

bicyclo[4.4.0]decane
(path of 4 atoms in each direction)

bicyclo[5.3.0]decane
(path of 5 atoms and path of 3 atoms)

Substituents and functional groups are indicated in the usual ways. Fused ring systems are always numbered larger before smaller, and numbered in such a way as to give the highest priority functional group the lower position number.

## Type 4. Two rings with more than two common atoms—bridged ring system

Two rings that share more than two common atoms are called bridged rings. Bridged rings share the same designation of ring system as Type 3 in which there are three paths between the two bridgehead atoms. The longer path is counted first, then the medium, then the shortest. Numbering starts at a bridgehead, continues around the largest ring, through the other bridgehead and around the medium path, ending with the shortest path numbered from the original bridgehead atom. (In these structures, bridgeheads are marked with a dark circle for clarity.)

bicyclo[2.1.1]hexane
(paths of 2 atoms, 1 atom, and 1 atom)

bicyclo[2.2.2]octane
(three paths of 2 atoms)

bicyclo[3.2.1]octane
(paths of 3 atoms, 2 atoms, and 1 atom)

5,5-dibromo-
bicyclo[2.1.1]hexane

bicyclo[2.2.2]oct-5-en-2-one

8,8-dimethyl-
bicyclo[3.2.1]octan-1-ol

## VI. Replacement Nomenclature of Heteroatoms

The term "heteroatom" applies to any atom other than carbon or hydrogen. It is common for heteroatoms to appear in locations that are inconvenient to name following basic rules, so a simple system called "replacement nomenclature" has been devised. The fundamental principle is to name a compound as if it contained only carbons in the skeleton, plus any functional groups or substituents, and then indicate which carbons are "replaced" by heteroatoms. The prefixes used to indicate these substitutions are listed here *in decreasing priority and listed in this order in the name:*

| Element | Prefix |
|---------|--------|
| O | oxa |
| S | thia |
| N | aza |
| P | phospha |
| Si | sila |
| B | bora |

Example

nonan-1-ol

2-thia-8-aza-4-sila-6-boranonan-1-ol

In the above example, note that the (imaginary) compound no longer has nine carbons, even though the name still includes "nonan". The heteroatoms have replaced carbons, but the compound is named as if it still had those carbons.

Where the replacement system is particularly useful is in polycyclic compounds. Shown below are three examples of commercially available and synthetically useful reagents that use this system.

parent hydrocarbon

reagent

abbreviation

bicyclo[2.2.2]octane

1,4-DiAzaBiCyclo[2.2.2]Octane
(upper case added to explain abbreviation)

DABCO

bicyclo[5.4.0]undec-7-ene

1,8-DiazaBicyclo[5.4.0]Undec-7-ene
(upper case added to explain abbreviation)

DBU

**685**

Appendix 1, Summary of IUPAC Nomenclature, continued

bicyclo[3.3.1]nonane

9-BoraBicyclo[3.3.1]Nonane
(upper case added to explain abbreviation)

9-BBN

## VII. Designation of Stereochemistry; Cahn-Ingold-Prelog system

Is this alkene *cis* or *trans*?

How can we distinguish
this structure from its
mirror image?

Compounds that exhibit stereoisomerism, whether geometric isomers around double bonds, substituent groups on rings, or molecules with asymmetric tetrahedral atoms (which are almost always carbons), require a system to designate relative and absolute orientation of the groups. The terms *cis/trans*, D/L in carbohydrates and amino acids, and $d / l$ for optically active compounds, are limited and cannot be used generally, although each still is used in appropriate situations. For example, *cis/trans* still is used to indicate relative positions of substituents around a ring.

A system developed by chemists Cahn, Ingold, and Prelog uses a series of steps to determine group priorities, and a definition of position based on the relative arrangement of the groups. In alkenes, the system is relatively simple:

high          high

low          low

This arrangement is defined
as Z = *zusammen*, "together",
from both high priority groups
on the same side of the C=C.

high          low

low          high

This arrangement is defined as
E = *entgegen*, "opposite", from
the two high priority groups on
opposite sides of the C=C.

As with alkenes, the orientation around an asymmetic carbon can be only one of two choices. In three dimensions, clockwise and counterclockwise are the only two directions that are definite, and even that description requires a fixed reference point. To designate configuration, the lowest (fourth) priority group is always placed farthest away from the viewer (indicated by a dashed line), and the group priorities will follow 1 to 2 to 3 in either a clockwise or a counterclockwise direction.

1 to 2 to 3 is
clockwise, defined
as the *R* = *rectus*
configuration.

1 to 2 to 3 is
counterclockwise,
defined as the *S* =
*sinister* configuration.

The only step remaining is to determine the priority of groups, for which there is a carefully defined set of rules.

Rule 1. Consider the first atom of the group, the point of attachment. Atoms with higher atomic number receive higher priority. Heavier isotopes have higher priority than lighter isotopes.

$$I > Br > Cl > F \qquad O > N > C > H \qquad {}^{14}C > {}^{13}C > {}^{12}C$$

Rule 2. If the first atoms of two or more groups are the same, go out to the next atoms to break the tie. One high-priority atom takes priority over any number of lower-priority atoms.

Rule 3. Treat multiple bonds as if they were all single bonds; one will be to the real atom, the others will be to imaginary atoms.

This is the hardest rule to put into practice. This example of an alkyne shows stepwise how to accomplish this. In the pictures, imaginary atoms (ones that did not start out in the structure, we had to invent them) are indicated by italics.

started with 3 bonds to carbon, so have to replace them

all atoms are real

halfway there—had to add two fake carbons

final—had to add four fake carbons

becomes

Added a fake O to the real C, and a fake C to the real O.

becomes

Added two fake Ns to the C, and two fake Cs to the N.

Examples applying the Cahn-Ingold-Prelog system

comparing F and Cl

comparing I and Br

E

E

(E)-3-methylpent-3-en-2-one

Z

(Z)-3-chloropent-3-en-2-one

Often, the hard part of applying the R/S system to asymmetric carbons is orienting the molecule so that the fourth priority group is farthest away. Sometimes it is easier to put the fourth priority group toward you, then take the opposite of what the direction of 1 to 2 to 3 says.

top view

flip
Br up so
F goes back

Putting the F (4th priority) pointing up toward you, the groups 1 to 2 to 3 appear counterclockwise, but we have to take the opposite, so instead of S, it is R.

# Appendix 2: Summary of Acidity and Basicity

One of the most important properties of molecules is their acidity or basicity. This section deals with protic acids, called Bronsted-Lowry acids. Similar statements can be made about Lewis acids but they are not the focus of this Appendix.

For the general chemical equation of an acid donating a proton, an equilibrium constant can be calculated, signified as $K_a$. As with pH, the $pK_a$ is defined as the $-\log_{10}K_a$.

$$H-A \rightleftharpoons H^+ + A^- \qquad K_a = \frac{[H^+][A^-]}{[HA]} \qquad pK_a = -\log_{10}K_a$$

conjugate acid    conjugate base

The way the above chemical equation and equilibrium expression are written is a common simplification where water is omitted, and the acid is a proton, $H^+$, rather than hydronium ion, $H_3O^+$. Often, organic acids and bases have no water present, so this is a reasonable simplification.

Some important comments about $pK_a$ values and dissociation of acids:

A) Values of $pK_a$ are measures of equilibrium constants, described further below. Values between 0 and 15.7 are measured by titration in water solution and are known accurately, to within 0.1 pK unit and sometimes better. Values outside of this range cannot be measured in water because of the leveling effect of water, and there is no universally accepted method for measuring these $pK_a$ values. This lack of a single standard of measurement means that the values below 0 and above 15.7 should be considered relative, not absolute. If your instructor says the $pK_a$ of methane is 46 and this book says it is 50, those should be considered the same value within experimental variation.

B) Acidity is a thermodynamic property, and the acid equilibrium constant, $K_a$, is a measure of the relative concentrations of species in the protonated and unprotonated form. As most organic acids are weak acids, meaning they are present mostly in the protonated form at equilibrium, the $K_a < 1$. Since the $pK_a = -\log K_a$, the $pK_a$ values will be greater than 0, with the larger $pK_a$ representing a weaker acid. If this is not clear, review text section 1-13.

C) Acids do not spontaneously spit out a proton! Despite our way of writing ionization equilibria as shown on the next page, acids do not give up a proton unless a base comes by to take the proton away. The reactions as drawn in the table should be considered half-reactions, just as the reactions in the electromotive series were half-reactions for balancing oxidation-reduction reactions in general chemistry.

## I. Acidity trends with the periodic table

The nonmetal elements, other than the noble gases, are shown here with their corresponding hydrides, that is, the compounds of the elements combined with hydrogen. Their $pK_a$ values are also listed.

Two trends in acidity become obvious:

1) acidity increases (lower $pK_a$) left to right across the periodic table, and
2) acidity increases top to bottom down the periodic table.

*Why?* *The fundamental principle of determining acid strength is this:*

# Acid strength is determined by stability of the conjugate base!

| IVA | VA | VIA | VIIA |
|---|---|---|---|
| C [6] | N [7] | O [8] | F [9] |
| $CH_4$ | $NH_3$ | $H_2O$ | HF |
| $pK_a$ ~50 | $pK_a$ ~35 | $pK_a$ 15.7 | $pK_a$ 3.2 |
| | P [15] | S [16] | Cl [17] |
| | $PH_3$ | $H_2S$ | HCl |
| | $pK_a$ ~29 | $pK_a$ 7.0 | $pK_a$ –7 |
| | | Se [34] | Br [35] |
| | | $H_2Se$ | HBr |
| | | $pK_a$ 3.9 | $pK_a$ –9 |
| | | Te [52] | I [53] |
| | | $H_2Te$ | HI |
| | | $pK_a$ 2.6 | $pK_a$ –10 |

Appendix 2 continued, Summary of Acidity and Basicity

Looking at the conjugate bases of the second-row elements, what correlates with this trend in stability?

*increasing stability of conjugate base*

$$CH_3^- \quad < \quad NH_2^- \quad < \quad OH^- \quad < \quad F^-$$

*increasing ELECTRONEGATIVITY*

The first stabilizing factor is **having the negative charge on the more electronegative atom.**

The second trend shows that acidity increases top to bottom, down the periodic table. What correlates with this trend in elements? SIZE. Why should size of the anion have anything to do with stability? **A charged species is more stable when the charge is more spread out**, and in a larger ion, the electron density is lower than in a small ion. Note that this is opposite of the electronegativity argument above: it appears that spreading out charge is more important than just electronegativity.

Other important factors in stabilizing conjugate bases will be introduced later. Now, let's turn to predicting acid-base reactions and their equilibrium position, meaning "which side is favored" at equilibrium.

## II. Predicting equilibrium position
We can use the table on p. 691 to make predictions about equilibrium position in acid-base reactions.

1) *A base will deprotonate any acid stronger than its conjugate acid.* This is the most useful principle of predicting acid-base reactions. On the table, this means that any base, hydroxide for example, can react with any acid more acidic than the conjugate acid of itself, water in our example. So hydroxide is a strong enough base to pull the proton from any of these: bicarbonate ion, a phenol, carbonic acid, a carboxylic acid, or a sulfonic acid. We can also predict that hydroxide is NOT a strong enough base to react with any acid above water on the table; for example, a mixture of hydroxide with an alkyne will favor the reactants at equilibrium, with only a small amount of products.

*reactants favored*

$$HO^- \; + \; RC\equiv C-H \; \longleftrightarrow \; H_2O \; + \; RC\equiv C\overset{\ominus}{:}$$

pK$_a$ 25         pK$_a$ 15.7

*weaker*    *weaker*         *stronger*    *stronger*
*base*      *acid*            *acid*      *base*

2) Another way of predicting the position of an equilibrium is to assign "stronger" and "weaker" to the acid and base on each side of the equation, using the table to determine which is stronger and which is weaker. *Equilibrium will always favor the weaker acid and base.* This method will always give the same answer as the principle in #1 above.

To lead into the next section, look again at the table on p. 691 and notice two things: a) with only a couple of exceptions, all the acidic protons are on either oxygen or carbon; and b) generalizations can be made about the acidity of functional groups. Learning to correlate acidity with functional group is important in predicting reactivity of the functional group.

Appendix 2 continued, Summary of Acidity and Basicity

## Approximate pK$_a$ Values of Organic Compounds

| | | pK$_a$ | | | |
|---|---|---|---|---|---|
| *weaker acid* | alkane | $\approx 50$ | $R-\overset{\mid}{\underset{\mid}{C}}-H$ | $\rightleftharpoons$ H$^+$ + | $R-\overset{\mid}{\underset{\mid}{C}}{:}^{\ominus}$ | *stronger base* |

$$\text{alkane} \quad \approx 50 \qquad R-\overset{|}{\underset{|}{C}}-H \;\rightleftharpoons\; H^+ + R-\overset{|}{\underset{|}{C}}{:}^{\ominus}$$

$$\text{alkene} \quad \approx 45 \qquad =C\overset{\diagup}{\underset{\diagdown H}{}} \;\rightleftharpoons\; H^+ + =C{.}^{\ominus}$$

$$\text{amine} \quad 35\text{-}40 \qquad -\overset{..}{\underset{|}{N}}-H \;\rightleftharpoons\; H^+ + -\overset{..}{\underset{|}{N}}{:}^{\ominus}$$

$$\approx 35 \qquad H-H \;\rightleftharpoons\; H^+ + H{:}^{\ominus}$$

$$\text{alkyne} \quad \approx 25 \qquad RC\equiv C-H \;\rightleftharpoons\; H^+ + RC\equiv C{:}^{\ominus}$$

$$\text{ketone and ester} \quad 20\text{-}25 \qquad -\overset{\displaystyle O}{\overset{\|}{C}}-\overset{|}{\underset{|}{C}}-H \;\rightleftharpoons\; H^+ + -\overset{\displaystyle O}{\overset{\|}{C}}-\overset{|}{\underset{|}{C}}{:}^{\ominus}$$

$$\approx 18 \qquad R-\overset{R}{\overset{|}{\underset{|}{C}}}-O-H \;\rightleftharpoons\; H^+ + R-\overset{R}{\overset{|}{\underset{|}{C}}}-\overset{..}{\underset{..}{O}}{:}^{\ominus}$$

alcohol

$$\approx 17 \qquad R-\overset{H}{\overset{|}{\underset{|}{C}}}-O-H \;\rightleftharpoons\; H^+ + R-\overset{H}{\overset{|}{\underset{|}{C}}}-\overset{..}{\underset{..}{O}}{:}^{\ominus}$$

cannot be measured in water solution

$$\approx 16 \qquad R-\overset{H}{\overset{|}{\underset{|}{C}}}-O-H \;\rightleftharpoons\; H^+ + R-\overset{H}{\overset{|}{\underset{|}{C}}}-\overset{..}{\underset{..}{O}}{:}^{\ominus}$$

- - - - - - - - - - - - - - - - - - - - - - - - - - - - - - - - - - - - - - - - - - -

measured in water solution

$$15.7 \qquad H_2O \;\rightleftharpoons\; H^+ + HO^-$$

$$10.3 \qquad HCO_3^- \;\rightleftharpoons\; H^+ + CO_3^{2-}$$

$$\text{phenol} \quad \approx 10 \qquad Ar-O-H \;\rightleftharpoons\; H^+ + Ar-\overset{..}{\underset{..}{O}}{:}^{\ominus}$$

$$6.4 \qquad H_2CO_3 \;\rightleftharpoons\; H^+ + HCO_3^-$$

$$\text{carboxylic acid} \quad 4\text{-}5 \qquad R-\overset{\displaystyle O}{\overset{\|}{C}}-O-H \;\rightleftharpoons\; H^+ + R-\overset{\displaystyle O}{\overset{\|}{C}}-\overset{..}{\underset{..}{O}}{:}^{\ominus}$$

$$\text{sulfonic acid} \quad <0 \qquad RSO_2-O-H \;\rightleftharpoons\; H^+ + RSO_2-\overset{..}{\underset{..}{O}}{:}^{\ominus}$$

*stronger acid*                                                         *weaker base*

Appendix 2 continued, Summary of Acidity and Basicity

## III. Correlation of Acidity with Functional Group

A. Oxygen Acids

$$\underset{\substack{\text{pK}_a < 0 \\ \text{sulfonic acid}}}{\overset{\displaystyle O \atop \displaystyle \underset{\displaystyle O}{\overset{\displaystyle \|}{R-S-O-H}}}{}} \;>\; \underset{\substack{\text{pK}_a\;4\text{--}5 \\ \text{carboxylic acid}}}{\overset{\displaystyle O \atop \displaystyle R-C-O-H}{}} \;>\; \underset{\substack{\text{pK}_a\;10.0 \\ \text{phenol}}}{\bigcirc\!\!-O-H} \;>\; \underset{\substack{\text{pK}_a\;15.7 \\ \text{water}}}{HO-H} \;>\; \underset{\substack{\text{pK}_a\;16\text{--}19 \\ \text{alcohol}}}{\overset{\displaystyle R \atop \displaystyle \underset{\displaystyle R}{R-C-O-H}}{}}$$

Sulfonic acids are the strongest of the oxygen acids but are not common in organic chemistry. Carboxylic acids, however, are everywhere and are considered the strongest of the common organic oxygen acids. (Note that "strong" and "weak" are relative terms: acetic acid, $pK_a$ 4.74, was a "weak" acid in general chemistry in comparison to sulfuric and hydrochloric acids, but acetic acid is a "strong" acid in organic chemistry relative to the other oxygen acids.) Phenols having OH groups on benzene or other aromatic rings are still stronger acids than water.

Why are phenols, carboxylic acids, and sulfonic acids stronger acids than water? *Because their anions are stabilized by resonance in which the negative charge is delocalized.* (Refer to text sections 1-13, 10-6, and 20-4, especially Figure 20-1.) Let's look at that statement in more detail.

Remember that acidity is a thermodynamic property; that is, acidity equilibrium depends on the difference in energy between the reactants and products. The more the anion is stabilized by resonance, the lower in energy it is, and the less positive the ΔG, as shown on the reaction energy diagram:

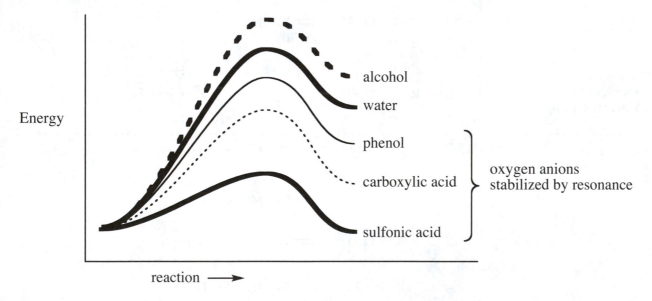

Why are alcohols weaker acids than water? There are two effects that contribute, both of which are consistent with the trend that 1° alcohols ($pK_a$ 16) are slightly stronger than 2° alcohols ($pK_a$ 17), which are slightly stronger than 3° alcohols ($pK_a$ 18). Alkyl groups are mildly electron-donating in their inductive effect (more about this later) and destabilize the anion, as shown in the energy diagram above. Second, the more crowded the anion is, the less it can be stabilized by hydrogen bonding with the solvent.

## B. Carbon Acids

When we think of "acids", we do not usually think of protons on carbon, yet carbon acids and the carbanions that come from them are of tremendous importance in organic chemistry.

"Unstabilized" carbon acids are those that do not have any substituent to stabilize the anion. Alkanes, having only $sp^3$ carbons, are the weakest acids with $pK_a$ around 50. The vinyl carbon in a carbon-carbon double bond is $sp^2$ hybridized with the electrons of the anion slightly closer to the positive nucleus, leading to some stabilization of the anion. This type of stabilization is particularly important in alkynes with sp hybridized carbons.

$$R-\overset{|}{\underset{|}{C}}-H \;\rightleftharpoons\; R-\overset{|}{\underset{|}{C}}\!\!:^{\ominus} \qquad\qquad =\!\overset{/}{\underset{\mathbf{H}}{C}} \;\rightleftharpoons\; =\!C\!:^{\ominus} \qquad\qquad RC\!\equiv\!C-H \;\rightleftharpoons\; RC\!\equiv\!C\!:^{\ominus}$$

| $sp^3$ hybridized | $sp^2$ hybridized | sp hybridized |
|---|---|---|
| $pK_a$ 50 | $pK_a$ 45 | $pK_a$ 25 |

## C. Carbon Acids Alpha to Carbonyl (This topic is described in detail in text section 22-2B.)

Look at these huge differences in $pK_a$ when the acidic group is next to a carbonyl.

$O-H \Rightarrow$  $RCH_2O-H$  $pK_a$ 16–18

$$\underset{}{R-\overset{\displaystyle O}{\overset{\|}{C}}-O-H} \quad pK_a\ 4\text{–}5$$

$N-H \Rightarrow$  $RCH_2NH-H$  $pK_a$ 35–40

$$R-\overset{\displaystyle O}{\overset{\|}{C}}-\overset{\displaystyle H}{\underset{}{N}}-H \quad pK_a\ 16$$

$C-H \Rightarrow$  $RCH_2CH_2-H$  $pK_a$ 50

$$R-\overset{\displaystyle O}{\overset{\|}{C}}-\overset{\displaystyle H_2}{C}-H \quad pK_a\ 20$$

**Hydrogens alpha to carbonyl are unusually acidic because of resonance stabilization of the anionic conjugate base.**

## D. Acidities of Acyl Functional Groups

In addition to the significant variation in the acidity of alpha hydrogens depending on which atom the H is bonded to, what is on the other side of the carbonyl also has a dramatic influence. In this case, the stabilization is more important on the starting material, not on the conjugate base. See the energy diagram on p. 694.

| aldehyde | ketone | ester | amide |
|---|---|---|---|
| $\overset{\mathbf H}{\underset{}{\mathbf H_2C}}-\overset{\mathbf O}{\overset{\|}{\mathbf C}}-\mathbf H$ | $\overset{\mathbf H}{\underset{}{\mathbf H_2C}}-\overset{\mathbf O}{\overset{\|}{\mathbf C}}-\mathbf{CH_3}$ | $\overset{\mathbf H}{\underset{}{\mathbf H_2C}}-\overset{\mathbf O}{\overset{\|}{\mathbf C}}-\overset{..}{\underset{..}{\mathbf O}}\mathbf{CH_3}$ | $\overset{\mathbf H}{\underset{}{\mathbf H_2C}}-\overset{\mathbf O}{\overset{\|}{\mathbf C}}-\overset{..}{\mathbf N}\mathbf{(CH_3)_2}$ |
| $pK_a$ 17 | $pK_a$ 20 | $pK_a$ 25 | $pK_a$ 30 |
| no stabilization of starting material | mild stabilization of starting material by weak electron donation from $CH_3$ | | |
| | | $\overset{\mathbf H}{\underset{}{\mathbf H_2C}}-\overset{\overset{\ominus}{\mathbf O}}{\underset{}{\mathbf C}}=\overset{\oplus}{\underset{..}{\mathbf O}}\mathbf{CH_3}$ | $\overset{\mathbf H}{\underset{}{\mathbf H_2C}}-\overset{\overset{\ominus}{\mathbf O}}{\underset{}{\mathbf C}}=\overset{\oplus}{\mathbf N}\mathbf{(CH_3)_2}$ |
| | | significant resonance stabilization of starting material | strongest resonance stabilization of starting material |

Appendix 2 continued, Summary of Acidity and Basicity

least
stabilization
aldehyde
ketone
ester
amide ← greatest stabilization of starting material, largest ΔG

Energy

reaction →

## E. Carbon Acids Between Two Carbonyls (This topic is described in detail in text section 22-15.)

While a hydrogen alpha to one carbonyl moves into the $pK_a$ 20-25 range for ketones and esters respectively, a hydrogen between two carbonyls (cyano and nitro are similar to a carbonyl electronically) is more acidic than water. The increased resonance stabilization of the conjugate base is largely responsible, but there are subtle variations depending on the type of functional group as noted at the bottom of p. 693. Look at the enormous influence of the nitro group.

$pK_a$ 13.5          $pK_a$ 11.2          $pK_a$ 10.2          $pK_a$ 9.0

$pK_a$ 10.2          $pK_a$ 5.8          $pK_a$ 3.6

## IV. Correlation of Basicity with Functional Group

The bulk of this Appendix is on acidity because many more functional groups are acidic than are basic. Basically (oooh, sorry), only one functional group is basic: amines. There is variation among aliphatic, aromatic, and heteroaromatic amines; these are covered thoroughly in text sections 19-5 and 19-6. One point in the text, just before Table 19-3, deserves emphasis: for any conjugate acid-base pair:

$$pK_a + pK_b = 14$$

Appendix 2 continued, Summary of Acidity and Basicity

This simple algebraic relationship is very useful:

Sample problem.  Is triethylamine (pK$_b$ 3.24) a strong enough base to deprotonate phenol (pK$_a$ 10.0)?

We need to calculate either the pK$_a$ of the conjugate acid of triethylamine or the pK$_b$ of the conjugate base of phenol to see which is stronger and weaker.  Then we can say with certainty which side of the equilibrium will be favored.

| | | | | |
|---|---|---|---|---|
| pK$_a$ 10.0 | pK$_b$ 3.24 | | pK$_b$ = **4.0**<br>(14 − 10.0) | pK$_a$ = **10.76**<br>(14 − 3.24) |
| *stronger*<br>*acid* | *stronger*<br>*base* | | *weaker*<br>*base* | *weaker*<br>*acid* |

Aha!  Products are favored at equilibrium, so the correct answer to the question is "Yes, triethylamine is a strong enough base to deprotonate phenol."

Try this for fun:  How weak must a base be before it does NOT deprotonate phenol?  What algebraic rule can you formulate to predict whether any combination of acid and base will favor products or reactants?

## V.  Substituent Effects on Acidity

So far, we have focused on acidities of different functional groups.  Let's turn to more minor, more subtle, structural changes to see what effect substituents will have on the acidity of a group.  Primarily, we imply *electronic* effects as opposed to *steric* effects, but this Appendix will conclude with a discussion of how steric and electronic effects can work together.

A.  Classification of Substituents—Induction and Resonance

Substituent groups can exert an electronic effect on an acidic functional group in two different ways: through sigma bonds, where this is called an *inductive effect*, or through p orbitals and pi bonds which is called a *resonance effect*.  Groups can also be electron-donating or electron-withdrawing by either of the mechanisms, so there are four possible categories for groups.  Note that a group can appear in more than one category, even in conflicting groups!

a)  Electron-donating by induction:  only alkyl groups (abbreviated R) have electrons to share by induction;

b)  Electron-withdrawing by induction:  every group that has a more electronegative atom than carbon is in this category; some examples:  F, Cl, Br, I, OH, OR, NH$_2$, NHR, NR$_2$, NO$_2$, C=O, CN, SO$_3$H,  CX$_3$ where X is halogen;

c)  Electron-donating by resonance:  groups that have electron pairs to share:  F, Cl, Br, I, OH, OR, NH$_2$, NHR, NR$_2$;

d)  Electron-withdrawing by resonance:  NO$_2$, C=O, CN, SO$_3$H.

B.  Generalizations on Electronic Effects on Acidity (refer to text section 20-4B)

Electric charge is the key to understanding substituent effects.  An acid is always more positive than its conjugate base; in other words, the conjugate base is always more negative than the acid.  Electron-donating and electron-withdrawing groups will have opposite effects on the acid-base conjugate pair.

Electron-donating groups stabilize the more positive acid form and destabilize the more negative conjugate base.  From the diagram, it is apparent that electron-donating groups widen the energy gap between reactants and products, making $\Delta G$ more positive, favoring reactants more than products.  In essence, this weakens the acid strength.

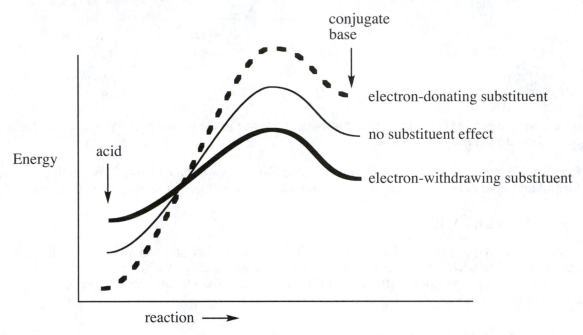

Electron-withdrawing groups destabilize the more positive acid form and stabilize the more negative conjugate base, narrowing the energy gap between reactants and products, making $\Delta G$ less positive. Products are increased in concentration at equilibrium which we define as a stronger acid.

*Electron-withdrawing groups increase acid strength; electron-donating groups decrease acid strength.*

C.  Inductive Effects on Acidity

Text section 20-4B gives a thorough explanation of the inductive effect of electron-withdrawing groups on simple carboxylic acids, from which three generalizations arise:

A)  *Acidity increases with stronger electron-withdrawing groups. (See solution to problem 20-33.)*
B)  *Acidity increases with greater number of electron-withdrawing groups.*
C)  *Acidity increases with closer proximity of the electron-withdrawing group to the acidic group.*

We don't usually look to aromatic systems for examples of inductive effects, because the pi system of electrons is ripe for resonance effects.  However, in analyzing the resonance forms of phenoxide on the next page, it becomes apparent that the negative charge is never distributed on the meta carbons.  Meta substituents cannot exert any resonance stabilization or destabilization; at the meta position, substituents can exert only an inductive effect.  The series of phenols demonstrates this phenomenon, consistent with aliphatic carboxylic acids.

Appendix 2 continued, Summary of Acidity and Basicity

The delocalized negative charge does not increase electron density at the meta position.

| OH CH₃ | OH H | OH OCH₃ | OH Cl | OH NO₂ |
|---|---|---|---|---|
| pK$_a$ 10.09 *donating* | pK$_a$ 10.00 | pK$_a$ 9.65 *withdrawing* | pK$_a$ 9.02 *withdrawing* | pK$_a$ 8.39 *withdrawing* |

D.  Resonance Effects on Acidity—benzoic acids and phenols (review the solution to problem 20-45)

Resonance effects can be expressed with placement of substituents at ortho or para positions, but ortho has the complication of steric effects, so just para substitution is shown here.

*Electron-withdrawing substituents* increase the acidity of benzoic acids and phenols:

| pK$_a$ 10.00 | pK$_a$ 8.05 | pK$_a$ 7.95 | pK$_a$ 7.15 | pK$_a$ 4.20 | pK$_a$ 3.55 | pK$_a$ 3.42 |
|---|---|---|---|---|---|---|

*Electron-donating by resonance but electron-withdrawing by induction:*

There is a group of substituents that donate by resonance but withdraw by induction: alkoxy groups and halogens are the most notable examples, and the acidity data provide an insight into which effect is stronger.

| pK$_a$ 4.20 | pK$_a$ 4.09 | pK$_a$ 4.47 |
|---|---|---|

*meta*-Methoxybenzoic acid is a stronger acid than benzoic acid, consistent with electron-withdrawing by induction, which is expressed at the meta position.  But the para isomer is *weaker* than benzoic acid; electron donation by resonance has not only compensated for the inductive effect (which is still operative at the para position) but has decreased the acidity even further.  Thus, the donating effect by resonance must be stronger than the withdrawing effect by induction for the methoxy group.

See another example on the next page.

697

Appendix 2 continued, Summary of Acidity and Basicity

*meta*-Fluorophenol is a stronger acid than phenol, consistent with electron-withdrawing by induction which is expressed at the meta position. The para isomer is still stronger than phenol; electron donation by resonance has not compensated for the inductive effcct (which is still operative at the para position). Thus, the donating effect by resonance must be weaker than the withdrawing effect by induction for the fluoro group.

$pK_a$ 10.00     $pK_a$ 9.28     $pK_a$ 9.81

*Studying substituent effects on acidity is the standard method of determining whether a group is donating or withdrawing by induction and resonance.*

E. Proximity Effects of Substituents

$pK_a$ 10.0     $pK_a$ 8.05     $pK_a$ 9.19     $pK_a$ 9.90

A     B     C     D

Three effects influence the $pK_a$ values of these substituted phenols. In **C**, the acetyl group at the meta position is electron-withdrawing by induction only. In **B**, the acetyl group at the para position exerts both resonance and inductive effects, both of which are electron-withdrawing, making the acid stronger. In theory, substituents at the ortho position should be like para, exerting both resonance and inductive effects; in fact, the inductive effect should be stronger because of closer proximity to the acidic group. So we would predict **D** to be a stronger acid than **B**, yet it is not. What other effect is operating?

Structure **E** shows that because of the proximity of the acetyl group to the OH, *intramolecular hydrogen bonding* is possible. Hydrogen bonding stabilizes the starting material, lowering the energy of the starting material and making ΔG more positive. Intuitively, it should be apparent that a hydrogen held between two oxygens will be more difficult to remove by a base. Also, after the proton has left as shown in structure **F**, the negative charge on the phenolic oxygen is close to the partial negative charge on the oxygen of the carbonyl, destabilizing product **F**, raising its energy, also making ΔG more positive. The proximity of the acetyl group influences both sides of the equation to make the acid weaker.

intramolecular hydrogen bond

E     F     + H⁺

Here are two more examples where intramolecular hydrogen-bonding influences acidity.

$pK_1$ 4.6; $pK_2$ 9.3     $pK_1$ 2.75; $pK_2$ 13.4     $pK_1$ 3.02; $pK_2$ 4.38     $pK_1$ 1.94; $pK_2$ 6.23

Appendix 2 continued, Summary of Acidity and Basicity

F. Steric Inhibition of Resonance

Another type of proximity effect arises when the placement of a substituent interferes with the orbital overlap required for resonance stabilization. This can be seen clearly in the acidity of substituted benzoic acids and in the basicity of substituted anilines.

Let's analyze this series of carboxylic acids.

COOH
|
H
$pK_a$ 3.75

COOH
$pK_a$ 4.20

COOH
CH$_3$
$pK_a$ 4.34

COOH
CH$_3$
$pK_a$ 4.27

COOH
$pK_a$ 3.46

COOH
H$_3$C     CH$_3$
$pK_a$ 3.21

Formic acid, $pK_a$ 3.75, serves as the reference carboxylic acid. Benzoic acid is weaker because the phenyl group is electron-donating by resonance, stabilizing the protonated form. Methyl substituents are known to be electron-donating by induction, strengthening the electron-donating effect, making the meta- and para-substituted acids weaker than benzoic acid.

Then come the anomalies. Alkyl groups are electron-donating by induction and should weaken the acids, but the ortho-*tert*-butyl and the 2,6-dimethylbenzoic acids are not only stronger than benzoic acid, they are stronger than formic acid! Something has happened to turn the phenyl group into an electron-withdrawing group.

Phenyl is electron-donating by resonance but electron-withdrawing by induction, so what has happened is that the ortho substituents have forced the COOH out of the plane of the benzene ring so that there is no resonance overlap between the benzene ring and the COOH orbitals. The COOH "feels" the benzene ring as simply an inductive substituent. Resonance has been "inhibited" because of the steric effect of the substituent.

CH$_3$
O
C
O—H
CH$_3$

COOH group is not parallel with the plane of the benzene ring— no resonance interaction.

This three-dimensional view down the C-C bond between the COOH and the benzene ring shows that COOH is twisted out of the benzene plane.

The same phenomenon is observed in substituted anilines. Anilines are usually much weaker bases than aliphatic amines because of resonance overlap of the nitrogen's lone pair of electrons with the pi system of benzene. When that resonance is disrupted, the aniline becomes closer in basicity to an aliphatic amine. (See problem 19-44(c).)

CH$_3$
N:
CH$_3$
$pK_b$ 8.94

CH$_3$
CH$_3$
N:
CH$_3$
CH$_3$
$pK_b \approx$ 6-7
(estimated)

More examples on the next page.

**699**

Appendix 2 continued, Summary of Acidity and Basicity

Examples of steric inhibition of resonance:

Amide—not basic because of
resonance sharing of N lone
pair with carbonyl.

Strong base similar to aliphatic amine;
geometry of bridged ring prevents
overlap of N lone pair with carbonyl.

----------------------------------------------------------------------

pK$_b$ 8.9

3° aromatic amine

pK$_b$ 3.4

3° aliphatic amine

pK$_b$ 6.2

3° amine, and the N is
bonded to a benzene, but the
bridged ring system prevents
overlap of N lone pair with
benzene.

----------------------------------------------------------------------

pK$_b$ 8.9

$(CH_3CH_2)_2N$:   :$N(CH_2CH_3)_2$
$H_3CO$         $OCH_3$

pK$_b$ − 2.3 (yes, negative!)

Not only is steric inhibition of
resonance important in this example,
but so is intramolecular hydrogen-
bonding in the protonated form. Draw
a picture.

Students: This is an EXAMPLE of the type of reaction summary that you should prepare for each chapter in which reactions are covered. Some students find lists more helpful, some prefer this "starburst" format. Each instructor will choose different reactions to emphasize; what is presented here is not an exhaustive summary. The most important admonition is that the main benefit to you comes when YOU PREPARE YOUR OWN SUMMARY.

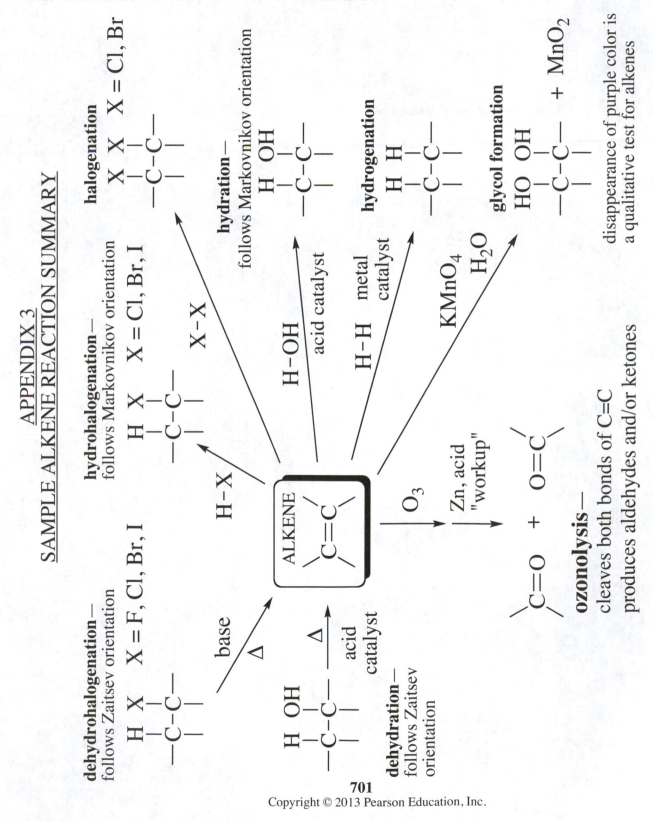

APPENDIX 3
SAMPLE ALKENE REACTION SUMMARY